# Earth Science:
# Earth's Weather, Water, and Atmosphere

# Earth Science:
# Earth's Weather, Water, and Atmosphere

## Volume 1

Editors

**MARGARET BOORSTEIN, PH.D.**
*Long Island University*

**RICHARD RENNEBOOG, M.SC.**

Salem Press
A Division of EBSCO Publishing
Ipswich, Massachusetts

**Library of Congress Cataloging-in-Publication Data**

Earth science. Earth's weather, water, and atmosphere / editors, Margaret Boorstein, Ph. D., Long Island University, Richard Renneboog, M. Sc.
     volumes cm
  Includes bibliographical references and index.
  ISBN 978-1-58765-989-8 (set) – ISBN 978-1-58765-985-0 (set 4 of 4) – ISBN 978-1-58765-986-7 (volume 1) – ISBN 978-1-58765-987-4 (volume 2)  1. Meteorology. 2. Hydrology. 3. Atmosphere. I. Boorstein, Margaret. II. Renneboog, Richard. III. Title: Earth's weather, water, and atmosphere.
  QC863.E27 2013
  551.5–dc23

                  2012027577

# CONTENTS

# PUBLISHER'S NOTE

Salem Press's *Earth Science: Earth's Weather, Water, and Atmosphere* provides a two-volume introduction to the major topics of study in climate, bodies of water, and atmosphere. These volumes provide a comprehensive revision and update to an earlier edition with the same title, which was published by Salem Press in 2001.

The essays in this collection cover a wide range of subject areas, including the components of the atmosphere, with important developments in the study of greenhouse effects and weather patterns, the study of waters within and on the surface of the earth, and the impact of sedimentology in the formation of Earth surfaces. The coeditors of the volume have reviewed each article for scientific authority and have ensured each article's currency.

Designed for high school and college students and their teachers, these volumes provide hundreds of expertly written essays supplemented by illustrations, charts, and useful reference materials, resulting in a comprehensive overview of each topic. Librarians and general readers alike will also turn to this reference work for both foundational information and current developments.

Each essay topic begins with helpful reference information, including a summary statement that explains its significance in the study of the earth and its processes. *Principal Terms* define key elements or concepts related to the subject, and the text is then organized following informative subheadings that guide readers to areas of particular interest. The background and history of each subject are provided and detail important contextual information on the topic. An annotated *Bibliography* closes each essay and refers the reader to external sources for further study that are of use to both students and nonspecialists. Finally, a list of *Cross-References* directs the reader to other subject-related essays within the set. At the end of every volume, several appendices are designed to assist in the retrieval of information, including a *Glossary* that defines key terms contained in each set, and tables such as *Atmospheric Pressure, Oil Spills Timeline, Bodies of Water Data Sheet,* and *Major Weather Events.*

Salem Press's *Earth's Weather, Water, and Atmosphere* is part of a series of Earth science books that includes *Physics and Chemistry of the Earth, Earth's Surface and History,* and *Earth's Materials and Resources.*

Many hands went into the creation of these volumes. Special mention must be made of its coeditors, Margaret Boorstein, Ph.D., and Richard Renneboog, M.Sc., who played a principal role in shaping the reference work and its contents. Thanks are also due to the many academicians and professionals who communicated their expert understanding of Earth science to the general reader; a list of these individuals and their affiliations appears at the beginning of the volume. The contributions of all are gratefully acknowledged.

# INTRODUCTION

The planet Earth is a complex set of living and non-living systems connected by the cycling of matter and flows of energy. The solid earth, or the lithosphere; the gaseous layer surrounding the solid earth, or the atmosphere; and the waters of the earth, or the hydrosphere, are all connected with one another and with the biosphere, or living things. Presented in alphabetical order, the articles in *Earth Science: Earth's Weather, Water, and Atmosphere* discuss and explain important components of these four realms, with a particular emphasis on the influences that people, industry, and commerce around the globe have on Earth's systems.

The evolution of the atmosphere is explained, along with its present structure and internal movements. The relationship between the earth and sun, including revolution and rotation, are discussed, as well as how the changing orientation of the earth's axis with respect to the sun over the course of a year creates seasons. An article on the greenhouse effect describes its necessity to life on Earth. Other articles provide insight into how human activities have interfered with natural balances, leading to global warming, climate change, and ozone depletion. Such distinctions are important to acknowledge so that individuals, through their formal governments, economic activities, and informal alliances, can make decisions about adaptation, mediation, or inaction.

A number of articles examine motions of matter and transfers of energy within the atmosphere. The dramatic consequences of energy transfer are discussed in the articles about hurricanes, tornadoes, lightning and thunder, and monsoons. El Niño/Southern Oscillation (ENSO) has been shown, through decades of gathering detailed weather statistics combined with modern measurement technologies, to have a worldwide impact on climate and weather. Essays about more mundane features of the atmosphere, such as clouds, winds, and climate basics, provide an understanding of their properties and importance to the atmospheric system. The government of the United States devotes human and fiscal resources to advance scientific research as well as improve weather forecasting. Anyone in the United States with access to the Internet can see satellite imagery; interpret and analyze local, regional, and national forecasts; and read discussions of the models that contributed to their creation.

Those articles devoted to hydrology focus on the study of waters that move within the earth's crust and on its surface. The movements of groundwater and the properties of aquifers—layers of rocks or sediment in which water is stored and flows—are discussed, along with consequences of their depletion and deterioration. Because freshwater is vital not only to human survival, but also to functioning economic systems, several articles discuss surface water, wells, water tables, and water quality. The technologies and environmental impact of desalination, dams and flood control, and floods also are important to societies around the world. Waterfalls, frequently spectacular in their beauty and haunting in their long-term erosion, can be and are often harnessed to produce electricity. Yet, erecting power lines to transport electricity from remote waterfall locations involves all sorts of economic, political, and environmental choices. Watersheds, or areas drained by a stream, need to be understood and monitored closely because so many competing uses depend on the waters that flow within them. Neither freshwater nor the intricacies of its movements and the consequences of its contamination are taken for granted today.

Oceans cover approximately two-thirds of the surface of the earth. It is said that less is known about the ocean floor than is known about space. Federal government agencies, including the National Aeronautics and Space Administration (NASA) and the National Oceanic and Atmospheric Administration (NOAA), have been expanding that awareness, especially in the last decade. Articles in this volume explain how heat is transported through wind-driven surface currents and thermal and salinity-driven deep ocean currents. The interactions of water and the tectonics of the ocean floor are discussed in articles about turbidity currents, seamounts, and hydrothermal vents. The danger of tsunamis to human life has always existed. The article about the tsunamis of December 26, 2004, and March 11, 2011, describes the unimaginably wide-ranging damage that occurred, as well as the resulting possible courses of action to mitigate casualties and damage in the future. Current ocean-atmospheric interactions are important influences on climate today. From the perspective of geologic eons, oceans serve as long- and short-term storehouses of carbon and other elements that are part of nutrient cycles vital to life on Earth. With recent human interference in these natural processes have come unforeseen effects on current atmospheric and oceanic processes. As global warming intensifies, sea level will be of greater concern not only to those populations that have to be evacuated and the areas to which they move, but to those who are dependent on the natural resources lost to rising seas.

More extensive discussions of specific oceans and seas provide descriptions of their origins and current geology and geography. The story of the Aral Sea in the former Soviet Union serves as a morality tale. Once the fourth largest body of freshwater in the world, it decreased greatly in size and deteriorated into a body of water with a higher saline concentration than the oceans. In contrast, Hudson Bay has been left relatively untouched, but that situation may change as demand for mineral resources and water power increases. The Atlantic and Indian Oceans, along with their climate influences and economic importance, have fascinating geologic features and marine ecologies. The Arctic Ocean illustrates the severity of the intensification of the greenhouse effect. As Arctic ice melts at alarming rates, less sunlight is reflected and more is absorbed by the ocean waters, thereby accelerating global warming. Yet, the economic benefits to countries and commercial interests include easier winter shipping and the opening of once-ice-blocked harbors. The global community must decide between apparent short-term advantages and long-term dangers. The Gulf of Mexico is a rich fishery and source of petroleum. As illustrated by the August 2010 oil-well blowout and spill, proper safety and environmental precautions have not always been balanced with economic expediency. Hurricanes intensify in the Gulf, again leading to questions about the possible impact of global warming on human populations and economies.

Lakes and rivers are also covered herein. The enormity of freshwater flow and the influences on plant and animal life are explained for the Amazon, Ganges, Yellow, Mississippi, and Nile Rivers, as are

consequences of misguided interference by people. The Great Lakes, which hold approximately 20 percent of the freshwater in the world, are important for fishing, recreation, and freshwater supply. They have also facilitated the economic growth of both the United States and Canada, with four of the five lakes sharing a border that features both countries' manufacturing belts and agricultural core. Contamination of the waters by excessive fertilizer and pesticide use, combined with industrial runoff, contributed to significant declines and contamination of the commercial and recreational fisheries. Cooperation by both countries and the neighboring states and provinces have alleviated these problems, but fish contamination remains.

Finally, sedimentology, including the influence of running and standing water on sediments and resulting landforms, is presented. Depositional systems such as alluvial systems and deltas create valuable farmland but, in turn, are related to streams or rivers overflowing their banks. Evaporites and deep-sea sedimentation both result in valuable resources for human use. The growing recognition of the ecological value of reefs has resulted in global-scale efforts to lessen atmospheric warming as well as regional attempts at physical protection. The challenges of weathering and erosion from careless farming, as well as the threat of beach erosion, will continue as long as people eat and fish and place a premium on living along the beautiful coast.

Our country and our world need an informed and scientifically literate public that understands environmental processes and can evaluate human solutions, whether proposed by governments, official and nonofficial agencies, or ordinary people. *Earth Science: Earth's Weather, Water, and Atmosphere* will help in that task as it provides basic knowledge and explanations and analysis of the interconnections of the solid earth, the atmosphere, the waters, and living inhabitants.

*Margaret Boorstein, Ph.D.*
*C. W. Post College of Long Island University*

# CONTRIBUTORS

**Arthur L. Alt**
*College of Great Falls*

**George K. Attwood**
*Maharishi International University*

**Michael P. Auerbach**
*Marblehead, MA*

**Dennis G. Baker**
*University of Michigan*

**Victor R. Baker**
*University of Arizona*

**Anita Baker-Blocker**
*Applied Meteorological Services*

**Elizabeth K. Berner**
*Yale University*

**Rachel Leah Blumenthal**
*Somerville, MA*

**Richard J. Boon**
*Independent scholar*

**Margaret Boorstein**
*Long Island University*

**Alan Brown**
*Livingston University*

**Joseph I. Brownstein**
*Atlanta, GA*

**Byron D. Cannon**
*University of Utah*

**Robert E. Carver**
*University of Georgia*

**Dennis Chamberland**
*Independent scholar*

**Dennis W. Cheek**
*Rhode Island Department of Education*

**John H. Corbet**
*Memphis State University*

**Robert G. Corbett**
*Illinois State University*

**Ralph D. Cross**
*University of Southern Mississippi*

**Loralee Davenport**
*Mississippi University for Women*

**Ronald W. Davis**
*Western Michigan University*

**Dave Dooling**
*Independent scholar*

**Walter C. Dudley, Jr.**
*University of Hawaii at Hilo*

**Dean A. Dunn**
*University of Southern Mississippi*

**John W. Foster**
*Illinois State University*

**Dell R. Foutz**
*Mesa State College*

**Joyce Gawell**
*Del Valle High School*

**Soraya Ghayourmanesh**
*Nassau Community College*

**Daniel G. Graetze**
*University of Washington, Seattle*

**Gina Hagler**
*Washington, DC*

**Jasper L. Harris**
*North Carolina Central University*

**Sara A. Heller**
*College of Charleston*

**William Hoffman**
*Independent scholar*

**Robert M. Hordon**
*Rutgers University*

**Stephen Huber**
*Beaver College*

**Samuel F. Huffman**
*University of Wisconsin—River Falls*

**Micah L. Issitt**
*Independent scholar*

**Pamela Jansma**
*Jet Propulsion Laboratory*

**Albert C. Jensen**
*Central Florida Community College*

**Jeffrey A. Joens**
*Florida International University*

**Karen N. Kähler**
*Independent scholar*

**Christopher Keating**
*Angelo State University*

**Gary G. Lash**
*State University of New York College, Fredonia*

**M. Lee**
*Independent scholar*

**Joel S. Levine**
*NASA Langley Research Center*

**Leon Lewis**
*Appalachian State University*

**W. David Liddell**
*Utah State University*

**Donald W. Lovejoy**
*Palm Beach Atlantic College*

**David W. Maguire**
*C. S. Mott Community College*

**Carl Henry Marcoux**
*University of California, Riverside*

**Sergei A. Markov**
*Austin Peay State University*

**Michael W. Mayfield**
*Appalachian State University*

**Joseph M. Moran**
*University of Wisconsin—Green Bay*

**Otto H. Muller**
*Alfred University*

**John E. Mylroie**
*Mississippi State University*

**Bruce W. Nocita**
*University of South Florida*

**Edward B. Nuhfer**
*University of Wisconsin—Platteville*

**David L. Ozsvath**
*University of Wisconsin—Stevens Point*

**Donald F. Palmer**
*Kent State University*

**C. Nicholas Raphael**
*Eastern Michigan University*

**Richard Renneboog**
*Independent scholar*

**James L. Sadd**
*Occidental College*

**Neil E. Salisbury**
*University of Oklahoma*

**Panagiotis D. Scarlatos**
*Florida Atlantic University*

**Elizabeth D. Schafer**
*Independent scholar*

**Kenneth J. Schoon**
*Indiana University Northwest*

**Rose Secrest**
*Independent scholar*

**John F. Shroder, Jr.**
*University of Nebraska at Omaha*

**J. H. Shugart**
*Kennesaw State University*

**R. Baird Shuman**
*University of Illinois at Urbana-Champaign*

**John Brelsford Southard**
*Massachusetts Institute of Technology*

**Joseph L. Spradley**
*Wheaton College*

**Kenneth F. Steele**
*Arkansas Water Resources Research Center*

**Robert J. Stern**
*University of Texas at Dallas*

**Dion Stewart**
*Adams State College*

**Toby Stewart**
*Independent scholar*

**Keith J. Tinkler**
*Brock University*

**Leslie V. Tischauser**
*Prairie State College*

**Kenrick Vezina**
*Lowell, MA*

**Nan White**
*Maharishi International University*

**Thomas A. Wikle**
*Oklahoma State University*

**Lisa A. Wroble**
*Redford Township District Library*

**Jay R. Yett**
*Orange Coast College*

# COMMON UNITS OF MEASURE

*Notes:* Common prefixes for metric units—which may apply in more cases than shown below—include giga- (1 billion times the unit), mega- (one million times), kilo- (1,000 times), hecto- (100 times), deka- (10 times), deci- (0.1, or one tenth), centi- (0.01, or one hundredth), milli- (0.001, or one thousandth), and micro- (0.0001, or one millionth).

| UNIT | QUANTITY | SYMBOL | EQUIVALENTS |
|---|---|---|---|
| Acre | Area | ac | 43,560 square feet<br>4,840 square yards<br>0.405 hectare |
| Ampere | Electric current | A *or* amp | 1.00016502722949 international ampere<br>0.1 biot *or* abampere |
| Angstrom | Length | Å | 0.1 nanometer<br>0.0000001 millimeter<br>0.000000004 inch |
| Astronomical unit | Length | AU | 92,955,807 miles<br>149,597,871 kilometers<br>(mean Earth-sun distance) |
| Barn | Area | b | $10^{-28}$ meters squared<br>(approximate cross-sectional area of 1 uranium nucleus) |
| Barrel<br>(dry, for most produce) | Volume/capacity | bbl | 7,056 cubic inches; 105 dry quarts; 3.281 bushels, struck measure |
| Barrel<br>(liquid) | Volume/capacity | bbl | 31 to 42 gallons |
| British thermal unit | Energy | Btu | 1055.05585262 joule |
| Bushel<br>(U.S., heaped) | Volume/capacity | bsh *or* bu | 2,747.715 cubic inches<br>1.278 bushels, struck measure |
| Bushel<br>(U.S., struck measure) | Volume/capacity | bsh *or* bu | 2,150.42 cubic inches<br>35.238 liters |
| Candela | Luminous intensity | cd | 1.09 hefner candle |
| Celsius | Temperature | °C | 1 degree centigrade |
| Centigram | Mass/weight | cg | 0.15 grain |
| Centimeter | Length | cm | 0.3937 inch |
| Centimeter, cubic | Volume/capacity | cm³ | 0.061 cubic inch |
| Centimeter, square | Area | cm² | 0.155 square inch |
| Coulomb | Electric charge | C | 1 ampere second |

| UNIT | QUANTITY | SYMBOL | EQUIVALENTS |
|---|---|---|---|
| Cup | Volume/capacity | C | 250 milliliters<br>8 fluid ounces<br>0.5 liquid pint |
| Deciliter | Volume/capacity | dL | 0.21 pint |
| Decimeter | Length | dm | 3.937 inches |
| Decimeter, cubic | Volume/capacity | dm³ | 61.024 cubic inches |
| Decimeter, square | Area | dm² | 15.5 square inches |
| Dekaliter | Volume/capacity | daL | 2.642 gallons<br>1.135 pecks |
| Dekameter | Length | dam | 32.808 feet |
| Dram | Mass/weight | dr *or* dr avdp | 0.0625 ounce<br>27.344 grains<br>1.772 grams |
| Electron volt | Energy | eV | $1.5185847232839 \times 10^{-22}$ Btu<br>$1.6021917 \times 10^{-19}$ joule |
| Fermi | Length | fm | 1 femtometer<br>$1.0 \times 10^{-15}$ meter |
| Foot | Length | ft *or* ' | 12 inches<br>0.3048 meter<br>30.48 centimeters |
| Foot, cubic | Volume/capacity | ft³ | 0.028 cubic meter<br>0.0370 cubic yard<br>1,728 cubic inches |
| Foot, square | Area | ft² | 929.030 square centimeters |
| Gallon (British Imperial) | Volume/capacity | gal | 277.42 cubic inches<br>1.201 U.S. gallons<br>4.546 liters<br>160 British fluid ounces |
| Gallon (U.S.) | Volume/capacity | gal | 231 cubic inches<br>3.785 liters<br>0.833 British gallon<br>128 U.S. fluid ounces |
| Giga-electron volt | Energy | GeV | $1.6021917 \times 10^{-10}$ joule |
| Gigahertz | Frequency | GHz | — |
| Gill | Volume/capacity | gi | 7.219 cubic inches<br>4 fluid ounces<br>0.118 liter |

| Unit | Quantity | Symbol | Equivalents |
|------|----------|--------|-------------|
| Grain | Mass/weight | gr | 0.037 dram<br>0.002083 ounce<br>0.0648 gram |
| Gram | Mass/weight | g | 15.432 grains<br>0.035 avoirdupois ounce |
| Hectare | Area | ha | 2.471 acres |
| Hectoliter | Volume/capacity | hL | 26.418 gallons<br>2.838 bushels |
| Hertz | Frequency | Hz | $1.08782775707767 \times 10^{-10}$ cesium atom frequency |
| Hour | Time | h | 60 minutes<br>3,600 seconds |
| Inch | Length | in or " | 2.54 centimeters |
| Inch, cubic | Volume/capacity | in³ | 0.554 fluid ounce<br>4.433 fluid drams<br>16.387 cubic centimeters |
| Inch, square | Area | in² | 6.4516 square centimeters |
| Joule | Energy | J | $6.2414503832469 \times 10^{18}$ electron volt |
| Joule per kelvin | Heat capacity | J/K | $7.24311216248908 \times 10^{22}$ Boltzmann constant |
| Joule per second | Power | J/s | 1 watt |
| Kelvin | Temperature | K | -272.15 degrees Celsius |
| Kilo-electron volt | Energy | keV | $1.5185847232839 \times 10^{-19}$ joule |
| Kilogram | Mass/weight | kg | 2.205 pounds |
| Kilogram per cubic meter | Mass/weight density | kg/m³ | $5.78036672001339 \times 10^{-4}$ ounces per cubic inch |
| Kilohertz | Frequency | kHz | — |
| Kiloliter | Volume/capacity | kL | — |
| Kilometer | Length | km | 0.621 mile |
| Kilometer, square | Area | km² | 0.386 square mile<br>247.105 acres |
| Light-year<br>(distance traveled by light in one Earth year) | Length/distance | lt-yr | 5,878,499,814,275.88 miles<br>$9.46 \times 1012$ kilometers |
| Liter | Volume/capacity | L | 1.057 liquid quarts<br>0.908 dry quart<br>61.024 cubic inches |

| Unit | Quantity | Symbol | Equivalents |
|---|---|---|---|
| Mega-electron volt | Energy | MeV | — |
| Megahertz | Frequency | MHz | — |
| Meter | Length | m | 39.37 inches |
| Meter, cubic | Volume/capacity | $m^3$ | 1.308 cubic yards |
| Meter per second | Velocity | m/s | 2.24 miles per hour<br>3.60 kilometers per hour |
| Meter per second per second | Acceleration | $m/s^2$ | 12,960.00 kilometers per hour per hour<br>8,052.97 miles per hour per hour |
| Meter, square | Area | $m^2$ | 1.196 square yards<br>10.764 square feet |
| Metric. See unit name | | | |
| Microgram | Mass/weight | mcg *or* µg | 0.000001 gram |
| Microliter | Volume/capacity | µL | 0.00027 fluid ounce |
| Micrometer | Length | µm | 0.001 millimeter<br>0.00003937 inch |
| Mile<br>(nautical international) | Length | mi | 1.852 kilometers<br>1.151 statute miles<br>0.999 U.S. nautical mile |
| Mile<br>(statute or land) | Length | mi | 5,280 feet<br>1.609 kilometers |
| Mile, square | Area | $mi^2$ | 258.999 hectares |
| Milligram | Mass/weight | mg | 0.015 grain |
| Milliliter | Volume/capacity | mL | 0.271 fluid dram<br>16.231 minims<br>0.061 cubic inch |
| Millimeter | Length | mm | 0.03937 inch |
| Millimeter, square | Area | $mm^2$ | 0.002 square inch |
| Minute | Time | m | 60 seconds |
| Mole | Amount of substance | mol | $6.02 \times 10^{23}$ atoms or molecules of a given substance |
| Nanometer | Length | nm | 1,000,000 fermis<br>10 angstroms<br>0.001 micrometer<br>0.00000003937 inch |

| UNIT | QUANTITY | SYMBOL | EQUIVALENTS |
|------|----------|--------|-------------|
| Newton | Force | N | 0.224808943099711 pound force<br>0.101971621297793 kilogram force<br>100,000 dynes |
| Newton-meter | Torque | N·m | 0.7375621 foot-pound |
| Ounce (avoirdupois) | Mass/weight | oz | 28.350 grams<br>437.5 grains<br>0.911 troy or apothecaries' ounce |
| Ounce (troy) | Mass/weight | oz | 31.103 grams<br>480 grains<br>1.097 avoirdupois ounces |
| Ounce (U.S., fluid or liquid) | Mass/weight | oz | 1.805 cubic inch<br>29.574 milliliters<br>1.041 British fluid ounces |
| Parsec | Length | pc | 30,856,775,876,793 kilometers<br>19,173,511,615,163 miles |
| Peck | Volume/capacity | pk | 8.810 liters |
| Pint (dry) | Volume/capacity | pt | 33.600 cubic inches<br>0.551 liter |
| Pint (liquid) | Volume/capacity | pt | 28.875 cubic inches<br>0.473 liter |
| Pound (avoirdupois) | Mass/weight | lb | 7,000 grains<br>1.215 troy or apothecaries' pounds<br>453.59237 grams |
| Pound (troy) | Mass/weight | lb | 5,760 grains<br>0.823 avoirdupois pound<br>373.242 grams |
| Quart (British) | Volume/capacity | qt | 69.354 cubic inches<br>1.032 U.S. dry quarts<br>1.201 U.S. liquid quarts |
| Quart (U.S., dry) | Volume/capacity | qt | 67.201 cubic inches<br>1.101 liters<br>0.969 British quart |
| Quart (U.S., liquid) | Volume/capacity | qt | 57.75 cubic inches<br>0.946 liter<br>0.833 British quart |
| Rod | Length | rd | 5.029 meters<br>5.50 yards |

| UNIT | QUANTITY | SYMBOL | EQUIVALENTS |
|---|---|---|---|
| Rod, square | Area | rd² | 25.293 square meters<br>30.25 square yards<br>0.00625 acre |
| Second | Time | s or sec | 1/60 minute<br>1/3,600 hour |
| Tablespoon | Volume/capacity | T or tb | 3 teaspoons<br>4 fluid drams |
| Teaspoon | Volume/capacity | t or tsp | 0.33 tablespoon<br>1.33 fluid drams |
| Ton<br>(gross or long) | Mass/weight | t | 2,240 pounds<br>1.12 net tons<br>1.016 metric tons |
| Ton<br>(metric) | Mass/weight | t | 1,000 kilograms<br>2,204.62 pounds<br>0.984 gross ton<br>1.102 net tons |
| Ton<br>(net or short) | Mass/weight | t | 2,000 pounds<br>0.893 gross ton<br>0.907 metric ton |
| Volt | Electric potential | V | 1 joule per coulomb |
| Watt | Power | W | 1 joule per second<br>0.001 kilowatt<br>$2.84345136093995 \times 10^{4}$ ton of refrigeration |
| Yard | Length | yd | 0.9144 meter |
| Yard, cubic | Volume/capacity | yd³ | 0.765 cubic meter |
| Yard, square | Area | yd² | 0.836 square meter |

# COMPLETE LIST OF CONTENTS

## Volume 1

# Volume 2

# CATEGORY LIST OF CONTENTS

# Earth Science:
# Earth's Weather, Water, and Atmosphere

# A

## ACID RAIN AND ACID DEPOSITION

*Acid rain is rain that is more acidic than would be natural as a result of reactions with pollutive, acid-forming gases, such as sulfur dioxide and nitric oxides. Lakes, forests, soils, and human structures in the eastern part of the United States and southeastern Canada have been damaged by acid rain and deposition of sulfuric and nitric acid aerosol on terrestrial objects.*

### PRINCIPAL TERMS

- **acid deposition:** the depositing of acidic materials on the ground surface through the action of precipitation
- **acid rain:** rain composed of water having a lower-than-normal pH due to having dissolved and reacted with airborne contaminants to produce acidic materials
- **alkaline:** having a pH greater than 7 due to a lower concentration of hydrogen ($H^+$) ions than are in neutral water
- **bicarbonate:** a negatively charged ion, as $HCO_3^-$, that effectively neutralizes excess hydrogen ions in natural waters, reducing acidity
- **cap and trade legislation:** legislation that places limits on the emission of acid-producing materials, such as sulfur dioxide, while allowing emitters of excess amounts to purchase and utilize the unused allowances of those whose emissions are below the legislated limit
- **limestone:** a rock containing calcium carbonate that reacts readily with acid rain and tends to neutralize it, being chemically eroded in the process
- **neutralization:** the adjustment of the concentration of hydrogen ions in solution in order to achieve neutral pH
- **nitric acid:** an acid formed in rain from nitric oxide gases in the air
- **nitric oxide gases:** gases formed by a combination of nitrogen and oxygen, particularly nitrogen dioxide and nitric oxide
- **pH:** a measure of the hydrogen ion concentration, which determines the acidity of a solution; the lower the pH, the greater the concentration of hydrogen ions and the more acidic the solution
- **sulfur dioxide:** a gas whose molecules consist of one sulfur atom and two oxygen atoms, formed by the combustion of sulfur in the presence of oxygen
- **sulfuric acid:** an acid formed as the primary component of acid rain by reaction of sulfur dioxide gas with liquid water in the atmosphere

### DEFINITION AND CAUSE

Acid rain is rain that is more acidic than it would be normally, usually because it has reacted with acid-forming pollutive gases. The acidity of rain is measured in pH units, which are the negative logarithmic values of the concentrations of hydrogen ions in solution. Pure water, which is neutral, has a pH of 7, reflecting the natural concentration of hydrogen ions in pure water of $1 \times 10^{-7}$ moles per liter; any solution with a pH greater than 7 is basic, or alkaline, and any solution with a pH less than 7 is acidic. The lower the pH, the more hydrogen ions there are and the more acidic the solution is. The natural acidity of rain is determined by its reaction with carbon dioxide gas in the atmosphere, a reaction that produces carbonic acid. Carbonic acid partly dissociates to produce hydrogen ions and bicarbonate ions. As a result of this reaction, pure rain should be moderately acidic, with a pH of 5.7. Any rain with a pH less than 5.7 is called "acid rain" and has reacted with acidic atmospheric gases other than carbon dioxide. Reaction of water with sulfur dioxide, for example, produces sulfuric acid in rain, and reaction of water with nitrogen dioxide produces nitric acid in rain. In some cases, acid rain has been observed to have a pH value as low as 2.4, which is as acidic as vinegar.

In addition to acid rain, there is "dry deposition," which occurs without rain and deposits acidic nitrate and sulfate particles and sulfuric and nitric acid aerosols from the atmosphere. The acidic particles are trapped by vegetation or settle out, and the gases are taken up by vegetation. "Acid deposition" usually refers to dry deposition of acids.

1

Acid rain was first recognized in Scandinavia in the early 1950's. It was discovered that acid rain (with a pH from 4 to 5) came from winter air masses that were carrying pollution into Scandinavia from industrial areas in Central and Western Europe. Rain became more acidic over the next twenty years, and the area of Europe receiving acidic rains increased. By the mid-1970's, most of northwestern Europe was receiving acid rain with a pH of less than 4.6. As a result of the discovery of acid rain in Europe, scientists began measuring the acidity of North American rain. Initially, around 1960, acid rain was concentrated in a bull's-eye-shaped area over New York, Pennsylvania, and New England. By 1980, however, most of the United States east of the Mississippi River and southeastern Canada was receiving acid rain (pH less than 5.0), and the central bull's-eye was receiving very acidic rain having a pH less than 4.2. The greatest increase in acidity of rain was in the southeastern United States.

The primary cause of acidity in U.S. and European rains is sulfuric acid, which comes from pollutive sulfur dioxide gas produced by the burning of sulfur-containing fossil fuels, particularly coal, but also oil and gas. In the United States, much of the sulfur dioxide gas is produced in the industrial area of the Midwest, However, sulfur dioxide gas—and the resulting sulfuric acid—can be transported for a distance of 800 kilometers to the northeast by the prevailing winds in the atmosphere before precipitating as acid rain in the northeastern United States and southeastern Canada. To reduce the acidity of rain in the East, then, the emissions of sulfur dioxide gas in the Midwest would have to be reduced. Another source of sulfur dioxide gas is smelters that process ores, such as that in Sudbury, Ontario, located north of Lake Huron in Canada, which is one of the largest sources of sulfur emissions in the world. This smelter's high exhaust stack spreads sulfuric acid aerosols over an extensive area hundreds of kilometers downwind. The original intent of building high exhaust stacks was to reduce local air pollution, but the net effect has been to spread the pollution over much larger areas. Acid rains are even found in Alaska, where sulfuric acid particles have been transported from the contiguous United States.

Nitric acid is a secondary cause of acid rain (contributing about 30 percent of the acidity), but it is one that is increasing. Nitric acid comes from the nitrogen oxide gases, nitrogen dioxide, and nitric oxide, which are produced by the burning of fossil fuels. In contrast to sulfur dioxide, 40 percent of the pollutive nitrogen oxide gas comes from vehicles and most of the remainder from power and heating. The production of nitrogen oxides therefore tends to be concentrated in urban areas. Nitric acid is an important component of acid rain in Los Angeles, for example, because air pollution from vehicle exhaust tends to become trapped in this area.

Some acid rain results from natural causes. Reduced sulfur gases, such as hydrogen sulfide and dimethyl sulfide, are produced by organic matter decay and converted to sulfur dioxide and sulfuric acid in the atmosphere. This process results in naturally acid rain. Volcanoes are another natural source of sulfur dioxide gas. Nevertheless, about 75 percent of the sulfur dioxide gas produced in the United States comes from the burning of fossil fuels. Naturally acid rain (with a pH less than 5.5) is uncommon, falling chiefly in remote areas such as the Amazon basin and some oceanic areas.

There are natural factors that work to reduce or neutralize the acidity of rain in certain areas. Windblown dust, particularly that containing limestone particles, tends to make rains in arid areas of the western United States less acidic by reacting with the acid to produce a rainwater solution with a pH of 6 or more. In addition, the presence of ammonia gas produced in agricultural areas by animal waste, fertilizers, and the decomposition of organic matter will reduce or counteract the acidity of rain on a local scale.

## EFFECTS OF ACID RAIN AND ACID DEPOSITION

The detrimental effects of dry deposition and acid rain include the corrosion and chemical erosion of structures and buildings made of susceptible materials, changes in soil characteristics, increases in the acidity of lakes, and other biological effects, particularly in high-altitude forests. The corrosive effects of airborne acids are particularly obvious on limestone, a rock composed of calcium carbonate, which reacts easily with acid rain. In many New England cemeteries, tombstones made of marble, a form of limestone, have been badly corroded, although older tombstones made of slate, which is less affected by acid rain, are intact. Limestone components of buildings and other structures are similarly corroded.

The effect of acid rain on soils depends on their composition. Alkaline soils that contain limestone have the ability to neutralize acid rain. Even in soils that do not contain limestone, several processes operate to neutralize acid rain, though such processes invariably alter the chemical nature of the soil. Cation exchange occurs, whereby hydrogen ions from the rainwater are exchanged for metal ions, such as calcium or magnesium, on the surface of clays and other minerals. This exchange removes excess hydrogen ions from soil solutions, rendering them less acidic. Another neutralization process involves the release of soil aluminum into solution and the accompanying uptake of hydrogen ions. This process occurs by dissolution of aluminum bound to clays and organic compounds. Frozen soils and sandy soils containing mostly quartz, which does not react with acid rain, have little ability to neutralize acid rain.

Lakes in certain areas have become acidic (with a pH less than 5) from the deposition of acid rain. Lakes in granitic terrain are most affected by acid rain because the surrounding bedrock has little or no ability to react with or neutralize the excess hydrogen ions in the water. Areas with many acid lakes include the Adirondack Mountains in New York, the Pocono Mountains in Pennsylvania, the Upper Peninsula of Michigan, Ontario, Nova Scotia, and Scandinavia. Generally, the deposition of highly acidic rain having a pH less than 4.6 over a long period of time is required. The effect is enhanced in bodies of water that are maintained by watershed drainage rather than by freshwater springs. Lake waters that have a tendency to become acidified initially have little ability to neutralize acid rain because they are low in carbonate and bicarbonate ions, which come predominantly from limestone. Such lakes are described as being poorly buffered. (Buffering is the resistance to changes in pH upon the addition of acid or base.) The soil in the drainage area surrounding acid lakes does not neutralize acid rain adequately before it reaches the lake because of a lack of limestone and clay minerals or because the soil cover is thin or lacking altogether. In addition, some lakes, although not usually acidic, may have periods of elevated acidity due to the runoff of snowmelt, which collects acid precipitation stored in the snow. This runoff gives a sudden large pulse of highly acidic water to the lake. In certain areas, such as Florida, acid lakes result partly due to causes other than acid rain, such as the presence of organic

acids produced by the decay of vegetation in poorly drained areas and nitric acid formed from nitrate-based fertilizer runoff. The gradual acidification of lakes results in the death of fish populations because of reproductive failure, as well as other changes in the organisms living in the lake. A reduction in the number of species occurs at all levels of the food chain. In some cases, snowmelt acidity has been identified as the cause of a massive, instantaneous fish kill in lakes.

Rivers are also known to become acidic. Eastern U.S. rivers show high concentrations of sulfate and a low pH in cases where the soil cannot neutralize the acid rain it receives. Certain acid rivers are caused by acidic drainage from mine dumps rather than by acid rain. Acid rivers rich in organic matter are found in the eastern United States coastal plain and in the Amazon basin. These rivers have naturally high concentrations of dissolved organic acids.

Acid rain and acid deposition are implicated in the decline and death of certain forests, particularly evergreen forests at high elevations. These forests receive very acidic precipitation from the accumulation of clouds at the mountaintops. It is thought that acid rain does not actually kill the forests but rather provides a stress that causes them to become less resistant and die from other causes. The actual stress provided by acid rain is still being studied. Possible stresses include loss of nutrients from soil and leaves through leaching, destruction of beneficial soil microorganisms, and increased susceptibility to frost damage during winter.

Efforts have been made to reduce the acidity of rain, particularly by controlling sulfur emissions. Power plants have been required to reduce the sulfur content of coal that they burn, thus lowering the amount of sulfur dioxide that is produced. Sulfate concentrations in rain in the northeastern United States have been reduced by this method. Nitrogen oxide emissions from cars have also been reduced through stricter emissions controls and improved engineering design for more efficient combustion. In some cases, acid lakes have been treated with limestone to temporarily neutralize their acidity, but the only permanent solution is a reduction in the acidity of the rain that they receive.

## STUDY OF ACID RAIN AND ACID DEPOSITION

The acidity of rain can be measured directly by an electronic device called a pH meter. A pair of electrodes is inserted into a solution, and the electrical potential, or voltage, is measured between them. This voltage is directly related to the concentration of hydrogen ions in the solution—that is, to the acidity of the solution. To monitor and measure the acidity of rainwater, networks have been constructed to collect rain samples over large geographical areas. The acidity of rainwater over the course of the entire year must be measured because pH varies between rainfalls, both seasonally and according to whether the air masses that produce the rain have passed over significant sources of pollution. The pH of rainwater and other forms of precipitation is also measured over a period of years. In addition to the concentration of hydrogen ions, the concentration of other ions, such as sulfate from sulfuric acid and nitrate from nitric acid, is measured in the rainwater samples. Such measurements give evidence of the source of the acidity—that is, which proportion is attributable to sulfuric acid and which to nitric acid. The pH levels of samples collected over a large geographical area are plotted on a map, and contours are drawn through equal values of the pH. Such maps show which areas are receiving the most acid rain. The amount of sulfate and nitrate being deposited by rain is also plotted separately. Meteorologists also use information about a storm system's path as it moves across the country. Such atmospheric systems transport pollutive gases from one area to another. Combining deposition patterns on maps with information about the path followed by a storm shows where the gas residues in rainwater may be coming from and suggests sources of the acidity.

Computers have been used to predict where acid rain will fall and how acidic it will be, given the sources and amounts of sulfur and nitrogen emissions, particularly from power plants and smelters, and the weather patterns. Predictions of this type require a detailed knowledge of the atmospheric chemistry by which sulfur dioxide is converted to sulfuric acid and the oxides of nitrogen are converted to nitric acid. This type of modeling is necessary to predict how much reduction in the acidity of rain in a distant area will result from a given reduction in a power plant sulfur source, for example.

The effects of acidity on soils have been the subject of study for many years. Laboratory experiments can demonstrate how soil clays and other minerals react to acid rain, including which chemical species

are taken up and which are released. In addition, soil solutions and minerals are collected and analyzed from actual field areas affected by acid rain. Ideally, such an analysis should be carried out over a period of time to determine whether any changes in the soil solution chemistry are occurring. From a knowledge of the soil chemistry, it is possible to predict how long a soil can receive acid rain before it loses particular nutrients or the ability to neutralize the excess acidity.

Measuring the acidity and chemical composition of lakes in various areas over long periods and sampling their fish populations and other biota enables scientists to see increases in lake acidity and to correlate the increases with changes in the populations of affected species. In some areas, lakes have been artificially acidified so that the changes in their chemistry and biological populations can be observed. Apparently, acidic lake water inhibits reproduction in fish and other creatures, in addition to destroying the organisms that they use for food. Computer models of acid rain falling on susceptible drainage areas of lakes are made in order to predict how the drainage area reacts to acid rain and how much reduction in acid rain would be necessary to lower the acidity of the lake to the point where it would support fish. In badly affected areas such as the Adirondacks, it may be necessary to reduce the acidity by half.

To study forest decline, surveys of present forest conditions are compared with historical records for the same areas. For example, in high-elevation areas in New England and the Adirondacks, more than half of the red spruce died between 1965 and 1990. Tree rings, which record annual growth, show reduced growth in certain forests. It is known that acid rain causes changes in the soil, such as the release of aluminum, which is toxic to root tissues and so prevents the uptake of essential nutrients. In addition, acid rain causes the loss of certain nutrients from the soil, such as sodium, calcium, and magnesium. Another effect of acid rain is the reduction of the numbers of microorganisms in the soil. Yet, because acid rain from nitric acid contains nitrogen, a plant nutrient, it may fertilize the soil if there is a deficiency of soil nitrogen. One problem in studying forests receiving acid rain is determining which of the many changes occurring are contributing most to forest damage. It is often difficult to distinguish between the stresses of acid rain and other stresses, such as those caused by drought, cold, and insects. Field studies in this area may involve artificial acidification of forest environments in order to determine which mechanisms are important.

*Elizabeth K. Berner*

## FURTHER READING

Ahrens, C. Donald. *Essentials of Meteorology: An Invitation to the Atmosphere.* Belmont, Calif.: Brooks/Cole Cengage Learning, 2012. Discusses various topics in weather and the atmosphere. Covers topics such as tornadoes, thunderstorms, acid deposition, air pollution, humidity, and cloud formation.

Berner, Elizabeth K., and Robert A. Berner. *The Global Water Cycle: Geochemistry and Environment.* Englewood Cliffs, N.J.: Prentice-Hall, 1986. Designed for college freshmen. Discusses acid rain and its effects. Various chapters describe the formation and global distribution of sulfur and nitrogen gases, from both natural and human-made causes. Covers the formation of acid rivers and acid lakes.

Brimblecombe, Peter, et al., eds. *Acid Rain: Deposition to Recovery.* Dordrecht: Springer, 2010. Compiles articles from *Water Air, & Soil Pollution: Focus,* Volume 7 (2007). Discusses acid rain from various perspectives. Covers ecological impact, wet versus dry deposition, soil chemistry, and surface water quality.

Hill, Marquita K., *Understanding Environmental Pollution.* Cambridge, England: Cambridge University Press, 2010. Examines numerous pollution problems and conditions. Chapter 6 deals specifically with acid deposition.

Jenkins, Jerry C., et al. *Acid Rain in the Adirondacks: An Environmental History.* Ithaca, N.Y.: Cornell University Press, 2007. Provides an account of the history of acid rain research conducted in the Adirondacks region. Begins with research from the 1950's through the 1990's. Includes multiple case studies. Final chapter summarizes conclusions drawn from the research. Offers chapter summaries, appendices, a bibliography, and an index.

Klaassen, G. *Acid Rain and Environmental Degradation: The Economics of Emission Trading.* Brookfield, Vt.: Edward Elgar, 1996. Examines the pros and cons of the industrial practice of trading the right to emit acid-rain-causing pollutants. Provides illustrations, index, and a lengthy bibliography.

Lane, Carter N. *Acid Rain. Overview and Abstracts.* Hauppauge, N.Y.: Nova Science Publishers, 2003. Compiles articles by several authors and provides an informative overview that describes the causes, effects, and ramifications of acid rain in a nontechnical manner.

Larssen, Thorjørn, et al. "Acid Rain in China." *Environmental Science & Technology* 40 (2006): 418-425. Describes the increase in air pollution and acid rain in China since the late 1970's. Discusses the impact of economic growth in the country, particularly the effects of increased coal combustion and emissions pollution. Discusses the current and future effects of acid rain on Chinese ecosystems.

Likens, Gene E., R. F. Wright, J. F. Galloway, and T. F. Butler. "Acid Rain." *Scientific American* 241 (October, 1979): 43-51. Describes the characteristics of acid rain and how it was recognized. Reviews the areas receiving acid rain in the United States and Europe. Describes the changes in acidity of rain over time and how they correlate with changes in sulfur and nitrogen emissions. Discusses the chemistry of acid lakes.

McCormick, John. *Acid Earth: The Politics of Acid Pollution.* 3d ed. London: Earthscan, 1997. Reviews the causes and environmental consequences of acid rain and acid deposition. Describes efforts to curb the drift of acid-rain-causing gases across state and national boundaries. Appropriate for the high school reader. Includes illustrations, maps, index, and references.

Mills, K. H., S. M. Chalanchuk, and D. J. Allan. "Recovery of Fish Populations in Lake 223 from Experimental Acidification." *Canadian Journal of Fisheries and Aquatic Sciences* 57 (2000): 192-204. Discusses the effects of increased acidity on multiple fish species in Canadian lakes. Describes the impact of acid rain on ecosystems.

Mohnen, Volker A. "The Challenge of Acid Rain." *Scientific American* 259 (August, 1988): 30-38. Discusses the formation of acid rain and how it is possible to predict where acid deposition will occur from known sulfur emissions. Describes the effects of acid rain on soils, forests, and lakes. Evaluates several ways of changing power-plant technology to reduce sulfur emissions. Written for a general audience.

Rose, John, ed. *Acid Rain: Current Situation and Remedies.* Langhorne, Pa.: Gordon and Breach Science Publishers, 1994. Describes the causes of acid rain, the techniques used to determine acid rain levels, and possible solutions to acid rain and air pollution problems. A good introduction to the field, written for college-level readers. Includes illustrations, bibliographical references, and an index.

Sommerville, Richard C. J. *The Forgiving Air: Understanding Environmental Change.* 2d ed. Boston: American Meteorological Society, 2008. Focuses on the various consequences of air pollution, including the depletion of the ozone layer, climatic changes, the greenhouse effect, and acid rain. Intended as an introduction for the layperson. Bibliography and index.

Wyman, Richard L., ed. *Global Climate Change and Life on Earth.* New York: Chapman and Hall, 1991.

Compiles articles on various aspects of climate changes and the effects of these changes.

**See also:** Air Pollution; Atmosphere's Evolution; Atmosphere's Global Circulation; Atmosphere's Structure and Thermodynamics; Atmospheric Properties; Auroras; Cosmic Rays and Background Radiation; Earth-Sun Relations; Freshwater and Groundwater Contamination Around the World; Global Warming; Greenhouse Effect; Industrial Metals; Nuclear Winter; Oil and Gas Distribution; Ozone Depletion and Ozone Holes; Plate Tectonics; Radon Gas; Rainforests and the Atmosphere

# AIR POLLUTION

*Air pollution is generated from both natural and human-made sources. Natural sources include pollen from plants, gases and particulate matter from volcanoes and decomposing organic matter, and windblown dust. Human-made sources include industrial and automobile emissions and airborne particles associated with human-induced abrasion.*

## PRINCIPAL TERMS

- **acid rain:** precipitation having elevated levels of acidity relative to pure water
- **atmosphere:** the layer of mixed gases that surrounds Earth
- **carbon dioxide:** $CO_2$, one of many minor gases that are natural components of the atmosphere; the product of the complete oxidation of carbon
- **greenhouse effect:** the environmental process that results when heat energy is absorbed and retained in the atmosphere by various gases and is not radiated out into space
- **inversion:** an unusual atmospheric condition in which temperature increases with altitude
- **off-gassing:** the spontaneous emission of entrained or entrapped gases from within natural and artificial sources
- **oxides of nitrogen:** several gases that are formed when molecular nitrogen is heated with air during combustion, primarily NO and $NO_2$
- **oxides of sulfur:** gases formed when fuels containing sulfur are burned, primarily $SO_2$
- **ozone:** a highly reactive compound composed of three atoms of oxygen, as $O_3$
- **photochemical oxidants:** pollutants formed in air by primary pollutants undergoing a complex series of reactions driven by light energy
- **photochemical reaction:** a type of chemical reaction that can occur in polluted air driven by the interaction of sunlight with various pollutant gases

### Earth's Atmosphere

Air pollution results from the unusual addition of gases, solids, and liquids to the atmosphere. The concentration of pollutants depends on prevailing atmospheric conditions as well as emission rates. Once pollutants are put into the atmosphere, it is impossible to control them to any significant degree. Thus emissions at the local level contribute to regional and global air pollution problems, such as smog and photochemical oxidants, acid precipitation, the depletion of the ozone layer, and global

warming associated with the intensification of the greenhouse effect. Although there are many air pollutants, the major ones are usually associated with burning, particularly the burning of fossil fuels and oil-based products. They are generally unburned hydrocarbons, oxides of sulfur and nitrogen, carbon dioxide, carbon monoxide, various photochemical oxidants and reactive compounds, and particulate matter from many different sources.

The atmosphere is a mixture of gases, aerosols, and particulate matter surrounding Earth. The concentration of some of the gases in clean air is fairly constant both spatially and temporally. Consequently, these gases are referred to as stable or permanent gases. Nitrogen and oxygen, the two most abundant permanent gases, account for 78 percent and 21 percent of the total atmosphere by volume, respectively. Gases that experience noticeable temporal and spatial variations are termed variable gases. The two most abundant of these are water vapor and carbon dioxide. The average concentration of carbon dioxide is about 0.034 percent. It varies seasonally in response to the growth cycle of plants, daily in response to plant photosynthesis, and spatially in response to the burning of fossil fuels. Water vapor is also highly variable. Some variable gases have natural origins and tend to have relatively high concentrations in urban areas. They are methane, carbon monoxide, sulfur dioxide, nitrogen dioxide, ozone, ammonia, and hydrogen sulfide.

The atmosphere is stratified according to its vertical temperature gradient. From the ground surface up, the major layers are the troposphere, the stratosphere, the mesosphere, and the thermosphere. The troposphere contains the bulk of atmospheric gases and, under normal conditions, is characterized by a fairly uniform temperature decline from the surface upward. The uppermost limit of the troposphere is called the tropopause, a transition zone between the troposphere and stratosphere where temperatures stabilize with increasing altitude. The troposphere extends up to about 10 kilometers. The next layer encountered is the stratosphere, which extends from about 12.5 kilometers up to about 45 kilometers above the surface. In its lower layer, the temperature gradient is somewhat stable. At an elevation of about 30 kilometers, however, the temperature starts to increase. Located within the stratosphere about 24 to 32 kilometers above the earth's surface is a zone with a relatively high concentration of ozone, a triatomic form of oxygen. This zone is called the ozone layer. It is important because the ozone absorbs most of the incoming ultraviolet rays emitted by the sun, preventing them from reaching the surface where they would have harmful effects on plant and animal life.

The two uppermost layers, the mesosphere and the thermosphere, have a distinctive temperature gradient. In the mesosphere, temperatures decline steadily with altitude, a condition that continues until its transition zone with the thermosphere, called the mesopause, is reached. At latitudes of 50 degrees north and higher, mesospheric clouds known as noctilucent clouds are sometimes seen during summer. These clouds may be anthropogenic in origin. The last layer, the thermosphere, slowly gives way to outer space and has no defined upper limit.

## ATMOSPHERIC INVERSIONS AND SMOG

Vertical and horizontal mixing of air is necessary to dilute pollutants in the atmosphere. Under normal conditions, temperatures decline with altitude in the troposphere. This decline in temperature with altitude is referred to as the thermal or environmental lapse rate. The warmer air near the surface rises, mixes with the air above it, and is dispersed upward by winds. This dilution process is important in reducing the concentration of pollution near the surface. Conversely, the vertical mixing of air is inhibited when the temperature profile in the troposphere inverts, developing a temperature or thermal inversion. When an inversion exists, a layer of warmer air becomes sandwiched between two layers of cooler air above and below, and the warmer, less dense air does not rise as it would normally. Pollutants can then accumulate below the warmer air as vertical mixing is prevented.

Conditions for temperature inversions develop when the earth readily radiates heat energy from its surface on clear nights or when air subsides and warms adiabatically from compression. On cool, clear nights, the earth readily radiates heat energy to space, cooling the surface. Air near the surface is, in turn, cooled by conduction, while the air above it is still relatively warm. This condition is referred to as a radiation inversion. Radiation inversions are common during autumn and are usually short-lived, as the rising sun in the morning heats the air near the surface, causing the inversion to dissipate as the day advances. Less frequent but more persistent subsidence-type inversions can occur when cooler air subsides in high-pressure systems or in valleys, as cooler, denser air descends along adjacent mountain slopes. Subsidence inversion episodes may last for days, allowing pollutants to concentrate to excessive levels, causing eye irritation, respiratory distress, reduced visibility, corrosion of materials, and soiling of clothes.

The atmosphere has inherent self-cleansing mechanisms. Pollutants are removed from the atmosphere through fallout due to gravitational settling, through rainout in condensation and precipitation processes, through washout as waterdrops and snowflakes accumulate pollutants as they fall to Earth, and through chemical conversion. Solar radiation, winds, and atmospheric moisture are important meteorological factors in these removal processes.

Chemical reactions between two or more substances in the atmosphere produce secondary pollutants, which are those created from other pollutants that have been released directly into the atmosphere from identifiable sources. Smog is a product of such reactions. Stability in the atmosphere that accompanies inversions provides favorable conditions for smog to develop. Smog is produced by chemical and photochemical reactions involving primarily sulfur oxides, hydrocarbons, and oxides of nitrogen. Smog that is characterized by sulfur oxides is called sulfurous smog and is associated with the burning of fuels having a relatively high sulfur content. This type of smog is common in undeveloped countries. Photochemical smog develops when oxides of nitrogen and various hydrocarbons undergo photochemical reactions to produce ozone and other chemical oxidizers. Sunlight promotes the reactions, and automobile exhaust is a primary source of nitrogen oxides. This is the type of smog typically encountered in large cities and urban centers such as Los Angeles and Mexico City.

## ACID PRECIPITATION

While smog is a relatively localized phenomenon closely associated with urban areas, acid precipitation is a more widespread phenomenon. Its effects are observed in national parks, agricultural regions, forested areas, and lakes and other bodies of water as well as in urban centers. Acid rain develops when oxides of sulfur or nitrogen combine with water vapor in the atmosphere to form sulfuric and nitric acids that fall back to Earth in precipitation. Once released into the atmosphere, these oxides and the compounds formed from them can travel great distances before returning to Earth in precipitation or as dry particulates, as much as 1,000 to 2,000 kilometers over three to five days. This long-range transport allows time for chemical reactions to convert pollutant gases into components of acid precipitation. Evidence suggests that the pH values in precipitation have been dropping, becoming more acidic, for some years. Wet precipitation is not the only way pollutants find their way to the surface. Diffusion and settling enable acidic gases and particles to find their way to the ground even under dry conditions. It is now widely accepted that both wet and dry deposition can be traced to human activity.

Much evidence has been gathered documenting the damaging effects of acid precipitation. These effects include damage to wildlife in lakes and rivers, reduction of forest productivity, damage to agricultural crops, and deterioration of human-made materials. Acid precipitation, also suspected of promoting the release of heavy metals from soils and pipelines into drinking water supplies, has different effects on different ecological systems and is most damaging to aquatic ecosystems. Acidity in precipitation at a given time depends not only on the type and quantity of pollutants being produced but also on the prevailing and immediate atmospheric conditions. Stagnant air, resulting from upper-level inversions, tends to cause higher levels of acidity. Furthermore, prevailing and local atmospheric systems are associated with the spread of acid precipitation over broader areas. Higher exhaust stacks, while minimizing levels of airborne pollutants locally, simply disperse the pollutants over larger areas, thus increasing their residence time in the atmosphere.

## OZONE DEPLETION AND GLOBAL WARMING

It is now realized that the impact of air pollution is more far-reaching than the troposphere. Evidence indicates that pollutants making their way up to the stratosphere are causing the ozone layer to break down or dissipate. Even though ozone constitutes a very small portion of the atmosphere, only about one part per million, it absorbs almost all of the ultraviolet rays from the sun, preventing them from reaching Earth's surface. Research findings from satellite-based monitoring systems have shown that there has been a breakdown of the ozone shield over the Antarctic, where a hole has been identified in the ozone layer. More recent research and satellite-based observation has shown that the ozone layer also appears to be thinning over the Arctic. While early laboratory studies showed that oxides of nitrogen could attack ozone, attention later focused on chlorofluorocarbons (CFCs) as being responsible for the decline in ozone. These compounds were widely used as refrigerants in common household appliances and air conditioning systems, propellants in aerosol sprays, agents for producing foam, and cleansers for electronic products, due to their low boiling points, facile compressibility, and very low chemical reactivity. Behaving much like inert gases, they do not degrade readily in the troposphere, but eventually make their way into the stratosphere. Laboratory studies have shown that when the CFC molecules come in contact with ozone and ultraviolet light, they enter into a complex series of gas-phase reactions by which they are converted into more reactive gases, such as chlorine. Since these gases tend to linger in the atmosphere for many years, it is believed that even though the use of CFCs was discontinued, the ozone layer would continue to disintegrate for several years afterward.

Further evidence suggests that CFCs not only destroy the ozone but also trap heat energy radiated from the ground and contribute to heating the atmosphere. The trapping of sensible heat energy in the atmosphere by gases is called the greenhouse effect. One of the most important gases contributing to the greenhouse effect, however, is not a chlorofluorocarbon but carbon dioxide. Carbon dioxide moves in a continuous cycle throughout the environment. It provides a link between the organic and inorganic components of the environment. Reacting with water and solar energy through photosynthesis in plants, it forms glucose that is subsequently passed through the food chain as a source of energy required by essentially all animal species.

Carbon dioxide is given off by plants and animals to the atmosphere during respiration. When plant and animal remains decay, carbon dioxide is passed back to the atmosphere and hydrosphere through the most natural processes of decomposition. When fossil fuels are burned, however, those natural processes are short-circuited, and large amounts of carbon dioxide are released directly into the atmosphere.

About 0.03 percent of the total atmosphere is carbon dioxide. Molecules of carbon dioxide in the atmosphere absorb and retain infrared radiation as heat, in much the same way that the glass walls of a greenhouse reflect heat back into the structure rather than allowing it to escape. While it is transparent to shortwave radiation from the sun, carbon dioxide absorbs strongly in the sensible heat or longwave radiation band. It is hypothesized that an increase in atmospheric carbon dioxide causes a decrease in outgoing longwave radiation and thus an increase in the atmospheric temperature.

The consequences of rising global temperatures will greatly alter the earth's surface. As the atmosphere warms, polar ice and glaciers will begin to melt and the rising oceans could flood many of the world's coastal regions, devastating low-lying countries. As shorelines rise, saltwater intrusion will contaminate the drinking-water supplies of many cities worldwide. Agricultural regions of the middle-latitude countries will migrate farther northward, increasing the length of the growing seasons in Canada and Russia.

## STUDY OF AIR POLLUTION

Methods of studying air pollution include controlled laboratory experiments, simulations in fluid-modeling facilities, computer simulations, and mathematical modeling. Controlled laboratory experiments are conducted in laboratories where gases are mixed to determine how they react. Laboratory experiments usually do not provide the definitive answer to what is actually occurring in the ambient environment because many variables cannot be replicated. These studies suggest what should be further studied and monitored in the natural environment. Some laboratory studies have simulated atmospheric conditions in a controlled environment, such as a biosphere where the impact of pollution on plants

can be determined by introducing the pollutants at various levels. Simulation studies may also include gathering data from fluid-modeling facilities, where the environment is replicated using miniature models and atmospheric conditions are controlled. These studies often contribute to an understanding of the dispersion and deposition of air pollutants.

Monitoring the atmosphere is an essential component of air pollution studies involving computer simulations and mathematical modeling. These types of studies rely largely on data sources or values from the ambient environment and are constrained by difficulties of measuring ambient levels. Sometimes, vessels containing samples of air are collected and returned to the laboratory for analysis, but continuous monitoring devices that are placed in the ambient environment are more common.

Many of the monitoring devices involve a colorimetric or photometric technique. Air to be analyzed is isolated and subjected to conditions in which carefully selected gas phase reactions can produce specific compounds from the pollutants that are present. The reaction product is then analyzed by photometric techniques, in which the concentration of light-absorbing substances is indicated by the light intensity that reaches the photometer. Particulate matter can be measured by fairly simple collectors that may use adhesive coated paper or filtration, and measuring the increase in weight resulting from the trapped particles. Another method involves passing a known volume of air through filter paper and measuring the intensity of light passing through it. The intensity of light indicates the scattering and absorptive properties of aerosols; it is expressed as a coefficient. Instruments may be located at the surface, mounted on airplanes, or allowed to ascend in balloons. Acidity in precipitation is determined by standard measures of acidity using a pH indicator.

## SIGNIFICANCE

Many industry officials continue to downplay the threat of pollution to the atmosphere and dispute the extent of damage caused by pollution. Yet as greater amounts of pollutants are released into the atmosphere, it becomes increasingly difficult to control their levels and reverse any resulting damage. The depletion of the ozone layer and increasing levels of smog and acid rain are all very real and threatening manifestations of air pollution. Efforts to protect the atmosphere must be made on a worldwide basis, as gases cannot be confined to political boundaries. Air pollution may be lowered by reducing emissions or by extracting pollutants from the atmosphere through natural means. Thus any plans to reduce air pollution should center on one or both of these approaches.

*Jasper L. Harris*

**FURTHER READING**

Bandy, A. R., ed. *The Chemistry of the Atmosphere: Oxidants and Oxidation in the Earth's Atmosphere.* Cambridge, England: Royal Society of Chemistry, 1995. Compiles essays and lectures first presented at the Priestley Conference in association with the Royal Society of Chemistry and the American Chemical Society at Bucknell University. Covers such topics as atmospheric chemistry, the ozone layer, and the causes and solutions to air pollution. Somewhat technical. Suitable for readers with some background in the subject.

Bolius, David. *Paleoclimate Reconstructions Based on Ice Cores: Results from the Andes and the Alps.* Berlin: SVH-Verlag, 2010. Presents a study of climate change using samples taken from ice cores in the Andes and Alps. Includes sampling methodology and chemical analysis. Documents the history of anthropogenic air pollution. Best suited for graduate students, professional geologists, and paleoclimatologists.

Brimblecombe, Peter. *Air Composition and Chemistry.* 2d ed. New York: Cambridge University Press, 1996. Provides a thorough account of atmospheric chemistry and the techniques and protocols involved in determining the makeup and properties of air. Written at the college level, but accessible to laypersons. Includes illustrations, maps, bibliography, and index.

Brimblecombe, Peter, et al., eds. *Acid Rain: Deposition to Recovery.* Dordrecht: Springer, 2010. Compiles articles from *Water Air, & Soil Pollution: Focus.* Discusses acid rain from various perspectives. Covers human impact, ecological impact, wet versus dry deposition, soil chemistry, and surface water quality.

Haerens, Margaret. *Global Viewpoints: Air Pollution* Farmington Hills, Mich.: Greenhaven Press, 2011. Written and compiled as a resource for students. Presents the problem of air pollution in a global context, in which regional air pollution effects are but components.

Hopler, Paul. *Atmosphere: Weather, Climate, and Pollution.* New York: Cambridge University Press, 1994. Offers a wonderful introduction to the study of Earth's atmosphere and its components. Chapters discuss the causes and effects of global warming, ozone depletion, acid rain, and climatic change. Illustrations, color maps, and index.

Larssen, Thorjørn, et al. "Acid Rain in China." *Environmental Science & Technology* 40 (2006): 418-425. Discusses the increase in air pollution and acid rain in China since the late 1970's. Discusses the trend of economic growth in the country and its impact on rates of coal combustion and emissions pollution. Discusses the current and future effects of acid rain on Chinese ecosystems.

McGranahan, Gordon, and Frank Murray, eds. *Air Pollution and Health in Rapidly Developing Countries* London: Earthscan Publications, 2003. Provides a thorough examination of the effects of air pollution on human health in the context of countries undergoing rapid technological development.

Ostmann, Robert. *Acid Rain: A Plague upon the Waters.* Minneapolis, Minn.: Dillon Press, 1982. Provides insight into the political history and development of the acid rain problem. Addresses the effects of acid rain on society from the perspective of a concerned citizen.

Sportisse, Bruno. *Fundamentals in Air Pollution: From Processes to Modelling.* New York: Springer, 2009. Discusses issues related to air pollution, such as emissions, the greenhouse effect, acid rain, urban heat islands, and the ozone hole. Well organized and clearly explained topics. Accessible to the layperson.

Stolarski, Richard S. "The Antarctic Ozone Hole." *Scientific American* 258 (January, 1988): 30-36. An excellent review of research efforts pertaining to ozone depletion in the Antarctic. A very understandable explanation of the chemical reactions associated with CFCs and ozone depletion. Presents an argument for natural processes as a major contributor to the ozone hole.

Tiwary, Abhishek, and Jeremy Colls. *Air Pollution: Measurement, Modelling and Mitigation.* New York: Routledge, 2010. A college-level textbook. Provides a detailed introduction to current methods of the measurement, modeling, and mitigation of air pollution using numerous charts and graphs. Some mathematical background expected for examples.

Vallero, Daniel. *Fundamentals of Air Pollution.* 4th ed. Burlington, Mass.: Academic Press, 2008. A good explanation of the physical aspects of air pollution. Offers an interdisciplinary approach, drawing from meteorology, chemistry, physics, engineering, medicine, and the social sciences. Some attention is given to the political aspects of air pollution problems.

Wyman, Richard L., ed. *Global Climate Change and Life on Earth.* New York: Chapman and Hall, 1991. A collection of articles on various aspects of climate change and the effects of these changes.

**See also:** Acid Rain and Acid Deposition; Atmosphere's Evolution; Atmosphere's Global Circulation; Atmosphere's Structure and Thermodynamics; Atmospheric Properties; Auroras; Cosmic Rays and Background Radiation; Earth-Sun Relations; Freshwater and Groundwater Contamination Around the World; Global Warming; Greenhouse Effect; Hydrologic Cycle; Nuclear Winter; Ozone Depletion and Ozone Holes; Precipitation; Radon Gas; Rainforests and the Atmosphere; Soil Chemistry; Weathering and Erosion

# ALLUVIAL SYSTEMS

*Alluvial systems include a variety of different depositional systems, excluding deltas, that form from the activity of rivers and streams. Much alluvial sediment is deposited when rivers top their banks and flood the surrounding countryside. Buried alluvial sediments may be important water-bearing reservoirs or petroleum-containing strata.*

## PRINCIPAL TERMS

- **braided river:** a relatively shallow river with many intertwined channels; its sediment is moved primarily as riverbed material
- **ephemeral stream:** a river or stream that flows briefly in response to nearby rainfall; such streams are common in arid and semiarid regions
- **floodplain:** the relatively flat valley floor on either side of a river that may be partly or wholly occupied by water during a flood
- **longitudinal bar:** a midchannel accumulation of sand and gravel with its long axis oriented roughly parallel to the river flow
- **meandering river:** a river confined essentially to a single channel that transports much of its sediment load as fine-grained material in suspension
- **oxbow lake:** a lake formed from an abandoned meander bend when a river cuts through the meander at its narrowest point during a flood
- **point bar:** an accumulation of sand and gravel that develops on the inside of a meander bend
- **saltation:** the movement of non-suspended sediment particles in the direction of fluid flow as particles impact against other particles
- **transverse bar:** a flat-topped body of sand or gravel oriented transverse to the river flow

## SEDIMENT

Deposits of silt, sand, and gravel produced by the activity of rivers and streams are called alluvial sediments. Sediment in rivers is moved primarily as either suspended load or bed load. The suspended load is the finest portion of sediment, composed of silt and clay particles, and is carried suspended within the flow itself by fluid turbulence. Material moved along the bottom of the river by rolling, sliding, and bouncing (saltation) is called the bed load, and it makes up the coarse fraction of a river's sediment. Rivers can be divided into four categories based on their morphology: braided, meandering, anastomosing, and straight. Straight rivers are rare, usually appearing only as portions of one of the other river types, and anastomosing rivers can be considered a special type of meandering river.

Several criteria are used to characterize alluvial systems. They include grain size, dominant mode of sediment transport (suspended load versus bed load), and migrational pattern of the river channel. Alluvial sediments can be broadly divided into three interrelated depositional settings: braided rivers, alluvial fans, and meandering rivers.

## BRAIDED RIVERS

Braided rivers have low sinuosity, which is defined as the ratio of the length of the river channel to the down-valley distance. They are characterized by relatively coarse-grained sediment transported as bed load. Fine-grained sediments make up a minor portion of the deposits. The main channel is internally divided into many subchannels and bars, which give the river a braided pattern. River bars are ridge-like accumulations of sand and gravel formed in the channel or along the banks, where deposition is induced by a decrease in velocity. Transverse bars are flat-topped ridges, oriented transversely to the flow, that grow by down-current additions and migration of sediment. Longitudinal bars are midchannel sand and gravel accumulations oriented with their long axes roughly parallel to current flow. During low-water stages, the braided pattern is very apparent, and water occupies only the subchannels. It is only during high-water stages that the entire braided channel has water in it, at which time the braided appearance may no longer be evident. The bars that occur in these rivers form as a result of high sediment loads and fluctuations in river discharge.

Braided rivers form in regions where sediment is abundant, water discharge fluctuates (but may be high), and, usually, vegetation is sparse. Some braided rivers and streams have flowing water in them sporadically, with long periods of dryness in between. These streams are called ephemeral, or intermittent. Braided rivers tend to have relatively high gradients. As such, they commonly occur as the upper reach of a river that may become a meandering river downstream as the sediment's grain size and the gradient both decrease.

## ALLUVIAL FANS

Alluvial fans are deposits that accumulate at the base of a mountain slope. There, mountain streams encounter relatively flat terrain and lose much of their energy, subsequently depositing the sediment that they were moving. The effect is similar to that of delta formation, which occurs as a flowing stream enters a body of relatively still water and loses the energetic motion necessary to maintain its suspended sediment load. This type of stream shifts the position of its channel over time, as sediment continues to be deposited, and builds a fan-shaped accumulation of debris that is coarse-grained near the mountain front and becomes progressively finer-grained away from the highland. Alluvial fans are best developed and observed in arid or semiarid climates, where vegetation is sparse and water flow is intermittent. Large quantities of sediment may be moved during short-term flash-flood events. There are also humid-climate fans, such as the enormous Kosi fan in Nepal, which measures 150 kilometers in its longest dimension.

Both arid and humid fans are built at least in part by deposition in a braided stream environment. Stream discharge, and therefore sediment deposition, is discontinuous on arid fans. Braided streams may operate over the fan's entire surface or predominantly on the outer regions, away from the region where the stream leaves the confines of the mountain valley. Arid fans are also built by deposition from mudflows and debris flows. These flows differ from braided stream flows in that they contain a lower proportion of water and a much higher proportion of debris. Humid fans have braided stream systems operating continuously on their surface. Their overall deposits are similar to braided stream deposits formed in other sedimentary environments.

## MEANDERING RIVERS

Meandering rivers have a greater sinuosity than braided rivers and are usually confined to a single channel. They have a lower gradient and therefore are typically located downstream from braided rivers. Sediment in meandering rivers is moved mostly as fine-grained suspended load. Several different types of sedimentary deposit result from the activity of meandering rivers. The coarsest material available to

the river is moved and deposited within the deepest part of the channel. These gravelly deposits are thin and discontinuous. Point bars develop on the inside curve of a meander bend and are a major site of sand deposition, although silt and gravel may also be components of point-bar deposits. Deposition takes place on point bars because of flow conditions in the river as water travels around the bend. Erosion takes place on the bank opposite the point bar, and in this way, meanders migrate. When a river floods and tops its banks, water and finer-grained sediment spill out of the channel and onto the surrounding valley floor, which is called the floodplain. On the bank directly adjacent to the river channel, large amounts of sediment are deposited to form natural levees, which are elongated narrow ridges parallel to the channel. Levees help to confine the river but are still topped during major floods. As the river spreads out across the floodplain, silt and clay are deposited.

Meandering rivers are constantly shifting their location within the river valley. In this way, very thick alluvial deposits may accumulate through time. New channels may be created between meanders such that old meander bends are cut off and isolated. These isolated meander bends are termed "oxbow lakes" and tend to fill quickly with sediment.

A special type of meandering river is an anastomosing river. It is characterized by a system of channels that do not migrate as much as meandering river channels and that are separated by large, permanent islands.

## TRANSPORT AND DEPOSITION

Alluvial sediments require running water for transport and deposition. This water may be available year-round, such that rivers and streams are constantly active. Sporadic stream discharge produces alluvial sediments in arid and semiarid environments. Alluvial sediments are also associated with glaciers. Large amounts of sediment with a wide range in grain size are deposited directly by glaciers. Streams fed by glacial meltwaters are important and effective agents of transport and deposition of this sediment. Most streams associated with glaciers are bed load streams and therefore have a braided pattern. Their sediment is usually quite coarse-grained.

Alluvial systems form broad, interconnected networks of drainage that feed water and sediment from highlands or mountainous regions to lowlands and eventually to the sea. These drainages form recognizable patterns that are controlled by the type of rock; by the type of deformation, if any, that the earth's crust in the region has undergone; and by the region's climate. As river systems age, the landscape changes and evolves. In this way, the surface of the land is sculpted by rivers, in concert with other surface processes.

**STUDY OF ALLUVIAL SYSTEMS**

The study of alluvial systems can be divided roughly into two categories: the study of modern river systems and the study of sedimentary rocks interpreted to have formed in some type of alluvial system. The study of modern alluvial processes and the sedimentary deposits that they generate is crucial to the understanding of such deposits in the geological record of rock strata. By understanding modern rivers—the ways in which sediment is moved and deposited, the changes through time in channel shape and location, and the characteristics of the deposits in relation to specific physical conditions—geologists can begin to interpret ancient alluvial sediments.

Modern alluvial systems are studied in a number of ways. It is important to know as much as possible about the flow of water itself, so measurements are made of flow velocity, depth, and width. The channel shape and configuration are also measured. Samples of river sediment, both suspended load and bed load, are collected, and estimates are made of how much sediment is moved by a river. It is also important to look at recent alluvial sediments not now directly associated with the river system. That may be done by digging trenches or collecting samples from floodplains or other alluvial sediments. This type of study looks at the last few hundred or few thousand years of river activity and is critically important to the understanding of these systems. The geologist must study not only what is happening today but also how the system has evolved. In this way, knowledge becomes a predictive tool that greatly enhances the overall understanding of the phenomenon.

It was not until the 1960's that geologists began to realize how large has been the contribution of alluvial systems to sedimentary rocks. This realization came about as a direct result of studies of modern alluvial systems that, in turn, allowed the correct identification of ancient alluvial deposits. The study of ancient alluvial deposits takes many forms and provides some

information that cannot be gathered from modern deposits. For example, modern deposits, for the most part, have only their uppermost surface exposed, except along the bank and in erosion gullies. Ancient accumulations, in contrast, have been transformed into rock such as shale, sandstone, and conglomerate and commonly are parts of uplifted and dissected terrains, including mountain ranges. These exposures of alluvial deposits allow the three-dimensional architecture of alluvial systems to be studied. Geologists look carefully at vertical changes and associations in the types and abundances of sediment that form as a result of alluvial processes.

It has been found that both meandering and braided rivers commonly form cycles of sedimentation that begin with relatively coarse-grained debris and progress to fine-grained debris. These cycles result mainly from the shifting and migration of the channel system. The coarser-grained base represents the channel and bar deposits, and the finer-grained top represents the overbank floodplain deposits.

The study of ancient alluvial deposits also provides clues to the geologic evolution of a region. The specific mineral composition of sedimentary rocks can indicate the nature of the terrain from which the sediment was derived. This, in turn, contributes to an understanding of the history that led to the generation of the ancient river system that produced the alluvial deposit.

**SIGNIFICANCE**

Alluvial systems operate over much of the earth's surface. They move and deposit enormous quantities of sediment each year and are both a boon and a hindrance to humankind. River valleys and floodplains are desirable regions for agricultural development because of the fertile soil typically found there. Rivers, however, naturally flood about every 1.5 years. This flooding often causes huge losses in property, crops, and sometimes even lives. Flooding associated with alluvial fans can be highly energetic and can occur almost without warning, usually in response to heavy precipitation over a short period. Water levels can build very quickly in narrow valleys with little vegetation and produce a rushing wall of water.

Many different economically valuable materials are found in alluvial deposits. Because alluvial deposits are relatively coarse-grained, they have spaces between the grains that may contain usable fluids,

15

such as water, oil, and natural gas. Many important aquifers are found in alluvial deposits. Petroleum typically originates in marine deposits, but it commonly migrates and may form reservoirs in alluvial deposits. Most sandy alluvial deposits are composed predominantly of quartz grains; however, concentrations of a number of different minerals and ores, including gold and diamonds, occur in alluvial deposits. Another important economic resource in alluvial deposits is the sand and gravel itself. This material is used for road construction and in the manufacture of cement and concrete.

The deposition of sediment from river systems represents the wearing away of the land. Much alluvial sediment eventually makes its way to the ocean, where it undergoes reworking by marine processes and deposition on continental shelves or perhaps in the deep sea. Through time, as continents move relative to one another, these sediments may be subducted into the mantle layer below a continental mass, or become compressed, folded, and uplifted in the formation of major mountain chains. The Appalachian Mountains in the eastern United States and the Himalaya in northern India and China are only two examples of mountains composed in part of sedimentary rocks, some of which are alluvial in origin.

*Bruce W. Nocita*

**FURTHER READING**

Charlton, Ro. *Fundamentals of Fluvial Geomorphology.* Abingdon: Routledge, 2007. Examines the manner in which river systems respond to environmental change and discusses the importance of this understanding for successful river management. Provides a comprehensive overview of river channel management methodologies.

Christopherson, Robert W., and Mary-Louise Byrne. *Geosystems: An Introduction to Physical Geography.* Toronto: Pearson Education Canada, 2006. Suitable for senior high school and college-level students as well as general readers. Provides a comprehensive overview of alluvial systems in the context of physical geography, with numerous references to online resources throughout.

Darby, Stephen, and David Sear. *River Restoration: Managing the Uncertainty in Restoring Physical Habitat.* Hoboken, NJ: John Wiley & Sons, Ltd., 2008. Theoretical and philosophical issues with habitat restoration begin this publication and provide a strong foundation for decision making. Later chapters address logistics, planning, mathematical modeling, and construction stages of restoration. Post-construction monitoring and long-term evaluations round out this publication to provide a full picture of the habitat restoration process. This text is highly useful for anyone involved in the planning and implementing of habitat restoration.

Davis, Richard A., Jr. *Depositional Systems: A Genetic Approach to Sedimentary Geology.* 2d ed. Englewood Cliffs, N.J.: Prentice-Hall, 1992. Offers good introductory chapters on alluvial systems and on related subjects, such as sediment transport. Written for the college-level reader.

Foresman, Timothy, and Alan H. Strahler. *Visualizing Physical Geography.* New York: John Wiley & Sons, 2012. Uses a unique approach to presenting the concepts of physical geography by heavily integrating visuals from *National Geographic* and other sources with the narrative to vividly illustrate the manner in which physical geographic processes are interconnected.

Harvey, Adrian M., Anne E. Mather, and Martin R. Stokes. *Alluvial Fans: Geomorphology, Sedimentology, Dynamics.* Geological Society Special Publication 251. London: Geological Society Publishing House, 2005. Collects papers by experts and researchers in the field. Focuses on resolving how large-scale geological controls such as tectonics and environmental change are reflected in the dynamics of alluvial fan formation. A wide range of rivers are examined. Characteristics of alluvial fans, processes affecting the alluvial fans, and the resulting ecological impacts are discussed.

Pavlopoulos, Kosmas, and Niki Evelpidou. *Mapping Geomorphological Environments.* New York: Springer, 2009. This text contains a chapter discussing fluvial environments and processes. Glacial formations and the waterfalls of glacial state park are also discussed. Many other geologic formations are described and mapping methodologies are examined.

Reading, H. G., ed. *Sedimentary Environments: Processes, Facies, and Stratigraphy.* 3d ed. Cambridge, Mass.: Blackwell Science, 1996. Probably the most comprehensive text available on sedimentary environments. Much of the material is technical, but the text has excellent figures and photographs and will not overwhelm the careful reader.

Savenije, Hubert H.G. *Salinity and Tides in Alluvial Estuaries.* San Diego: Elsevier, 2005. This book is provides clear descriptions of alluvial fan processes. Tidal movements, deposition dynamics, and estuary salinity are discussed. The text has a number of equations, without much explanation of the variables, so a background in hydraulics and water dynamics is needed. This book is well suited for researchers and students looking for examples of the applications of physics in alluvial systems.

Schumm, Stanley A. *Active Tectonics and Alluvial Rivers.* New York: Cambridge University Press, 2002. Suitable for college-level readers, this source is packed with information on river systems and tectonics.

Tarbuck, Edward J., and Frederick K. Lutgens. *Earth: An Introduction to Physical Geology.* 10th ed. Upper Saddle River, NJ: Prentice Hall, 2010. An introductory textbook suitable for high school students. Contains sections on channel deposits (bars), floodplain deposits, alluvial fans, and various types of river. Includes many diagrams and photographs. Review questions and a list of key terms conclude the chapter.

Walker, R. G., ed. *Facies Models: Response to Sea level Change.* 2d ed. Tulsa, Okla.: Society of Economic Paleontologists and Mineralogists, 1994. An excellent compilation on sedimentary systems. Several chapters are devoted to alluvial systems. Suitable for college students.

**See also:** Beaches and Coastal Processes; Deep-Sea Sedimentation; Deltas; Desert Pavement; Drainage Basins; Floodplains; Lakes; Reefs; River Bed Forms; River Flow; Sand; Sediment Transport and Deposition; Weathering and Erosion

# AMAZON RIVER BASIN

*The Amazon River Basin is the largest in the world, covering an area of more than 4,810,000 square kilometers. It is home to more plant and animal species than any other region in the world and contains thousands of species that have not yet been discovered or named.*

## PRINCIPAL TERMS

- **basin:** the region drained by a river system, including all of its tributaries
- **delta:** the area at the mouth of a river that is built up by deposits of soil and silt
- **fauna:** the animal population of a region
- **flora:** the plant species of a region
- **tributaries:** rivers that flow into larger rivers
- **várzea:** the part of the rainforest that is flooded for up to six months per year

### GEOGRAPHY

The Amazon River of South America is the world's largest river, though, at 6,400 kilometers, not quite the longest. Africa's Nile River is 6,650 kilometers long, about 200 kilometers longer than the Amazon. The amount of water flowing down the Amazon is, however, about sixty times greater than that of the Nile. The Amazon carries more water than the Nile, the Mississippi, and China's Yangtze River combined, a total that adds up to more than one-fifth of all the river water in the world. Its basin—the region drained by a river and its tributaries—includes parts of Brazil, Bolivia, Peru, Ecuador, and Colombia and measures more than 7 million square kilometers, an area about three-fourths the size of the continental United States. More than a thousand tributaries flow into the Amazon, and the basin contains about 22,680 kilometers of navigable water.

There are two different types of tributaries, so-called black-water and white-water rivers. The black-water rivers, such as the Urubu, the Negro, and the Uatama, come from very old geologic areas that are poor in salts, very acidic, and filled with sediments. They are avoided by fishermen because they have small fish populations. The water is actually tea-colored because of heavy concentrations of tannic acid. The black-water rivers have few nutrients and carry much decaying vegetation from river banks. The decaying matter, chiefly leaves and branches, consumes much of the oxygen in the water. Because of low oxygen levels, these rivers support few forms of life;

they are so muddy that little light can penetrate and few plants can live in them, further reducing the food supply for fish. Indians living along these rivers call them "starvation rivers" because little food can be obtained from them. On the other hand, they also support fewer insects and mosquitoes, making the areas around them more livable for human beings. The white-water, or clear-water, tributaries come from areas of little erosion. The waters from these streams, such as the Xingu and the Tapajos, are clear enough that the bottom is visible in many places.

The Amazon begins high in the Andes Mountains in Peru, less than 160 kilometers from the Pacific Ocean, at an elevation of about 5,200 meters. For its first 970 kilometers, the Amazon headstream, known as the Apurímac River, drops through the mountains at a rate of about 5 meters per kilometer. It passes through grasslands used for grazing sheep and alpaca and by Cuzco, the old capital of the Inca Empire and the region's largest city. After passing through the Montana, an area of dense forests and steep valleys, the Apurímac flows into the Urubamba, forming a river called the Ucayali. It then heads north for about 325 kilometers, where the stream is joined by several smaller tributaries coming from Peru and Ecuador.

Where the Ucayali joins the Maranon, about 160 kilometers above the city of Iquitos, the middle course of the Amazon begins. There, the elevation is about 105 meters, and the river is more than 3 kilometers wide. It then travels through 405 kilometers of unpopulated territory before it enters Brazil. For the next 1,296 kilometers, the Amazon becomes a slow-moving, meandering stream, forming many huge lakes. There the Río Negro, named for its very dark waters, enters from the north. The Río Negro is 6.5 kilometers wide and carries water from Colombia and Venezuela. A few hundred miles downstream, the Madeira, the Amazon's longest tributary, enters from the south. Its waters come from as far away as La Páz, the capital of Bolivia, to the Paraguay River in far southern Brazil. Below the Madeira, the Amazon narrows to little more than 0.5 kilometer wide, but it is about 130 meters deep. It still has about 800 kilometers to go before it reaches the ocean.

The Xingu is the last major river to enter the Amazon, and its junction is considered the place where the delta—the triangular part of a river formed by soil deposits just before it enters the sea—begins. The Xingu is about 1,620 kilometers long and rises in the Brazilian highlands. The Amazon Delta stretches along the Atlantic coast for about 325 kilometers to the north and inland for about 485 kilometers. The delta contains the island of Marajó. The city of Belem is in the delta and is the official port of entry for the Amazon basin.

### CLIMATE AND FLORA

Most of the basin has a rainy tropical climate. Annual rainfall in the delta averages about 215 centimeters per year, with 180 centimeters in the middle region, and 300 to 400 centimeters per year in the upper course. The middle course and the delta have a rainy season that lasts from December through May, during which as much as 45 centimeters of rain can fall every month. During this time, the floodplain, called the *várzea*, is soaked and becomes a giant lake that can be anywhere from 50 to 200 kilometers wide. Above the mouth of the Madeira, the Amazon flows through a different type of floodplain called the *terra firme*. There the soil is generally very poor in nutrients and unsuitable for agriculture. Even in the *várzea*, however, less than 2 percent of the floodplain can be used for farming. Mostly, there is a tropical rainforest of light woods, brush, wild cane, and grasses. The soil is poorly drained and often clayey. Cattle are grazed on Marajó Island, and farmers have brought in water buffalo, but little else can be grown or raised in the region.

There are more than thirty thousand flowering plant species in the basin. That number represents about one-third of all the flowering plants found in South America and three times the number found in all of Europe. The amount of vegetation in various areas of the river is influenced by the rise and fall of the water. Some plants are drowned during floods, while others thrive. During periods of low water, the floodplain has little vegetation. There are, however, grasses and wild rice that grow at low water. The grasses become floating islands during flood season, sometimes reaching sizes of more than 0.6 kilometer in length and several hundred meters wide. The islands are home to passion fruit, morning glories, and giant water lilies more than 0.6 meter in diameter.

Ants, spiders, and grasshoppers also live on the islands. The root zone becomes a home for insects and an important source of food for fish. Many fish also eat the fruit of the plants. At low water, the islands are trapped in bushes and trees and the plants begin to rot, turning the water a deep, murky brown.

In the *várzea*, forest trees and bushes must adapt to survive during the flood season, which reaches its peak from May until July. During this period, plants become waterlogged and the forests, or *igapo*, constantly are under water from 1 to 2 meters deep. In some cases, the shoreline is swamped all year, and many plants and trees will die. Young plants are often under water from seven to ten months per year. There, growth is also limited by lack of light, as bigger trees block out the light. In many cases, it takes seedlings ten to twenty years to reach a size that will enable them to survive. There are also other problems; for example, *kapok*, or silk-cotton trees, have been among the most successful plants in the *várzea*. This species has been almost entirely eliminated by loggers cutting down trees to be made into boxes.

### FAUNA

Approximately nine hundred species of birds live in the Amazon, or about 10 percent of the total number of bird species on the planet. Some 50 percent of these species are found in this region and nowhere else. The bird population includes hummingbirds, macaws, owls, parakeets, parrots, toucans, and many others. Most species favor a particular level of rainforest, either on the ground or in the lower, middle, or upper level of trees. Only a few kinds of birds move between levels. The Amazon is the richest bird region in the world. There is also an enormous population of another kind of flying animal, bats, with more than fifty different species represented, making it the most diverse group of mammals in the region.

There are no crocodiles in the Amazon, but there is a smaller relative, the caiman. The black caiman can reach a length of 5.5 meters. Unfortunately, it is endangered due to overhunting. Its shiny black skin, which is used for handbags and expensive shoes, is a sought-after commodity for hunters.

There are believed to be somewhere between 1,300 and 2,000 species of fish in the river. In comparison, the Congo River in Africa has an estimated 560 species of fish, and the Mississippi River has fewer than 300. There are so many different species in the

Amazon because there were no mass extinctions in the river caused by glaciers or ice ages. The huge size of the basin also provides for numerous kinds of different environments, each of which can be uniquely exploited by different species. There are rapids, waterfalls, lakes, and streams, each with different types of water and vegetation. Both freshwater and saltwater species are found. Lungfish, which have survived from the Paleozoic era 150 million years ago, are common.

About 43 percent of the fish are characoids, including piranhas and neons, which are familiar to many aquarium owners. Piranhas are a diverse group of twenty species; they are red-bellied and can be very dangerous. They hunt in schools of up to one hundred fish, and are constantly in search of food. Normally, they are found in the floodplain lakes of the *várzea*. Various kinds of catfish and twenty species of stingrays are found in deeper water. One species, the piraiba catfish, is the largest fish in the river, reaching a length of 4.5 meters and a weight of 230 kilograms. The Amazon River is also home to many freshwater dolphins, which can reach lengths of 2.4 meters.

The phylum Arthropoda is well represented in the Amazon region, with 1 million or more species in the area, a large number of them still unclassified. They are without a doubt the most successful animals in the rainforest. About 90 percent are insects (jointed invertebrates, or animals without backbones), 9 percent are spiders, and 1 percent are other, more obscure kinds of creatures. Insects include flies, beetles (the most diverse group), ants, scorpions, millipedes, centipedes, and symphylans. The latter are centipede-like and live on the floodplain floor. The floor-dwelling species generally migrate upward into the trees when the rainy season begins. In the middle and upper tree levels are spiders, moths, butterflies, and various other winged species. The most abundant species at all levels are the ants, many of them adapted to living on one particular kind of tree or plant at one specific level. Millions more live on the ground, where huge numbers of mollusks, segmented worms, and flatworms also thrive.

Several types of lizards live in the basin, iguanas being the best known. Iguanas are typically green in color but can quickly change color to blend in with their background. They have long toes and sharp nails that enable them to climb trees, and most live at the tops of trees. Iguanas are vegetarians and can reach a size of more than 0.3 meter in length. Another kind of lizard, the teiid, can reach lengths of 1.5 meters. There are only a few frog species in the region, primarily because they have difficulty competing with fish for food and are easy prey for piranhas and other flesh-eating fish. There are more toads than frogs in the Amazon, as toads can more easily hide in the leaf litter that covers the ground.

The Amazon is home to the world's largest rodent, the capybara, which looks something like a large guinea pig and lives in groups of a dozen or more along the riverbanks, feeding on grasses and other plants. The capybara is being rapidly reduced in numbers, however, because it is intensively hunted for food by native peoples. Other rodents include the toro, a spiny rat that is a common night hunter; tree-dwelling porcupines; anteaters; leaf-eating sloths; five species of opossums; and several species of armadillos. Most of these groups migrated into the basin from the north when Central America became connected to South America via Panama from 3 million to 4 million years ago.

There are some hoofed mammals in the basin, but most species of this type have become extinct. The tapir is the oldest hoofed mammal in the basin and the only one that enters the river, usually for bathing and to eat fish. Monkeys are abundant in the region, with more than forty species found. There are two main groups. Marmosets and tamarins form one group. They live in small troupes in the trees and are most active during the day. Cebids, the other major group, include capuchins and howlers. They are much larger in size than the species of the first group, have relatively large brains, and resemble African apes.

There are 160 species of snakes in the Amazon, most living on the floodplain or on the riverbank. Only the coral snakes and vipers are poisonous. The area's pit vipers are among the most dangerous snakes in the world. They hunt at night, reach lengths of 1 meter, and live on a diet of small animals. The anaconda is the largest snake in the basin, reaching a length of 6 meters or more. It is a constrictor and therefore not poisonous, but kills by squeezing its victims to prevent them from breathing so that they die from suffocation. The anaconda is most active at night, grabbing birds, rodents, turtles, or caimans and coiling around them to first kill and then to break bones so that they are easier to swallow and digest. Boa constrictors use the same method of capturing their food, and are common in the water and on the shore.

Manatees, the largest animals in the Amazon, weighing 450 kilograms or more, are found in large numbers. These relatives of the elephant eat grasses and plants and live in the water. They can stay under water for more than an hour before coming to the surface to breathe. Other animals include coatis, which resemble raccoons; several kinds of deer; and peccaries, or wild hogs. Human beings were the last species to enter the basin, perhaps about ten thousand years ago.

*Leslie V. Tischauser*

**FURTHER READING**

Bonotto, Daniel Marcus, and Ene Gloria da Silveira. *The Amazon Goldruch and Environmental mercury Contamination.* Hauppauge, N.Y.: Nova Science Publishers, 2009. Reviews the extent to which *garimpo* gold mining methods that have been used in the Brazilian Amazon for more than thirty years have affected the environment and the lives of the *garimperos*.

Davis, Wade. *One River: Explorations and Discoveries in the Amazon Rain Forest.* New York: Simon & Schuster, 1996. A history of human exploration of the region. Includes much information on local peoples and the impact of modern economic development. Well written, with many photographs and a useful index.

Fraser, Lauchlan H., and Paul A. Keddy. *The World's Largest Wetlands: Ecology and Conservation.* Cambridge, England: Cambridge University Press, 2005. Presents the views of leading experts on the characteristics and conservation of wetlands.

Freitag, Bob, et al. *Floodplain Management: A New Approach for a New Era.* Washington, D.C.: Island Press, 2009. Each chapter presents a different case study that focuses on a new topic in flood control. Strategies of floodplain management revolve around the natural processes and dynamics of rivers. The text discusses a multiple approaches, which vary as widely as the locations in which they are used. A technical text best suited for engineers and hydrologists taking part in floodplain management.

Goulding, Michael, Ronaldo Barthem, and Efrem Ferreira. *The Smithsonian Atlas of the Amazon.* The Smithsonian, 2003. Filled with hundreds of excellent images, this book provides a comprehensive guide to the Amazon River. Including topics ranging from the thirteen tributaries to the mixing of freshwater and saltwater at the river's end. More of an information resource rather than an atlas; deforestation and species diversity are also discussed.

Hecht, Susanna B., and Alexander Cockburn. *The Fate of the Forest: Developers, Destroyers and Defenders of the Amazon.* Chicago: University of Chicago Press, 2010. Examines the history of the destruction of the Amazon rainforest, and reviews the efforts now being undertaken to save and restore the area.

McClain, Michael E., Reynaldo L. Victoria, and Jeffrey Edward Richey. *The Biogeochemistry of the Amazon Basin.* New York: Oxford University Press, 2001. Addresses many questions about the geochemical and biochemical processes that occur throughout the Amazon Basin. Articles are compiled to include biogeochemistry, nutrient cycling, land development and conservation. Topics such as biomass, trace elements, organic matter and nutrients levels are evaluated in regard to the Amazon River Basin. Written by experts in a wide variety of environmental fields.

Moran, Emilio. *Developing the Amazon.* Bloomington: Indiana University Press, 1981. Presents an economist's view of development in the river basin. Pays some attention to the ecological losses caused by development. Written mainly for specialists in development. Contains some excellent maps and descriptions of the region's geography.

Wohl, Ellen. *A World of Rivers.* Chicago: University of Chicago Press, 2011. The Amazon, Ob, Nile, Danube, Ganges, Mississippi, Murray-Darling, Congo, Chang Jiang, and Mackenzie rivers each have a chapter in this book. Figure 1.1 contains more straightforward and organized information than some full textbooks. Natural history, anthropogenic impact, and the future environment of these ten great rivers are discussed. The bibliography is organized by chapter.

**See also:** Ganges River; Great Lakes; Hydrothermal Vents; Lake Baikal; Mississippi River; Nile River; Rainforests and the Atmosphere; Reefs; Tropical Weather; Yellow River

# AQUIFERS

*Aquifers are the source of water for approximately 40 percent of the U.S. population. The identification, conservation, and protection of aquifers are important to the future of drinking water supplies in North America and elsewhere in the world.*

## PRINCIPAL TERMS

- **aquifer:** any porous, permeable geologic structure or rock formation that contains water
- **cone of depression:** the depression, in the shape of an inverted cone, of the groundwater surface that forms near a pumping well
- **confined aquifer:** an aquifer that is completely filled with water and whose upper boundary is a confining bed; also called an artesian aquifer
- **confining bed:** an impermeable layer in the geologic stratigraphy that prevents vertical water movement
- **groundwater:** water found below the ground surface in the zone of saturation
- **permeability:** the ability of rock, soil, or sediment to transmit a fluid
- **porosity:** the ratio of the volume of void space in a given geologic material to the total volume of that material
- **unconfined aquifer:** an aquifer whose upper boundary is the water table; also called a water table aquifer
- **water table:** the upper surface of the zone of saturation
- **zone of saturation:** a subsurface zone in which all void spaces are filled with water

### THE WATER TABLE

To understand aquifers, one must understand how water occurs beneath the earth's surface. The world's water supply is constantly circulating through the environment in a never-ending process known as the hydrologic cycle. Natural reservoirs within this cycle include the oceans, the polar ice caps, underground water, surface water, and the atmosphere. Water on the land surface that is able to infiltrate the ground becomes underground water. "Groundwater recharge" occurs when the infiltration reaches the water table.

Underground water exists in three different subsurface zones: the soil moisture zone, the intermediate vadose zone, and the zone of saturation. The soil moisture zone is found in soil directly beneath the land surface and contains water not confined below a rock stratum so that it is available to plant roots. This zone is generally not saturated unless a prolonged period of rainfall or snowmelt has occurred. Water in this zone is held under tension by the attractive forces between soil particles and water molecules, or surface tension forces. The depth of this zone corresponds to the depth to which plant roots can grow.

Water able to infiltrate through the soil moisture zone may pass into the intermediate vadose zone, also called the zone of aeration, before reaching the water table. The vadose zone is always unsaturated, since the pore spaces between particles contain both water and air, and the water it contains is held under tension. The thickness of a vadose zone depends on how close the water table is to the surface.

The water table forms the uppermost surface of the zone of saturation and is characterized by a water pressure equal to atmospheric pressure. It may be only a short distance below the land surface in humid regions and hundreds of meters below the surface in desert environments. In general, the water table mimics the land surface topography. If the water table intersects the land surface, the result is a lake, swamp, river, or spring. Below the water table, in the zone of saturation, geologic materials are completely saturated, and the water pressure increases with depth. Water contained within the zone of saturation is known as groundwater. When groundwater occurs in a particular type of geologic formation known as an aquifer, it can feed a well.

### POROSITY

An aquifer can be functionally defined as any earth material—rock, soil, or sediment—that yields a significant quantity of groundwater to a well or spring. The definition of "a significant quantity" varies according to the intended use; what constitutes an aquifer for an individual homeowner may be quite different from what constitutes an aquifer for a municipal supply. For a geologic formation to be useful as an aquifer, it must be able both to hold and to transmit water.

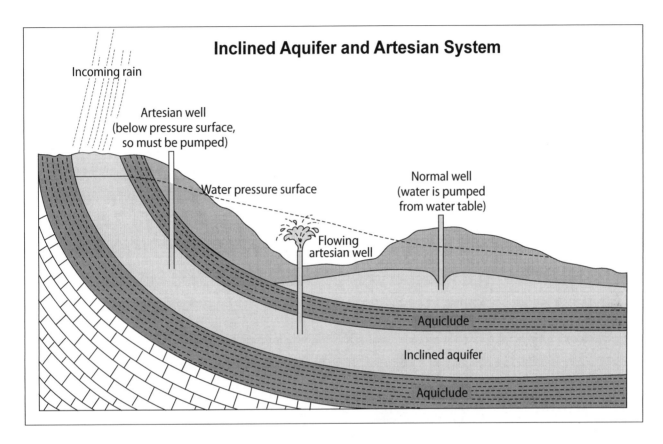

The ability of geologic materials to hold, or store, water is known as porosity. Simply stated, porosity is the volume of void space present divided by the total volume of a given rock or sediment. This proportion is usually expressed as a percentage. The higher this ratio is, the more void space there is to hold water. There are various types of porosity. Unconsolidated materials (soil and sediment) have pore spaces of varying sizes between adjacent grains referred to as intergranular porosity. The ratio of pore space to total volume depends on several factors, including particle shape, sorting, and packing. Loosely packed sediments composed of well-sorted, spherical grains are the most porous. Porosity decreases as the angularity of the grains increases because such particles are able to pack more closely together. Similarly, as the degree of sorting decreases, the pore spaces between larger grains become filled with smaller grains, and porosity decreases. Values of porosity for unconsolidated materials range from 10 percent for unsorted mixtures of sand, silt, and gravel to about 60 percent for some clay deposits. Typical porosity values for uniform sands are between 30 and 40 percent.

Rocks have two main types of porosity: pore spaces between adjacent mineral grains and voids caused by fractures. Rocks formed from sedimentary deposits, such as shale and sandstone, may have significant intergranular porosity, but it is usually less than the porosity of the sediments from which they were derived because of the compaction and cementation that take place during the process of transforming sediments into rock. Therefore, although sandstone porosities may be as high as 40 percent, they are commonly closer to 20 percent because of the presence of natural cements that partially fill available pore spaces. Igneous and metamorphic rocks are composed of tightly interlocked mineral grains and therefore have little intergranular porosity. Virtually all void space in such rocks is a result of fractures (joints and faults). For example, granite, a dense igneous rock, usually has a porosity of less than 1 percent, but it may reach 10 percent if the rock is fractured.

There are additional types of porosity that occur only in certain kinds of rocks. Limestone, a rock that is soluble in water, can develop solution conduits along fractures and bedding planes. In the extreme case, solution weathering may lead to the development of a cave, which has 100 percent porosity. The overall porosity of solution-weathered limestone sometimes reaches 50 percent. Rocks created by volcanic eruptions may contain void space in the form of trapped air bubbles (called vesicles), shrinkage cracks developed during cooling (called columnar joints), and tunnels created by flowing lava (called lava tubes). In extreme cases, the porosity of volcanic rocks may exceed 80 percent.

## PERMEABILITY

The presence of void space alone does not constitute a good aquifer. It is also necessary for groundwater to be able to move through the geologic material in question. The ability of porous formations to transmit fluids, a property known as permeability, depends on the degree to which the void spaces are interconnected. Some high-porosity materials, such as clay, shale, and pumice, do not make good aquifers because the void spaces are largely isolated from one another. Materials that have high permeability include sand, gravel, sandstone, and solution-weathered limestone. Rocks with low porosities, such as shale, quartzite, granite, and other dense, crystalline rocks, have low permeabilities, unless they are significantly fractured.

Groundwater moves along tortuous paths through the available void space in a given porous formation. Regardless of the material's permeability, groundwater flows much more slowly than surface water in a river. The velocity of stream flow may be measured in meters per second, whereas groundwater velocities commonly range between 1 meter per day and less than 1 meter per year, averaging about 17 meters per year in rocks. Underground rivers are uncommon, occurring only in cavernous limestone or lava tubes in volcanic terrane.

The geologic materials that make good aquifers are those that have both high porosity and high permeability. The response of any given aquifer to a pumping well will also depend, however, on its position beneath the surface and its relationship to the water table. Aquifers near the ground surface usually have the water table as their upper boundary. The thickness of these aquifers therefore changes as the water table rises or falls. An aquifer under these conditions is an unconfined aquifer, or water table

aquifer. These types of aquifer are the easiest to exploit for a water supply, but they are also the easiest to contaminate. It is therefore important to delineate the extent of unconfined aquifers and to take measures to protect them from various forms of pollution.

Because the water table is free to fluctuate in unconfined aquifers, the amount of water supplied to a well reflects the gravity drainage of water from void spaces. The volume of water available from aquifer storage, or the "specific yield," therefore approaches the upper limit set by porosity. Some groundwater is unable to drain from void spaces under the influence of gravity because it is tightly held by surface tension forces; this retained water, known as "specific retention," forms a thin film around individual grains. The highest values of specific yield occur in coarse-grained, permeable aquifers, such as sand, gravel, and sandstone.

## CONFINED AQUIFERS

Some aquifers, usually found at depth or those known as "inclined" aquifers, are completely filled with groundwater and bounded at the top by an impermeable layer called a confining bed, or aquiclude. The water in these confined, or artesian, aquifers is under pressure because of the weight of overlying formations and the fact that the confining bed does not allow groundwater to escape. If a well intrudes into such an aquifer, the water level will usually rise above the upper boundary of the confining bed, creating an artesian well. In some cases, water may rise above the land surface at the point where the well is placed. This condition is known as a flowing artesian well. Water will flow freely out of such wells as long as the aquifer remains under pressure. Many of the Great Plains states (Kansas, Nebraska, and the Dakotas) are underlain by important shale layers. The original pressure in these aquifers was quite high because the sandstone beds are upwarped along the eastern front of the Rocky Mountains and Black Hills, where they receive groundwater recharge.

Confined aquifers supply water to a well not through the gravity drainage of void spaces but through compression of the aquifer as water pressure is reduced during pumping. The volume of water available from storage in a confined aquifer, or the "specific storage," is only a small fraction of the total volume and is therefore always much less than porosity. When confined aquifers are pumped to the extent that they become dewatered, the accompanying pressure reduction can lead to extensive aquifer compression and land surface subsidence. This problem is most serious in cases where the confining bed is composed of clay because the loss of fluid pressure beneath the clay causes water to be squeezed out of the clay layer by the weight of overlying formations. Once a clay layer is compressed in this way, it will not be able to reabsorb water, even if the surrounding materials become saturated again.

## STUDY OF AQUIFERS

The first step in studying aquifers is to utilize data that have already been collected, such as geologic maps. These maps show the distribution of various geologic formations on the land surface and are therefore valuable tools for delineating the outcrop patterns of potential aquifers. If cross-sections are also available, they can aid in the estimation of potential aquifer thicknesses and the identification of possible confined aquifers. Geologic maps, however, serve only as a preliminary tool in aquifer study. Any interpretations made from maps need to be verified by field descriptions and, if possible, pumping tests.

Topographic maps can be used to make generalizations about the groundwater flow system. Springs, lakes, streams, and swamps may indicate areas of groundwater discharge. Because the water table is usually a subdued version of the land surface, it may be possible to infer groundwater flow directions from the local topography. That can provide a clue as to where recharge areas occur. Other indications of recharge areas are topographic high points and a general lack of surface-water features. For groundwater recharge to occur, however, permeable materials must be exposed at the land surface.

Reliable estimates of aquifer properties require fieldwork. Often, samples are taken from the field for the purpose of determining aquifer properties, such as grain-size distribution or permeability, in the laboratory. Such tests, however, are performed only with small samples and may not be representative of the overall aquifer unit. That is particularly true in fractured rock structures, where the movement of groundwater may be very difficult to predict. In such situations, the injection and monitoring of tracer dyes has proved helpful in understanding groundwater flow.

The most direct way to study aquifers is by boring holes and installing wells. By drilling, geologists are able to discover the exact nature of the subsurface materials. Detailed drilling logs are kept of the different layers and the depths at which they were encountered. Using their knowledge of geologic materials, properly trained geologists can predict which formations will constitute the best aquifers.

Once wells have been installed, additional information about the aquifer can be learned by conducting pumping tests (pumping the wells at known rates for extended periods). Because water moves relatively slowly through an aquifer, the pumping of a well removes groundwater faster than it can be replaced. The resulting water-level drawdown is called a cone of depression; it is a zone of dewatering near the pumping well that resembles an inverted cone. The exact shape of this cone is a function of the pumping rate and the aquifer properties. Therefore, certain aquifer characteristics, such as permeability, are discovered by studying the cone of depression created by a known pumping rate. To identify the shape of the cone, monitoring wells must be placed near the pumped well to detect drawdown at various distances.

The prediction of long-term well yields requires not only an understanding of aquifer properties but also a knowledge of the groundwater recharge rates. To determine the amount of water in a particular drainage basin that is available for groundwater recharge, it is necessary to develop a hydrologic budget for that basin. Hydrologic budgets attempt to account for all water inputs and losses from the basin in question. Inputs include precipitation, surface-water inflow, groundwater inflow, and water imported by human activities (as for irrigation). Losses include surface-water runoff, groundwater outflow, evaporation, transpiration by plants, and water exported from the basin by human activities. The difference between these inputs and losses equals the amount of water gained or lost by the groundwater reservoir.

## SIGNIFICANCE

The term "aquifer" is generally not precise because its definition depends somewhat on the intended use. For an individual homeowner who requires only 7 to 20 liters per minute from a well, fractured bedrock may serve as a suitable source of water. By comparison, only a few geologic materials are capable of delivering the larger water supply demanded by a municipality or industrial plant—often more than 1,500 liters per minute. In the case of high-capacity well requirements, the term "aquifer" is restricted to highly permeable geologic materials, such as unconsolidated sands and gravels, sandstone, and solution-weathered limestone.

Most private wells draw water from unconfined aquifers, which have the water table as an upper boundary. The saturated thicknesses of these aquifers can fluctuate as the water table rises or falls. Therefore, shallow wells completed in unconfined aquifers can be pumped dry as the water table declines during periods of prolonged drought. Because of their proximity to the land surface, these aquifers are also susceptible to groundwater contamination. Confined aquifers do not comply to a water table because they are fully saturated and capped with an impermeable confining bed. The natural groundwater quality in confined aquifers is not necessarily superior to that found in unconfined aquifers, but the presence of a low-permeability confining bed may help to protect these aquifers from being contaminated by materials descending from the surface.

An understanding of aquifer characteristics is important to the proper use of groundwater. The determination of maximum sustainable well yields requires a knowledge of how the aquifer stores and transmits water. Such information is also needed to estimate how close adjacent wells can be to one another without causing interference through the formation of overlapping cones of depression. Pumping water out of an aquifer at rates exceeding the rate of groundwater recharge will eventually cause the depletion of that aquifer as a valuable water supply. Many arid regions in the western United States are facing this problem because of years of groundwater mismanagement. The overpumping of confined aquifers is especially troublesome because if these aquifers become dewatered, they compress, leading to subsidence of the land surface. This has been especially noted in California, where groundwater withdrawal has resulted in subsidence of 9 meters or more in the San Joaquin Valley and the formation of deep, chasm-like cracks in the level grounds of Rogers Lake.

Delineating the extent of aquifers and particularly of their recharge zones is critical to the development of policies that protect groundwater supplies from pollution. In the case of confined aquifers, recharge zones may be restricted to areas where the confining bed is absent, so contaminants entering the ground through a relatively small zone could affect a large number of downgradient wells. Moreover, because groundwater moves very slowly through the subsurface, it takes a long time to be renovated once contaminated.

*David L. Ozsvath*

## FURTHER READING

Ahmed, S., R. Jayakumar, and Abdin Saleh. *Groundwater Dynamics in Hard Rock Aquifers: Sustainable Management and Optimal Monitoring Network Design.* Dordrecht: Springer, 2008. Presents the findings of pilot research on the structure and functioning of an aquifer in a granitic formation in order to characterize the system and its properties with regard to geophysical, geological, and remote-sensing applications.

Batu, Vedat. *Applied Flow and Solute Transport Modeling in Aquifers.* Boca Raton, Fla.: CRC Press, 2006. Discusses the dynamics of aquifer water flow, flow conditions, solute transport, and sorption principles. Covers aquifer modeling fundamentals, both numerical and analytical approaches. Contains a list of references and indexing. Best suited for aquifer researchers and advanced graduate students with a background in fluid dynamics or groundwater analysis.

_____. *Aquifer Hydraulics: A Comprehensive Guide to Hydrogeology Data Analysis.* New York: Wiley, 1998. Offers a detailed look at the role aquifers play in groundwater flow. Well illustrated and well indexed. Accompanied by a CD-ROM.

Fetter, Charles W. *Applied Hydrogeology.* 4th ed. Upper Saddle River, N.J.: Prentice Hall, 2000. Emphasizes the practical aspects of understanding groundwater occurrence and movement. Contains a detailed discussion of the influence that geologic conditions have on groundwater occurrence, with special emphasis on groundwater regions in the United States. Addresses regional groundwater movement within aquifers and contains helpful illustrations of the principles discussed. Includes a glossary of important terms. Suitable for college-level readers.

Gilbert, Janine, Dan L. Danielopol, Jack A. Stanford, et al., eds. *Groundwater Ecology.* San Diego, Calif.:

Academic Press, 1994. Compiles ecological essays written by experts. Examines the problem of groundwater contamination and pollution, as well as possible solutions. Color illustrations and maps.

Glennon, Robert. *Unquenchable. America's Water Crisis and What to Do About It.* Washington, D.C.: Island Press, 2009. A serious analysis of the problems with fresh, potable water in the United States, most of which is drawn from aquifers. Offers some hard-hitting suggestions for amending what is potentially the greatest crisis facing the nation.

Gorelick, Steven M., et al., eds. *Groundwater Contamination: Optimal Capture and Contamination.* Boca Raton, Fla.: Lewis, 1993. Looks at pollution levels in groundwater and surface water. Provides a thorough description of water quality measures, standards, and procedures. Illustrations, index, and bibliography.

Guymon, Garry L. *Unsaturated Zone Hydrology.* Upper Saddle River, N.J.: Prentice-Hall, 1994. Focuses on the movement of water through the unsaturated zone and into aquifers. Uses a quantitative modeling approach and combines research and expertise form a variety of disciplines.

Hamblin, Kenneth W., and Eric H. Christiansen. *Earth's Dynamic Systems.* 10th ed. Upper Saddle River, N.J.: Prentice Hall, 2003. Although not limited to the topic of groundwater, this widely used introductory textbook has a good discussion of groundwater occurrence and movement. The color figures are helpful to readers unfamiliar with subsurface geology. Suitable for both high school and college-level readers.

Kasenow, Michael. *Aquifer Test Data: Analysis and Evaluation.* Highlands Ranch, Colo.: Water Resources Publications, 2006. Focuses on aquifer test analysis. Discusses methods of evaluation, data collection, and well-field analysis. Includes a CD-ROM with specific capacity tables and theisian tables. Provides a well-rounded overview of basic testing topics and specific aquifer types and analyses.

Misstear, B. D. R., David Banks, and Lewis Clark. *Water Wells and Boreholes.* New York: John Wiley & Sons, 2006. Suitable for postgraduate students and researchers. Brings scientific background information and practical advice together using world case studies.

Nielson, Kurt Ambo. *Fractured Aquifers: Formation Evaluation by Well Testing.* Vancouver: Trafford Publishing, 2007. Presents state-of-the-art methods for analyzing well-test results from aquifers flowing in fractured rock structures, including vertical and horizontal fracturing and dual-porosity behavior.

Nonner, Johannes C. *Introduction to Hydrogeology.* London: Taylor & Francis Group, 2006. Covers all major fields in hydrogeology, as well as the occurrence, behavior, and properties of groundwater. Well illustrated, with numerous examples of groundwater cases from around the world.

Randolph, John. *Environmental Land Use Planning and Management.* Washington, D.C.: Island Press, 2004. Describes basic principles and strategies of land-use planning and management. Discusses various land features, types, and environmental issues, such as soils, wetlands, forests, groundwater, biodiversity, and runoff pollution. Case studies and specific examples are provided.

Todd, David K. *Groundwater Hydrology.* 3d ed. New York: John Wiley & Sons, 2004. Emphasizes the practical aspects of groundwater occurrence and movement. Introduces the reader to groundwater utilization and its relationship to the hydrologic cycle. Discusses aquifer types and the occurrence of groundwater in the United States. Suitable for college-level readers.

**See also:** Air Pollution; Artificial Recharge; Atmosphere's Evolution; Atmosphere's Global Circulation; Atmosphere's Structure and Thermodynamics; Climate; Dams and Flood Control; Floods; Freshwater and Groundwater Contamination Around the World; Groundwater Movement; Groundwater Pollution and Remediation; Hurricanes; Hydrologic Cycle; Monsoons; Precipitation; Salinity and Desalination; Saltwater Intrusion; Storms; Surface Water; Tornadoes; Waterfalls; Water Quality; Watersheds; Water Table; Water Wells; Weather Forecasting

# AR4 SYNTHESIS REPORT

*Released in 2007, the* AR4 Synthesis Report *combines the efforts of the three working groups of the Intergovernmental Panel on Climate Change into one document that assesses the state of research on climate change and makes possible recommendations for future action and research.*

## PRINCIPAL TERMS

- **about as likely as not:** something that is 33 to 66 percent likely to be true, as used in the context of the *AR4 Synthesis Report*
- **climate change:** as defined by the Intergovernmental Panel on Climate Change, a change in the climate that persists for roughly ten years or longer.
- **exceptionally unlikely:** less than 1 percent likely to be true
- **extremely likely:** greater than 95 percent likely to be true
- **extremely unlikely:** less than 5 percent likely to be true
- **glacial lakes:** lakes formed by the melting of a glacier; an indicator used in estimating average annual temperatures
- **greenhouse gas:** any of the gases in the earth's atmosphere, including carbon dioxide and methane, that absorb heat, thereby affecting atmospheric temperatures
- **ice core record:** use of samples from ice sheets to ascertain temperature, atmospheric and volcanic activity, and precipitation in past years
- **likely:** greater than 66 percent likely to be true
- **mitigation:** with respect to greenhouse gases, strategies to prevent emissions from passing current baseline levels
- **more likely than not:** greater than 50 percent likely to be true
- **radiative forcing:** a measurement of how much the greenhouse gas in a given area will impact the solar radiation
- **tree-ring data:** data harvested from tree rings that indicate historical climates; thicker rings indicate the presence of light and nutrients, which allow more growth
- **unlikely:** less than 33 percent likely to be true
- **very likely:** greater than 90 percent likely to be true
- **very unlikely:** less than 10 percent likely to be true
- **virtually certain:** greater than 99 percent likely to be true

## The IPCC and *AR4 Synthesis Report*

The *AR4 Synthesis Report (AR4)* is the Fourth Assessment Report of the Intergovernmental Panel on Climate Change (IPCC), an international group originally founded by two United Nations agencies, whose mandate is to review and assess the latest research findings on Earth's climate. The *AR4* is a synthesis of reports published by three separate working groups of the IPCC in 2007. Previous reports had been published in 1990, 1995, and 2001. The three working groups are tasked with evaluating, respectively, the scientific basis for climate change, the impact of climate change, and appropriate responses to mitigate the impact.

The *AR4*, like the three individual reports from the working groups, does not comprise independent research; rather, it comprises an evaluation of existing research for a large-picture evaluation of climate change. The report evaluates the events of climate change and the drivers and impacts of and responses to climate change. Finally, the report, which evaluates the uncertainties of predictions of climate change, is divided into six sections.

In the 2001 *Third Assessment Report* (commonly called *TAR*), links were established between changes in the climate and their impact. One of the goals established for the *AR4* report was to determine ways to reverse climate trends, proposing solutions that would reduce emissions and, therefore, mitigate future impacts.

The *AR4* also includes specific definitions (provided in the Principal Terms) for terms encompassing the uncertainty of statements in the report. As with any other area of science, the discussions in the report have some level of uncertainty, so the IPCC felt it was important to clearly define terms for scientists across different disciplines to explain the level of confidence related to conclusions in the report.

The levels of uncertainty include a qualitative and a quantitative scale. The qualitative scale is simple terminology explaining the level of agreement and the quantity of evidence (using terms such as "high agreement" and "medium evidence").

The quantitative terms are more specific, referring to a numerical range of certainty based on statistics and the expertise in the field. When expressing confidence in a data set, a model, or analysis, confidence is expressed on a scale of 1 to 10, with descriptors of confidence: very high (9), high (8), medium (5), low (2), and very low (1). When evaluating a full body of evidence for a prediction or statement of causality, the report uses the scale listed in its glossary.

## CHANGES IN CONDITIONS

The *AR4* begins by stating that the warming of the global climate is "unequivocal." The authors attribute this to three primary observations: temperature changes on land and water, a rise in sea level, and the melting of snow and ice cover.

Many of the observations on which current assessments are based began in 1970, when the availability of powerful remote-sensing satellites made such data gathering and analysis possible. While this relatively small data set yields insufficient data for some conclusions, it also illustrates that the years between *TAR* and *AR4* have seen a significant increase in the availability of climate data. However, when it comes to observations on a local scale, there is still a profound difference in data gathering and analysis, as developing countries tend not to have the resources to monitor environmental changes in the same way.

Among the strongest effects found for global warming, the authors cite the number and size of lakes formed from melted glaciers, the number of predators at the top of a food chain in ocean ecosystems, and avalanches or other instances of instability in mountains or other regions covered by heavy snow. The authors state with high confidence that there have been increased water levels in lakes and rivers that have glaciers as their source earlier in a given year, a factor that has had an impact on global water temperature and quality. Additionally, plants and animals have begun sprouting and migrating earlier in the season. Similarly, oceanic plant life has increased because of higher temperatures.

The authors state with medium confidence that the temperature changes have affected crop planting times and have led to increases in the rates of certain human ailments, such as infectious diseases and heat-related health problems (in Europe, for example). Ultimately, the conclusion is that of the seventy-five studies encompassing twenty-nine thousand data sets, 89 percent are consistent with the idea that the earth is warming.

The authors note that some events and features of the environment have not appeared to change substantially or consistently in the period studied, although this could be a result of a lack of sufficient data. The ice coverage of Antarctica on the whole, and the surface temperatures on that continent, do not seem to have changed significantly. The authors also cite a lack of evidence to determine if the number of extreme weather events such as tornadoes and tropical cyclones has changed.

## CAUSES OF CHANGE

In the second section of the *AR4* report, the researchers explore potential causes of global temperature increases. The major observation is that the atmosphere is filled with historically high levels of the greenhouse gases carbon dioxide ($CO_2$), nitrous oxide ($N_2O$), and methane ($CH_4$). These greenhouse gases lead to radiative forcing and, therefore, higher temperatures on Earth.

In reviewing the causes of climate change, the authors conclude that global warming is "very likely" because of greenhouse gas increases and that these gases have "likely" impacts on the warming of all six inhabited continents. Furthermore, they note, the warming "likely" has had an impact on many of the changes seen in local systems. In *TAR*, the human connection to global warming had been regarded as "likely."

Despite their strong statements on the findings, the authors also note that there are some limits to the data, including a lack of data from developing parts of the world, limits on how much is available over time in some areas where averages can be thrown off by extreme weather events, and means to attribute changes in smaller geographic areas, where adaptation may reduce the effects observed.

## EFFECTS OF CLIMATE CHANGE

In discussing the effects of climate change, the authors state there is "high agreement and much evidence" that higher temperatures will continue as long as greenhouse gas emissions-levels remain high. While some aerosols (for example, sulfur dioxide) have the effect of cooling the temperature, their levels are expected to drop.

Under current projections, temperatures are expected to rise 0.4 degree Celsius (0.7 degree Fahrenheit) through about 2030. It is very likely, researchers conclude, that the environmental changes observed through the twentieth century will continue to a greater degree during the twenty-first century, even without any significant growth in greenhouse gas emissions.

Some climate models are predicting that those changes will continue beyond the twenty-first century, even without an increase in emissions, because of radiative forcing's impact on temperatures under current conditions. One major concern is water supply, as glaciers are often a source of freshwater for many parts of the world. The disappearance of some glaciers, combined with a growing world population, could have severe effects.

## RESPONSES TO CLIMATE CHANGE

In the second half of *AR4*, the authors discuss sustainable development. In light of the need for development in much of the world, of primary concern is developing areas without exacerbating negative climate change. Humans are making changes to ameliorate climate change and its effects, but more changes are needed. Without these changes, humans will not be able to adapt to new conditions on Earth.

Although the impact of global warming is known, policy solutions to avoid global warming have proved challenging. Along with concerns about greenhouse gas emissions, there are concerns about development in many parts of the world that would result in increases in greenhouse gas emissions.

A primary concern is mitigation. Because of the higher emissions related to development, the strategy is to immediately attack the problem of reducing carbon emissions in developed parts of the world. By mitigating now, the hope is that emissions levels will be constant. Without this, global warming will worsen and more adverse effects from climate change will arise in the coming years.

Technologies available now or expected to be available by 2030 can help in this regard, with the hope that greenhouse gas levels will remain at their present levels or become lower. The expectation is that some of these steps will lower economic costs, with the highest cost being $100 (U.S. dollars) per ton of carbon dioxide emitted (or the equivalent).

The authors note that economic policies not directly tied to climate change may still have significant effects, and those effects remain poorly understood. There exist a number of other uncertainties about climate change. Solutions will depend on the specific challenges of a given region and on the resources available there, whether these resources are financial or governmental and institutional. Future patterns of use, and especially the technologies that may develop, will be key to whether the challenge of development can be met without harming Earth's climate.

*Joseph I. Brownstein*

## FURTHER READING

Bernstein, Lenny, et al. *Climate Change 2007: Synthesis Report*. Geneva: IPCC, 2008. The complete *AR4 Synthesis Report*, which summarizes the findings of the three IPCC working groups.

Curry, Judith A., and Peter J. Webster. "Climate Science and the Uncertainty Monster." *Bulletin of the American Meteorological Society* 92, no. 12 (2011): 1667-1682. In this article, which led to some controversy, the authors discuss how uncertainty is handled in the IPCC report.

Intergovernmental Panel on Climate Change. *Climate Change 2001: IPCC Third Assessment Report*. Geneva: IPCC Secretariat, 2001. A number of changes took place since the last report from the IPCC, both in methods and the strength of conclusions. This early report provides some insight into how climate change science and research has evolved.

_____. Working Group I. *Climate Change 2007: The Physical Science Basis*. New York: Cambridge University Press, 2007.

_____. Working Group II. *Climate Change 2007: Impacts, Adaptation, and Vulnerability*. New York: Cambridge University Press, 2007.

_____. Working Group III. *Climate Change 2007: Mitigation of Climate Change*. New York: Cambridge University Press, 2007. These three reports make up the most recent IPCC review of available information on climate change and subsequent recommendations.

**See also:** Climate; Climate Change Theories; Climate Modeling; Global Energy Transfer; Global Warming; Greenhouse Effect; Hydrologic Cycle; Impacts, Adaptation, and Vulnerability; IPCC; Long-Term Weather Patterns; Observational Data of the Atmosphere and Oceans; Ocean-Atmosphere Interactions; Ozone Depletion and Ozone Holes; Recent Climate Change Research; Remote Sensing of the Atmosphere; Remote Sensing of the Oceans; Satellite Meteorology; Severe and Anomalous Weather in Recent Decades; Satellite Meteorology; World Ocean Circulation Experiment

# ARAL SEA

*The Aral Sea was once the fourth largest body of freshwater on Earth. Located in the desert of the central Asian republics of Kazakhstan and Uzbekistan, its waters once supported productive fisheries. As the result of unwise water management practices, the Aral Sea has since lost about 60 percent of its surface area, is saltier than the world's oceans, and has markedly changed the local climate and ecology.*

## PRINCIPAL TERMS

- **climate moderation:** a change in the climate or average weather of a region that reduces the extremes of heat and cold
- **ecological backlash:** the unanticipated ecological effect of what appears, at first, to be a harmless activity
- **irrigation:** the relocation and application of water by means of ditches or pumping as an aid to crop production
- **salinity:** a measure of the quantity of dissolved salts in water
- **water equilibration:** a condition in a lake or other body of water in which the water lost by evaporation is equal to the water added by rainfall or runoff

## OVERVIEW

Tectonic movements during the Tertiary period (from about 65 million years to about 3 million years ago) formed large basins in parts of central Asia. In time, these basins filled with water. As the climate and rainfall varied over millions of years, the number and sizes of these water-filled basins also varied. The Caspian Sea and Aral Sea (actually large freshwater lakes) are remnants of much larger, ancient bodies of water. The Aral Sea formed in a depression in the earth's surface at the beginning of the Pleistocene epoch (the last Ice Age) about 1.6 million years ago.

Aquatic vegetation eventually established a presence in the water-filled basins. The nearby wetlands supported lush growths of plants to form extensive, highly productive marshes. Farther back from the marshes, shrubs and trees developed into riparian forests. Beyond the influence of the waters, shrubby grasslands and sparse vegetation gave way to near-desert conditions.

Flocks of migratory waterfowl, including ducks, geese, and egrets, fed among the lush vegetation, nested, and raised their young. Deer, wild pigs, muskrats, and a host of small mammals also occupied the extensive wetlands and meadows. Fertilized by the abundant aquatic vegetation, shoals of sturgeon, carp, roach, pike, perch, bream, and other fish provided a ready supply of food for humans from the time they first ventured into the area.

## HUMAN DEVELOPMENT

During the early decades of the twentieth century, human settlement increased around the shores of the Aral Sea, and the harvest of the rich fisheries increased. In addition to the catch consumed by local populations, large volumes of fish were packed and shipped across the continent. To satisfy the demand for Aral Sea fish as food, the waters yielded an annual harvest of fifty thousand tons of fish. Trawlers, some as long as 15 meters, plied the lake waters, gathering up Lake Aral's fish.

In addition, cattle grazed on the lush grasslands surrounding the Aral Sea, and crops were grown and harvested to feed the local human population. Although the region was surrounded by the sparse vegetation and desert, the availability of water from the Aral Sea and its riverine systems made agriculture possible.

The area surrounding the Aral Sea enjoyed what has been described as a desert-continental climate with cold winters, hot summers, and sparse rainfall. The climate was moderated somewhat by the heat capacity of the volume of water in the Aral Sea. During the summer, the waters absorbed heat from the air and slowly released it in the winter. This had the effect of reducing the extremes of seasonal temperatures. Although the Aral Sea never became a popular resort area, its waters were used by villagers around its shores for swimming and boating during the warm season.

In addition, the waters of the Aral Sea and the rivers that supplied it, the Amu Darya and the Syr Darya, were an important source of potable water and water for domestic and farming uses. Both humans and livestock, as well the abundant wildlife of the region, depended on these waters for life.

**COTTON AT A PRICE**

In the early days of the Soviet Union following World War I, the communist authorities developed a plan designed to make the new nation self-sufficient in cotton. The program, which involved irrigating vast, near-desert regions in central Asia and planting and growing cotton, began in the mid-1920's and appeared to be successful. By the 1950's, vast volumes of water were being diverted from the two rivers that fed the Aral Sea. The waters flowed through hundreds of kilometers of canals that extended from the rivers to the surrounding desert to irrigate the fields of new cotton. Poor or hasty planning, however, led to the construction of canals that were open to the searing heat of the sun, resulting in great losses of water to evaporation. In addition, the canals were unlined, and much water was lost by percolation into the sandy desert soils.

Despite these problems, the irrigated cotton project yielded rich dividends, and the Soviet Union joined the United States and China as a world leader in cotton exports. However, the ecological backlash of the project became apparent during the early 1960's. The first sign of trouble was the shrinking of the Aral Sea as billions of liters of water were diverted from the feeder rivers to irrigate the cotton crops. Between 1950 and 1988, irrigated cotton fields were expanded to cover more than 7 million hectares of land. The fishermen and cattle workers in the Aral Sea region were forced to become cotton pickers.

**LOSS OF WATER VOLUME**

The impact of the huge withdrawal of water from the feeder rivers of the Aral Sea soon became apparent as an increasing amount of sea bottom was exposed. Details about the dismal state of the Aral Sea were largely unknown outside the Soviet Union. The communist government maintained a wall of silence about what many eventually called one of the world's worst environmental disasters, perhaps second only to the nuclear power plant accident at Chernobyl in 1986. However, in the mid-1980's, under the principle of *glasnost* (openness) practiced by Soviet president Mikhail Gorbachev, the world soon learned the truth about the Aral Sea.

What had been the world's fourth largest lake—slightly bigger than Lake Huron in North America—had, by the 1980's, lost 80 percent of its volume. This loss was largely caused by evaporation. With its riverine input

diverted to irrigate cotton, there was insufficient water to resupply the lake basin. More than 3 million hectares of seabed, now exposed to the drying sunlight and wind, became an expanse of white salt. The mineral content of the remaining Aral Sea waters increased dramatically until it was three times the salinity of the world's oceans. The resulting brine concentration killed off all twenty-four of the Aral Sea's native species of fish. With the collapse of the once-thriving fisheries and related economic adversities, 100,000 workers were displaced.

The loss of water volume caused both immediate and long-term problems for the Aral Sea, the surrounding region, and the local people. Not only were the fish gone, but also the environment of the sea became inhospitable to birds and other wildlife. Deer and small mammals deserted the sea edges where the lush grasses and shrubs had been replaced by growing layers of crusting salts. Even the cotton harvest slowly declined, with the plants producing fewer and smaller bolls. The shrinking of the Aral Sea, which formerly had moderated the local climate, resulted in a harsher climate, a shorter growing season, and frequent dust storms. The salts and chemical residues blowing off the dry seabed reduced the soil's fertility.

The long-term impact on human populations in the Aral Sea region has extended well beyond the loss of the fisheries and jobs. Many of the people in the region have said that they fear they are slowly being poisoned. Runoff from the irrigation system supplies the local drinking water, which has become a brew of pesticides, defoliants, and fertilizers. Worse, domestic sewage only partially treated in the old, Soviet-era equipment is also part of the drinking-water supply. The frequent wind storms blow clouds of salt, dust, and agricultural chemicals from the exposed, dry seabed, resulting in a high incidence of respiratory problems. The windborne pollutants, coupled with the contaminated drinking water, have been blamed for many digestive upsets, typhoid, dysentery, and birth defects. Populations close to the Aral Sea report infant mortality rates as high as one hundred deaths per one thousand births (10%), the highest in the former Soviet Union. Further, life expectancy at birth tends to be lower than the average in more economically developed countries.

**SIGNIFICANCE**

The story of the Aral Sea is not unfamiliar. A number of environmental disasters have been induced by those who elevate the need for economic gain above sound environmental planning. Some of those disasters have come about simply because the needed information was not available, while others have occurred because available information was ignored by planners.

The Aswan Dam across the Nile River in Egypt was built in the early 1960's to provide electricity to the city of Cairo and to furnish irrigation water to the lower Nile basin. The completion of the dam caused a number of ecological backlashes. Nutrient-rich silt, which provided annual enrichment of the local farmland, was suddenly trapped behind the dam and could no longer flow downriver. The silt had also enriched the nearby waters of the Mediterranean Sea, thereby boosting its productive sardine fishery. Deprived of this natural fertilizer, farmers were forced to import chemical fertilizers at great expense. The sardine fisheries collapsed. Furthermore, Lake Nasser, the reservoir that formed behind the Aswan Dam, provided a habitat for snails carrying the organism that causes schistosomiasis, a disease with a significant fatality rate in humans.

Other examples of environmental backlash include the unregulated dumping of organic household waste (sewage) in many communities, which often results in the pollution of drinking water; the introduction of non-native species to new areas, which sometimes causes the extinction of native plants and animals; and the widespread use of agricultural insecticides, which may accumulate in food chains and lead to the deaths of larger animals that prey on the target insects.

The restoration of the Aral Sea may prove to be beyond human technology and capabilities. In the long run, the most feasible course of action may be one of inaction to allow the natural processes of ecological succession to eventually restore the area to some semblance of its natural state. Certainly the greatest benefit to be gained from the example of the Aral Sea is that of a lesson learned. While it is difficult, if not impossible, to foresee all or most of the consequences of human action in the natural environment, at least as much attention should be paid to a soundly produced environmental impact plan as is paid to engineering and economic plans.

*Albert C. Jensen*

## FURTHER READING

Breckle, Siegmar-W., Walter Wucherer, Liliya A. Dimeyeva, and Nathalia P. Ogar. *Aralkum: A Man-Made Desert: The Desiccated Floor of the Aral Sea (Central Asia)* Berlin: Springer-Verlag, 2012. Provides a comprehensive examination of the ecological catastrophe that is the Aralkum, the new desert on the desiccated floor of the Aral Sea that has resulted from human environmental mismanagement.

Ellis, William S. "The Aral: A Soviet Sea Lies Dying." *National Geographic* (February, 1990): 70-93. Describes the destruction of the once-productive Aral Sea and the impact on the local people. Suitable for high school readers. Includes detailed color photographs.

Imeson, Anton. *Desertification, Land Degradation and Sustainability.* Hoboken, N.J.: Wiley, 2012. Explores research from the first decade of the twenty-first century. Discusses desertification, management, and mitigation on the global and local scale, and explores the cultural and physical impact. Well-suited for undergraduates and graduate students with an environmental studies, geology, ecology, or economics background.

Kostianoy, Andrey G. *The Aral Sea Environment,* Berlin: Springer-Verlag, 2010. Reviews information gathered by specialists from Russia, Uzbekistan, France, Germany, and the United States. Offers an overall examination of the Aral Sea system from paleohistorical records through its present, ever-worsening state. Provides an analysis of the runoff waters and deltas of the Amudarya and Syrdarya Rivers.

Kotlyakov, V. M. "The Aral Sea Basin: A Critical Environmental Zone." *Environment* 23 (1): (January, 1991). Describes the role of Soviet scientists in covering up possible adverse environmental impacts of the diversion of water from the Aral Sea. Suitable for high school readers.

Mainguet, Monique. *Aridity: Drought and Human Development.* Berlin: Springer-Verlag, 2010. Discusses global and local aridity and drought. Describes changes in vegetation and hydrology. Focuses on anthropogenic impact. Examines hydrology, desertification, and the concept of decadence. Includes references and multiple indexes.

Morgan, Michael D., Joseph M. Moran, and James H. Wiersma. "The Vanishing Aral Sea: Can It Be

Saved?" *Environmental Science: Managing Biological and Physical Resources.* Dubuque, Iowa: Wm. C. Brown, 1993. Places the Aral Sea disaster in the context of managing water and aquatic food resources. Suitable for high school students.

Nicholson, Sharon E. *Dryland Climatology.* New York: Cambridge University Press, 2011. Covers the geomorphology, hydrology, ecology, and climatology of dry land. Discusses adaptation, microhabitats, and desertification. Written for graduate students, professional researchers, and scientists.

Nihoul, Jacques C. J., Peter O. Zavialov, and Philip P. Micklin. NATO Scientific Affairs Division. *Dying and Dead Seas: Climatic Versus Anthropic Causes.* Dordrecht: Kluwer Academic Publishing, 2004. The first comprehensive study of dead and dying seas. Presents an analytical look at past, present, and future roles of climatic and anthropic causes.

Pearce, Fred. "Poisoned Waters." *New Scientist 2000* (October 21, 1995): 29-33. Examines the impact on the human population in the area around the Aral Sea, especially human health. Includes photographs. Suitable for high school and college students.

Schneider, David. "On the Level: Central Asia's Inland Seas Curiously Rise and Fall." *Scientific American* (July, 1995): 14. Discusses the historic changes in water volume of the Aral Sea and the nearby Caspian Sea. Suitable for high school students.

Stone, Richard. "Coming to Grips with the Aral Sea's Grim Legacy." *Science* 284 (April 2, 1999): 30-33. Reports on the weather and the ecological environment of the Aral Sea and some possible efforts to ameliorate the conditions. Suitable for high school and college students.

Zavialov, Peter O. *Physical Oceanography of the Dying Aral Sea* Chichester: Praxis Publishing, 2005. Describes the past and background of the Aral Sea, its present crisis state, and the attempts to foresee its future from a physical oceanography standpoint. Based on Russian scientific literature that was previously unavailable to international readers and researchers.

**See also:** Arctic Ocean; Atlantic Ocean; Black Sea; Caspian Sea; Deep Ocean Currents; Gulf of California; Gulf of Mexico; Hudson Bay; Indian Ocean; Mediterranean Sea; North Sea; Pacific Ocean; Persian Gulf; Red Sea; Surface Ocean Currents

# ARCTIC OCEAN

*The Arctic Ocean, the fourth largest of the world's oceans with an area of 12,257,000 square kilometers, lies completely within the Arctic Circle. Large segments of it remain frozen throughout the year. It has an average depth of 1,000 meters but in some parts is nearly 5,500 meters deep.*

## PRINCIPAL TERMS

- **brash:** splinters that become detached from ice floes and float in the Arctic Ocean
- **ice floes:** large formations of ice, usually 2.5 to 3.5 meters thick, that float in the waters of the Arctic Ocean
- **igloo:** a temporary Inuit structure made from blocks of dense snow
- **Inuit:** the native dwellers of the northern polar regions, whose name means "the People"; often referred to incorrectly as Eskimos, a word from a more southerly native language
- **phytoplankton:** tiny floating sea plants that are most plentiful in the presence of sunshine and rich nutrients
- **pingo:** a large, stable ice intrusion of the Arctic tundra terrain, appearing as a large, dome-shaped, earth-covered mound with cracks visible at the top, the core being solid ice
- **salinity:** the salt content of such substances as water and food
- **umiak:** a large boat, covered with animal skins, that the Inuit traditionally used when hunting marine mammals such as seals and whales

## LOCATION OF THE ARCTIC OCEAN

Lying wholly within the Arctic Circle, the Arctic Ocean was once viewed by geographers as a part of the Atlantic Ocean. It is now viewed as a discrete body of water with definite boundaries. As such, it is the fourth-largest ocean on Earth, smaller than the Pacific, the Atlantic, and the Indian Oceans, which are, respectively, the largest, second-largest, and third-largest oceans in the world. Waters from all oceans intermingle as tides and currents carry them along and disperse them. The Arctic Ocean is unique in that it is almost landlocked. Essentially circular, it extends from the North Pole south to about 80 degrees north latitude or, if one includes its smaller fringe seas, to about 70 degrees north latitude. The main landmasses that it touches are Canada, Alaska, Siberia, Greenland, Iceland, and Norway.

The Arctic Ocean spills about 60 percent of its water into the Atlantic Ocean between Greenland and Spitsbergen, a group of islands that belong to Norway. It is largely the surface waters that are exchanged because a high range of submerged mountains known as the Faeroe-Icelandic Ridge, which in some places breaks the surface and creates islands, blocks the exchange of the deepest water. Of all the major oceans, the Arctic, because of its unique pattern of temperatures, currents, and ice conditions, probably has the most independent existence. There is virtually no flow of water from the Arctic Ocean into the Pacific because land barriers and Earth's rotation prevent such a flow. Some 35 percent of the water that comes to the Arctic Ocean flows in from the Pacific. Most of the other water that enters it comes from the Norwegian Sea.

Crucial to the ecology of the Arctic Ocean is the Greenland Sea. Several large rivers flow from Canada and Siberia into the sea, bringing enormous quantities of water into its somewhat constricted basin. In this part of the world, evaporation is not great, so if the water that comes into the ocean were not expelled into the Greenland Sea, serious problems would ensue.

The water from the Greenland Sea creates a cold current, termed the East Greenland Current, that flows south along Greenland's east coast. The much weaker Labrador Current flows through Smith Sound and Baffin Bay. Yet another weak current flows from the Bering Strait. Water that does not flow out through the Greenland Sea is deflected by Greenland's northern shore. It forms a current that, off the northwest portion of the Arctic Archipelago, runs southwest and west, then turns again seeking an outlet, which creates a unique circular current in the Arctic Ocean.

This current explains why the part of the Arctic Ocean that is bounded by Siberia has less ice than the same ocean has in Greenland and in parts of the Arctic Archipelago, notably Ellesmere Island. Large ice floes tend to drift southward and westward, many of them melting before they have a chance to drift into the congested shipping lanes of the Atlantic Ocean. Icebergs that reach the Atlantic Ocean are usually brought there by the Labrador Current from western Greenland's fjords. Although some of the ice melts as it moves southward, the polar ice cap that covers part of the Arctic Ocean has not melted in recorded history.

It was once thought that the Arctic Ocean had a considerable effect upon the climate and ecology of all the other oceans. Researchers, however, have questioned this supposition, concluding generally that the Arctic Ocean is more affected by conditions in the world's other oceans than they are affected by conditions in the Arctic Ocean.

## DERIVATION OF THE NAME "ARCTIC"

In early times, the Arctic Ocean lay in what the ancients called "terra incognita," meaning "unknown land." The areas that it touched were thought to be incapable of sustaining life, although it is now recognized that the Arctic area is teeming with life of many varieties, from complex vertebrates such as humans, seals, and polar bears to simple microorganisms such as phytoplankton, which flourish in both the Arctic and the Antarctic.

The ancient Greeks named the Arctic after the astronomical constellation known as "the great bear." The Greek word for bear is *arktos*—hence "Arctic." The ancients observed that the great bear constellation appeared to revolve around the North Pole. Convinced that there must be another pole at Earth's other extremity, the Greeks coined the term "*antarktos*," or "opposite bear," which in time became the term "Antarctic."

In early times, the Arctic Ocean was often referred to as the Frozen Ocean because so much of it was permanently covered with ice extending from the polar ice cap. With winter temperatures typically as low as −33 degrees Celsius and below, it was generally thought that the ocean and the area surrounding it precluded human habitation, a fallacy that has since been disproved.

## PEOPLE OF THE POLAR REGIONS

Despite early conjecture that suggested that the polar regions were uninhabited, it is now clear that they have sustained human life for thousands of years. The Inuit have long been permanent residents of areas around the Arctic Ocean. These people are traditionally hunters who hunt the native animals, particularly seals, for food.

The Inuit are expert at fishing through the ice and take a great deal of their food from the Arctic Ocean, particularly in winter when they tend to dwell along the ocean's shore. They also hunt whales on a limited basis, which they once harpooned from "umiaks," large open boats made of animal skins on a wood-and-bone frame. Whales provided the Inuit with oil to fuel their lamps and with blubber—the large, fatty layers that preserve the whale's buoyancy and body warmth in the frigid Arctic waters. Eating the blubber helped sustain the Inuit during the long winters. In the present day, Inuit cultural traditions live on, but the traditional weapons and boats have largely given way to their modern counterparts, and snowmobiles now run where dogsleds and snowshoes were once the norm.

During the winter, the Inuit often build igloos, domed shelters made from blocks of densely packed snow, that are surprisingly comfortable inside. Caught out in the open with a storm approaching or simply traveling away from home for an extended period, a skilled person could construct an igloo for shelter in a matter of minutes. Sometimes a hole can be cut in the ice of the igloo's floor in order to catch fish protected from the driving winter winds.

In summer, most of the Inuit went inland to the tundra to hunt for game and to fish in the lakes. They gathered the wild berries and roots that grow there. They preserved the meat of some of the fish and game they caught by drying or smoking it so that they would have food during the harsh winter.

In the present day, the vast majority of Inuit live in simple frame houses in scattered villages throughout the north, and the art of building igloos is becoming a rare skill. Concerted action by tribal elders now seeks to preserve the traditional knowledge and pass it down to younger generations as a right of their heritage.

The modern people of the polar regions have been touched by technology, usually living in villages that have schools, stores, churches, and medical centers. They live in houses made from imported materials. Such houses usually have an air lock, an area between the door that opens to the outdoors and the second door that opens into the house so that the cold Arctic air does not penetrate the building's interior warmth.

Only sketchy and inaccurate information was available about the Arctic and its people until the nineteenth century, when exploration of the region began. Earlier European sailors had gotten to the region in their search for a short route to Asia. Most, however, found the area so forbidding that they turned back. In the early twentieth century, a now-classic documentary film titled *Nanook of the North* was shot on location and was the first real exposure to Inuit culture experienced by the rest of the world.

## EXTENT OF THE ARCTIC OCEAN

The Arctic Ocean occupies an area of more than 12 million square kilometers, about one-seventh the size of the Atlantic Ocean. Many scientists believe that global warming will gradually increase the size of the Arctic Ocean and will cause flooding in the land areas that touch it.

Several seas lie on the fringes of the Arctic Ocean, including the Barents, Beaufort, East Siberian, Greenland, Kara, and Laptev Seas, as well as Baffin Bay. These fringe seas extend 10 degrees farther south than the Arctic Ocean proper, reaching as far as 70 degrees north latitude. Although some geographers consider the polar regions to be those that cannot sustain the growth of trees, the polar circles that most geographers accept are generally calculated as being 66.33 degrees north and south of the equator, so that even the fringe seas are wholly within the Arctic Circle. As global warming progresses, it should be noted that the region that cannot support tree growth is shrinking accordingly, and boreal forest encroaches farther and farther into the northlands.

The whole of the Arctic Ocean is roughly the size of Antarctica, whose ocean is very different from its northern counterpart in that it surrounds a continent, whereas the Arctic Ocean is surrounded by landmasses. That more life thrives in the Arctic than in Antarctica is attributable to the fact that the landmasses around the Arctic Ocean are warmed by the ocean's currents, with much of the water being quite shallow because of the continental shelves. Nevertheless, much of the Arctic Ocean is frozen all year long, with its water temperature hovering between −1.1 and +1 degree Celsius.

The northern and southern polar regions are generally defined as being those points on Earth that experience at least one day each year when the sun does not set. This phenomenon is true at both the North and South Poles in June and December of each year, respectively.

## APPEARANCE OF THE ARCTIC OCEAN

Looking down upon the Arctic landscape from above, one is impressed by how much water and ice constitute the landscape. The land below seems solidly frozen, rimmed with ice from the sea, much of which is covered with ice. The view of the frozen sea from the air is dramatic beyond imagining. At times it is not punctuated for hundreds of kilometers by any stretches of water. As one moves out from the land, the ocean waters, if they are visible at all, appear dark, but they gradually moderate into a lighter green, their surfaces studded by huge chunks of ice called brash that have splintered off from larger ice floes.

If one could look below the surface to the ocean floor, it would present the appearance of a warped, rifted surface with great irregularities. In some places, it sinks to depths of 5,500 meters. Freshwater ice formations, similar to the pingos that mark the Arctic tundra, are present in some of the shallower water, each moored soundly to the bottom on which they rest.

It may seem odd that freshwater islands exist in a saltwater ocean. This peculiarity is explained by the fact that ice resulting from the freezing of saltwater becomes less and less saline in its constitution as it ages. Ice loses one-half of its salinity in its first year of being frozen; eventually, it reaches the point where it has virtually no salinity, so that such ice can be returned to its liquid state and drunk with no ill effects.

Beneath its surface, the Arctic Ocean consists of two basins. They are believed to have developed separately more than 100 million years ago when the tectonic plates of Earth's surface drifted apart. The Eurasian Basin resulted when the sea floor spread in a line along the Nansen Cordillera, a range of submerged mountains that constitutes the northernmost part of the Mid-Atlantic Range. In time, this movement pushed a narrow portion of Asia away from the mainland. Known as the Lomonosov Ridge, this sliver lies directly beneath the North Pole. At roughly the same time, the land that is now Alaska moved away from North America and left a basin, known as the Canada Basin, on the other side of the Lomonosov Ridge. This basin is also sometimes referred to as the Amerasian Basin.

## CONTINENTAL SHELVES

The Arctic Ocean contains the widest continental shelf in the world, between 490 and 1,780 kilometers in width around the Eurasian basin, stretching north from Siberia toward the North Pole. A similar shelf, in the Amerasian Basin, which is from 97 to 200 kilometers wide, extends north from North America. Its exposed portions form the Arctic Archipelago, which consists of Wrangel Island, the Franz Josef Archipelago, the New Siberian and Lyakhov Islands, Severnaya Zemlya, Novaya Zemlya, and Spitsbergen.

The continental shelf that extends north from North America eventually drops off into a deep, oval basin reaching south from the North Pole to the Bering Sea. This and parts of the Greenland Sea east of Greenland are the deepest parts of the Arctic Ocean. It was once thought that no form of life could exist at such depths, but it has now been established that in some of the deepest parts of the oceans, hot water, sometimes as hot as 367 degrees Celsius, flows out of deeply submerged structures called hydrothermal vents that are comparable in some ways to volcanoes. Life exists in many forms around such vents.

Nothing in nature remains the same forever. Cores of ice taken from the Arctic reveal that the area once had a temperate climate. The fossils found in the ice cores suggest that all sorts of vegetation grew where now a wholly different kind of vegetation exists in the harsh climate of the Arctic Circle. Evidence also substantiates the theory that the polar climate is still changing, once again becoming warmer. This warming has implications for the entire world: If the polar ice caps begin to melt at a rapid rate, low-lying areas of the world may become flooded and uninhabitable. Studying the Arctic Ocean and the circumpolar regions will raise awareness that everything in nature is connected and that no alteration in nature, whether natural or human-made, is without consequences.

*R. Baird Shuman*

## FURTHER READING

American Museum of Natural History. *Ocean.* New York: Dorling Kindersley Limited, 2006. Discusses the geology, circulation, climate, and physical characteristics of the ocean. Covers marine biology and ocean chemistry. Includes a discussion of icebergs and polar ocean circulation. An excellent starting point for anyone learning about oceans and marine ecology. Includes images on each page, an extensive index, a glossary, and references.

Ballard, Robert D., and Malcolm McConnell. *Explorations: My Quest for Adventure and Discovery Under the Sea.* New York: Hyperion, 1995. Presents fascinating information about the very deep sea, correcting many previously held notions about it.

Bischof, Jens. *Ice Drift, Ocean Circulation and Climate Change.* Chichester: Praxis Publishing, 2000. Presents and discusses the concept of ice rafting, in which large pieces of Arctic ice floes break away and drift like rafts on the ocean currents, their movement adding to our knowledge of past and present conditions of oceanic circulation.

Broad, William J. *The Universe Below: Discovering the Secrets of the Deep Sea.* New York: Simon and Schuster, 1997. Provides useful information about the very deep sea. Explores the notion that there is little life in the very deep sea. Demonstrates that all sorts of life that have hitherto been undetected inhabit the abyssal depths, sometimes enduring high temperatures from hot flows that would kill most known forms of life.

Byers, Michael. *Who Owns the Arctic? Understanding Sovereignty Disputes in the North.* Vancouver: Douglas & McIntyre Publishers, 2009. Covers many issues of Arctic sovereignty. An especially topical subject since global warming is rapidly freeing up access to Arctic Ocean resources.

Frolov, Ivan E. *The Arctic Basin: Results from the Russian Drifting Stations.* Chichester: Praxis Publishing, 2005. Discusses the results obtained by manned research stations on the drift ice of the high Arctic. Describes the meteorological, oceanographic, and geophysical observations.

Lemke, Peter, and Hans-Werner Jacobi. *Arctic Climate Change: The Acsys Decade and Beyond.* New York: Springer Science+Business Media, 2012. Addresses a number of major topics related to climate change in the Arctic brought on by increasing global temperatures with respect to the role of the Arctic in the global climate system.

Penny, Malcolm. *Seas and Oceans: The Polar Seas.* Austin, Tex.: Raintree Steck-Vaughn, 1997. Written specifically for the juvenile audience. Articulates clearly a great deal of information about the Arctic Ocean and the people who live near it.

Serreze, Mark C., and Roger Graham Barry. *The Arctic*

*Climate System.* New York: Cambridge University Press, 2005. Provides a comprehensive, up-to-date assessment of the Arctic climate system for researchers and advanced students.

Stein, Ruediger, and Robie W. Macdonald. *The Organic Carbon Cycle in the Arctic Ocean.* Berlin: Springer, 2004. Various topics are discussed in relation to the Arctic Ocean carbon dynamics. Covers dissolved organic matter, particulate organic carbon, productivity and growth rates, benthic carbon cycling, and organic carbon burial rates. Summarizes the Arctic carbon cycle and compares it to global cycling.

U.S. Department of the Interior, Minerals Management Service. *Programmatic Environmental Assessment: Arctic Ocean Outer Continental Shelf Seismic Surveys.* U.S. Department of the Interior Minerals Management Service, Alaska OCS Region, 2006. Provides an overview of seismic surveys and the exploration of the Alaskan continental shelf. Includes alternative scenarios for surveys and their evaluation. Addresses the environmental impact of such surveys.

**See also:** Aral Sea; Atlantic Ocean; Black Sea; Caspian Sea; Gulf of California; Gulf of Mexico; Hudson Bay; Ice Ages and Glaciations; Icebergs and Antarctica; Indian Ocean; Lakes; Mediterranean Sea; North Sea; Pacific Ocean; Persian Gulf; Red Sea; Water Quality; Weather Modification

# ARTIFICIAL RECHARGE

*Artificial recharge is the technique of capturing water that might otherwise go to waste, such as flood runoff or treated sewage effluent, and using it to replenish groundwater supplies by allowing it to infiltrate the soil or forcing it underground with recharge wells. Not only can this technique help to conserve drinking water resources, but it can also correct problems caused by excessive pumping, such as seawater invasion and land subsidence.*

## PRINCIPAL TERMS

- **groundwater:** the water contained in soil and rock pores or fractures below the water table
- **groundwater recharge:** the water that infiltrates from the ground surface downward through soil and rock pores to the water table, causing its level to rise
- **losing stream:** a stream that is located above the elevation of the water table and that loses water to the ground via infiltration through the stream bed; the opposite of a gaining stream
- **recharge well:** a well designed to pump surface water underground in order to recharge the groundwater; sometimes called an injection well
- **unsaturated zone:** the area of the soil or rock between the land surface and the water table, in which voids between the soil and rock particles contain both air and moisture; also called the vadose zone
- **water spreading:** an artificial recharge technique in which floodwaters are diverted from the stream channel and spread in a thin sheet over a flat land surface, allowing the water to infiltrate the ground
- **water table:** the upper surface of the saturated zone or groundwater, below which all soil and rock pores are filled with water

## NATURAL RECHARGE

Groundwater is one of the world's most vital resources. More than 20 percent of all freshwater used in the United States comes from groundwater sources, and demand has been steadily increasing over the years. Groundwater typically comes from rainfall and snowmelt, which infiltrates the soil and percolates slowly downward through a region of soil and rock known as the unsaturated or vadose zone. At some level below the surface, the small spaces in and between the soil and rock particles become completely filled with water. This area is the saturated zone, and the water it contains is by definition groundwater. The upper surface of the saturated zone is called the water table and is equivalent to the level of water in a well that might intersect it. When a water well is pumped, water is withdrawn from the soil and rock pores, and the level of the water table is lowered correspondingly.

Recharge of the groundwater occurs when more water from precipitation infiltrates the unsaturated zone. When this infiltrating water reaches the water table, the water table level will rise again. Natural recharge can also occur when a stream flows over permeable sands and gravels and water seeps downward through the stream bed. Streams that leak water into the ground in this way are called losing streams, and, for the most part, are found in arid climates. Many streams in humid climates actually gain water from the ground, and are accordingly referred to as gaining streams.

In many regions of the United States, the amount of rainfall is small enough that natural groundwater recharge occurs very slowly, even though groundwater may be abundant. In addition, humans often modify the land in such a way as to reduce natural recharge. For example, impervious coverings of the soil surface such as parking lots, streets, roof tops, and airports all prevent rainfall from infiltrating the ground. Small losing streams that might also have previously recharged the groundwater are frequently diverted into sewer systems and hastened out of an area. For a combination of these reasons, many cities that pump groundwater have experienced a severe and continuous drop in the level of the water table, which has not only threatened the water supply but also resulted in other undesirable side effects, such as subsidence (sinking) of the land surface or invasion of saltwater from the ocean into the fresh groundwater supply.

## ADVANTAGES OF ARTIFICIAL RECHARGE

Artificial recharge is a technique used to increase the amount of surface water moving into the ground. Surface water is manipulated by some method of construction such as building pits or basins, the use of permeable materials for ground covering, spreading it on bare ground surface, or injecting it directly into the ground via recharge wells. In this manner, natural recharge is supplemented, and water that might otherwise go to waste (such as floodwater) can be stored underground for later use. The advantages are numerous. Storing water inside the earth rather than in traditional surface reservoirs means that little or no water will be wasted as a result of evaporation. There is also no wastage of land because of the flooding of a reservoir and no expensive dam to build that later may threaten to break catastrophically. Underground reservoirs do not eventually fill with sediment as do surface reservoirs, and they are less vulnerable to contamination.

Artificially recharged groundwater can be used not only to replenish water supplies and raise the water table but also to increase the pressure of the underground water enough to prevent seawater from migrating farther into the ground or even to flush the intruding seawater out of an area where pumping wells are located. In some areas, severe lowering of the water table (or the underground water pressure) through overpumping has caused the soil structure to compress irreversibly, resulting in a lowering, or subsidence, of the actual ground surface. This subsidence is usually gradual, but in some areas with limestone bedrock, sinkholes can collapse suddenly. In either case, ground subsidence can result in substantial damage to buildings or even flooding of the area. Artificially recharged water does not and cannot restore the ground surface to its former levels, but it can substantially reduce or even halt further subsidence. Another use of artificial recharge is for the storage of energy. The demand for heating or air-conditioning is seasonal. During the summer, surplus hot water can be stored underground through recharge wells. This water has a different viscosity (flow behavior) from that of the cold regional groundwater and thus mixes with it surprisingly little. Even after a storage period of up to three months underground, as much as three-fourths of the stored heat can be recovered.

## SEWAGE EFFLUENT

The water that is used to recharge groundwater artificially normally comes from excess storm runoff collected by various means. One of the most attractive aspects of artificial recharge, however, is that treated sewage effluent can be used as the water source. The effluent can be disposed of in this way, and, if the operation is carefully designed and monitored, the waste water becomes cleansed and purified during the recharge process as a result of the very nature of groundwater movement: The slow percolation of water through the tiny soil and rock pores allows the earth itself to act as a filter for the water. Movement of sewage effluent through the unsaturated zone can remove all of the bacteria and viruses, along with the solid matter and a large proportion of any undesirable chemical contaminants. Artificial recharge of sewage effluent must be carefully planned and carried out so that the water does not percolate too rapidly or reach the water table before most of the contaminants have filtered out; otherwise, the groundwater will become polluted.

Artificial recharge of groundwater has been widely practiced for more than two hundred years throughout the world for a variety of purposes, but there are some difficulties with the method. One of the biggest problems in the United States is that there are often separate laws dealing with surface water and groundwater, as well as separate governing bodies or "owners." That makes the legal and economic aspects of the conjunctive use of surface water and groundwater extremely complicated. Artificial recharge may not be possible in some areas, or it may be too expensive in others. If it is not carried out properly, it can lead to groundwater contamination or other problems. In spite of these drawbacks, artificial recharge has been highly successful in restoring groundwater levels and reducing seawater intrusion and subsidence.

## RECHARGE PITS

The most common technique used for artificial recharge involves the excavation of pits or basins to collect local storm runoff or diverted stream flow. Economically, old gravel pits can also be used. The pits or basins must intersect soils or layers of rock that have a high permeability, such as sands and gravels, and be located well above the water table. When a storm occurs, floodwater is diverted to a basin or series of basins paralleling a naturally losing stream channel. This water then fills the basin and, over a few days or weeks, gradually infiltrates through the permeable sand or gravel at the bottom of the basin and works its way downward to the water table. The water table then rises directly below the basin, forming a mound shape, which grows upward and spreads outward as recharge of the groundwater occurs. This "mound" formation is exactly analogous to the formation of a zone of depression in the water table at a location such as a well when water is removed. Both reflect the differential between the rate at which water is added to or removed from the water table and the rate at which water flows within the water table.

Recharge pits and basins require continual maintenance, as the infiltration rate decreases sharply with time. Fine-grained material (clay and silt) suspended in the floodwater settles to the bottom and works downward into the uppermost soil pores, clogging them, as do subsequent algae and bacteria growth. To reopen the soil pores, it is necessary to allow the basin to dry out from time to time. This causes the clays to dry and crack and allows the organic matter to decompose. Sometimes it may even be necessary to till and scrape the basin floor. Clogging of the pit floor may not be problematic, however, if it is possible to construct the pit so that it is steep-sided and deep, as long as the pit's walls are sufficiently permeable.

Since 1935, Long Island, New York, has used recharge basins to divert storm water, which would otherwise run out to sea via sewers, to the groundwater. There are now more than three thousand recharge basins that dispose of approximately 230,000 cubic meters of water per day. These basins are unlined open pits about 3 meters deep and open to the underlying gravels. Their infiltration rate is high enough that almost all of the basins are dry within five days after a 2-centimeter rainfall.

## WATER SPREADING

Another common artificial recharge technique is known as water spreading. Small losing tributary streams in a relatively flat area are modified in such a way that floodwaters, instead of racing down the stream bed, are diverted and spread as a thin sheet over the land surface. The floodwaters then infiltrate permeable soils and recharge the groundwater over an extensive area. The changes to the stream bed are inexpensive and generally involve the construction of low check dams bulldozed from river-bottom material and perhaps reinforced by vegetation, wire, or rocks. An added benefit of this technique is that by trapping floodwaters harmlessly in the upstream areas, urban areas that are located downstream are

also better protected from flood damage. Some drawbacks to the technique are the large amounts of land required, ice buildup in winter, and the fact that the small dams are easily washed out by larger floods and must constantly be rebuilt. If the check dams were to be constructed of permanent materials, however, they might create a flood hazard upstream. Water spreading recharge techniques can be adapted to steeper terrains by constructing a series of shallow, flat-bottomed, closely spaced ditches or furrows near the losing stream channel. The exact configuration of the ditches (contoured, tree pattern, or trellis) can be adjusted to the local terrain, but the lowest ditches should flow back into the main stream channel to avoid any overflow flood hazard. Such ditches are somewhat more costly to construct than are check dams. They also require a lot of land area, and, like recharge basins, are subject to clogging.

Sometimes it is possible to avoid the construction costs of ditches and check dams entirely by using preexisting irrigation canals for water spreading. In this case, cropland is irrigated by floodwaters during the dormant or winter season as well as the growing season. Care must be taken with this method to ensure that the constant leaching (removal by downward percolation) of salts and nutrients from the topsoil does not adversely affect the quality of the groundwater or the crops during the growing season. If treated sewage effluent is used as the recharge water in this technique, although agricultural land is usually spray-irrigated from a surface water source, nutrients are actually added to the soil. This not only benefits the crops but also cleanses the recharged effluent. The problem with this technique is that irrigation can overwhelm the natural filtering characteristics of the unsaturated zone and allow the groundwater to become contaminated.

## RECHARGE WELLS

The artificial recharge techniques already discussed are practical only in areas where permeable soils on rock beds allow direct infiltration from the surface to the water table. In many areas, the layers of the earth bearing the groundwater (aquifers) are deeply buried below impermeable geological materials, and there is no direct access to the surface. It is still possible, however, to recharge these layers artificially using recharge wells (sometimes called injection wells). Unlike pumping wells, in which water is pumped from the groundwater to the surface, recharge wells allow surface water to be forced under pressure into the ground and underlying permeable rocks. Large volumes of water can be stored in this way and later pumped out as the need arises. The recharged freshwater can also be used to force seawater, which may have invaded the ground and contaminated drinking water supplies, back out to sea. An additional benefit of this particular technique is that, unlike recharge basins or water spreading, it does not require much land area, so it is practical to use in urban areas.

Recharge wells are not without their drawbacks. Compared with other artificial recharge techniques, they are expensive to construct and require continual maintenance. Their main problem, similar to recharge basins, is well clogging. Any fine-grained sediment in the recharge water will enter the rock pores and clog them. Dissolved air, also very abundant in the recharge water, will enter the pore spaces as well and prevent water infiltration. Bacteria growing in the water will coat the well sides and rock pores with clogging growths, and chemical reactions between the recharge and groundwater will also cause clogging substances, such as rust and carbonate salts, to grow. For these reasons, the recharge water must be carefully treated before it is injected, and the clogged wells must be pumped frequently to restore some of their original permeability. Finally, recharge wells allow direct access of the surface water to the ground, bypassing the filtering characteristics of the unsaturated zone. That means that if any contaminant finds its way into the recharge water, it may rapidly enter the ground and contaminate drinking water supplies.

Recharge wells have been used successfully in Los Angeles, California, to prevent contamination of drinking water supplies by invading seawater. More than ninety wells have been placed in a line about 15 kilometers long and parallel to the coast. Filtered, chlorinated wastewater is injected into four underlying, water-bearing layers at an average rate of 1,500 cubic meters per day per well. The wells create a ridge of pressurized water along the coast that separates the invading seawater from pumping wells farther inland. Water levels in the pumping wells can even be drawn below sea level with no danger of saltwater contamination.

*Sara A. Heller*

## FURTHER READING

Asano, Takashi, ed. *Artificial Recharge of Groundwater.* Boston, Mass.: Butterworth, 1985. A comprehensive and somewhat technical compendium of articles dealing with artificial recharge, grouped by topic. Emphasis is placed on recharge with reclaimed wastewater and case histories. Includes a short section on the legal and economic concerns of groundwater recharge. Bibliography, index. Suitable for the college-level reader.

D'Angelo, Salvatore, Thomas G. Richardson, Richard P. Arber, et al., eds. *Using Reclaimed Water to Augment Potable Water Resources.* 2d ed. Denver, Colo.: American Water Works Association, 2008. Collects lectures and essays compiled by the Water Environment Federation and the American Water Works Association. Focuses on the practice of recharging and recycling contaminated groundwater and other water sources.

Dillon, P. J., ed. *Management of Aquifer Recharge for Sustainability.* Boca Raton, Fla.: Taylor & Francis, 2002. A compilation of papers discussing groundwater management and aquifer recharging. Covers topics such as bank filtration, soil aquifer treatment, and rainwater harvesting.

Environmental and Water Resources Institute. *Standard Guidelines for Artificial Recharge of Ground Water.* Reston, Va.: American Society of Civil Engineers, 2001. Presents the steps to be taken in planning, designing, constructing, maintaining, operating, and closing an artificial recharge project. Includes a discussion of the economic, environmental, and legal considerations of such projects.

Healy, Richard W., and Bridget R. Scanlon. *Estimating Groundwater Recharge.* New York: Cambridge University Press, 2010. Provides a critical evaluation of the principles, theory, and assumptions used in estimating the rates of groundwater recharge. Geared to serious students of hydrogeology and related fields.

Moore, John E. *Field Hydrogeology.* 2d ed. Boca Raton, Fla.: CRC Press, 2012. Provides practical information on field investigation, aquifer testing, and groundwater quality testing. Discusses planning and report writing in detail. Offers case studies from around the world. Best suited for students of hydrology.

O'Hare, Margaret P., et al. *Artificial Recharge of Ground Water.* Chelsea, Mich.: Lewis, 1986. Covers the basic artificial recharge techniques, including chapters on how to evaluate an area for its recharge potential and case histories of several artificial recharge projects. Contains a lengthy annotated bibliography. Includes appendices with data tables on the recharge potential and needs of various geographic areas of the United States. Index. Suitable for the college-level reader.

Saether, Ola M., and Patrice de Caritat, eds. *Geochemical Processes, Weathering, and Groundwater Recharge in Catchments*. Rotterdam: A. A. Balkema, 1997. Looks at the geochemical cycles and processes associated with catchment basins. Deals with the artificial recharge of groundwater in such systems. Somewhat technical.

Steenvoorden, J. H. A. M., and Theodore A. Endreny, eds. *Wastewater Re-use and Groundwater Quality*. Wallingford, England: International Union of Geodesy and Geophysics, 2004. Presents a series of articles examining the methods, precautions, and consequences of returning waste water to the environment as a means of recharging groundwater.

Thompson, Stephen A. *Hydrology for Water Management*. Rotterdam: A. A. Balkema, 1999. Discusses hydrology, groundwater flow, and stream flow. Focuses on the management of water supplies in efforts to keep them free of pollutants and available to the largest number of people possible. Includes illustrations, maps, an index, and a bibliography.

Todd, David Keith. *Ground Water Hydrology*. 3d ed. New York: John Wiley & Sons, 2004. A classic textbook on groundwater that covers artificial recharge. Provides detailed descriptions of water spreading and recharge well techniques, in addition to some case studies. Nontechnical, easy to follow, with excellent illustrations and extensive references and index. Suitable for the college-level reader.

**See also:** Aquifers; Dams and Flood Control; Floods; Freshwater and Groundwater Contamination Around the World; Groundwater Movement; Groundwater Pollution and Remediation; Hydrologic Cycle; Precipitation; Salinity and Desalination; Saltwater Intrusion; Surface Water; Waterfalls; Water Quality; Watersheds; Water Table; Water Wells

# ATLANTIC OCEAN

*The Atlantic Ocean separates the North and South American continents from Europe and Africa. With an area, excluding its dependent seas, of about 83 million square kilometers, it is second in size only to the Pacific Ocean. The dependent seas add another 23 million square kilometers to the Atlantic's total size.*

## PRINCIPAL TERMS

- **continental drift:** the gradual movement of continental landmasses across the earth's surface driven by convection processes in the mantle
- **continental shelf:** the part of the sea floor that is generally gently sloping and extends beneath the ocean from adjacent continents
- **continental slope:** the part of the continental shelf that drops off sharply toward the ocean's floor
- **equator:** an imaginary line, equidistant from the North and South Poles, around the middle of the planet
- **estuary:** an area where the mouth of a river broadens as it approaches the sea, characterized by the mixing of freshwater and saltwater
- **lagoon:** a long, narrow body of saltwater that is separated from the ocean by a bank of sand
- **tectonic plates:** large segments of the earth's crust, affected by the movement of magma in the underlying mantle layer
- **tidal range:** the difference in water depth between high and low tides

### LOCATION AND ORIGIN

The Atlantic Ocean, with an average depth of 3,300 meters and a maximum depth of 8,380 meters, touches the eastern coastlines of North and South America and the western coastlines of Europe and Africa. The Atlantic Ocean is often divided into the North Atlantic and the South Atlantic, although the entire oceanic mass may properly be considered a single ocean. The North Atlantic extends from the equator north to the Arctic Ocean. The South Atlantic, that part of the ocean south of the equator that extends as far as Antarctica, meets the Pacific Ocean at Cape Horn, the southernmost tip of South America, and the Indian Ocean at the Cape of Good Hope, the southernmost tip of Africa.

More than 200 million years ago, Earth's land surface consisted of a single supercontinent, referred to as Pangaea, surrounded by an enormous sea, called Panthalassa. A large, shallow bay known as the Tethys

Sea also protruded into the supercontinent from Panthalassa. The landmass, like contemporary landmasses, was a large, solid rock plate that floated on the mantle layer of molten rock. Over the intervening millions of years, convection currents in the mantle drove Pangaea to break apart into several smaller segments and become the present-day continents. The magmatic currents have driven, and are still driving, the continental masses and tectonic plates across the face of the planet. Looking at a map of the modern world and regarding it as a huge jigsaw puzzle, it is easy to visualize how the western coast of Africa fits into the eastern coast of the Americas and how the coastlines of other geographical areas fit neatly into those facing them.

As Pangaea split apart, the Tethys Sea essentially became enclosed by land as continental drift continued over millions of years, forming part of the present-day Arctic Ocean and Mediterranean Sea. The Pacific Ocean is the remnant of Panthalassa that exists today. The main rift in Pangaea separated large tectonic plates that have since formed North and South America, Asia, Africa, Europe, India, and Australia. The ever-widening space between the separating continents filled with water to form what is now the Atlantic Ocean. Current data regarding the movement of the continents and tectonic plates reveal that the North and South American continental structure and the Europe-Africa-Asia-Australia chain continue to move away from each other as though pivoting about a point in the North Atlantic Ocean approximately at the location of Iceland, such that the greater the distance from that location, the faster the tectonic plate is moving. Thus Europe and the easternmost coast of Canada are separating at the slowest rate, while Australia continues to move in a northeasterly direction at a relatively fast rate.

Two major tectonic plates, each a segment of the Atlantic Ocean floor, are moving apart annually at the average rate of 1.3 centimeters from a volcanically active line that runs approximately down the middle of the Atlantic Ocean, called the Mid-Atlantic Ridge. Where volcanic activity has been intense, molten rock has cooled into volcanic cones that sometimes protrude above the water's surface to create islands. However, in the Atlantic, unlike in the Pacific (where islands are scattered throughout the ocean), more islands are found close to shore than toward the middle of the ocean. The rate of movement of the Atlantic Ocean sea floor away from the Mid-Atlantic Ridge is such that the entire sea floor is replaced about every 100,000 years, precluding the formation of mid-Atlantic islands. The few exceptions are actually the exposed peaks of underwater volcanic mountains, some of which, such as Mount Tiede in the Canary Islands, extend more than 3,600 meters above sea level.

Ancient people named the Atlantic Ocean after the mythological giant Atlas, who is said to have carried the world on his shoulders. Many ancients, believing that earth was flat, feared that anyone sailing far enough into the ocean would eventually fall over the edge of the earth into an abyss. This belief inhibited early explorers, although even in ancient times some skeptical adventurers sailed far into the ocean and returned safely to shore.

## CONTINENTAL SHELF AND
### COASTLINE

Close to shore, the Atlantic Ocean is generally shallow, seldom exceeding a depth of 545 meters. Waters cover a shelf of rock that, in most of the North Atlantic, extends between 65 and 80 kilometers from the coastline into the ocean. At the end of this shelf, the ocean drops precipitously to a depth of about 3,940 meters, gradually leveling off into what is designated the abyssal plain. The continental shelf off the coast of Africa is much narrower than the continental shelf off the coast of North America.

The coastlines that border the Atlantic Ocean have been forming for millions of years, shaped largely by the motion and activity of the water that laps the shore. The rockiest areas are found in the northern parts of the North Atlantic and in the South Atlantic. The level of the

ocean has risen considerably over the past 10,000 years as glaciers have melted. With the end of the Ice Age, huge quantities of water poured down rivers into the oceans. The point at which these rivers meet the ocean is called an estuary. Among the deepwater estuaries are the Chesapeake Bay in the United States and the Falmouth Estuary in Great Britain, both of which are sufficiently deep to accommodate large ships.

Less rocky portions of coastline are often characterized by long stretches of sandy beaches that have been formed over the years as the moving waters of the ocean have pulverized rocks and shells, transforming them into sand. Sometimes sandbars form near the coast. These may be long expanses of sand, often permanently covered by water, while others are below the water's surface only at high tide. Some sandbars that are permanently above water may be broad enough to be inhabited, as are the Outer Banks of North Carolina, although at times devastating floods occur in communities built on sandbars. Sandbars endanger ships that venture too close to them, sometimes causing them to founder. Such

*The Atlantic Ocean off Cape Cod, Massachusetts.* (PhotoDisc)

ships may float off a sandbar as the water rises at high tide, but in many cases this does not happen, and the ship, trapped on the sandbar, must be hauled off. Most seaports employ pilots whose duty it is to guide incoming ships through deep channels away from dangerous sandbars or rock formations beneath the surface of the water. Lagoons sometimes form between the coast and sandbars that are not permanently covered with water. These are salty bodies of water in which marsh grasses and other vegetation grow. Many of them are rich in the microorganisms that fish need for their survival. Extensive fish and bird populations cluster around lagoons.

No ocean is fed by more rivers than the Atlantic. These rivers have, over hundreds of thousands of years, carried silt and sand toward the ocean they feed. These deposits have, in many places, built up to form deltas, roughly triangular areas whose broad base fronts the ocean. Some deltas, particularly those of the Amazon and Niger Rivers, are hundreds of miles wide with channels running through them. Often in tropical areas, mangrove swamps flourish in the deltas, the roots of the mangrove trees being covered with saltwater at high tide to provide them with the nutrients they require for their growth. Fish and birds flourish in mangrove swamps.

## TIDES AND TRADE WINDS

Ocean tides are generally predictable. High tide and low tide each occur twice in any twenty-four-hour period. Tidal ranges—the depth variations between high tide and low tide—are most pronounced on continental shelves and in deepwater bays. In some parts of the world, the tidal range is dramatic. In the Canadian province of Nova Scotia, the Bay of Fundy has a tidal range of more than 12 meters, one of the largest in the world.

Gentler tides are observed as one approaches the equatorial regions, although there can be dramatic tidal activity in such areas during hurricanes and tropical storms, which have traditionally occurred in August through October in the Northern Hemisphere and February through April in the Southern Hemisphere. There is considerable concern at present that global climate change is causing more powerful hurricanes and extending the hurricane season. Hurricanes usually form in tropical areas but can move rapidly into more temperate zones where, if they strike land, they can result in injury, death, and substantial destruction. In some cases, such storms erode entire beaches and completely inundate waterside property. The Atlantic coastline in many places is shrinking rapidly as the result of water erosion from such storms.

Predictable wind currents, much like jet streams in the upper atmosphere, blow across the Atlantic Ocean. The trade winds near the equatorial areas blow in a westward direction. Such winds made it possible for early explorers to sail from European ports to the Americas. For their return trips, explorers depended upon winds to the north of the areas where the trade winds blow called the "westerlies," which blow from west to east. Along the equator is a narrow area where waters are calm and where there is virtually no wind. In this area, referred to as the doldrums, sailors can drift languidly for long periods of time with no wind to propel them.

## ATLANTIC CURRENTS

Ocean water is never still. Much ocean water moves in one direction in predictable flows, in essence forming rivers through the ocean. Such rivers are called ocean currents. The warm tropical currents near the equator push the waters of the Atlantic Ocean westward toward North and South America, moving clockwise in the Northern Hemisphere and counterclockwise in the Southern Hemisphere.

The northern equatorial current runs from just north of the equator along the northeastern coast of South America and on toward the southern coast of North America. There, in the ocean east of the United States, the Gulf Stream runs through the Atlantic, flowing in a northerly direction. The Labrador Current that flows from the Arctic Ocean meets the Gulf Stream off Newfoundland, causing temperature variations that result in high humidity and dense fogs. North of Labrador, the Gulf Stream divides into the North Atlantic Drift, which flows north toward Greenland, and the Canaries Current, which flows southeast through the Atlantic off the west coast of Africa. These clockwise currents flow in what is termed the North Atlantic Gyre.

South of the equator, the southern equatorial current moves counterclockwise off South America's east coast, proceeds south to the Brazilian Current, which then veers southeast to meet the Benguela Current that flows from just north of Antarctica to the Tropic of Capricorn. These counterclockwise currents are referred to as the South Atlantic Gyre.

The oceanic rivers that form the currents of the North and South American Gyres have a profound effect upon climate. The Gulf Stream, which brings warm, tropical waters north at the rate of some 130 kilometers per day, makes far northern areas such as Greenland and Iceland warm enough for human habitation. Its southeasterly branch gives Great Britain a more temperate climate than one would expect at such latitudes. The northern regions of the United States are considerably warmer than they would be if the Gulf Stream did not flow along their coasts. Even areas within the Arctic Circle feel the effects of the Gulf Stream, which keeps Russia's Arctic port of Murmansk free of ice throughout the year.

The South Atlantic Ocean's Benguela Current propels cold waters from the Antarctic Ocean north along Africa's southwestern coast, keeping its temperatures much more temperate than would be expected in such latitudes. Cold air, however, does not hold moisture well, so when warm, humid winds from Africa's southwestern coast strike the cold air over the Benguela Current, they cause rain to fall over the ocean. The result is that the whole southwestern coast of Africa is arid. The South Atlantic Ocean there is bordered by desert.

## FOOD CHAIN

The world's oceans are teeming with life. In a somewhat hierarchical arrangement called the food chain, the sea's larger creatures feed on the smaller ones. At the base of the food chain are microscopic organisms called plankton, which form the basis for life in oceans. Plankton, which exist in both animal and vegetable form, are the basic diet of many aquatic creatures and hence of all ocean life. Plankton cannot live without sunlight, so they are found close to the water's surface. Plankton depend upon nutrients borne in the water for sustenance, and such nutrients are usually found in coastal areas that are not very deep, such as mangrove swamps, where plankton often flourish. Some of the minerals the plankton require are washed into the ocean through estuaries that carry silt from inland areas. Other minerals come from deep in the sea, carried to the surface by ocean currents. In the Atlantic, the most abundant plankton are found in the extreme northern and southern reaches of the ocean, as well as off the west coast of Africa. These areas have a high concentration of the nutrients—abundant in deep, cool waters—on which plankton feed.

The presence of large amounts of plankton in these areas results in a wealth of other aquatic creatures, most of which live near the surface of the ocean where plankton are found, although exploration of the very deep oceans has revealed a remarkable amount of sea life in parts of the ocean so deep that it was previously thought that nothing could survive there.

The Atlantic has an intriguing population of creatures that range from among the smallest on Earth, such as plankton, to the largest, such as whales. Most of the fish caught by commercial fisherman for human consumption—notably sardines, smelt, cod, halibut, mackerel, hake, sole, and anchovies—swim in large schools near the ocean's surface to harvest the plankton and smaller fish that constitute their diets. Larger fish, such as sharks, whales, swordfish, and sailfish, are fast-moving and live on a diet of smaller fish. They are often found in the cooler waters of the Atlantic, where plankton are abundant enough to nourish the small fish on which larger fish feed.

Some of the sea's creatures are extremely mobile. Whales travel from the polar regions to areas thousands of kilometers away for spawning. Dolphins swim fast enough to cover 245 kilometers in a single day. Conversely, some shellfish, such as mussels, attach themselves to formations in the sea and are immobile, protected from predators by their thick shells. Some forms of sea life are protected by camouflage that allows them to blend in with the surrounding area. Others are able to emit noxious jets of fluid into the waters around them when they are threatened.

Considerable bird life is associated with the Atlantic Ocean. Most of the birds that depend upon the Atlantic for their food supply can fly, but some of them, such as the cormorant and the puffin, can swim as well. They dive into the water to catch their prey and can swim to catch it. Penguins, found in the Antarctic, cannot fly and can barely walk, but they swim as well as many fish. Cranes and other long-legged birds flourish in shallow waters, where they eat tiny sea creatures that they strain from the water and mud through their beaks.

In the middle of the North Atlantic is a large, relatively calm area of water known as the Sargasso Sea, whose surface is covered with huge fields of nutritious seaweed. The Sargasso Sea has become a spawning ground for eels that migrate to it during mating season from Europe and North America.

## SEASIDE SETTLEMENTS

Throughout history, humans have tended to settle beside the sea or along rivers. The transportation opportunities provided by such locations still make them desirable places for settlement. Such areas usually offer a temperate climate as well. The ancient cultures of Rome and Greece grew up on the shores of the Mediterranean and the Aegean, two of the dependent seas of the Atlantic Ocean. The Iberian Peninsula borders on the Atlantic to the west, as do France and Great Britain. Advanced civilizations have flourished on the shores of the Atlantic, particularly in Europe and Africa, since long before recorded history.

The islands of the Atlantic—products of violent volcanic activity that created large, hilly outcroppings—were slow to develop. The exceptions are the Canary Islands and Madeira, both of which became necessary stopover points for explorers sailing from Europe to the Americas. In recent times, considerable development of the Atlantic islands has taken place in such areas as the Caribbean. The Falkland Islands off the southeastern coast of South America have a stable, permanent population, as do such island enclaves as Prince Edward Island and Newfoundland in the north.

Shipping and commerce have been the backbone of the economy in most of the areas that have developed along the Atlantic coastline. Raw materials are brought into Atlantic port cities such as New York, Philadelphia, Rio de Janeiro, and Lisbon to feed the manufacturing industries of those countries. The fishing industry has flourished for many centuries along the Atlantic coastline. Tourism has also become a major economic factor in the more temperate and scenic regions on the Atlantic coast.

Trade patterns were substantially altered with the opening in 1896 of the Suez Canal, which connects the Mediterranean Sea with the Indian Ocean, and with the opening in 1914 of the Panama Canal, which links the Atlantic and Pacific Oceans. These strategic canals opened up a great deal of trade worldwide and overcame the necessity of plying the turbulent and dangerous waters around Cape Horn and the Cape of Good Hope in order to deliver goods to distant markets.

*R. Baird Shuman*

## FURTHER READING

American Museum of Natural History. *Ocean.* New York: Dorling Kindersley Limited, 2006. Geology, tides and waves, circulation and climate, physical characteristics of the ocean, and more are discussed in this all-encompassing text on oceans. Marine biology and ocean chemistry are covered, along with discussion of icebergs and polar ocean circulation. With so many topics covered, each section provides merely an overview. Regardless, this text is an excellent starting point for anyone learning about oceans and marine ecology. There are numerous images on each page, an extensive index, a glossary and references.

Ballard, Robert D., and Malcolm McConnell. *Explorations: My Quest for Adventure and Discovery Under the Sea.* New York: Hyperion, 1995. Relates a great of information about a leading oceanographer's explorations of the world's oceans, particularly about the Atlantic Ocean, its currents, its animal life, and its general place among the oceans of the world.

Bertness, Mark D. *Atlantic Shorelines: Natural History and Ecology.* Princeton, N.J.: Princeton University Press, 2006. The book is filled with useful images and graphs to support the well-organized text. The author combines the fields of ecology, geology, and hydrology into a seamless story of the Atlantic coastline. Written without technical jargon or intense mathematics, this book is ideal for undergraduates and can be understood by those with a limited science background.

Broad, William J. *The Universe Below: Discovering the Secrets of the Deep Sea.* New York: Simon and Schuster, 1997. One of the most compelling books about the very deep ocean. Explodes many myths about the deep ocean, pointing out that considerable life flourishes in its very deepest parts, even at volcanic underwater temperatures of more than 500 degrees Celsius.

Collins, Elizabeth. *The Living Ocean.* New York: Chelsea House, 1994. Provides crucial information about the Atlantic Ocean and about environmental factors that threaten it and its sea life.

Goni, G. J., and Paola Malanotte-Rizzoli. *Interhemispheric Water Exchange in the Atlantic Ocean.* Amsterdam: Elsevier B.V., 2003. Discusses the interhemispheric and intergyre exchange of heat, salt, and freshwater in order to improve understanding of the dynamics of the tropical Atlantic Ocean and how they affect the oceans globally.

Morozov, Eugene G., Alexander N. Demidov, Roman Y. Tarakanov, and Walter Zenk. *Abyssal Channels in the Atlantic Ocean: Water Structure and Flows.* Dordrecht: Springer, 2010. Written for oceanography specialists. Summarizes some thirty years of observational data regarding the movement of water in the Atlantic Ocean deeps.

Oceanography Course Team. *Ocean Circulation.* 2d ed. Oxford: Butterworth-Heinemann, 2001. Surface currents and deep water currents are discussed, with focus on the North Atlantic Gyre, Gulf Stream, and equatorial currents. The El Niño phenomenon is discussed as well as the great salinity anomaly. This book provides a good introduction to oceanography.

Stevenson, R. E., and F. H. Talbot, eds. *Oceans.* New York: Time-Life, 1993. Provides excellent coverage of the Atlantic Ocean. Written in direct, easily understandable prose. Numerous excellent illustrations.

Trujillo, Alan P., and Harold V. Thurman. *Essentials of Oceanography.* Upper Saddle River, N.J.: Prentice-Hall, 2010. Suitable for all readers. Uses information from many branches of the sciences to illustrate the relationship between ocean phenomena and other earth systems.

Ulanski, Stan L. *The Gulf Stream: Tiny Plankton, Giant Bluefin, and the Amazing Story of the Powerful River in the Atlantic.* Chapel Hill: University of North Carolina Press, 2008. Explores the science and history of the Gulf Stream and its effects on global climate. Well illustrated. Suitable for all readers.

Van der Voo, Rob. *Paleomagnetism of the Atlantic, Tethys and Iapetus Oceans.* New York: Cambridge University Press, 2004. Discusses paleomagnetism, the continents and supercontinents, and terrane displacement.

Waterlow, Julia. *The Atlantic Ocean.* Austin, Tex.: Raintree Steck-Vaughn, 1997. Directed toward the young adult market. Offers comprehensive and lucid coverage of the Atlantic Ocean. Thoroughly researched. Well illustrated, with a glossary of key terms.

**See also:** Aral Sea; Arctic Ocean; Black Sea; Caspian Sea; Gulf of California; Gulf of Mexico; Hudson Bay; Indian Ocean; Mediterranean Sea; North Sea; Pacific Ocean; Persian Gulf; Red Sea

# ATMOSPHERE'S EVOLUTION

*The chemical composition of the atmosphere has changed significantly over the 4.6-billion-year history of Earth. The composition of the atmosphere has been influenced by a number of processes, including the "outgassing" of volatile materials originally trapped in Earth's interior during its formation; the geochemical cycling of carbon, nitrogen, hydrogen, and oxygen compounds between the surface, the ocean, and the atmosphere; and the evolution of life.*

## PRINCIPAL TERMS

- **chemical evolution:** the synthesis of amino acids and other complex organic molecules—the precursors of living systems—by the action of atmospheric lightning and solar ultraviolet radiation on atmospheric gases
- **photosynthesis:** the biochemical synthesis of glucose and molecular oxygen from carbon dioxide and water by chlorophyll-containing organisms in the presence of sunlight
- **prebiotic:** relating to the period of time before the appearance of life on Earth
- **primordial solar nebula:** an interstellar cloud of gases and dust that condensed by the action of gravitational forces to form the bodies of the solar system about 5 billion years ago
- **solar ultraviolet radiation:** biologically lethal solar radiation in the spectral interval between approximately 0.1 and 0.3 micron (1 micron = 0.0001 centimeter)
- **T Tauri stars:** a class of stars that exhibits rapid and erratic changes in brightness
- **volatile outgassing:** the release of the gases and liquids, such as argon, water vapor, carbon dioxide, and nitrogen sulfur, trapped within Earth's interior during its formation

## VOLATILE OUTGASSING

About 5 billion years ago, a cloud of interstellar gas and dust, called the primordial solar nebula, began to condense under the influence of gravity. This condensation led to the formation of the sun, moon, Earth, the other planets and their satellites, asteroids, meteors, and comets. The primordial solar nebula was composed almost entirely of hydrogen gas, with a smaller amount of helium, still smaller amounts of carbon, nitrogen, and oxygen, and still smaller amounts of the rest of the elements of the periodic table. About the time that the newly formed Earth attained its approximate present mass, gases that were released from the planet's interior could be retained by Earth's gravity instead of escaping into space, thus forming a gravitationally bound atmosphere. It is believed that the atmospheres of the other terrestrial planets, Mars and Venus, also formed in this manner. The release of gases and other volatiles in this manner is called volatile outgassing. The period of extensive volatile outgassing may have lasted for many tens of millions of years. The outgassed volatiles or gases had roughly the same chemical composition as present-day volcanic emissions: 80 percent water vapor by volume, 10 percent carbon dioxide by volume, 5 percent sulfur dioxide by volume, 1 percent nitrogen by volume, and smaller amounts of hydrogen, carbon monoxide, sulfur, chlorine, and argon.

The water vapor that outgassed from the interior eventually reached the saturation point, which is controlled by the atmospheric temperature and pressure. Once the saturation point was reached, the atmosphere could not hold any additional gaseous water vapor. Any new outgassed water vapor that entered the atmosphere would have precipitated out of the atmosphere in the form of liquid water. The equivalent of several cubic kilometers of liquid water released from Earth's interior in gaseous form precipitated out of the atmosphere and formed the oceans. Only small amounts of water vapor remained in the atmosphere, ranging from a fraction of a percent to several percent by volume, depending on atmospheric temperature, season, and latitude.

The outgassed atmospheric carbon dioxide, being somewhat water soluble, dissolved in the newly formed oceans and subsequently formed carbonic acid through its reaction with water. Once formed, carbonic acid can dissociate into ions of hydrogen, bicarbonate, and carbonate. The carbonate ions reacted with ions of calcium and magnesium in the ocean water, forming first insoluble carbonate salts, which precipitated out of the ocean and accumulated as seafloor carbonate sediments, eventually accumulating in sufficient quantities to form beds of carbonate rock. Most of the outgassed atmospheric carbon dioxide formed carbonates, leaving only trace amounts of gaseous carbon dioxide in the atmosphere (about 0.035 percent by volume). Sulfur dioxide, the third most abundant component of volatile outgassing, was chemically transformed into other sulfur compounds, including sulfuric acid and other sulfates in the atmosphere. Eventually, the sulfates formed atmospheric aerosols and gravitated out of the atmosphere to settle on the surface.

The fourth most abundant outgassed compound, nitrogen, is almost completely chemically inert in the atmosphere and thus was not readily transformed in large quantities, as was sulfur dioxide. Only minor amounts of nitrogen would be converted to various oxides by the action of lightning, to be trapped as nitrogen oxide salts in minerals. Unlike carbon dioxide, nitrogen is relatively insoluble in water and, unlike water vapor, does not condense out of the atmosphere. For these reasons, nitrogen built up in the atmosphere to become its major constituent (78.08 percent by volume). Accordingly, outgassed volatiles led to the formation of Earth's atmosphere, oceans, and the earliest, prebiotic carbonate rocks.

## CHEMICAL EVOLUTION

It has been demonstrated in laboratory experiments that molecular nitrogen, carbon dioxide, and water vapor in the early atmosphere would have been acted upon by solar ultraviolet radiation and atmospheric lightning. In the process, molecules of formaldehyde and hydrogen cyanide were chemically synthesized in the early atmosphere. These molecules were precipitated and diffused out of the atmosphere into the oceans. In the water, formaldehyde and hydrogen cyanide entered into chemical reactions that eventually led to the chemical synthesis of amino acids—the building blocks of proteins in living systems. The synthesis of amino acids and other compounds from nitrogen, carbon dioxide, and water vapor in the atmosphere is called chemical evolution. Chemical evolution preceded and provided the material for biological evolution.

For many years, it was thought that the early atmosphere was composed of ammonia, methane, and hydrogen rather than of carbon dioxide, nitrogen, and water vapor. Experiments show, however, that ammonia and methane are chemically unstable and are readily destroyed by both solar ultraviolet radiation and chemical reaction with the hydroxyl radical, which is formed from water vapor. In addition,

ammonia is very water soluble and is readily removed from the atmosphere by precipitation. Hydrogen, the lightest element, is readily lost from a planet by gravitational escape. Thus, an early atmosphere composed of methane, ammonia, and hydrogen would be very short lived, unless these gases were produced at a rate equal to their destruction or loss rates (an equilibrium state). These gases are also known to be extremely efficient "greenhouse gases," even more effective than carbon dioxide; their presence in the primordial atmosphere in any substantial amount would have maintained an extraordinarily high atmospheric temperature, by which many of the materials that were formed would thermally decompose. Today, methane and ammonia are very minor components of the atmosphere, at concentrations of 1.7 parts per million by volume and 1 part per billion by volume, respectively. Both gases are produced by microbial activity at the ground surface, and methane is released during coal mining and oil production, and from seafloor accumulations of methane hydrate. Clearly, microbial activity and microbes were nonexistent during the prebiotic phase of the planet. The atmospheres of the outer gas giant planets—Jupiter, Saturn, Uranus, and Neptune—all contain quantities of hydrogen, methane, and ammonia. It is believed that the atmospheres of these planets, unlike the atmospheres of the terrestrial planets—Earth, Venus, and Mars—are captured remnants of the primordial solar nebula resulting from the greater ability of the gravitational fields of those large planets to capture such light materials, preventing them from being drawn toward the sun. Because of the outer planets' great distance from the sun and their very low temperatures, hydrogen, methane, and ammonia are stable and long-lived constituents of their atmospheres. This is not true of hydrogen, methane, and ammonia in Earth's atmosphere.

Some have suggested that at the time of its formation, Earth may have also captured a remnant of the primordial solar nebula as its very first atmosphere. Such a captured primordial solar nebula atmosphere would have been composed of mostly hydrogen (about 90 percent) and helium (about 10 percent), the two major elements of the nebula. Even if such an atmosphere had surrounded the very young Earth, it would have been very short lived. As the young sun went through the T Tauri phase of its evolution, very strong solar winds (the supersonic flow of protons and electrons from the sun) associated with that phase would have quickly dissipated this remnant atmosphere. In addition, there is no geochemical evidence to suggest that early Earth ever possessed a primordial solar nebula remnant atmosphere.

## EVOLUTION OF ATMOSPHERIC OXYGEN

There is microfossil evidence for the existence of fairly advanced anaerobic microbial life on Earth by about 3.8 billion years ago. The ability to carry out photosynthesis evolved in one or more of these early microbial species. Through photosynthesis, the organism utilizes water vapor and carbon dioxide in the presence of sunlight and chlorophyll to form glucose and molecular oxygen. The glucose molecules are subsequently used by the organism for food and for biopolymerization into starches and celluloses. The production of oxygen by photosynthesis was a major event on Earth and eventually transformed the composition and chemistry of the early atmosphere as oxygen built up to become the second most abundant constituent of the atmosphere (20.90 percent by volume). It has been estimated that atmospheric oxygen reached only 1 percent of its present atmospheric level 2 billion years ago, 10 percent of its present atmospheric level about 550 million years ago (at the beginning of the Paleozoic), and its present atmospheric level as early as 400 million years ago.

The evolution of atmospheric oxygen had important implications for the evolution of life. Because molecular oxygen is a very effective oxidizing agent and would have been harmful to existing anaerobic life-forms, the presence and buildup of oxygen required the evolution of respiration and aerobic organisms. Accompanying and directly controlled by the buildup of atmospheric oxygen were the origin and evolution of atmospheric ozone, which is chemically formed from oxygen. The evolution of atmospheric ozone resulted in the shielding of Earth's surface from biologically lethal solar ultraviolet rays. The development of the atmospheric ozone layer and its accompanying shielding of Earth's surface permitted early life to evolve such that it could leave the safety of the oceans and go ashore for the first time in the history of the planet. Prior to the evolution of the atmospheric ozone layer, early life was restricted to a depth of several meters below the ocean surface. At this depth, the ocean water offered shielding from solar ultraviolet radiation. Theoretical computer calculations indicate that atmospheric ozone provided sufficient shielding from biologically lethal ultraviolet radiation for the evolution of non-marine organisms once oxygen reached about one-tenth of its present atmospheric level.

## VENUS AND MARS

Calculations indicate that the atmospheres of Venus and Mars also formed as a consequence of volatile outgassing of the same gases that led to the formation of Earth's atmosphere—water vapor, carbon dioxide, and nitrogen. In the case of Venus and Mars, however, the outgassed water vapor may never have existed in the form of liquid water in quantities comparable to those on Earth. Because of Venus's closer distance to the sun (108 million kilometers versus 150 million kilometers for Earth), its lower atmosphere was too hot to permit the outgassed water vapor to condense out of the atmosphere. Thus, the outgassed water remained in gaseous form in the atmosphere and, over geological time, was decomposed by solar ultraviolet radiation into molecular hydrogen and oxygen. The very light hydrogen gas quickly escaped from the atmosphere of Venus, and the heavier oxygen combined with surface minerals to form a highly oxidized surface. In the absence of liquid water on the surface of Venus, the outgassed carbon dioxide remained in the atmosphere and built up to become the overwhelming constituent of Venus's atmosphere, about 96 percent by volume. The outgassed nitrogen accumulated to make up about 4 percent by volume of the atmosphere of Venus. The present-day carbon dioxide and nitrogen atmosphere of Venus is massive—its atmospheric surface pressure is about 90 atmospheres (compared to the surface pressure of Earth's atmosphere of only 1 atmosphere). If the outgassed carbon dioxide in Earth's atmosphere had not been dissipated via dissolution in the oceans and carbonate formation, the planet's surface atmospheric pressure would presumably be about 70 atmospheres, with carbon dioxide accounting for about 98 to 99 percent of the atmosphere and nitrogen about 1 to 2 percent. Thus, the atmosphere of Earth would closely resemble that of Venus. The carbon dioxide-rich atmosphere of Venus causes a very significant greenhouse temperature enhancement, giving the surface of Venus a temperature of about 750 kelvins, which is hot enough to melt lead. The surface temperature of Earth is only about 288 kelvins, a range at which water can exist in equilibrium between all three phases, as solid, liquid, and gas.

Like Venus, Mars has an atmosphere, though extremely thin, composed primarily of carbon dioxide (about 95 percent by volume) and nitrogen (about 3 percent by volume). The total atmospheric surface pressure of Mars is only about 7 millibars (1 atmosphere is equivalent to 1,013 millibars). There may be large quantities of outgassed water in the form of ice or frost below the surface of Mars, but in the absence of liquid water, the outgassed carbon dioxide has remained in the atmosphere. The smaller mass of the atmosphere of Mars compared to the atmospheres of Venus and Earth may be attributable to the smaller mass of the planet and, accordingly, the smaller mass of gases that could have been trapped in the interior of Mars during its formation. In addition, the amounts of gases trapped in the interiors of Venus, Earth, and Mars during their formation apparently decreased with increasing distance from the sun. Venus appears to have trapped the greatest amounts of gases and was the most volatile-rich planet. Earth trapped the next greatest amounts, and Mars trapped the smallest amounts.

## STUDY OF EARTH'S ATMOSPHERE

Information about the origin, early history, and evolution of Earth's atmosphere comes from a variety of sources. Information on the origin of Earth and other planets is based on theoretical computer simulations, with ever-increasing empirical data input from celestial observation. These computer models simulate the collapse of the primordial solar nebula and the formation of the planets. Astronomical observations of what appears to be the collapse of interstellar gas clouds and the possible formation of planetary systems have provided new insights into the computer modeling of this phenomenon. Information about the origin, early history, and evolution of the atmosphere is based on theoretical computer models of volatile outgassing, the geochemical cycling of the outgassed volatiles, and the photochemistry of the outgassed volatiles. The process of chemical evolution, which led to the synthesis of organic molecules of increasing complexity—the precursors of the first living systems on the early Earth—is studied in laboratory experiments in which mixtures of gases simulating Earth's hypothetical

early atmosphere are energized by solar ultraviolet radiation and atmospheric lightning. The resulting products are analyzed by chemical techniques. A key parameter affecting atmospheric photochemical reactions, chemical evolution, and the origin of life was the flux of solar ultraviolet radiation on the early Earth. Astronomical measurements of the ultraviolet emissions from young sun-like stars have provided important information about the probable ultraviolet emissions from the sun during the early history of the atmosphere.

Geological and paleontological studies of the oldest rocks and the earliest fossil records have provided important information on the evolution of the atmosphere and the transition from an oxygen-deficient to an oxygen-rich atmosphere. Studies of the biogeochemical cycling of the elements have provided important insights into the later evolution of the atmosphere. Thus, studies of the origin and evolution of the atmosphere are based on a broad cross-section of science, involving astronomy, geology, geochemistry, geophysics, and biology as well as atmospheric chemistry.

## SIGNIFICANCE

Studies of the origin and evolution of Earth's atmosphere have provided new insights into the processes and parameters responsible for global change. Understanding the history of the atmosphere provides a sound basis for better understanding its future. Today, several global environmental changes are of national and international concern, including the depletion of ozone in the stratosphere and increasing global temperatures caused by the buildup of greenhouse gases in the atmosphere. The study of the evolution of the atmosphere has provided new insights into the biogeochemical cycling of elements between the atmosphere, biosphere, land, and ocean. Understanding this cycling is a key to understanding environmental problems. Studies of the origin and evolution of the atmosphere have also provided new insights into the origin of life and the possibility of life outside Earth.

*Joel S. Levine*

## FURTHER READING

Ackerman, Steven A., and John A. Knox. *Meteorology: Understanding the Atmosphere.* 3d ed. Sudbury, Mass.: Jones and Bartlett Learning, 2012.

Provides an overview of the atmosphere and atmospheric phenomena, beginning from the evolution of the early terrestrial atmosphere. Suitable for university-level readers.

Ahrens, C. Donald. *Essentials of Meteorology: An Invitation to the Atmosphere.* Belmont, Calif.: Brooks/Cole Cengage Learning, 2012. Covers various topics in weather and the atmosphere. Discusses topics such as tornadoes and thunderstorms, acid deposition and other air pollution topics, humidity and cloud formation, and temperature.

Bandy, A. R., ed. *The Chemistry of the Atmosphere: Oxidants and Oxidation in the Earth's Atmosphere.* Cambridge, England: Royal Society of Chemistry, 1995. Collects essays and lectures first presented at the Priestely Conference at Bucknell University. Covers such topics as atmospheric chemistry, the ozone layer, and the causes and remedies of air pollution. Somewhat technical. Suitable for the reader with some background in the subject.

Berner, Robert A. *The Phanerozoic Carbon Cycle: $CO_2$ and $O_2$.* New York: Oxford University Press, 2004. Discusses climate and atmosphere of the Paleozoic, Mesozoic, and Cenozoic eras. Also covers aspects of weathering and erosion on the carbon cycle. Suited to undergraduates. Contains references for each chapter and indexing.

Brimblecombe, Peter. *Air Composition and Chemistry.* 2d ed. New York: Cambridge University Press, 1996. Provides a thorough account of atmospheric chemistry and the techniques and protocol involved in determining the makeup and properties of air. Appropriate for readers with some background in chemistry and mathematics. Illustrations, maps, bibliography, and index.

Hobbs, Peter V. *Introduction to Atmospheric Chemistry.* Cambridge, England: Cambridge University Press, 2000. Provides a sound introduction to atmospheric chemistry, beginning with the early evolution of the planet's atmosphere. Written at the college level.

Holland, H. D. *The Chemical Evolution of the Atmosphere and Oceans.* Princeton, N.J.: Princeton University Press, 1984. A comprehensive and technical treatment of the geochemical cycling of elements over geological time and the coupling between the atmosphere, ocean, and surface. Covers the origin of the solar system, the release and recycling of volatiles, the chemistry of the early atmosphere and ocean, the acid-base balance of the atmosphere-ocean-crust system, and carbonates and clays.

Levine, Joel S., ed. *The Photochemistry of Atmospheres: Earth, the Other Planets, and Comets.* Orlando, Fla.: Academic Press, 1985. A series of review papers dealing with the origin and evolution of the atmosphere, the origin of life, the atmospheres of Earth and other planets, and climate. Compares the origin, evolution, composition, and chemistry of Earth's atmosphere with the atmospheres of the other planets. Contains two appendices that summarize atmospheric photochemical and chemical processes.

Lewis, John S., and Ronald G. Prinn. *Planets and Their Atmospheres: Origin and Evolution.* New York: Academic Press, 1983. A comprehensive, textbook treatment of the formation of the planets and their atmospheres. Begins with a detailed account of the origin and evolution of solid planets via coalescence and accretion in the primordial solar nebula. Discusses the surface geology and atmospheric composition of each planet.

Marshal, John, and R. Alan Plumb. *Atmosphere, Ocean and Climate Dynamics: An Introductory Text.* Burlington, Mass.: Elsevier Academic Press, 2008. An excellent introduction to atmospheres and oceans. Discusses the greenhouse effect, atmospheric structure, oceanic and atmospheric circulation, and climate change. Suitable for advanced undergraduates and graduate students with some background in advanced mathematics.

Voronin, P., and C. Black. "Earth's Atmosphere as a Result of Coevolution of Geo- and Biospheres." *Russian Journal of Plant Physiology* 54 (2007): 132-136. Covers the evolution of the atmosphere's composition and factors altering the gas composition. Provides background content on photosynthesis and chemolithotrophy. Highly technical. Appropriate for readers with a strong chemistry or geology background.

Yung, Yuk Ling, and William B. DeMore. *Photochemistry of Planetary Atmospheres.* New York: Oxford University Press, 1999. Examines the atmospheric chemistry of all of the planets in the solar system, with a focus on photochemical processes. Intended for the advanced reader. Illustrations, bibliography, and index.

**See also:** Acid Rain and Acid Deposition; Air Pollution; Atmosphere's Global Circulation; Atmosphere's Structure and Thermodynamics; Atmospheric Properties; Auroras; Climate; Climate Change Theories; Cosmic Rays and Background Radiation; Earth-Sun Relations; Global Warming; Greenhouse Effect; Ice Ages and Glaciations; Long-Term Weather Patterns; Nuclear Winter; Observational Data of the Atmosphere and Oceans; Ozone Depletion and Ozone Holes; Radon Gas; Rainforests and the Atmosphere

# ATMOSPHERE'S GLOBAL CIRCULATION

*The general circulation of Earth's atmosphere involves the large-scale movements of significant portions of air in the atmosphere. Variations in surface temperatures produce pressure gradients that combine with the Coriolis force to circulate most of the air in the atmosphere. This involves the Hadley circulation, which moves air in the Northern and Southern Hemispheres in three huge convection cells each, and the Walker circulation along the equatorial belt, which produces the El Niño phenomenon when it oscillates.*

## PRINCIPAL TERMS

- **adiabatic:** the effect of changing the temperature of a gas or other fluid solely by changing the pressure exerted on it, without the input or removal of heat energy
- **anticyclone:** a general term for a high-pressure weather system that rotates clockwise in the Northern Hemisphere and counterclockwise in the Southern Hemisphere
- **Coriolis force:** a non-Newtonian force acting on a rotating coordinate system; on the Earth, this causes objects moving in the Northern Hemisphere to be deflected toward the right and objects moving in the Southern Hemisphere to be deflected toward the left due to Earth's rotation
- **cyclone:** a general term for a low-pressure weather system that rotates counterclockwise in the Northern Hemisphere and clockwise in the Southern Hemisphere
- **geostrophic wind:** a wind resulting from the balance between a pressure gradient force and Coriolis force; the flow produces jet streams and is perpendicular to the pressure gradient force and the Coriolis force
- **pressure gradient force:** a wind-producing force caused by a difference in pressure between two different locations

## DRIVING FORCES

The general circulation of the atmosphere operates as a heat engine driven by the uneven distribution of solar energy over the surface of the planet. Atmospheric circulation transfers some of this energy from regions where it is abundant to regions where it is scarce. This energy transfer reduces the difference in surface temperature between the equatorial regions and the polar regions, between the oceans and the continents, and between continental interiors and coastal regions. The existence of an atmosphere, with water vapor, carbon dioxide, and other greenhouse gases, keeps the average temperature at the surface considerably warmer than it would otherwise be. The motions of the atmosphere cool the warmer regions and warm the cooler regions, smoothing out the extremes of temperatures.

The sun is so far from Earth that its rays of light can be thought of as being parallel when they strike the planet. Because Earth is a sphere, the areas warmed by this light vary with latitude. During the equinoxes, the sun is directly over the equator. A beam of sunlight with a square cross-section of 1 meter on each side is perpendicular to Earth's surface and will illuminate an area of 1 square meter at the equator. However, as the distance from the equator increases, the angle of incidence of that beam of sunlight increases in accord with the circumference of the planetary surface. At a latitude of 45 degrees north (near Ottawa, Ontario, for instance), that same 1 square meter beam of sunlight will be spread out over 1.4 square meters of the surface. At a latitude of 60 degrees north (near Anchorage, Alaska), it will be spread out over 1.7 square meters.

The 23.5-degree tilt of Earth's axis alters this in a cyclic manner over the course of a year. At the Northern Hemisphere summer solstice, 45 degrees north has a 1 square meter of sunlight spread out over 1.07 square meters, and 60 degrees north has it spread out over 1.24 square meters. At the Northern Hemisphere winter solstice, 45 degrees north has its 1 square meter of sunlight spread out over 2.73 square meters, while 60 degrees north has its spread out over 8.83 square meters. The intensity of solar radiation at these two latitudes differs by a factor of 1.16 in the summer and 1.22 in spring and fall, but 3.23 in the winter. Other choices for latitudes would yield similar results. This helps to explain why the temperature contrasts between northern and southern states in the United States is so much greater during the winter months than during the summer ones. It also explains some of the seasonal differences in general atmospheric circulation.

## MAJOR CONVECTION CELLS

Air above a warm surface absorbs heat and expands, in accordance with the gas law equation of physical chemistry, PV = nRT. This equation specifies the direct relationship between the temperature, pressure, and volume of a gas, so that as temperature increases, the pressure the gas exerts against its surroundings and the volume that it occupies also increase. Because the mass of the heated air has not changed, the expanded air is less dense than the air surrounding it, and it will rise, for the same reason that a hot air balloon rises. The rising air displaces air above it, pushing that air away with a pressure gradient force.

To understand how temperature gradients at the surface produce pressure gradient forces aloft, consider two adjacent columns of air that initially contain the same amount of air, one of which is warmer than the other. The warmer one expands and reaches higher above the surface. Because the mass of air in both columns is initially the same, the air pressure at the surface is also the same beneath both columns. Next, consider the air pressure halfway up the cooler column. This level is beneath one-half of the air in the cool column, but it is beneath more than one-half of the air in the warm column. The air pressure at this elevation in the warm column is therefore greater than the air pressure at this elevation in the cool column. This difference in pressure is a pressure gradient and will cause air to flow aloft from the warm column toward the cool column. As it does so, the total mass in the warm column will decrease, and the total mass in the cool column will increase; therefore, the air pressures at the surface will no longer remain the same. Beneath the cool column, the air pressure at the surface will be greater than beneath the warm column, and this pressure gradient will cause air to flow along the surface from the cool column toward the warm column. Eventually, a steady-state flow will result with an elevation that separates the two directions of air flow. Not surprisingly, above this elevation the total mass in each column will be the same.

If Earth were not rotating, displacements of the air initially above the equator would be toward the poles aloft and toward the equator at the surface; because it is rotating, however, an additional effect called the Coriolis force must be considered. A point on the equator is a distance of one Earth radius away from Earth's rotation axis, whereas a point at one of the

poles is directly on the rotation axis. As a parcel of air moves to the north from the equator, it is moving closer to the rotation axis. It inherited a certain angular momentum from when it was on the equator, and conservation of this angular momentum requires it to move somewhat to the east. In contrast, a parcel moving to the south in the Northern Hemisphere is moving farther from the rotation axis, and conservation of its angular momentum requires it to move a bit to the west. In either case, in the Northern Hemisphere, the Coriolis force causes objects to move to the right of the direction that they would normally be moving; in the Southern Hemisphere, it causes things to move to the left. This effect is only significant when there is very little frictional interaction between the moving object and Earth, so it is extremely important in the circulation of the oceans and atmosphere.

Air being displaced from its location over the equator will initially go due north or due south. The Coriolis force deflects it more and more until, in the vicinity of 30 degrees north or south latitude, it will be moving due east. At high altitudes, this is a geostrophic wind, a wind from the west (called a westerly) resulting from a poleward-directed pressure gradient force being deflected 90 degrees by the Coriolis force. The fastest elements of this flow form the subtropical jet stream. No longer moving toward the poles, this high-altitude air is now considerably cooler and denser than it was when first heated at the surface of Earth. As a consequence, it descends. As it gets lower, the additional air pressure it experiences causes it to warm up. Bicycle pumps illustrate this adiabatic effect, as they get hot when the air within them is pressurized without the input of heat energy.

The amount of water vapor that can be contained in air varies with temperature. This is commonly observed as water condenses out of cooler air at night to produce dew or frost. The air rising over the equator is initially warm and heavily laden with water. As it cools, this water vapor condenses, eventually producing the intense rainfall essential for equatorial rainforests. Losing its entrained water vapor warms the air, enhancing its ascent. Later, after moving to the northeast or southeast, this air descends, warming up so that it once again absorbs water vapor. The regions of Earth's surface near 30-degree latitudes are characterized by intense evaporation that produces deserts such as the Sahara and the Kalahari in Africa.

Because the descending air at 30-degree latitudes is denser than average, the air pressure at the surface beneath it will be higher than average. Similarly, the air pressure beneath the rising air column at the equator is lower than average. This difference in pressure causes winds to blow across the surface, from 30 degrees north or south latitude toward the equator. The Coriolis force also affects this flow so that these winds are deflected; by the time they reach the equator, they are coming out of the east. These easterly winds are called the trade winds, a name given because they were favorable for sailing ships bound to the East for trade.

## HADLEY CELLS

The movements of air already described connect to form two circulating cells, one in the Northern Hemisphere and one in the Southern Hemisphere. These are called Hadley cells, named after George Hadley, who discovered them in 1735.

The convection responsible for the Hadley cells is not seen as clearly elsewhere on Earth. However, four additional cells, two in each hemisphere, have long been a part of the theoretical development of meteorology. One, called the Ferrel cell, after William Ferrel, who proposed it in 1856, also descends at about 30 degrees latitude and presumably ascends at latitudes of about 60 degrees. The other, also proposed by Ferrel and called the Polar cell, ascends at latitudes of about 60 degrees and descends at the poles. The surface winds produced by these cells return air to the region around 60 degrees. This returning air is deflected by the Coriolis force, producing westerlies between 30 degrees and 60 degrees, and Polar easterlies at higher latitudes. Much of the general circulation of Earth's atmosphere can be explained by this six-cell model, and it continues to be used in many elementary meteorology and earth-science courses and textbooks. As a theoretical tool, it is useful and easily grasped, and it yields insights about global systems that are generally accurate. Air does descend at the poles and at 30 degrees latitude, and returns to the vicinity of 60 degrees latitude, but the situation there is not as simple as the rising heated air at the equator. The boundary between the Polar cell and the Ferrel cell, called the Polar front, has opposing surface winds, not converging ones; high-altitude winds over this front do not simply diverge as they do above the equator.

## FRONTS AND CYCLONES

During World War I, dirigibles were used to drop bombs on locations in England. To avoid assisting this tactic, the global system of weather data gathering was suspended. Subsequently, Norway's fishing fleet was put at risk from the weather, and scientists there developed theoretical models to make up for the lack of distant weather data. Led by Vilhelm Bjerknes and Halvor Solberg, this group of meteorologists was called the Bergen school. Relying on more closely spaced but effectively synchronous data, they developed the concept of fronts and proposed the cyclone model for storm genesis and evolution.

The cyclone model grew out of observations of storm systems in the middle latitudes. Such a system has a low-pressure region near its center and a pattern of winds moving in a concentric fashion around this low. This is the result of surface winds trying to move into the low-pressure region but being deflected by the Coriolis force. In the Northern Hemisphere, these cyclonic winds move around the low in a counterclockwise direction, whereas in the Southern Hemisphere, they move clockwise. Often, observations revealed a consistent pattern of temperature gradients, precipitation bands, and surface wind configurations. The entire storm system usually moved from west to east and evolved in similar ways from its initial genesis, to being fully developed with maximum winds, to fading out and disappearing. As this evolution occurred, the interactions between air masses of different temperatures and with different moisture content followed reasonably consistent patterns.

The meteorologists of the Bergen school saw that the shear zone between the westerlies of the warm, moist air to the south and the easterlies of the cold, dry air to the north was an important element in generating these storms. A line connecting the various cyclonic disturbances in the middle latitudes defined the Polar front—not as a simple, smooth surface, but one with major excursions to the north and south, much like a meandering river.

Air above the poles is cooler and more compressed than the warmer air in the middle latitudes. This produces a pressure gradient aloft that is directed toward the pole. This would cause poleward movement, except that the Coriolis force deflects such movement, again producing a westerly geostrophic wind, the fastest part of which is called the Polar jet stream. This is the jet stream referred to on weather maps and in forecasts in the United States and Canada. As already described, the temperature gradients at the surface are greater at higher latitudes, and hence the pressure gradients and velocities of the Polar jet stream are greater than those for the subtropical jet stream. In addition, because the surface temperature gradients are greater in winter than in summer, the velocity and significance of the Polar jet stream are also greater in the winter.

As the Polar jet stream races around the globe at velocities of about 125 kilometers per hour in the winter and 60 kilometers per hour in the summer, instabilities develop that deflect it into a meandering path. As a meander develops, the range of temperatures over which it moves increases, causing higher winds and even greater meandering. Eventually, portions of its path may be nearly north-south, bringing warmer air to higher latitudes and cooler air to lower ones. This diminishes the temperature gradients, causing the meanders to shrink until nearly east-west flow is reestablished. Called Rossby waves, after Carl-Gustav Rossby, who described them in 1939, these meanders form a path that resembles a very blunt, rounded star with three to six points centered on the pole. Moving along such a path, the jet stream speeds up at some places and slows down at others. Speeding up decreases air pressure aloft, while slowing down increases it. Cyclonic disturbances tend to form beneath places where the air pressure aloft is reduced and to move along tracks that lie beneath the jet stream. As the disturbances evolve, fronts develop, precipitation occurs, and warm air is transported to higher altitudes and then to higher latitudes.

## EL NIÑO/SOUTHERN OSCILLATION

The surface components of the Hadley cells move toward the equator. The Coriolis force deflects the flow, so that by the time these winds reach the equator, they are coming out of the east. Ocean currents, driven by their own geostrophic flows, are quite similar, with major east-to-west flows near the equator. The water moved by these ocean currents is warm surface water, and its transport to the west results in a buildup of such waters in the western part of the Pacific and, to some extent, the Atlantic. With warmer sea-surface temperatures to the west and upwelling of cooler water on the eastern side of the Pacific, yet another temperature gradient exists of sufficient scale to influence general circulation patterns.

The convection cell in this case is called a Walker cell, named after Sir Gilbert Walker, who identified it in 1924. The conditions needed for its development are neither constant nor periodic. Every three to five years, because of factors not yet well understood, this circulation breaks down. Generally coupled with a decrease in the strength of the trade winds, the Walker cell reverses its orientation: Instead of ascending air in the west, with sufficient rainfall to support equatorial rainforests in Indonesia, the air ascends over the eastern regions of the Pacific, bringing sometimes intense rainfall to regions that are otherwise deserts in Peru. Called El Niño by oceanographers and the Southern Oscillation by atmospheric scientists (often abbreviated as ENSO), this reversal has dramatic effects on weather patterns.

Usually lasting between twelve and eighteen months, the ENSO has been identified in historic and geologic data sets. Droughts in Africa, floods in the American West, and other phenomena appear to be related to the sea-surface temperature in the equatorial Pacific. Certainly one of the more interesting aspects of the ENSO is its effect on the other aspects of general atmospheric circulation.

## SIGNIFICANCE

Understanding general atmospheric circulation is essential for accurate, useful weather predictions. Knowledge of this circulation permits meteorologists to estimate how various parcels of air will move, how pressures will change, and how and where precipitation will occur. These estimates, in turn, permit them to project further into the future and improve the accuracy of their predictions.

By recognizing which variables are most likely to alter the general atmospheric circulation, and being able to guess how these variables might have been different in the past, geologists and climatologists can make more informed models about ancient weather patterns and climate evolution. The Himalayan Mountains and Tibetan Plateau are comparatively recent features on Earth, for example, and their presence has dramatically altered circulation patterns. Better understanding of past climates will help assess the influence of anthropogenic inputs, such as carbon dioxide and chlorofluorocarbons (CFCs), and should serve to guide public policy.

*Otto H. Muller*

## FURTHER READING

Ahrens, C. Donald. *Essentials of Meteorology: An Invitation to the Atmosphere.* 6th ed. Belmont, Calif.: Cengage Learning, 2012. Written in a style that promotes visualization of the concepts being discussed. Treatment of general atmospheric circulation is straightforward and easily grasped. Presents few equations, but includes lists of terms and questions for thought.

Barry, Roger G., and Richard J. Chorley. *Atmosphere, Weather, and Climate*. 9th ed. London: Routledge, 2010. Covers the subject of general atmospheric circulation in a thorough but not technically challenging style. Presents a vast number of figures and black-and-white line drawings to illustrate its points. Suitable for advanced high school students.

James, Ian N. *Introduction to Circulating Atmospheres*. Cambridge, England: Cambridge University Press, 1994. Presents many equations and technically challenging concepts. Includes illustrations that help to explain the concepts. Develops the theoretical basis for much of terrestrial meteorology. Gives considerable attention to atmospheres on other planets.

Lutgens, Frederick K., Edward J. Tarbuck, and Dennis Tasa. *The Atmosphere: An Introduction to Meteorology*. 11th ed. Upper Saddle River, N.J.: Prentice Hall, 2009. A very readable textbook with a wealth of color illustrations, including photographs and satellite images. Discussions are concise, easily understood, and internally consistent, although sometimes not as thorough as they might be.

Marshal, John, and R. Alan Plumb. *Atmosphere, Ocean and Climate Dynamics: An Introductory Text*. Burlington, Mass.: Elsevier Academic Press, 2008. An excellent introduction to atmospheres and oceans. Covers the greenhouse effect, convection and atmospheric structure, oceanic and atmospheric circulation, and climate change. Suitable for advanced undergraduates and graduate students with some background in advanced math.

Monmonier, Mark S. *Air Apparent: How Meteorologists Learned to Map, Predict, and Dramatize Weather*. Chicago: University of Chicago Press, 1999. An approachable treatment of meteorology cleverly presented as a history of weather cartography. Includes discussions of television channels, Web sites, and NOAA radar and radio stations. Traces much of the history of the subject since the end of the nineteenth century. Discusses general circulation patterns throughout the text. Clearly written, making this book valuable reading for anyone interested in meteorology.

Schneider, Tapio, and Adam H. Sobel, eds. *The Global Circulation of the Atmosphere*. Princeton, N.J.: Princeton University Press, 2007. Collects papers from the 2004 conference at California Institute of Technology. Covers large-scale atmospheric dynamics, storm-tracking dynamics, and tropical convection zones. A strong understanding of advanced mathematics is required. Best suited for graduate students and professionals.

Vallis, Geoffrey K. *Atmospheric and Oceanic Fluid Dynamics: Fundamentals and Large-scale Circulation*. New York: Cambridge University Press, 2006. Begins with an overview of the physics of fluid dynamics to provide foundational material on stratification, vorticity, and oceanic and atmospheric models. Discusses topics such as turbulence, baroclinic instabilities, wave-mean flow interactions, and large-scale atmospheric and oceanic circulation. Best suited for graduate students studying meteorology or oceanography.

Wells, Neil. *The Atmosphere and Ocean: A Physical Introduction*. 3d ed. New York: John Wiley & Sons, 2012. The atmosphere and oceans are both fluids circulating on a rotating planet, intimately and profoundly influencing each other. Treats atmospheric and oceanic interactions in a readable, yet thorough, manner. Presents numerous quantitative concepts and equations with graphs or figures that make them understandable to readers with little technical background.

Williams, James Thaxter. *The History of Weather*. Commack, N.Y.: Nova Science Publishers, 1998. A very easy-to-read treatment of the development of meteorology. Provides qualitative insights into how the observational and theoretical approaches to understanding weather sometimes competed but generally complemented and augmented each other.

**See also:** Acid Rain and Acid Deposition; Air Pollution; Atmosphere's Evolution; Atmosphere's Structure and Thermodynamics; Atmospheric Properties; Auroras; Climate Change Theories; Cosmic Rays and Background Radiation; Earth-Sun Relations; Geochemical Cycles; Global Energy Transfer; Global Warming; Greenhouse Effect; Nuclear Winter; Observational Data of the Atmosphere and Oceans; Ocean-Atmosphere Interactions; Oceans' Origin; Ozone Depletion and Ozone Holes; Radon Gas; Rainforests and the Atmosphere; Remote Sensing of the Atmosphere

# ATMOSPHERE'S STRUCTURE AND THERMODYNAMICS

*An atmosphere is a layer of gases that surrounds a planetary surface. Earth's current atmosphere is a complex, dynamic system that interacts closely with and controls the surface environment. Atmospheric thermodynamics involves the process by which energy from the sun is absorbed and deposited on Earth, including its oceans and atmosphere. Early life-forms substantially altered Earth's atmosphere, and humans continue to do so.*

## PRINCIPAL TERMS

- **adiabatic:** characterizing a process in which no heat is exchanged between a system and its surroundings
- **air drainage:** the flow of cold, dense air downslope in response to gravity
- **chlorofluorocarbons (CFCs):** chemicals in which chlorine and fluorine replace one or more of the hydrogen atoms in the molecular structure of the corresponding hydrocarbon
- **cosmic rays:** high-energy atomic nuclei and subatomic particles, as distinct from electromagnetic radiation
- **environmental lapse rate:** the general temperature decrease within the troposphere; the rate is variable but averages approximately 6.5 degrees Celsius per kilometer
- **exosphere:** the outermost layer of Earth's atmosphere
- **greenhouse effect:** a planetary phenomenon in which the atmosphere absorbs and retains more heat radiation than it passes back out into space
- **growing degree-day index:** a measurement system that uses thermal principles to estimate the approximate date when crops will be ready for harvest
- **heterosphere:** a zone of the atmosphere at an altitude of 80 kilometers, including the ionosphere, made up of rarefied layers of oxygen atoms and nitrogen molecules
- **homosphere:** a major zone of the atmosphere below the heterosphere whose chemical makeup is consistent with the proportions of nitrogen, oxygen, argon, carbon dioxide, and trace gases at sea level; includes the troposphere, stratosphere, and mesosphere
- **infrared radiation:** electromagnetic radiation with frequency in the range of $10^{13}$ to $10^{14}$ Hertz (Hz)
- **ionosphere:** the layer of ionized gases in Earth's atmosphere, starting about 50 to 100 kilometers above the surface of the planet (between the thermosphere and the exosphere)
- **latent heat:** the energy absorbed or released during a change of physical state
- **mesosphere:** the extremely rarefied atmospheric layer at altitudes from 50 to 80 kilometers above the surface, characterized by rapid decreases in temperature
- **net radiative heating:** the driving force for atmospheric thermodynamics, essentially the difference between heat entering the atmosphere due to solar heating and heat leaving the atmosphere as infrared radiation
- **photodissociation:** the condition in which light energy absorbed by a molecule is sufficient to dissociate the bonds between atoms in the molecule, typically caused by light in the ultraviolet range
- **radiational cooling:** the cooling of Earth's surface and the layer of air immediately above it by a process of radiation and conduction
- **stratosphere:** the atmospheric zone 20 to 50 kilometers above the surface that contains the functional ozone layer
- **temperature inversion:** a condition in which a region of warmer occupies a position above its normal location, causing air temperature to increase with increasing elevation from Earth's surface
- **thermosphere:** the atmospheric zone extending from 80 to 480 kilometers in altitude, and containing the ionosphere
- **troposphere:** the lowest atmospheric layer, extending from Earth's surface to an altitude of about 18 kilometers, containing 90 percent of the total mass of the atmosphere, marked by considerable turbulence and a decrease in temperature with increasing altitude
- **ultraviolet light:** electromagnetic radiation having a frequency in the range of $10^{15}$ to $10^{17}$ Hz

### ATMOSPHERIC CONTENT

"Atmosphere" usually refers to the layer of gases that covers Earth. Although most planets have atmospheres of some sort, Earth's atmosphere is unique among those known in this solar system in that it contains a substantial amount of oxygen and supports the equilibrium existence of water in solid, liquid, and vapor phases.

The atmosphere of Earth contains 78 percent nitrogen and 21 percent oxygen; the remainder consists of 0.9 percent argon, 0.03 percent carbon dioxide, and traces of hydrogen, methane, nitrous oxide, and inert gases. In addition, the atmosphere carries varying amounts of water vapor and aerosol particles such as dust and volcanic ash, depending on local and global events. As early as 3.5 billion years ago, primitive algae-like life-forms emerged and fed on the carbon dioxide by using photosynthesis to metabolize carbon dioxide and water molecules to form simple sugars and more complex polysaccharides. Photosynthesis over time gradually brought about a drastic change in the ratios of gases in the atmosphere, eventually producing the current carbon dioxide concentration of only 0.03 percent and free molecular oxygen concentration of 20 percent.

## A DYNAMIC SYSTEM

Gases are compressible, and they absorb or transmit varying amounts of electromagnetic radiation. Because of these characteristics, the atmosphere is not static but is instead a highly dynamic system. Phenomena such as weather and climate are short- and long-term events involving the exchange of energy and transport of mass within the atmosphere and also between the solid Earth, liquid oceans, and space.

Links between solar activities (especially sunspot cycles), weather, and climate have been sought for decades, but associations between them remain inconclusive. Interactions between the oceans and the air, however, are more substantial, and therefore more readily discernible. Winds drive waves and affect ocean currents; in return, the oceans act as a heat source or sink for the atmosphere. The most famous interaction is the El Niño/Southern Oscillation event. El Niño ("the child"), which usually happens around Christmas, is an upwelling of cold water off the Pacific coast of South America that occurs every two to seven years. Besides having a disastrous effect on the fishing industry, it is associated with changes in circulation and precipitation patterns in the atmosphere over the Pacific basin.

## TEMPERATURE

The atmosphere has no clear upper boundary, but is generally considered to extend to an altitude of about 300 kilometers, at which it responds more to electromagnetic effects and acts less like a fluid body. It can be described in three major characterizations: temperature, chemistry, and electrical activity.

Temperature changes with altitude and, as there is less overlying gas with increasing altitude, pressure and exposure to radiation also affect temperature. The lowest region of the atmosphere, enclosing virtually all life and weather on Earth, is the troposphere, extending to an altitude of 8 to 18 kilometers. The name is taken from the Greek word *tropo*, meaning "turn," and refers to the fact that this region turns with the solid earth. The bottom of the troposphere is the boundary layer, where the atmosphere interacts directly with the surface of the planet. The boundary layer is often turbulent, as moving air masses (winds) encounter and flow around or over obstructions and exchange heat with the ground or water. Temperature and pressure in the troposphere decrease at about 2 degrees Celsius per kilometer until the top of the troposphere, the tropopause, is reached. Life becomes increasingly difficult to maintain with altitude (humans may require additional oxygen above 4 kilometers and must wear pressure suits above 10.6 kilometers). Some 90 percent of the mass of the atmosphere is contained below the tropopause.

At the tropopause, temperature reaches a minimum of about –50 degrees Celsius, then rises again as one enters the stratosphere, to peak at about 15 degrees Celsius at the stratopause at an altitude of about 50 kilometers. Above this level, in the mesosphere, temperature declines again to a low of –60 degrees Celsius at the mesopause, at an altitude of 85 kilometers. Only 1 percent of the atmosphere is in the mesosphere and above; 99 percent lies below. Particles from atomic nuclei to meteors generally are destroyed in the mesosphere. Nuclei, or cosmic rays, encountering gas molecules in the mesophase will be shattered into secondary and tertiary particles. Most meteors are heated by friction and vaporize when they encounter the mesosphere. Above the mesosphere, the thermosphere (the hot atmosphere) extends to approximately 300 kilometers in altitude, and temperatures soar to between 500 and 2,000 degrees Celsius, depending on solar activity. Because the atmosphere is so thin, however, the total heat present is minuscule. Finally, beyond the thermosphere is the exosphere (outer layer), which extends from 300 kilometers to the solar wind (also called the interplanetary medium).

## SOLAR AND INFRARED RADIATION

Atmospheric thermodynamics can be described as the process by which energy from sunlight is absorbed by matter on Earth's surface, in the oceans, and in the atmosphere. This energy must be returned to space in the form of infrared radiation in order for the planet to maintain its stable cyclic range of ambient temperatures. Even though solar radiation and infrared radiation must be approximately balanced on a global level, they are frequently out of balance on a local level, accounting for the ambient weather changes seen on a day-by-day basis.

Absorption of solar energy is most concentrated on Earth's surface, particularly in tropical regions, whereas most of the infrared radiation going out to space originates in the middle troposphere and is more evenly distributed between equatorial and polar regions. Time-averaged temperature distribution is maintained by the system by transporting heat from regions where solar heating dominates over infrared cooling to regions where radiative cooling is able to dominate. In this way, the atmosphere transports heat from the ground to the upper troposphere, and the atmosphere and the oceans work together to transport heat toward the Arctic and Antarctic poles from the equatorial belt.

## NET RADIATIVE HEATING

Net radiative heating, the difference between heat gained by absorption of solar energy and heat lost due to infrared reradiation, is the driving force for atmospheric thermodynamics. Solar radiation reaches the top of the atmosphere, where about 30 percent of it is reflected back into space. The remaining radiation is absorbed. About 69 percent of solar absorption happens at the surface, and about 40 percent of that leaves the surface as net infrared radiation. This serves to leave a net surface radiative heating of about 150 watts per square meter. On a global and annual average, the net radiative heating of Earth's surface is equivalent to the net radiative cooling of the atmosphere; normal thermodynamics within the atmosphere thus maintains an appropriate balance

between the two. A prolonged imbalance between these two systems would result in global warming, which could melt the polar ice caps and raise the levels of the oceans.

Over the oceans, most of the energy from net radiative heating of the surface is used in the process of water evaporation. Energy is removed from the surface as latent heat; when water vapor condenses, it deposits almost all of its latent heat into the air rather than into the condensed water, which returns to the surface as precipitation. The result is that heat is transported from the surface to the air, where the potential for condensation takes place. On a global average, about 70 percent of the net radiative heating of the surface is removed by latent heat flux, and the remaining 30 percent leaves the surface by conduction of the sensible heat to the overlying air. The atmosphere thus experiences diabatic heating or cooling from four ongoing processes: latent heating, sensible heating at the surface, solar absorption, and infrared heating or cooling.

## TEMPERATURE INVERSION

The general temperature decrease within the troposphere is called the "environmental lapse rate." The occurrence of shallow layers where temperatures actually reverse this normal pattern and get warmer with increasing height is known as a "temperature inversion." Temperature inversions are often seen in cities located at high altitudes with a low partial pressure of oxygen, and in basins surrounded by mountains that serve to block winds and trap industrial pollutants. A "greenhouse effect" inversion causes a reversal of an ecosystem's normal atmospheric temperature gradient, thermally altering harmful airborne chemical compounds and enhancing their negative effects on organisms living below. A notable example is the strong eastern wind that blows toward Denver, Colorado, and traps a brown cloud of pollutants against the Rocky Mountains. This requires the daily broadcast of air-quality reports on radio and television and leads to the frequent calls for weak and elderly residents to stay indoors during hotter parts of the day. Trapped carbon monoxide, coming mainly from automobile tailpipes via the incomplete combustion of gasoline, combines quickly with hemoglobin, the oxygen-carrying compound in the blood in humans and animals. By taking up binding sites on the hemoglobin molecule, carbon monoxide impairs oxygen delivery to the tissues. Elderly persons with heart disease are at special risk through any restrictions in oxygen delivery.

The thermosphere, which does not have well-defined limits, exhibits another increase in temperature as a result of the absorption of very short-wavelength solar energy by atoms of oxygen and nitrogen. Although temperatures rise to values in excess of 1,000 degrees Celsius in this outermost layer, it is difficult to compare the temperature in this layer with that seen on Earth's surface. Temperature is directly proportional to the average speed at which molecules are moving, and because gases within the thermosphere are moving at very high speeds, ambient temperature remains very high. However, the gases in this region are so sparse that only a very small number of these fast-moving air molecules collide with foreign bodies, thus causing only a very small amount of energy to be transferred. For this reason, the temperature of a satellite orbiting Earth in the thermosphere is determined by the amount of solar radiation it absorbs and not directly by the temperature of the surrounding environment. Thus, if an astronaut in the space shuttle were to expose his or her hand to ambient space, it would not feel hot.

Ground inversions occur frequently and commonly extend upward for 100 meters or more. They can develop from several different causes. Their primary cause is radiative cooling, which occurs under clear skies at night. During the day, the ground stores thermal energy from solar radiation. After sunset, the ground surface radiates the stored heat energy, thereby cooling the ground surface. Energy from the relatively warm air is physically conducted into the radiationally cooled surface of the ground as the two are in contact, thereby cooling the air immediately above it. Only about the first 100 meters experiences a temperature decrease as a result of radiative cooling. Such induced temperature inversions are more likely to occur on nights with clear skies than on nights with cloudy skies. Calm wind conditions or light breezes also are more conducive to developing a ground inversion. Ground inversions are more likely to develop and to last longer in cold climates because snow and ice reflect sunlight, and the small amount of heat absorbed is utilized in the melting process, thus cooling the surface rapidly and producing an inversion.

A second mechanism for the formation of an inversion is the result of a phenomenon called air drainage. On cold nights over rolling topography, denser cold air responds to gravity and moves downslope to collect in local depressions. Continued cooling can cause inversions to extend over larger areas both vertically and horizontally if the vertical cooling extends above the summits of the rolling terrain. Evidence of the initial development of a ground inversion often is heavy dew or frost. Frequently, ground fogs occur in association with inversions because of cooling of the air. This is particularly true in an air drainage situation where fogs first appear in depressions at the surface.

A third way in which ground inversions are formed is through the movement of a warm air mass into a region. A warm current of air may move over a cool ground surface or a cooler layer of surface air. The lower portions of the air mass are cooled, and stable or nonturbulent conditions result, producing an inversion.

The frequency of ground inversions varies across the United States (frequency being expressed as a percentage of the total time a region has inversions). Most ground inversions occur at night in winter. Summer inversions are less frequent than winter inversions, but they do occur.

## CHEMISTRY

From a chemical standpoint, the atmosphere is divided into two major realms: the homosphere and the heterosphere. The homosphere—which overlaps the troposphere, stratosphere, and mesosphere—has a chemical makeup essentially identical to sea-level proportions of nitrogen, oxygen, and trace gases, even though the absolute numbers of atoms and molecules drop sharply. With increasing altitude, some important differences start to appear. Ozone becomes an important constituent of the atmosphere in this realm. Ozone is formed in the stratosphere by short-wavelength ultraviolet sunlight splitting apart oxygen molecules by photodissociation. These free oxygen atoms then form ozone with oxygen molecules. Molecular nitrogen also is dissociated.

Although the stratosphere and mesosphere (sometimes treated together as the middle atmosphere) are quite tenuous compared to the troposphere, gases in them form an optically dense layer that absorbs or reflects short-wavelength ultraviolet and X-ray radiation that would be damaging to life on the surface. Ozone is especially important with regard to the absorption of ultraviolet radiation. Nevertheless, ozone is quite fragile and can be destroyed by chlorofluorocarbons (Freons and related compounds) used for several years as spray can propellants and refrigeration system coolants. Studies indicate that these gases migrate upward in the atmosphere and chemically remove thousands of times their own mass in ozone molecules before they are broken down after several decades or even centuries. This is believed to have led to the formation of an ozone hole over the South Polar region, and the appearance of a similar, but less pronounced, effect over the Arctic. Atomic oxygen becomes more common in the mesosphere. In the heterosphere, the gas mixture changes drastically, and hydrogen and helium become dominant.

## ELECTRICAL ACTIVITY

From an electrical standpoint, there are the regions designated as the neutral atmosphere and the ionosphere. The neutral atmosphere, below 50 kilometers in altitude, is largely devoid of electrical activity other than lightning, which might be regarded as localized "noise." Above 50 kilometers, atoms and molecules are ionized largely by sunlight (ultraviolet radiation in particular) and, to a lesser extent, by celestial X-ray sources, collisions with other atoms, and geomagnetic fields and currents. Although the ionosphere as a whole is electrically neutral, it comprises positive (ion) and negative (electron) elements that conduct currents and respond to magnetic disturbances. The ionosphere starts at about 50 to 100 kilometers in altitude and extends outward to more than 900 kilometers as it gradually merges with the magnetosphere and its components. It is sometimes called the Heaviside layer after Oliver Heaviside (1850-1925), who predicted a layer of radio-reflecting ionized gases. The ionosphere is divided into C, D, E, and F layers, which in turn are subdivided (F$_1$, for example).

The ionosphere is one of the most active regions of the atmosphere and one of the most responsive to changes in solar activity. Ions and electrons in the ionosphere form a mirrorlike layer that reflects radio waves. Radio waves are absorbed by the lower (D) region of the ionosphere (which also reaches down into the mesosphere). The D-layer dissipates at night in the absence of solar radiation, allowing radio waves to be reflected by the F-layer at higher altitudes, thus causing radio "skip." These effects vary at different wavelengths. Intense solar activity can alter the characteristics of the ionosphere and make it unreliable as a radio reflector, either through the input of high-energy radiation or by the injection of particles carried by the solar wind.

## ATMOSPHERIC LIGHT DISPLAYS

Such particle injections would go unnoticed but appear as the aurora borealis and aurora australis (the Northern and Southern Lights, respectively). Earth's magnetic field shields the planet from most charged radiation particles. At the polar regions, however, where the magnetic field lines are vertical (rising from the surface), the environment is magnetically open to space. Many charged particles from space or from the solar wind are "funneled" into the polar regions along the lines of force of the planetary magnetic field. When the particles strike the atmosphere, they surrender their energy as light with spectral lines unique to the electrochemical interactions taking place. These auroral displays generally take place at 120 to 300 kilometers in altitude, with some occurring as high as 700 kilometers.

The aurora is the best-known atmospheric light display. Other "dayglow" and "nightglow" categories are caused by lithium, sodium, potassium, magnesium, and calcium at altitudes from about 60 to 200 kilometers. These metals may be introduced by meteors as they are vaporized upon entering the atmosphere. A layer of hydroxyl radicals causes an infrared glow at about 100 kilometers, and dull airglows are caused by poorly understood effects at 100 to 300 kilometers in altitude.

## JET STREAMS AND WAVES

Although the principal division of the atmosphere is vertical, there are horizontal differences related to latitude and to weather. Two major phenomena that affect the atmosphere are the jet streams and waves. The jet stream is a high-speed river of air moving at about 10 to 20 kilometers in altitude and at 100 to 650 kilometers per hour. Its location plays a major role in the movements of larger air masses that make up weather fronts in the troposphere.

More than twenty wave phenomena take place in the atmosphere in response to different events. The three principal categories are gravity, Rossby, and acoustic. Gravity waves are not associated with relativity but with vertical oscillations of large air masses causing ripples, like a bottle bobbing in a pond. Rossby (or planetary) waves are associated with the wavelike distribution of weather systems. Acoustic waves are related to sound.

## ATMOSPHERIC STUDY TOOLS

The earliest types of instruments used for atmospheric study remain among the most important. Barometers, thermometers, anemometers, and hygrometers provide the most immediate records of atmospheric change and warnings of impending events. Vertical profiles of atmospheric conditions are obtained by transporting such instruments up into the air using balloons and suborbital rockets. The term "sounding rocket" comes from the earliest days of atmospheric study, when scientists were "sounding" the ocean of air just as they would the ocean of water: Small charges were attached to balloons, and the time sound took to reach the ground was a crude measure of atmospheric density. Balloon-borne instrument packages continued to be called radiosondes ("radio sounders").

Instrumentation carried aboard spacecraft is of a different nature. Many of the most revealing devices have been spectrometers of various types that analyze light reflected, emitted, or adsorbed by the atmosphere. Absorption of ultraviolet light by the atmosphere led to the discovery of the hole in the Antarctic ozone cover; drops in absorption meant that ultraviolet light was passing through, rather than being returned to space (typically, such measurements also require observation of the solar ultraviolet output). Optical instruments usually are most effective when they view the atmosphere "edge-on" so as to increase the brightness of the signal (somewhat like viewing a soap bubble at the edges). Atmospheric studies can be difficult when viewing straight down, because the weak signals from airglow and other effects are washed out by the brighter glow of Earth or the stellar background. Special techniques can be employed. The U.S. space shuttle has twice carried sensors designed to monitor carbon monoxide pollution in the atmosphere. Gas cells containing carbon monoxide at different pressures acted as filters that blocked all signals but the wavelengths corresponding to carbon monoxide at the same pressure (that is, altitude) as that in the cell.

The most powerful tools used in studying the atmosphere have been the weather satellites deployed to observe the atmosphere from geostationary orbit (affording continuous views of half a hemisphere) and from lower polar orbits. Images from these satellites reveal the circulation of the atmosphere by the motion of clouds. Other sensors (called sounders) provide temperature profiles of the atmosphere at various altitudes.

The most extensive analyses of the atmosphere have been carried out by the Atmosphere Explorers, the Orbiting Geophysical Observatories, the Atmosphere Density Explorers, and the Dynamics Explorers. Operating in the upper reaches of the atmosphere, these spacecraft have enabled determination of the structure and composition of the atmosphere and the changes it experiences with seasons and solar activity. The more sensitive chemical assays, however, have been conducted by instruments carried aboard the manned Spacelab 1 and 3 missions of the NASA space shuttle program. An Imaging Spectrometric Observatory carried on Spacelab 1 produced highly detailed emission spectra of the atmosphere between 80 and 100 kilometers in altitude. Atmospheric Trace Molecules Observed by Spectroscopy (ATMOS), on Spacelab 3, measured the altitude ranges of some thirty chemicals and identified five, such as methyl chloride and nitric acid, in the stratosphere, where previously they were only suspected.

## COMPARATIVE PLANETOLOGY

Comparative planetology analyzes the differences and similarities between and among the planets. Earth, Venus, and Mars are used most often in comparative atmospheric studies. These three "terrestrial" planets are similar in size and in general chemistry but totally different in environment, largely because of their different atmospheres. Venus has a dense atmosphere composed largely of carbon dioxide and topped by clouds of sulfuric acid, which has led to surface temperatures of 900 degrees Celsius and to normal atmospheric pressures ninety times greater than those of Earth. The circulation pattern, though, is unaltered by precipitation and oceans and thus can be used as a model in studying Earth. Efforts to understand how Venus became a "runaway greenhouse" have suggested a similar scenario for Earth. Mars, in contrast, has a tenuous atmosphere composed of carbon dioxide and traces of water vapor and oxygen. Studies of Mars focus on how its climate and atmosphere evolved and whether it was once Earthlike.

## ATMOSPHERIC ALTERATION BY LIFE-FORMS

The atmosphere as it currently exists is a relatively recent phenomenon brought about by the gradual alteration of the environment by life-forms. Awareness of this global alteration is helping humans understand the effects they are having on the environment over a relatively short time span—essentially since the onset of the Industrial Revolution. The widespread use of fossil fuels and the burning of forests to clear land for agriculture have converted the carbon that plants spent billions of years converting into solid carbon compounds back into gaseous carbon dioxide. Furthermore, the plants that were "sinks," or absorbers, of carbon dioxide are available in lesser quantities to liberate oxygen. Sulfur compounds are naturally introduced by volcanoes, biological decay, and oceanic processes, but large quantities have been added by industrial processes, including coal burning. One product, sulfur dioxide, combines with water vapor at low altitudes to form sulfuric acid. At high altitudes, it can also alter the ozone layer and the terrestrial radiation balance. In addition, the ratios of nitrogen compounds are altered by combustion and by widespread use of nitrogen-based fertilizers; these products, too, have an adverse effect on ozone.

The immediate concern is not that the oxygen supply will be depleted, although that is a credible, long-term possibility, but that the increased amounts of carbon dioxide in the atmosphere will cause a greenhouse effect. In the greenhouse effect, long-wavelength (infrared) radiation emanating from the soil or ground is absorbed by the atmosphere and retained. Glass serves this purpose for a greenhouse by reflecting the radiation back into the interior of the structure and so increasing the interior temperature. Carbon dioxide has the same effect in Earth's atmosphere, but it functions by absorbing and retaining the infrared energy rather than by reflecting it back toward the ground. Other human-made gases that enhance the greenhouse effect are nitrous oxide, methane (which is also produced naturally), and chlorofluorocarbons (which also deplete ozone, thus allowing more radiation to enter). Because so little is known about causes and effects in this field, there are uncertainties in predicting what will happen. It is expected, however, that increases in the carbon dioxide content of the atmosphere will raise global temperatures and that such a rise in temperature will shift weather patterns and cause large portions of the polar ice caps to melt, thus flooding coastal regions.

## AGRICULTURAL APPLICATIONS

There are many practical applications of the knowledge of thermodynamic data, one of which is regularly utilized in agriculture to estimate the approximate date when crops will be ready for harvest. The growing degree-day index estimates the number of growing degree-days for a particular crop on any given day as the difference between the daily mean temperature and the minimum temperature required for growth of a particular crop. For example, the minimum growing temperature for corn is 10 degrees Celsius (50 degrees Fahrenheit), which means that on a day when the mean temperature is 24 degrees Celsius (75 degrees Fahrenheit), the number of growing degree-days for sweet corn is estimated at fourteen. Starting with the onset of the growth season, the daily growing degree-day values are added. If 1,111 growing degree-days are needed for corn to mature in a particular region, the corn should be ready to harvest when the total number of growing degree-days reaches that figure.

*Daniel G. Graetzer and Dave Dooling*

## FURTHER READING

Ahrens, C. Donald. *Essentials of Meteorology: An Invitation to the Atmosphere.* Belmont, Calif.: Brooks/Cole Cengage Learning, 2012. Discusses various topics in weather and the atmosphere, including tornadoes and thunderstorms, acid deposition and other air pollution topics, humidity and cloud formation, and temperature.

_____. *Meteorology Today: An Introduction to Weather, Climate, and the Environment.* 9th ed. Belmont, Calif.: Thomson Brooks-Cole, 2009. A thorough and useful text designed for college-level students taking an introductory course on atmospheric science, yet readable for anyone with an interest in meteorology.

American Meteorological Society. "Policy Statement of the American Meteorological Society of Global Climate Change." *Bulletin of the American Meteorological Society* 72 (1991): 57-59. A review of current knowledge and a call for more research into the nature of climate variability.

Burroughs, W. J. *Watching the World's Weather.* New York: Cambridge University Press, 1991. A very readable text with helpful explanations regarding weather changes in all parts of the world.

Campbell, I. M. *Energy and the Atmosphere: A Physical-Chemical Approach.* New York: John Wiley & Sons, 1986. An excellent text on the physical and chemical nature of the atmosphere and the effects of pollution.

Curry, Judith A., and Peter Webster. *Thermodynamics of Atmospheres and Oceans.* San Diego, Calif.: Academic Press, 1999. Offers a look at the effects of the interaction between oceans and the atmosphere on weather patterns and climatic changes. Provides good insight into the role that atmospheric thermodynamics plays in meteorology. Illustrations, maps, and index.

Garstang, Michael, and David R. Fitzjarrald. *Observations of Surface to Atmosphere Interactions in the Tropics.* New York: Oxford University Press, 1999. Looks at the atmospheric and oceanic systems of tropical environments. Suitable for readers with some background in meteorology. Illustrations and maps.

Grotjahn, R. *Global Atmospheric Circulations: Observations and Theories.* Oxford, England: Oxford University Press, 1993. Provides a useful presentation of old concepts and new theories regarding atmospheric thermodynamics. Suitable for advanced readers.

Lutgens, Frederick K., Edward J. Tarbuck, and Dennis Tasa. *The Atmosphere: An Introduction to Meteorology.* 11th ed. Upper Saddle River, N.J.: Prentice Hall, 2010. Contains excellent chapters on the atmosphere, meteorology, and weather patterns.

Marshal, John, and R. Alan Plumb. *Atmosphere, Ocean and Climate Dynamics: An Introductory Text.* Burlington, Mass.: Elsevier Academic Press, 2008. An excellent introduction to atmospheres and oceans. Discusses the greenhouse effect, convection and atmospheric structure, oceanic and atmospheric circulation, and climate change. Suited for advanced undergraduates and graduate students with some background in advanced mathematics.

Stacey, Frank D., and Paul M. Davis. *Physics of the Earth.* 4th ed. New York: Cambridge University Press, 2008. Discusses Earth's atmosphere in Chapter 2. An appendix provides information on thermodynamics. Well organized, with additional mathematics and physics concepts geared to graduate students. Includes many appendices and student exercises.

Vallis, Geoffrey K. *Atmospheric and Oceanic Fluid Dynamics: Fundamentals and Large-scale Circulation.* New York: Cambridge University Press, 2006. Begins with an overview of the physics of fluid dynamics to provide foundational material on stratification, vorticity, and oceanic and atmospheric models. Discusses topics such as turbulence, baroclinic instabilities, wave-mean flow interactions, and large-scale atmospheric and oceanic circulation. Best suited for graduate students studying meteorology or oceanography.

Van Ness, H. C. *Understanding Thermodynamics.* Mineola, N.Y.: Dover Publications, 1983. Discusses the abstract properties of thermodynamics, making them more understandable. Covers topics quickly.

Williams, J. *The "USA Today" Weather Book.* 2d ed. New York: Random House, 1997. An excellent guide to the interpretation of weather reports. Gives numerous applied examples of thermodynamic principles.

Yung, Yuk Ling, and William B. DeMore. *Photochemistry of Planetary Atmospheres.* New York: Oxford University Press, 1999. Examines the atmospheric chemistry of all of the planets in the solar system, with a focus on photochemical processes. Intended for the advanced reader. Illustrations, bibliography, and index.

Zdunkowski, Wilfred, and Andreas Bott. *Thermodynamics of the Atmosphere: A Course in Theoretical Meteorology.* New York: Cambridge University Press, 2004. Written for graduate students and researchers in meteorology and related sciences. Assumes a significant background in mathematics in the discussion of thermodynamic principles as they relate to atmospheric phenomena.

**See also:** Acid Rain and Acid Deposition; Air Pollution; Atmosphere's Evolution; Atmosphere's Global Circulation; Atmospheric and Oceanic Oscillations; Atmospheric Properties; Auroras; Barometric Pressure; Climate; Clouds; Cosmic Rays and Background Radiation; Earth-Sun Relations; Floods; Global Warming; Greenhouse Effect; Nuclear Winter; Observational Data of the Atmosphere and Oceans; Ocean-Atmosphere Interactions; Ozone Depletion and Ozone Holes; Radon Gas; Rainforests and the Atmosphere; Remote Sensing of the Atmosphere; Storms; Surface Ocean Currents; Weather Forecasting

# ATMOSPHERIC AND OCEANIC OSCILLATIONS

*The oceans' effect on developing weather patterns has long been known and taken into consideration in weather prediction. Atmospheric and oceanic patterns fluctuate over a one- to twenty-year course. These fluctuations, or oscillations, create major climate change, such as the well-known El Niño Southern Oscillation (ENSO) in the tropical Pacific, which affects weather across the globe.*

## PRINCIPAL TERMS

- **conveyor belt current:** a large cycle of water movement that carries warm waters from the North Pacific westward across the Indian Ocean, around southern Africa, and into the Atlantic, where it warms the atmosphere, then returns to a deeper ocean level to rise and begin the process again
- **Ekman spiral:** water movement in lower depths that occurs at a slower rate and in a different direction from surface water movement
- **solar radiation:** transfer of energy from the sun to Earth's surface, where it is absorbed and stored
- **trade winds:** winds that blow steadily toward the equator; north of the equator, trade winds blow from the northeast, whereas south of the equator they blow from the southeast
- **upwelling:** the process by which colder, deeper ocean water rises to the surface and displaces surface water

### AIR AND OCEAN INTERACTION

Currents in the water and the air push heated water toward cooler climates, where the heat is released, helping to regulate temperature on northern continents. As heat is released, currents in the ocean and atmosphere return the cooled water to warmer climates, and the process continues in an endless cycle. These exchanges are an important way of recycling energy through the ocean-atmosphere system, which tends to balance land temperatures and climate response. Ocean currents, which keep water in constant motion, are affected by Earth's rotation, the sun's energy, wind, and the salinity (salt content) and temperature of the ocean. Just as there are a series of air streams, pressures, and currents in the various levels of the atmosphere, the ocean also contains a similar network of circulation patterns and pressure zones.

Some deep-sea currents, such as the Gulf Stream, push through the waters much as rivers flow through land. By coupling the sciences of meteorology and oceanography, several major ocean currents have been detected using peak evaporation levels caused by warm, dry air in the subtropics. Strong evaporation throughout the year off the coast of South Africa characterizes the presence of the Agulhas Current, strong evaporation off the eastern United States in January characterizes the Gulf Stream, and strong evaporation throughout the year in the northeast Atlantic characterizes the North Atlantic Drift. These currents all have generally higher surface temperatures in relation to overlying air, especially in winter, and are frequently accompanied by strong winds. Approximately 40 percent of the total heat transported from the Southern to the Northern Hemisphere is through the action of surface ocean currents.

Another type of ocean current, called an upwelling, brings deeper, colder waters to the surface. This colder water replaces warmer surface waters that are pushed away by the strong trade winds. Upwellings carry with them large amounts of nutrients that nourish the plankton, which make up the base of the food chain. Such currents are an excellent example of how atmospheric currents and ocean currents interact: Without the trade winds to push surface waters away, colder water could not surface.

Ocean currents called Ekman spirals also demonstrate how the atmosphere affects ocean waters. Winds drive surface water in the same direction that the wind blows. Water just below the surface moves more slowly due to friction and the moving water is slightly deflected to the right by the Coriolis effect. Therefore, water at lower depths moves even more slowly, and in the opposite direction of the flow of surface water. This causes the water to flow in a downward motion called the Ekman spiral. Because of the Ekman spiral, surface water actually flows at a 90-degree angle to the wind flowing out to sea along coastlines, thus creating the opportunity for upwellings.

### CLIMATE SHIFTS AND GLOBAL IMPACT

Changes in ocean temperature and air currents influence weather, and weather changes over time determine climate. As warm-water masses move toward the coasts, they bring with them atmospheric moisture that causes rainfall. When air currents shift, as when strong trade winds die down or reverse course, areas that normally get little rain may be flooded, and areas that normally get a lot of rain may have droughts.

The best-known climate shift has occurred for centuries. It was named El Niño (Spanish for "Christ child" or "the little boy") by Peruvian fisherman because it occurs during the Christmas season. Peru is known for its anchovy fisheries, and El Niño hampers this important harvest. Normally, strong trade winds push warmer surface waters away from Peru's coast so upwelling can occur. This upwelling brings rich nutrients from the lower ocean waters, and Peruvian fishermen reap strong harvests. During El Niño years, which cycle approximately every seven years and may last up to two years, trade winds weaken, and warmer surface waters remain along Peru's coast. Upwellings do not take place, surface waters heat up, and the anchovy harvest suffers.

The effects of El Niño do not end with the Peruvian anchovy fisheries, however. The phenomenon triggers a climatic ripple effect that disrupts weather patterns around the globe. Unusual numbers of storms may rage across North America. Drought conditions have been experienced in northeastern Brazil, southeastern Africa, and western Pacific islands. El Niño effects have also caused unusually wet springs in the eastern United States and elsewhere.

The 1982-1983 El Niño is estimated to have caused more than $8 billion in damage. The severe 1997-1998 El Niño, in turn, is thought to have caused more than $15 billion in damage. It was not until the 1982 El Niño, however, that meteorologists and oceanographers began to seriously study the phenomenon to learn ways to predict both its approach and the severe weather it often causes. Researchers learned that a fluctuating wind pattern known as the Southern Oscillation, which is the same fluctuation that causes the trade winds to weaken, triggers El Niño. Researchers also learned that the ocean-current pattern is dependent upon the wind pattern. Together, this climatic event is known as the El Niño Southern Oscillation (ENSO).

## ENSO

Meteorologists and oceanographers continue to learn all they can about ENSO and other atmosphere-ocean oscillations around the globe. The ENSO pattern was the first to be studied, and many weather forecasting models have been created to aid the study of atmosphere-ocean oscillations. These computer models have also helped researchers study ENSO historical trends for the past five hundred years. Many have come to believe that ENSO events may have contributed to plagues and other similar disasters throughout history.

ENSO patterns fluctuate between the extremes of heated tropical waters associated with El Niño and a colder weather-front pattern known as La Niña ("the little girl"). This oscillation pattern takes about seven years to cycle, and each extreme (hot or cold) in the pattern may last up to two years. ENSO researchers have also linked the onset of El Niño with other climate events in surrounding ocean waters. Just before El Niño begins warming the Pacific Ocean, the tropical Indian Ocean warms. Warming then appears in the tropical Pacific, triggering ENSO, which causes warm winds to blow over South America. About nine months after this occurs, the circulation of the tropical Atlantic changes and the waters there begin warming.

The study of the ENSO pattern has led researchers to discover a major ocean circulation pattern called the conveyor belt. This conveyor belt of water is thought to have existed for the several million years since the continents have occupied their current positions in the oceans. It serves to connect the major bodies of water making up the global ocean. Its significance was overlooked until oscillation patterns were studied. The conveyor belt current circulates heated ocean water from the tropical Pacific Ocean to the North Pacific Ocean, turning clockwise around the Pacific, to pass westward between Australia and Malaysia across the Indian Ocean, around the southern tip of Africa, and up into the North Atlantic Ocean. As the surface waters cool in the North Atlantic Ocean, they sink to a lower ocean level and return eastward across the Atlantic Ocean, around southern Africa, across the Indian Ocean, passing south of Australia, and back to the tropical waters of the Pacific Ocean.

Some researchers are concerned that if this conveyor belt current were to stop or slow down, heat would build up in the Southern Hemisphere, while the Northern Hemisphere would experience a severe drop in temperature. Such an occurrence may have caused the last ice age, and research findings continue to stress the importance of this conveyor belt current to maintaining global climate. Temperatures of surface water change in direct relation to changes in this conveyor belt current, the mass flow of which is estimated to equal that of the Amazon River a hundred times over, delivering a heat load to the upper North Atlantic equivalent to one-fourth of the solar energy that reaches the surface in that region.

## NAO and PADO

The study of the ENSO pattern and the use of computer modeling to forecast weather revealed other atmospheric-oceanic oscillations, such as the North Atlantic Oscillation (NAO), which fluctuates on a twenty-year time scale. The atmospheric behavior of NAO has long been known to meteorologists. It is typically a low-pressure, counterclockwise wind circulation that centers over Iceland. This weather pattern contrasts with a high-pressure, clockwise circulation near the Azores Islands off the coast of Portugal. Strong winds blow west to east between these two weather centers. NAO seesaws between these two centers, and strong winds drive heat from the Gulf Stream across Eurasia during high-index years, producing unusually mild winters there. At the other extreme, air pressure builds up over Iceland, weakening the warming winds and delivering bitter winters to Europe and Greenland.

Oceanographers have concluded the ocean must be the cause of the unique atmospheric patterns created by NAO. Because the ocean has a huge capacity for storing heat and reacts at a slower rate than the atmosphere, it must provide the input for atmospheric patterns operating in the same mode year after year. The source of the oceanic oscillator is a pipeline of warm water fed by the Gulf Stream. It takes twenty years to complete one cycle, thus setting the timing for the long-term swings of NAO. Researchers are unsure, however, how the temperature of the waters in the pipeline actually triggers NAO.

The discovery of the Pan-Atlantic Decadal Oscillation (PADO) resulted from a theory that tropical oscillations in the Atlantic actually extended beyond the tropical region. PADO covers an area of more than 11,000 kilometers extending from the southern Atlantic to Iceland. Meteorologists proposed the existence of PADO, claiming it fluctuates on a ten- to fifteen-year time scale. PADO consists of east-to-west bands of water spanning the Atlantic Ocean. The bands alternate between warmer and cooler water and are accompanied by changes in atmospheric circulation. An oscillation to the other extreme reverses the temperature variance. It is also believed that PADO is triggered by NAO.

The study of ENSO, NAO, and PADO has led many researchers to theorize that such oscillations are linked in a chain that circles the global ocean. The Arctic Oscillation, which reverberates from the far northern Atlantic Ocean to the far northern Pacific Ocean, is associated with the NAO, especially in winter. The two phases that characterize the Arctic oscillation correlate precisely to those of the NAO. Varying from day to day and decade to decade, the oscillation is a natural response of the atmosphere to the complex interactions of the ocean-atmosphere system.

The Pacific Decadal Oscillation (PDO) is believed to be another link in the chain of atmosphere-ocean oscillations. This high-latitude oscillation was first noted in 1977 when the northern Pacific Ocean cooled dramatically. An atmospheric low-pressure center off the Aleutian Islands intensified and shifted eastward. This brought more frequent storms to the West Coast of North America and warmed Alaska, while Florida experienced periodic winter freezes. A multitude of other environmental changes also resulted. Researchers have been able to pinpoint other shifts occurring in 1947 and 1925 and have also identified close ties between PDO and ENSO events.

## Significance

The greatest impact of the study of atmosphere-ocean oscillations has been the joining of two separate fields of science. As meteorologists and oceanographers research the manner in which atmospheric and oceanic patterns rely on each other and ultimately influence climate, they have become more dependent on coupled research.

Computer simulations, sometimes referred to as "oceans in a box," have become more sophisticated as researchers from both fields collaborate and combine data. Climate scientists began using computer simulations in the mid-1980's to study winds in the atmosphere to determine how they stir ocean currents and alter pressure patterns in the atmosphere that feed back on the ocean again. Joint research between oceanographers and climatologists extends knowledge about the complex interactions of atmosphere-ocean oscillations and their influence upon each other and the climate.

*Lisa A. Wroble*

**FURTHER READING**

Bigg, Grant R. *The Oceans and Climate.* 2d ed. Cambridge, England: Cambridge University Press, 2003. Written by a noted professor of environmental sciences. Details atmospheric and oceanic circulation patterns, and discusses their influence on meteorological developments. Describes the influence of the atmosphere and the ocean on each other and demonstrates how this interaction influences major ocean-atmosphere oscillations.

Christopherson, Robert W., and Mary-Louise Byrne. *Geosystems: An Introduction to Physical Geography.* Toronto: Pearson Education Canada, 2006. Presents a discussion of atmospheric and oceanic oscillations as fundamental systems of the global environment. Includes numerous references and links to online resources. Accessible to general readers.

Easterbrook, Don. *Evidence Based Climate Science: Data Opposing $CO_2$ Emissions as the Primary Source of Global Warming.* Burlington, Mass.: Elsevier, 2011. Presents an evidence-based analysis of the scientific data concerning climate change and global warming. Authored by eight of the world's leading climate scientists who refute the claims embraced by proponents of $CO_2$ emissions as the cause of global warming. Includes comprehensive citations and references, as well as an extensive bibliography.

Garrison, Tom S. *Oceanography: An Invitation to Marine Science.* Belmont, Calif.: Brooks/Cole, Cengage Learning, 2010. Discusses oceanic currents, circulation, and oscillations. Builds a story from the formation of oceans, through waves and tides, to economics and conservation of the ocean, its inhabitants, and its resources. Provides abundant diagrams that aid readers from the layperson to advanced undergraduates.

Houghton, John. *Global Warming: The Complete Briefing.* 4th ed. Cambridge, England: Cambridge University Press, 2009. Provides an overview of the evidence for global warming. Several chapters focus on the climate system and seasonal forecasting. Offers accessible coverage of the ENSO system. Includes discussions of how ENSO influences weather and the potential for improved weather modeling based on research of ENSO.

Marshal, John, and R. Alan Plumb. *Atmosphere, Ocean and Climate Dynamics: An Introductory Text.* Burlington, Mass.: Elsevier Academic Press, 2008. Offers an excellent introduction to atmospheres and oceans. Discusses the greenhouse effect, convection and atmospheric structure, oceanic and atmospheric circulation, and climate change. Best suited for advanced undergraduates and graduate students with some background in advanced mathematics.

McKinney, Frank. *The Northern Adriatic Ecosystem: Deep Time in a Shallow Sea.* New York: Columbia University Press, 2007. Covers the paleogeography of the Adriatic Sea. Discusses the succession of the ecosystem as the sea's geography changed. Discusses oceanography topics such as circulation and sedimentation. Topics are thorough and logically ordered, making this book accessible to undergraduates.

Michaels, Patrick J., and Robert C. Balling. *Climate of Extremes: Global Warming Science They Don't Want You to Know.* Washington, D.C.: Cato Institute, 2009. Presents a view of climate change as the natural and inevitable consequence of oceanic and atmospheric oscillations, instead of the apocalyptic result of anthropogenic effects.

Sarachik, Edward S., and Mark A. Cane. *The El Niño-Southern Oscillation Phenomenon.* New York: Cambridge University Press, 2010. Offers a comprehensive discussion of ENSO and other oceanic-atmospheric processes. Covers research measurements, models, and predictions of future occurrences. Provides many diagrams, appendices, a reference list, and index.

TAO Project Office of NOAA/Pacific Marine Environmental Laboratory. *Upper Ocean Heat Content and ENSO.* Seattle: National Oceanic and Atmospheric Administration (NOAA) Web site (http://www.pmel.noaa.gov/tao). Researchers at NOAA and other research institutions work together to track and model current oscillations, including ENSO and NAO. More information about their findings and other ocean-atmosphere interactions are available through links from their Web site.

Vallis, Geoffrey K. *Atmospheric and Oceanic Fluid Dynamics: Fundamentals and Large Scale Circulation.* New York: Cambridge University Press, 2006. Begins with an overview of the physics of fluid dynamics to provide foundational material on stratification, vorticity, and oceanic and atmospheric models. Discusses topics such as turbulence, baroclinic instabilities, wave-mean flow interactions, and large-scale atmospheric and oceanic circulation. Best suited for graduate students studying meteorology or oceanography.

Woods Hole Oceanographic Institution (WHOI) Web site (http://www.whoi.edu/index.html). Offers general information about the ocean, research, education, and resources, including video animations. WHOI researchers work to understand the complexities of the ocean; atmosphere-ocean interaction is just one component of their research.

**See also:** Atmosphere's Global Circulation; Atmosphere's Structure and Thermodynamics; Climate; Clouds; Cyclones and Anticyclones; Drought; Geochemical Cycles; Global Energy Transfer; Greenhouse Effect; Hurricanes; Lightning and Thunder; Monsoons; Observational Data of the Atmosphere and Oceans; Remote Sensing of the Atmosphere; Remote Sensing of the Oceans; Satellite Meteorology; Seasons; Storms; Tornadoes; Van Allen Radiation Belts; Volcanoes: Climatic Effects; Weather Forecasting; Weather Forecasting: Numerical Weather Prediction; Weather Modification

# ATMOSPHERIC PROPERTIES

*The atmosphere is a layer of gaseous elements that surrounds the earth and differentiates the environment of the earth from outer space. The atmosphere retains gases produced by chemical reactions and protects the surface of the earth from both solar and cosmic radiation, thereby enabling life on Earth. The troposphere contains the greatest density of gases in the atmosphere while the stratosphere contains the ozone layer, which protects the earth from cosmic radiation.*

## PRINCIPAL TERMS

- **atmosphere:** gaseous "envelope" surrounding the earth that contains all gases produced by terrestrial sources
- **circulation cell:** cyclic pattern of air movement within the atmosphere
- **convection:** the vertical transport of atmospheric properties
- **Coriolis effect:** illusion of deflection observed when a body moves through the atmosphere with regard to an individual situated on the moving surface of the earth
- **magnetosphere:** outer region of Earth's ionosphere where the movement of particles is dominated by Earth's magnetic field
- **ozone:** form of oxygen containing three joined oxygen atoms responsible for blocking much of the solar radiation that hits Earth's atmosphere
- **stratosphere:** uppermost region of the atmosphere able to support life; extends from 10 to 50 kilometers (6 to 31 miles) above Earth's surface
- **thermosphere:** outer region of the atmosphere between 80 and 800 kilometers (50 to 497 miles) from the surface where temperature increases with increasing altitude because of bombardment by solar radiation
- **topography:** the relief features or surface configuration of a certain area
- **troposphere:** the lowest level of Earth's atmosphere extending to approximately 10 kilometers above sea level

## COMPOSITION OF THE ATMOSPHERE

The atmosphere is a gaseous layer surrounding the earth that is generated by chemical activity on the earth's surface and retained by the planet's gravitational field. The atmosphere has no exact upper limit but gradually dissipates with increasing distance from the surface, blending into outer space. The outermost layer of the earth's atmosphere is known as the exosphere, which stretches from approximately 500 to more than 1,000 kilometers (310 to 620 miles) from the earth's surface and gradually blends into outer space.

The earth's atmosphere is largely made up of nitrogen ($N_2$), which accounts for between 78 and 79 percent of the total gases found beneath the exosphere. Oxygen ($O_2$) is the next most abundant type of gas, typically representing between 20 and 21 percent of the remaining gases. The remaining 1 percent of the atmosphere is composed mostly of argon (Ar), which accounts for approximately 0.93 percent, and carbon dioxide ($CO_2$), which is present at levels of up to 0.031 percent.

The remaining atmosphere consists of minute quantities of trace gases, including hydrogen ($H_2$), helium (He), neon (Ne), ozone ($O_3$), and methane ($CH_4$). Though the trace gases are present only in minute quantities, many of them are essential for life on Earth. Ozone, for instance, makes up only 0.000002 to 0.000007 percent of the atmosphere, but it is the most important element, as it blocks solar radiation; this radiation would otherwise be deadly to life on Earth.

Oxygen and hydrogen combine to form water vapor ($H_2O$), which varies in concentration depending on circulation patterns. Water vapor can compose more than 4 percent of the atmosphere in extremely humid areas. Water vapor is responsible for all precipitation on Earth, making it essential for life outside the marine environment.

## STRUCTURE OF THE ATMOSPHERE

More than 80 percent of atmospheric gases are concentrated within the lower levels of the atmosphere, approximately 16 km (10 mi) above sea level. Most of this bottom 16 km is taken up by the troposphere, which extends to approximately 10 km (6 mi) above the surface and is the layer of the atmosphere most affected by the chemical processes generated by life on Earth.

The troposphere is in a constant state of turbulence, as chemical reactions contribute new gases and absorb other gases from the air. Changes in temperature and pressure combined with the gravitational pressures on Earth's surface cause weather patterns to emerge and thereby fuel cycles of exchange that move elements between the atmosphere and the earth. The troposphere is heated by a transfer of heat from the surface of the earth; temperatures decrease with increasing altitude. This pattern ends at the tropopause, which is the uppermost layer of the troposphere.

The stratosphere is the next lowest level of the atmosphere, extending from 10 km to approximately 50 km (6 to 31 mi) above Earth's surface. In comparison with the troposphere, the stratosphere is relatively free of turbulence and is stratified into distinct layers of different gaseous composition. At the lowest levels of the stratosphere, sufficient oxygen exists to support bacterial life. Commercial airliners also tend to fly in the lowest level of the stratosphere to avoid the turbulent weather patterns that dominate the troposphere.

The stratosphere contains a thin layer of ozone, in the region between 25 and 30 km (16 and 19 mi) above Earth's surface, which blocks solar radiation. Solar radiation therefore concentrates in the upper mid- and upper levels of the stratosphere. For this reason, temperature in the stratosphere increases with altitude until it reaches the topmost layer of the stratosphere, called the stratopause.

Above the stratopause is the mesosphere, which extends from 50 to 80 km (31 to 50 mi) above the earth. Within the mesosphere, the effect of solar radiation reflected by the ozone layer is reduced, and temperatures again begin to decrease with altitude. In addition, $CO_2$ trapped in this region further reduces the temperature, making the mesosphere the coldest region in the atmosphere. Temperatures can drop below -90 degrees Celsius (-194 degrees Fahrenheit) in the upper reaches of the mesosphere.

Above 80 km (50 mi), the mesosphere gradually blends into the ionosphere, which is the general term used for the atmospheric region between 80 and 800 km (497 mi) above sea level. This is the portion of the atmosphere that is most susceptible to solar and cosmic radiation and is therefore populated by ionized gas. Ionization occurs when streams of free electrons in cosmic radiation impact with atoms and cause them to exist in an ionized state where free electrons and ions dominate the environment.

The thermosphere is the region where spacecraft enter low Earth orbits and where electromagnetic fields create the visual effects known as auroras. The atmospheric atoms and molecules that exist in the thermosphere increase in temperature as they react with solar and cosmic radiation. The temperature of atoms in the thermosphere can reach more than 2,000 degrees Celsius (3,632 degrees Fahrenheit), but the concentration of atoms is so low that the region would still feel cold to any life-form. Past 300 km (186 mi) from the surface, the concentration of atoms is so low that it becomes difficult to measure any further increase in temperature, and the environment begins to resemble the vacuum of outer space.

Beyond the thermosphere is the exosphere, which is the region where atoms and molecules leaving Earth's atmosphere exit into space. While it is still possible to observe minute quantities of gas in the exosphere, the region is largely a vacuum; the extremely low density of atoms means that there is little temperature variation within the region.

### ATMOSPHERIC CIRCULATION

The circulation of atmospheric gas depends on a number of interrelated factors, including temperature, pressure, and the rotation of the earth. The most important factor in determining the movement of atmospheric gas is the uneven distribution of heat over the earth. This creates convection currents, which move air vertically through the atmosphere.

Because of the earth's shape and orientation relative to the sun, more solar radiation falls on the equator than on the poles. Warm air rises in the atmosphere and loses density as it cools, creating areas of low and high pressure. Coupled with variation in Earth's topography and with the contribution of ocean currents driven by thermal energy from beneath the crust, these cycles of warming and cooling give rise to cells of circulation that create wind and weather patterns and distribute solar energy across the planet.

The path of wind currents over the earth's surface is complicated by the Coriolis effect, which is an observational phenomenon caused by Earth's rotation. Because the earth is rotating, an object moving in a straight line above the surface of the earth will appear to veer to the right or left before landing. The observed variance is not actually a result of the object's movement but is caused by the earth's continued rotation while the object is moving, thereby altering the landing point with respect to the observed direction of travel.

Earth rotates in a counterclockwise direction; therefore, in the Northern Hemisphere the Coriolis effect appears to shift the path of a moving object to the right, while in the Southern Hemisphere movement appears to shift to the left. At the equator, there is no net movement to either direction, and an object will appear to follow a straight path to its destination. Because of this rotational effect, winds originating in the north and flowing south will arrive on Earth as if moving in a southwest direction, while winds originating in the south and heading north will appear to be traveling in a southeastern direction.

Three primary circulation systems act on the earth; these systems are known as circulation cells or wind belts. The first of these systems is the tropical or Hadley cell, which begins at the equator. Here warm air rises in response to solar heating until it meets the tropopause, where it begins to spread north and south toward the poles.

Two tropical cells exist, one to the north and one to the south of the equator, from 0 to 30 degrees in either direction. Because of the Coriolis effect, wind from the tropical cells flows either southeast or northeast from the cells, creating the trade winds, which were named because of their importance to merchant mariners. At the equator, there is virtually no net movement of wind, creating a dead zone called the doldrums.

At about 30 degrees east or west, the cooled air moving poleward descends to the earth and creates two more areas—the horse latitudes—with little directional air current. Most of the cooled air returns to the equator and is recycled within the tropical cell; however, some of the air pushes farther toward the pole and becomes part of what meteorologists call the midlatitude or Ferrel cell.

The midlatitude cell blends cold air from the polar region with warmer air flowing from the equatorial zone. This mixture of temperatures and pressures creates complex wind patterns. Because both the tropical winds and the polar winds blow in an easterly direction, the midlatitude winds are caught between the two and tend to move in a westerly direction, creating what meteorologists call the prevailing westerlies. While westerly winds are typical, changes in wind direction are common in this region, leading to a

high degree of temperature and pressure variation. This variability results in the temperate zone, which is marked by higher variation in seasonal temperatures and weather patterns than in either the tropical or polar regions.

At 60 degrees north and south, the midlatitude cells meet the polar cells, which behave similarly to the tropical cells. Here warm air rises again to the troposphere and moves poleward, eventually falling to the earth at the poles and recycling into the polar cell. Where the midlatitude and the polar cells meet, the winds of the polar cells tend to dominate, flowing over the weaker currents from the midlatitude cells. This causes intense shifts in pressure that translate into strong winds and unpredictable changes in current direction. For this reason, the polar fronts are known for their potential to generate windstorms and cyclones.

## ELECTROMAGNETISM AND THE ATMOSPHERE

Earth generates a magnetic field because of the inherent magnetism, heat, and movement of material in the planet's core. The magnetic field stretches beyond the exosphere and reaches thousands of kilometers into space. The magnetic field plays a role in protecting the earth from solar radiation.

The sun is constantly releasing millions of tons of radiation, called the solar wind, into space. This solar wind permeates the entire heliosphere, which is the portion of space affected by the sun. All of the planets in the solar system are bombarded by energetic particles in the solar wind.

The interaction of the solar wind and the earth's magnetic field creates the magnetosphere, which is a field of charged magnetic particles that surrounds the earth. The magnetosphere is not spherical but contains a long tail stretching thousands of kilometers into space on the side facing away from the sun. Some of the charged particles in the solar wind penetrate the magnetosphere, moving eventually into the thermosphere, where they interact with atmospheric atoms to create the optical phenomena known as auroras.

Without the magnetosphere, the charged particles of the solar wind would destroy the ozone layer in the earth's stratosphere, thereby leaving the earth vulnerable to increased levels of solar radiation. Life on Earth therefore depends on both the shielding effects of the atmosphere and the magnetic field.

*Micah L. Issitt*

## FURTHER READING

Grotzinger, John, Thomas S. Jordan, and Frank Press. *Understanding Earth.* 5th ed. New York: W. H. Freeman, 2006. Comprehensive introduction to aspects of Earth systems science, including the formation of the lithosphere, oceanographic sciences and atmospheric sciences.

Holton, James R. *An Introduction to Dynamic Meteorology.* 4th ed. Burlington, Mass.: Academic Press, 2004. Introduction to the science of meteorology, including detailed descriptions of atmospheric chemistry and physics. Written for students focusing on environmental sciences.

Marshall, John, and R. Allan Plumb. *Atmosphere, Ocean, and Climate Dynamics: An Introductory Text.* Burlington Mass.: Academic Press, 2008. Covers the relationship between the oceans and the atmosphere and aspects of environmental chemistry and meteorology.

Moran, Joseph M. *Weather Studies: Introduction to Atmospheric Science.* Washington D.C.: American Meteorological Association, 2009. Introduction to the study of weather and weather patterns. Includes information about atmospheric composition and the development of atmospheric currents.

National Research Council. *The Atmospheric Sciences: Entering the Twenty-First Century.* Washington, D.C.: National Academy Press, 2010. Comprehensive description of atmospheric sciences, including weather forecasting, atmospheric chemistry, physics, and climate change.

Ruddiman, William F. *Earth's Climate: Past and Future.* 2d ed. New York: W. H. Freeman, 2007. Multidisciplinary introduction to atmospheric sciences and climate studies, including theories of ocean and atmosphere development.

Wallace, John M., and Peter V. Hobbs. *Atmospheric Science: An Introductory Survey.* 2d ed. Burlington, Mass.: Academic Press, 2006. University-level introduction to atmospheric sciences. Covers atmospheric chemistry, weather forecasting, and other atmospheric phenomena.

**See also:** Auroras; Barometric Pressure; Carbon-Oxygen Cycle; Climate Change Theories; Cyclones and Anticyclones; Earth-Sun Relations; Global Energy Transfer; Greenhouse Effect; Long-Term Weather Patterns; Nitrogen Cycle; Observational Data of the Atmosphere and Oceans; Rainforests and the Atmosphere; Recent Climate Change Research; Remote Sensing of the Atmosphere

# AURORAS

*Auroras are visual phenomena related to interactions between Earth's magnetic field and the solar radiation released by the sun. The aurora borealis of the Northern Hemisphere and the aurora australis of the Southern Hemisphere occur year round but are normally visible only at high latitudes.*

## PRINCIPAL TERMS

- **corona:** the outermost layer of the sun, which extends into space in an irregular pattern surrounding the main body of the star
- **coronal mass ejection:** larger than average burst of solar wind related to deformations and reconfigurations of the sun's magnetic field
- **electromagnetic wave:** a wave of energy consisting of oscillating electric and magnetic fields
- **electromagnetism:** relationship between electric energy and magnetic energy responsible for attraction between negatively and positively charged particles
- **heliosphere:** portion of space affected by the presence of the sun or another star
- **ionosphere:** portions of the upper atmosphere consisting of part of the mesosphere, thermosphere, and exosphere; characterized by gas ionization through exposure to solar radiation
- **magnetosphere:** outer layer of the earth's atmosphere constituted by the interaction between the earth's magnetic field and charged particles released by the sun
- **optics:** branch of physics that deals with the properties and characteristics of light
- **plasma:** state of matter similar to a gas in which a portion of the molecules have become ionized, giving rise to matter built from free ions and electrons
- **solar wind:** charged particles emanating from the sun that extend to the end of the heliosphere

## INTRODUCTION

Auroras are complex visual phenomena caused by the interaction of solar radiation with the earth's atmosphere and magnetic field. Auroras are primarily visible only at high latitudes near the magnetic North and South Poles of the earth.

In the Northern Hemisphere, the phenomenon is called the northern lights or the aurora borealis, which is derived from the name of the Roman goddess of the dawn combined with the Greek word *boreas*, which refers to the northern wind. In the Southern Hemisphere, these phenomena are called the aurora australis or the southern lights.

Auroras are some of the oldest geophysical phenomena witnessed and recorded in human history, and they have been incorporated into many legends and mythologies. The connection between auroras and solar activity was first proposed by English physicist Richard C. Carrington 1859, but the hypothesis was not generally accepted until nearly a half-century later as advances in technology allowed scientists to gain a better understanding of Earth's electromagnetic field and the properties of solar radiation.

Auroras usually appear as curtain- or sheet-like waves of color pulsating along invisible lines in the sky. Typical colors include blue, green, red, and purple, though other combinations are possible. The colors and shapes of auroras are now understood as a product of the shape of Earth's magnetic field and the composition of gases in the atmosphere.

## OPTICS AND ELECTROMAGNETISM

Optics is a branch of physics concerned with the behavior of light and its interaction with other types of matter. Light is a type of electromagnetic radiation, which is a form of energy that exists as an oscillating wave combining electric and magnetic forces. The relationship between electricity and magnetism causes the attraction between negatively charged and positively charged particles. The adhesion of electrons and protons within an atom is one example of the electromagnetic force in action.

There exist many types of electromagnetic radiation, and only a small fraction can be detected by optical means. Visible light is the portion of the electromagnetic spectrum that can be viewed by the human eye. Because electromagnetic energy exists as a wave, it can be defined partially by its wavelength, which is a measurement of the distance between oscillations of the wave.

Wavelengths are typically measured in nanometer (one billionth of a meter) increments. Visible light falls into the range of approximately 400 to 700 nanometers and makes up the visual spectrum. An optical phenomenon is an interaction involving light that falls within the visual spectrum and is therefore detectable by the human eye.

## EARTH'S MAGNETIC FIELD

In many respects, the earth functions as an electromagnet, generating an electromagnetic field from within the core that reaches thousands of kilometers into space. This field plays a critical role in governing the relationship between the earth and other astrological bodies in the solar system.

The poles of Earth's magnetic field are shifted several kilometers from the North and South Poles of Earth's rotation. The earth's magnetic field also shifts position and may move by more than 15 kilometers (9.3 miles) per year.

The absolute core of the earth is composed primarily of iron that, because of the extreme pressures generated by the earth's mass, has crystallized into solid form. The solid inner core is surrounded by the outer core, which is filled with iron and other metals that have melted into a liquid state because of the decay of radioactive elements within the core. Heat from radioactive decay generates convection currents within the outer core, which causes the iron and nickel to spin around the inner core. Furthermore, the rotation of the earth causes this spinning metal to align with the North and South Poles of the planet's rotational axis. The spinning core acts like an electromagnetic generator, creating waves of positive and negative energy that constitute the planet's magnetic field.

## THE MAGNETOSPHERE

The sun functions as a large nuclear reactor, generating enormous quantities of energy within its core. The sun is composed of plasma, a form of matter that is similar to gas but that contains ionized particles. Astrophysicists estimate than 99 percent of the observable universe is filled with plasma.

The sun comprises primarily hydrogen and helium, with small amounts of other gases. Because of the sun's mass, the material at its center is subjected to intense pressure and heat; these forces generate nuclear reactions that transform the gases of the sun into superheated plasma.

Because it is constantly generating electric and magnetic energy, the sun releases vast amounts of energy into space. Radiation from the sun passes through the solar system to the ends of the heliosphere, which is the area of space affected by the sun's properties. Radiation emanates from the outer layer of the sun's atmosphere, known as the corona. The wave of radiation emanating from the sun is called the solar wind, which consists of millions of charged particles (protons, electrons, and ionized atoms) moving rapidly through space at speeds of approximately 400 km (249 mi) per second.

As the solar wind hits the earth's magnetic field, it forms the magnetosphere, an "envelope" surrounding the earth that consists of solar radiation particles moving along the lines of positive and negative magnetic current generated by the earth. The magnetosphere blocks much of this solar radiation from reaching the inner atmosphere, where it would be harmful to most life on Earth.

The magnetosphere is shaped like a teardrop with the leading, semispherical end facing the sun and a long tail, called the magnetotail, trailing on the antisolar side of the earth. Where the solar wind first encounters the magnetic field, the charged particles collide and create a repulsive force that pushes from the earth. This force, called the bow shock, pushes most of the solar wind particles out and to the sides of the earth; the particles continue to flow into the deeper reaches of the solar system.

As these magnetically charged particles move along the magnetic field, they attach to and stretch the magnetic field from the planet, thereby creating the magnetotail. Physicists are uncertain about the length of the magnetotail, but it appears to extend for millions of kilometers into outer space.

## SOLAR FLARES AND AURORAS

In its typical state, the solar wind is too weak to penetrate the magnetic field and is therefore radiated into space along the magnetotail. However, periodically the sun erupts, producing massive energy discharges known as solar flares and coronal mass ejections. When a wave of solar wind produced by one of these phenomena strikes the magnetosphere, the magnetic field distorts and becomes linked with the magnetic field of the solar wind.

At this point, solar particles are drawn along the inside of the magnetosphere and into the magnetotail. However, because this strand of particles is still connected to the leading edge of the earth's magnetic field, the strand of particles stretches until it snaps.

The front part of this strand returns to Earth, while the trailing end is pulled through the magnetotail and into outer space.

The particles attached to the leading edge of the magnetic field are accelerated as the magnetic field snaps into its original shape. As Earth's magnetic field reconnects, sealing the hole in the magnetosphere, these accelerated particles are ejected into the atmosphere. As they draw closer to the earth, the particles are channeled into Earth's magnetic field lines, which are concentrated waves of magnetism that surround the earth in a series of lateral circles. Solar particles typically create auroras when they settle into the field lines located at the earth's poles.

## THE AURORA EFFECT

Auroras occur in the portion of the upper atmosphere known as the ionosphere. The lower edge of the auroral curtain may fall to as low as 60 km (100 mi) above sea level while the upper edge may stretch to more than 400 km (249 mi) above the surface. Light given off by an aurora is caused by collisions between electrons from the solar wind and atoms in the ionosphere.

When electrons collide with an oxygen atom, some of the energy from the electron is momentarily transferred to the oxygen atom's electrons. This places the oxygen in an excited state, which resolves when the oxygen atom releases extra energy in the form of light. Oxygen atoms emit either a greenish light or a dark red light depending on the time between the collision and the release of light.

When the electrons strike nitrogen atoms, a similar phenomenon will occur, giving off a red light. Additionally, some of the nitrogen atoms struck by charged electrons will undergo a different reaction in which one of the nitrogen atom's electrons is temporarily dislodged from the atom during the collision. As this electron rejoins with the nitrogen atom, excess energy is produced and released as a blue light.

Because the composition of gases in the atmosphere changes with altitude, the colors of the auroras are divided into levels of different colors. The topmost portion of the auroral curtain appears red because the atmosphere at high altitudes has high concentrations of oxygen and low concentrations of nitrogen. Because of this, oxygen and electron collisions at this level tend to produce red light. In the middle altitudes, oxygen and nitrogen are more evenly mixed, so the light produced from collisions at this level tends to be a mixture of green, blue, and red, ultimately giving rise to a greenish blue pattern.

At the lowest level of the aurora, molecules are so dense that oxygen atoms do not release light. Excited oxygen atoms will release light only if they have sufficient time for the excess energy to be released. If an excited oxygen atom strikes another oxygen atom while in its excited state, it will transfer excess energy to the other oxygen atom, rather than release light. Oxygen atoms are so dense at lower levels of the ionosphere that excited oxygen atoms generally collide with one another rather than release light. Therefore, the light typically observed at the lowest level of the auroral curtain is pinkish and composed of a combination of blue and red light given off by nitrogen atoms.

In some cases, the light present in an aurora may suddenly shift in color and pattern. This rapid shift results from what physicists call a substorm, which is a momentary disruption of the magnetic field that causes an additional pulse of radiation to be released from the magnetotail and injected into the ionosphere. The mass changes in color and shape from substorms are sometimes called auroral eruptions.

Auroras can appear on any day of the year, though they are typically best viewed at night and are generally only visible in the far Northern and Southern Hemispheres. Because the magnetic poles shift from year to year, the optimal location for viewing auroras will shift accordingly.

Solar flares and coronal mass ejections also occur throughout the year, but they tend to increase and decrease according to a regular eleven-year pattern called the solar activity cycle. During the high point of the cycle, called the maxima, solar flares and coronal mass ejections are more common and come with an increase in the potential for auroras to appear.

*Micah L. Issitt*

## FURTHER READING

Bone, Neil. *The Aurora: Sun-Earth Interactions.* New York: Wiley, 1996. Overview of scientific information regarding the study of auroras. Includes detailed descriptions of different types of auroras and how they form within the atmosphere.

Falck-Ytter, Harald. *Aurora: The Northern Lights in Mythology, History, and Science.* Hudson, N.Y.: Bell Pond Books, 1999. General-level text that covers the history, mythology, and science of auroras from a variety of perspectives. Contains basic information about the scientific study of auroras.

Liu, William, and Masaki Fujimoto, eds. *The Dynamic Magnetosphere.* New York: Springer, 2011. Detailed description of the magnetosphere and its role in atmospheric sciences. Written for advanced students of atmospheric sciences and physics.

National Research Council. *The Atmospheric Sciences: Entering the Twenty-First Century.* Washington, D.C.: National Academy Press, 2010. Detailed description of research into atmospheric sciences. Discusses the hydrosphere and the atmosphere and solar interactions with the environment.

Pielou, Evelyn Chrystalla. *A Naturalist's Guide to the Arctic.* Chicago: University of Chicago Press, 1994. Overview of the landscape and physical chemistry of the Arctic environment. Contains a discussion of auroras and the phenomena that produce them.

Wallace, John M., and Peter V. Hobbs. *Atmospheric Science: An Introductory Survey.* 2d ed. Burlington, Mass.: Academic Press, 2006. Technical introduction to atmospheric science, with discussions of weather patterns and Earth-sun interactions.

**See also:** Atmospheric Properties; Earth-Sun Relations

# B

## BAROMETRIC PRESSURE

*Barometric pressure is a measure of atmospheric pressure as recorded by a scientific instrument known as a barometer. Barometric pressure readings have been made since the eighteenth century and remain useful today in forecasting weather. These readings allow meteorologists to identify the areas of high and low pressure that are integral parts of weather events.*

### PRINCIPAL TERMS

- **altimeter:** scientific instrument that measures the altitude of an object above a fixed level
- **aneroid barometer:** device that uses an aneroid capsule composed of an alloy of beryllium and copper to measure changes in external air pressure
- **atmospheric pressure:** force exerted on a surface by the weight of air above that surface; measured in force per unit area
- **barograph:** a graph that records atmospheric pressure in time
- **barometer:** device for measuring atmospheric pressure; some are water-based, some use mercury or an aneroid cell, and some create a line graph of atmospheric pressure
- **high-pressure area:** region in which the atmospheric pressure is greater than that in the areas around it; represented by H on weather maps
- **low-pressure area:** region where the atmospheric pressure is lower than that in surrounding areas; represented by L on weather maps
- **mercury barometer:** glass tube of a minimum of 84 centimeters (33 inches), closed at one end, with a mercury-filled pool at the base; the weight of the mercury creates a vacuum at the top of the tube; mercury adjusts its level to the weight of the mercury in the higher column
- **meteorology:** the study of changes in temperature, air pressure, moisture, and wind direction in the troposphere; the interdisciplinary scientific study of the atmosphere
- **water-based barometer:** also known as a storm glass or Goethe barometer, a device with a glass container and a sealed body half full of water; also has a spout that fills with more or less water depending upon atmospheric conditions and their forces

### BAROMETRIC PRESSURE DEFINED

Barometric pressure, also known as atmospheric pressure, is the measure of the amount of pressure the atmosphere exerts on the surface of the earth at a given point in time. Barometric pressure is the accumulated weight of the air, influenced by the force of gravity, above the point being measured.

When the temperature is cold and the air is dense, the air pressure will be higher than it is when the temperature is warm and the air is less dense. This occurs because the denser, heavier air exerts more downward pressure than warmer, lighter air.

An active interest in atmospheric pressure can be traced to Galileo Galilei and Evangelista Torricelli in the seventeenth century. Galileo wondered how long a straw could be and still remain useful when moving fluid up that straw. The question was based on actual observations of pumps used to bring water from beneath the ground. The pumps were effective up to about 10 meters (33 feet). Beyond that height, the pumps failed. Torricelli began exploring an understanding of this phenomenon.

Having observed that water could not reach a height of more than 10 m, Torricelli calculated that if atmospheric pressure were the cause, mercury, with a greater density, would not be able to attain a height of more than 74 centimeters (29 inches). Through experimentation, he had removed the air from the top of a glass straw by pumping it out; he correctly concluded that the force of the atmosphere had caused the water to rise to a height of 10 m and the mercury to rise to a height of 74 cm.

Torricelli noticed that the height of the column of mercury changed from day to day. He also noticed that if he traveled to areas that ranged in altitude, the column of mercury changed in height depending upon the altitude. He theorized that the weight of the atmosphere was changing but was unable to identify how or why.

Rene Descartes was next to investigate atmospheric pressure. In 1647, he added numbers to a Torricelli tube so that the readings from day to day could be accurately recorded. Blaise Pascal posited that air became thinner as one goes higher in the atmosphere and proved this in 1648. In 1666, Robert Boyle was the first to describe the Torricelli tube as a barometer.

By the eighteenth century, mercury barometers were an important part of the equipment of ocean-going vessels, in large part because of the work of Edmund Halley and John Locke in proving a correlation between barometer readings and weather conditions. To this day, barometric pressure readings play a vital role in predicting the weather. They help track the flow of air from high-pressure areas to low-pressure areas.

## BAROMETER TYPES AND MEASURING BAROMETRIC PRESSURE

All barometers determine the weight of the atmosphere, and the weight of gravity, pressing down upon them.

Widely used in the eighteenth century, mercury barometers are seldom used outside a lab today. Mercury barometers rely upon columns of mercury in glass tubes marked with a scale, and the height of the mercury column is noted. As the pressure of the atmosphere becomes greater, the column of mercury rises and the high pressure is noted. As the pressure of the atmosphere becomes less, the column of mercury falls and low pressure is noted. By recording readings at the same time and in the same place each day, the readings can be compared to note the trends in pressure. These trends can be used to forecast the weather.

Aneroid barometers came into wide use in the mid-nineteenth century but do not include mercury. Instead they include a bellows that contracts and expands with the pressure of the atmosphere. When the pressure is great, the pin measuring the barometric pressure moves to a higher position and indicates high pressure. When the pressure is less, the pin measuring the barometric pressure moves to a lower position and indicates low pressure. As is the case with the mercury barometer, the trend in readings can be used to forecast the weather.

Electronic barometers, in use today, rely upon internal sensors and electronic circuits to measure and display atmospheric pressure. These barometers are very accurate and often can display graphs of recent readings. Their displays are digital and generally include other readings, such as temperature, which are of use.

## INFLUENCE OF GRAVITY ON BAROMETRIC PRESSURE

Gravity plays a part in barometric pressure readings because gravity is part of the force of the atmosphere. If a person envisions himself or herself in what Torricelli referred to as an ocean of air, that person is at the bottom of that "ocean." The pressure is not felt because it comes from all sides and effectively cancels itself out. However, the force of gravity plays a part in the force of the atmosphere.

If a person goes to a location 1,524 m (5,000 ft) above sea level, that individual will find that the atmosphere is always lighter than it is at a location at or below sea level. Because of this, adjustments are made to bring all readings to sea level. In this way, forecasters can note meaningful differences between locations that vary in altitude. It also is useful to track changes in barometric pressure at one location because this allows the forecaster to note the trends for that specific location.

## BAROMETRIC PRESSURE AND THE WEATHER

Barometric pressure readings are integral to weather forecasting, as air flows from high-pressure areas into low-pressure areas because of atmospheric pressure. This flow of air is what brings a weather event.

The air flowing into low-pressure areas is doing so because of the force of gravity as it pulls upon the denser air. The air flowing from the high-pressure area does so in the form of winds. These winds are subject to the rotation of the earth and to the Coriolis force. Named for the nineteenth-century French engineer and mathematician Gaspard-Gustave Coriolis, the Coriolis force, or effect, describes the effect of an inertial force and demonstrates that an object appears to deviate from its path when viewed in a coordinate system.

In truth, the object is not deflected; however, the wind does travel counterclockwise around a low-pressure area in the Northern Hemisphere and clockwise around a high-pressure area in the Northern Hemisphere. (The reverse is true in the Southern Hemisphere.) The weather conditions experienced will be the result of the winds, temperatures, humidity, and other factors in play. By noting the trend in the barometric pressure, forecasters have an idea of what type of weather is developing and approaching and can tailor their observations to those expectations when creating their forecasts.

*Gina Hagler*

## FURTHER READING

Burch, David. *The Barometer Handbook: A Modern Look at Barometers and Applications of Barometric Pressure.* Seattle: Starpath, 2009. A thorough discussion of the uses and importance of the barometer. Includes worldwide average monthly pressures and their standard deviations. Provides a firm understanding of the importance of this simple instrument in weather prediction.

Dunlop, Storm. *The Weather Identification Handbook.* Guilford, Conn.: Lyons Press, 2003. An illustrated guide that provides methods of identification and descriptions of weather phenomena including clouds, cloud formations, optical phenomena, precipitation, and severe weather.

Frederick, John E. *Principles of Atmospheric Science.* Sudbury, Mass.: Jones and Bartlett, 2008. A complete introduction to atmospheric science. Presents the fundamental scientific principles and concepts related to the earth's climate system.

Gibilisco, Stan. *Meteorology Demystified.* New York: McGraw-Hill, 2006. An authoritative introductory work with information on meteorology and its underlying areas of science. Includes tornadoes, hurricanes, winter storms, and the impact of human activity on climate.

Middleton, W. E. K. *The History of the Barometer.* Baltimore: Johns Hopkins University Press, 1964. Recounts the development of the barometer from the sixteenth and seventeenth centuries through the mid-twentieth century. Includes the history of experiments and attendant controversies, as the barometer became a precision scientific instrument.

Randall, David A. *Atmosphere, Clouds, and Climate.* Princeton, N.J.: Princeton University Press, 2012. An overview of the major atmospheric processes. Covers the function of the atmosphere in the regulation of radioactive energy flows and the transport of energy through weather systems, including thunderstorm and monsoons. Examines obstacles in predicting the weather and climate change.

Salby, Murray L. *Physics of the Atmosphere and Climate.* New York: Cambridge University Press, 2011. Provides an integrated treatment of the earth-atmosphere system and its processes. Begins with first principles and continues with a balance of theory and applications as it covers climate, controlling influences, theory, and major applications.

Wallace, John M., and Peter V. Hobbs. *Atmospheric Science: An Introductory Survey.* Burlington, Mass.: Elsevier Science, 2006. An updated version of a widely respected earlier work. Includes an explanation of atmospheric phenomena within the context of the latest discoveries and technologies. Contains new chapters on atmospheric chemistry, the earth system, climate, and more.

**See also:** Atmospheric Properties; Climate Change Theories; Climate Modeling; Cyclones and Anticyclones; Greenhouse Effect; Long-Term Weather Patterns; Observational Data of the Atmosphere and Oceans; Recent Climate Change Research; Remote Sensing of the Atmosphere; Remote Sensing of the Oceans; Severe and Anomalous Weather in Recent Decades; Tropical Weather

# BEACHES AND COASTAL PROCESSES

*The shoreline is the meeting place for the interaction of land, water, and atmosphere, and rapid changes are the rule rather than the exception.*

## PRINCIPAL TERMS

- **beach:** an accumulation of loose material, such as sand or gravel, that is deposited by waves and currents at the border of a body or stream of water
- **longshore current:** a slow-moving current between a beach and the breakers, moving parallel to the beach; the current direction is determined by the wave refraction pattern
- **longshore drift:** the movement of sediment parallel to the beach by a longshore current
- **oscillatory wave:** a wind-generated wave in which each water particle describes a circular motion; such waves develop far from shore, where the water is deep
- **tsunami:** a low, rapidly moving wave created by a disturbance on the ocean floor, such as a submarine landslide or earthquake
- **wave base:** the depth to which water particles of an oscillatory wave have an orbital motion; generally the wave base is equal to one-half the distance between successive waves
- **wave refraction:** the process by which the vertical angle of a wave moving into shallow water is changed or bent

## BEACH COMPOSITION

The processes that create, erode, and modify beaches are many. Marine processes, such as waves, wave refraction, currents, and tides, work concurrently. This suggests that a beach is a sensitive landform, and indeed it is. Generally a beach is a sedimentary deposit made by waves and related processes. Beaches are often regarded as sandy deposits created by wave action; however, beaches may be composed of broken fragments of lava, sea shells, coral reef fragments, gravel, or even large stones. A beach is composed of whatever sediment is available. Beaches have remarkably resilient characteristics. They are landforms made of loose sediment and are constantly exposed to wave and current action. On occasion, the coastal processes may be very intensive, such as during a hurricane or tropical storm. Yet, in spite of the intensity of wave processes, these rather

thin and narrow landforms, although perhaps displaced, restore themselves within a matter of days. Coastal scientists are inclined to believe that the occurrence and maintenance of beaches are related to their flexibility and rapid readjustment to the varying intensity of persistent processes.

Sediment deposited on beaches is often derived from the continents. Rivers are one source of beach sediment, and the coast is another. As rivers erode the land, the sediment they carry ultimately finds its way to the shore. Because the sediment may be transported several tens or hundreds of kilometers, it is refined and broken down even further to finer-sized particles. Once the river reaches the sea, the sediment is distributed by longshore currents along the shoreline as a beach. Beaches also occur where the shoreline is composed of cliffs, such as along the Pacific coast of North America. Here, waves erode the sea cliffs, and the sediment is deposited locally as a beach. Under these conditions, the beach deposit is most often gravelly because the sediment is transported only a short distance and has not had an opportunity to break up into finer-sized sediment such as sand. Some island beaches are made up entirely of shell fragments.

## WAVES

The most obvious mechanical energy source working on beaches is waves. Waves approaching a beach are generally created by winds at sea. As wind velocities increase, a wave form develops and radiates out from the storm. An oscillatory motion of the water occurs as the wave form moves across the water surface. It is important to note that the water movement within a wave is not the same as the movement of a wave form. In a wave created in deep water that is approaching a beach, the water particles move in a circular orbit, and little forward movement of the water occurs. The water at the surface moves from the top of the orbit (the wave crest) to the base of the orbit (the wave trough) and then back up. Thus the water particles move in an oscillatory wave; this motion continues down into the water. Although the size of the orbits in the water column decreases, motion occurs to a depth referred to as the wave base. At this point, the depth of the wave base is less than the depth of the water. As wave crests and troughs move into shallow water, the water depth decreases to a point where it is equal to the wave base. From this point, the orbital motion is confined because of the shallowness of the water and takes on an elliptical path. The ellipse becomes more confined as the waves enter ever shallower water and eventually becomes a horizontal line. At this point, there is a net forward movement of water in the form of a breaker.

As a wave enters shallow water, many adjustments occur, such as a change in the orbital path of water particles already described and a change in the velocity of the wave form. Since the wave base is "feeling the bottom" in shallower water, friction occurs, slowing the wave down. As seen from an airplane, waves entering shallow water do so at an angle, not parallel to the shoreline. Therefore, one part of the wave enters shallow water and slows down relative to the rest of the wave. Thus a part of the wave crest is feeling bottom sooner than the rest of the wave, causing the breaking action of the wave to appear to peel itself along the beachfront. Because the wave crests and troughs have different velocities, the wave refracts or bends. In so doing, the wave crests and troughs try to parallel the shallow bottom topography that they have encountered.

The wave refraction is seldom completed, and the breaking wave surges obliquely up the slope of the beach and then returns perpendicular to the shoreline. The result is a current that basically moves water in one direction parallel to the beach in a zigzag pattern (a longshore, or littoral, current). It is a slow-moving current that is located between the breaking wave out in the water and the beach. Because longshore-current movement operates in shallow water and along the beach, it is capable of transporting sediment along the shoreline. "Longshore drift" refers to the movement of sediment along beaches. In a sense, the longshore current is like a river moving sand and other material parallel to the shore. Beaches are always in a state of flux. Although they appear to be somewhat permanent, they are constantly being moved in the direction of the longshore current. Along any shoreline, several thousand cubic meters of sediment are constantly in motion as longshore drift. Along the beaches of the eastern United States, about 200,000 cubic meters of sediment is transported annually within a longshore drift system.

A different type of ocean wave is one that is generated on the ocean floor rather than by the wind. Such waves are properly termed tsunamis. Although they popularly have been coined "tidal waves," they are completely unrelated to tides or the movement of planets. Some type of submarine displacement, such as the creation of a volcano, a landslide on the sea floor, or an earthquake beneath the ocean bottom, displaces a volume of water, which triggers waves. Tsunami waves are low, subdued forms traveling thousands of kilometers over the ocean surface at extremely high velocities, exceeding 800 kilometers per hour under the proper conditions. As a tsunami approaches shallow water or a confined bay, its height increases. There is no method for direct measurement of the heights of tsunami waves. However, in 1946, a lighthouse at Scotch Cap, Alaska, located on a headland 31 meters above the Pacific Ocean, was destroyed by waves caused by a landslide-generated tsunami.

The changing character of a beach is very dynamic because the properties of waves are variable. During storm conditions or when more powerful waves strike a beach during the winter season, for example, the beach commonly is eroded, has a steeper slope, and is composed of a residue of coarser sediment, such as gravel, that is more difficult to remove. During fair-weather or summer conditions, however, beaches are redeposited and built up.

## STUDY OF BEACHES AND COASTAL PROCESSES

The study of beaches and coastal processes is not particularly easy because of constant wave motion and changes in the beach shape. Scientists have, however, developed field methods as well as laboratory techniques to study these phenomena. To study wave motion and related current action, several techniques have been devised that include tracers, current meters, aerial photography, and satellite-based monitoring. Two types of tracers are commonly used to determine the direction and velocity of sand movement: radioactive isotopes and fluorescent coatings to produce luminophors. (Luminophors are sediment particles coated with selected organic or inorganic substances that glow under certain light conditions.) In the former type, a radioactive isotope such as gold, chromium, or iridium is placed on the surface of the grains of sediment. Alternatively, grains of glass may be coated with a radioactive element. The radioactive sediment can then be detected relatively easily and quickly using just a Geiger counter. Both techniques can trace the direction and abundance of sediment along the seashore. Bright-colored dyes or current meters can also be used to document the direction of longshore currents. Surveying instruments are used to determine the high and low topography of beaches and adjacent sand bars. By measuring beach topography before and after a storm, for example, scientists can record immediate changes. In this way, they can document the volume and dimensions of beach erosion or deposition that has taken place during a storm. Measurement of wave height and other wave characteristics can be done with varying degrees of sophistication. Holding a graduated pole in the water and visually observing wave crest and troughs is both the simplest and the cheapest method. To achieve more refined measurements, however, scientists place pressure transducers or ultrasonic devices on the shallow sea floor to record pressure differences or fluctuations of the sea surface. These more precise instruments also record other wave data for later study and analysis.

All these methods are detailed field techniques. By comparing aerial photographs, detailed maps, satellite pictures, and in some cases government studies in a coastal sector, changes over long periods of time may be detected. Finally, because wave motion cannot be controlled on a shoreline, wave tank studies are used to derive wave theories. Normally, an elongated glass-lined tank containing water from 0.3 to 1 meter in depth is used. A mechanical device called a wave machine displaces water at one end of the wave tank to create waves and sediment is introduced at the other end. The heights and other characteristics of the waves can be varied, as can the type of sediment, to form the beach. In this controlled way, the various beach and wave relationships can be studied.

## SIGNIFICANCE

More than 70 percent of the population of the United States lives in a shoreline setting, and thus an understanding of how waves interact with beaches is important. Shoreline property is prized for residential use and hence highly valuable. On many beaches, the great investments made in hotels and condominiums suggest that beaches are the most sought-after

environment on Earth, and shoreline frontage is sold by the foot or meter, not by the acre or hectare. Currently, however, such demand and investment are threatened with rising sea levels and continued coastal erosion.

Beaches represent the buffer zone that mitigates wave erosion. Waves generated in the open sea slow down in shallow water, and the beach deposit absorbs the impact of waves. Beaches are therefore constantly changing and are one of the most ephemeral environments on Earth. Thus a sound knowledge of longshore currents and beach development is necessary prior to nearshore or marine construction. Sea walls and groins, for example, interfere with waves and longshore currents and may cause considerable erosion in selected areas, while inducing abnormal sedimentation in others. Sea walls are often constructed perpendicular to a shoreline to slow longshore currents so that a beach can be deposited. Downcurrent, beyond the area of beach deposition, erosion will take place. Similarly, rivers, which are a major source of the sediment that creates beaches, are sometimes dammed, thus depriving beaches of sediment and resulting in their erosion. Unless planners, developers, and builders understand these processes, major damage can result: Failure to understand coastal processes has, on occasion, caused a property owner to sue a neighbor who caused beach erosion to take place.

*C. Nicholas Raphael*

## FURTHER READING

Bird, Eric. *Beach Management.* New York: John Wiley, 1996. Includes chapters on waves and on beaches. Covers fundamental concepts, although most examples are Australian. A good introduction for anyone with a high school-level earth background.

Cameron, Silver Donald. *The Living Beach.* Toronto: Macmillan Canada, 1998. Looks at the ecosystems and biological systems of North American beaches. Focuses on the relationships between beaches and the environment. Index and bibliography.

Davis, Richard Jr., and Duncan Fitzgerald. *Beaches and Coasts.* John Wiley & Sons, 2009. Provides an exhaustive overview of the world's different coasts and the processes that shape them, combining tectonics, hydrography, climate, and geology to explain the morphology of coasts and coastal processes.

Dean, Robert G., and Robert A. Dalrymple. *Coastal Processes with Engineering Applications.* Cambridge,

England: Cambridge University Press, 2002. Covers coastal engineering and oceanography theories and applications intended to mitigate coastal erosion. For advanced students in fields related to coast morphology.

Fall, J. A., et al. *Long-Term Consequences of the Exxon Valdez Oil Spill for Coastal Communities of South-central Alaska.* Alaska Department of Fish and Game, Division of Subsistence, Anchorage, Alaska. Technical Report 163. 2001. Covers social, economic, and environmental impact of the oil spill.

Haslett, Simon K. *Coastal Systems.* 2d ed. Abingdon: Routledge, 2009. Offers a concise introduction to the global environment of coasts in which ocean, land, and atmosphere all play a role in the physical and ecological evolution of coastlines.

Holthujsen, Leo H. *Waves in Oceanic and Coastal Waters.* New York: Cambridge University Press. The text is slightly above introductory level. It contains a chapter on observation techniques such as remote-sensing, altimetry, and wave buoys. The physics of water waves and linear wave theory are thoroughly discussed.

Kaufman, Wallace, and Orrin Pilkey. *The Beaches Are Moving: The Drowning of America's Shoreline.* Durham, N.C.: Duke University Press, 1983. Deals with the processes working in the coastal zone, such as winds, waves, and tides. Highlights the impact of rising sea levels and the modification and urbanization of the coast. Written in a nontechnical style. A narrative text suitable for all ages.

Komar, Paul D. *Beach Process and Sedimentation.* Upper Saddle River, N.J.: Prentice Hall, 1998. Extensive treatment of waves, longshore currents, and sand transport on beaches. Equations and mathematical relationships are presented and elaborated upon. College-level material. Appropriate for those interested in the specifics of coastal processes.

Leatherman, Stephen P. *Barrier Island Handbook.* 3d ed. College Park, Md.: Laboratory for Coastal Reasearch, University of Maryland, 1988. Based on actual field studies along the East Coast of the United States. Includes numerous photographs, diagrams, and tables. Most suitable for coastal managers and government employees; however, very readable and suitable for nonscientists as well as the general scientist. Emphasizes the dynamic

nature of beaches, recreation and construction impacts, and nearshore processes.

Lencek, Lena, and Gideon Boskar. *The Beach: The History of Paradise on Earth.* New York: Viking, 1998. A natural history of the environmental and social importance of coastal areas. Includes illustrations.

Masselink, G., and M. G. Hughes. *Introduction to Coastal Processes and Geomorphology.* London: Hodder Arnold Publication, 2003. The book is an excellent resource for students of geomorphology. Contains many illustrations and diagrams.

Micallef, Anton, and Allan Williams. *Beach Management: Principles and Practice.* London: Earthscan, 2009. Uses case studies from the United Kingdom, the United States, New Zealand, the Mediterranean, and Latin America. Provides a comprehensive coverage of beach management principles and practice.

Pethick, John. *An Introduction to Coastal Geomorphology.* Baltimore, Md.: Edward Arnold, 1984. A thorough survey of coastal processes. Discusses wave energy on the coast and its characteristics, the relationship between currents and the movement of beach material along the shore, and the landforms, such as beaches, mud flats, and estuaries. Most suitable for anyone needing an equation or a technical explanation of selected processes operating in a coastal zone.

Schwartz, Maurice L. *Encyclopedia of Coastal Science.* Dordrecht: Springer, 2005. Features contributions by well-known international specialists in their respective fields. Covers geomorphology, ecology, engineering, technology, oceanography, and human activities as they relate to coasts.

Thorsen, G. W. "Overview of Earthquake-Induced Water Waves in Washington and Oregon." *Washington Geologic Newsletter* 16 (October, 1988). Offers an introduction to the impact of a tsunami on the coasts of Washington and Oregon following an earthquake in March 1964. Discusses the tsunami in nontechnical language. Estimates the damage and economic losses. Presents maps and tidal gauge records. A good review of wave activity, intended for interested laypersons and teachers.

Walker, H. J. "Coastal Morphology." *Soil Science* 119 (January, 1975). A nontechnical overview of coastal landforms. Discussion includes beaches, deltas, and lagoons. Includes maps illustrating processes. Useful for a nonscientist interested in the causes and distribution of coastal features from a geographical perspective.

**See also:** Alluvial Systems; Bleaching of Coral Reefs; Dams and Flood Control; Deep-Sea Sedimentation; Deltas; Desert Pavement; Drainage Basins; Floodplains; Floods; Hurricanes; Lakes; Monsoons; Reefs; River Bed Forms; River Flow; Sand; Sediment Transport and Deposition; Surface Water; Tsunamis; Weathering and Erosion

# BLACK SEA

*The Black Sea is a nearly landlocked body of water that was formerly linked to its eastern neighbor, the Caspian Sea. Geological changes separated it from the Caspian and created two narrow waterways, the Bosphorus Straits and the Dardanelles Straits, connecting it with the Mediterranean. Some of its most interesting physical features make it unique among the world's small to medium-sized seas.*

## PRINCIPAL TERMS

- **Bosphorus and Dardanelles Straits:** straits leading to and from the Sea of Marmara; they are the only sea lanes linking the countries with shores on the Black Sea to the Mediterranean
- **Crimean Peninsula:** a large peninsula located on the north-central shore of the Black Sea; it is the most significant irregularity on the otherwise regular coastline
- **Danube River:** probably the most famous river that enters the Black Sea; the river is a vital means for international maritime access, via the Black Sea, for several European countries
- **Istanbul:** a major international city located in Turkey just at the entry point of the Bosphorus into the Sea of Marmara
- **Kerch Straits:** the strategic narrows that join the Black Sea to the Sea of Azov, located northeast of the Crimean Peninsula and sharing a similar ecology
- **Odessa:** the major Black Sea city of the Ukraine; its port, Ilichyovsk, is possibly the busiest maritime center on the Black Sea
- **Tethys Sea:** the much larger geological predecessor of both the Black Sea and the Caspian Sea to its east; named after the Titan and wife of the Greek god of the great outer sea, Oceanus

### GENERAL FEATURES

The Black Sea, which encompasses approximately 414,960 square kilometers of saline water, is located on the southern margin of the Eurasian landmass. Its shores include territories in several countries. From the southern and southwestern coast in Turkey, the shoreline proceeds in a clockwise direction through Bulgaria, Romania, Ukraine, Crimea, and Georgia.

The name Black Sea is a literal translation of the Turkish Kara Deniz. This name, suggesting a somewhat ominous image (particularly in comparison to the Turkish name White Sea, or Ak Deniz, referring to the Mediterranean), dates from the Turkish occupation of the southern shores of the Black Sea between the thirteenth and the fifteenth centuries. The sense of blackness refers not to any impression of the color of its waters, but rather to a presumed inhospitable environment. When the Turks captured the Greek-named Pontus Euxinus (meaning "hospitable sea") from the Byzantine Greeks, they adopted a much earlier Greek denomination for the same body of water, namely Pontus Axeinus (or "inhospitable sea").

The major ports of the Black Sea littoral states are Trabzon, Samsun, Sinop, and Zonguldak (Turkey); Burgas and Varna (Bulgaria); Odessa (Ukraine); Sevastopol (Crimea); and Batumi (Georgia). In the 1990's, Turkey began developing a new port between Trabzon and Batumi to accommodate major increases in the movement of goods, including petroleum, from the former southern Caspian Sea coastal territories of what had been the Soviet Union.

### GEOLOGICAL ORIGINS

Geologically, the Black Sea appears to be a basin left behind by the retreating ancient Tethys Sea, a process that took place over about 200 million years. This long geological evolution reached a significant stage about 50 million years ago as the upward thrust of the Anatolian landmass (Turkey) and Western Iran split the Caspian Sea basin off from the Black Sea. Further mountain-forming activity, including the upheavals that created the Pontic, Caucasus, Carpathian, and Crimean landmasses, also affected the basin that eventually held the waters of the Black Sea. Finally, the geology associated with the famous Black-Sea-to-Mediterranean connection via the Bosphorus and Dardanelles Straits, which lead into and out of the Sea of Marmara, appears to have been the most recent stage contributing to the current configuration of the Black Sea, occurring less than 10,000 years ago.

The total area of the Black Sea, including its immediately contiguous arm, the Sea of Azov, is about 423,800 square kilometers. At its widest, the sea extends more than 1,134 kilometers from west to east. Its narrowest north-to-south point, running from the tip of the Crimean Peninsula to the Turkish coast at the Kerempe Burmi Cape, is slightly more than 240 kilometers.

The sea is fed by a number of important freshwater rivers. The most famous of these, the Danube, enters the sea along the coast of Romania. The Dniester, which flows parallel to the northern fringes of the Carpathian Mountain range, ends at the delta that feeds the farthest northwest area of the Black Sea. The Dniester Delta forms part of the major port complex of Odessa in the Ukraine. To the east of the Dniester is the Dnieper, the last major river to enter the north coast of the Black Sea proper before its coastline is broken by the Crimean landmass and the waters and rivers associated with the Sea of Azov.

The depth of the sea varies; the deepest area, in the southern part of the basin about midway from east to west, measures more than 2,120 meters. The fact that the deeper zones of the Black Sea receive very low levels of ventilation (circulation of oxygen) means that these areas do not support any significant plant growth or fish populations. In the deep waters, a high concentration of hydrogen sulfide causes whatever oxygen that exists to ionize. These conditions mean that only certain species of bacteria can survive at lower depths. The beginnings of low oxygen levels, with consequential decreases in marine life, occur at depths of around 150 meters. Where waters are sufficiently ventilated, there are substantial numbers of fish, although only a few species are sought by commercial fisheries.

The most outstanding irregularity at surface level along the coast is the Crimean Peninsula, which juts southward into the Black Sea from the mainland just east of the Dnieper River Delta. The Crimean Peninsula is the site not only of the traditional seaside vacation town of Yalta (where the Western allies met with the Soviet Union's Joseph Stalin near the end of World War II) but also of the strategic port of Sevastopol, founded by Russian czars as a strategic maritime outpost for Russian access to the Mediterranean.

There are no major islands in the Black Sea. Most geographical accounts overlook the few smaller islands, mentioning only Zmiyini, located east of the Danube River delta, and Berezan, near the mouth of the Dniester.

The Kerch Strait, located to the east of the Crimean Peninsula where a city by that name stands, connects the Black Sea to the Sea of Azov. The latter, which forms part of the same residual basin complex left behind by the receding Tethys Sea, encompasses approximately 67,600 square kilometers. The Sea of Azov receives the waters of two major rivers, the Don and the Kuban.

With the one major exception of the Crimean Peninsula, the coastline of the Black Sea is quite regular, reflecting its geological origins as a residual basin. The shallow coastal shelf is seldom deeper than 90 meters. It can be very wide but generally extends out from the shore about 11 kilometers. If one takes the average of 11 kilometers and adds areas where the shelf is considerably wider, these shallow waters make up almost one-fourth of the total surface of the Black Sea.

## TIDAL PHENOMENA AND CURRENTS

There is little regularly predictable tidal movement in the Black Sea. When moderate tidal movements do occur, their levels can be affected by a sudden rise or fall of atmospheric pressures in different regions either over the sea or on the landmasses around it. Only rarely do such tides cause dangers for boats or pier structures along the shorelines.

Currents are strongest near the shores of the sea, where the water is not very deep. Such currents run an average of between 40 and 50 centimeters per second. In areas of deeper water, very slow currents are difficult to measure but generally do not run faster than 2.5 to 5 centimeters per second. A particular phenomenon of currents exists in the Bosphorus Straits zone leading from the Black Sea to the Sea of Marmara. Here, a current near the surface carries water from the Black Sea past the city of Istanbul into the Sea of Marmara and eventually into the Mediterranean. At deeper levels, the current moves in the opposite direction, carrying water with higher saline content that originates in the Mediterranean into the Black Sea.

The salinity level of the Black Sea is roughly one-half that of the world's major oceans. Salinity counts (relative proportions of salt to water) near the surface are between seventeen and eighteen parts of salt per thousand parts of water. In the cold band that lies beneath the first layer of Black Sea surface water, the salinity count increases rapidly, reading twenty-one parts per thousand. The proportion of salt to water continues to increase as one descends to deeper zones, but the rate of increase is much more gradual than the apparently sudden increase that occurs in the intermediate-level cold band. The highest salinity count in the deep water is about thirty parts salt to one thousand parts water. Measurements are higher in the Bosphorus Straits, but this phenomenon is linked to the transfer of Mediterranean water into the Black Sea via subsurface currents.

Because the Black Sea is not a very large body of water, its surface temperature varies considerably from season to season. Winter temperatures in the northern zones hover just below 0 degrees Celsius. Where the inflow of freshwater from major rivers is substantial, this means some freezing can occur near the coast. The saline content of most zones is sufficiently high, however, to prevent widespread freezing of Black Sea waters. In the southern half of the sea, winter temperatures are more moderate, reaching about 10 degrees Celsius in the southeast corner near the Turkish border with Georgia. Summer brings surface temperatures up to nearly 21 degrees Celsius. The warming effect, however, is limited to depths down to about 45 meters. Below this level, down to about 90 meters, there is a more or less constant cold band, where temperatures remain at about 7 degrees Celsius throughout the year. In fact, deeper waters may even be a bit warmer than the cold band, but this phenomenon stems from an insulating effect and is not tied to rising temperatures at the surface.

## NATURAL RESOURCES AND ECONOMIC POTENTIALS

Until the late twentieth century, perhaps the best-known natural resource exploited on an industrial scale in the Black Sea region was coal, shipped in large quantities from the Turkish port of Zonguldak. Caspian Sea petroleum transit shipments from the Black Sea port of Batumi were always a factor in the economic history of the Black Sea during the period of the Soviet Union, but their importance grew rapidly in the last decade of the twentieth century, as newly opened free international markets for Caspian Sea oil provided incentives for expanded construction of pipelines and port facilities in the southeast coastal region of the Black Sea.

Fishing is the most traditional form of economic activity in the Black Sea. It appears to have a limited capacity for expansion beyond levels reached by the middle of the twentieth century. The most common species of fish in the Black Sea are Mediterranean varieties, some of which actually move into and out of the Black Sea via the Sea of Marmara from season to season. Specialists have noted nearly two hundred different species of fish, but only about forty varieties are exploited commercially. The most important locally processed fish include horse mackerel and another species bearing the local name khamsa.

## SIGNIFICANCE

Study of the Black Sea region can be particularly useful for comparison with small to moderately large inland seas that exist around the world. The scale of ecological considerations in such circumstances is drastically different from what would apply in the study of large oceans. In this case, two comparable bodies of water, the Black Sea and the Caspian Sea, are located much in the same region but demonstrate quite different characteristics. On the one hand, for example, the Black Sea has not been seriously affected by large-scale petroleum development, something that has had major effects on the ecology of the Caspian Sea. On the other hand, the Black Sea has considerably more large urban port cities around its coasts. Examination of pollution levels and marine biology in the Black Sea can lead to a better understanding of its own environmental prospects as well as those of similar bodies of water in other areas of the world.

*Byron D. Cannon*

## FURTHER READING

Cooper, Bill, and Laurel Cooper. *Back Door to Byzantium: To the Black Sea by the Great Rivers of Europe.* London: Adlard Coles Nautical, 1997. A skillfully written travel book emphasizing the importance of several major waterways that allow the interior of Europe to connect with the Black Sea and the Mediterranean.

Dumont, Henri, Tamara A. Shiganova, and Ulrich Niermann, eds. *Aquatic Invasions in the Black, Caspian, and Mediterranean Seas.* Boston: Kluwer Academic Publishers, 2004. Provides an overview of the links between the Black Sea, Caspian Sea, and Mediterranean Sea. Discusses the physical connections and the resulting movement of invasive species due to shipping. Examines the invasion of jellies into the Caspian Sea. Provides a comparison to other invasive species, including invertebrates, plankton, and ctenophores.

King, Charles. *The Black Sea: A History.* Oxford: Oxford University Press, 2005. Provides a thorough overview of the history and importance of the Black Sea to the many nations that it borders.

Kostianoy, Andrey G., and Alexey N. Kosarev. *The Black Sea Environment.* Dordrecht: Springer, 2007. Presents a systematic description of physical oceanography, marine chemistry and pollution, marine biology, geology, meteorology, and hydrology of the Black Sea based on observational data. Aimed at specialists in physical oceanography, marine chemistry, pollution studies, and biology.

Kotlyakov, V., M. Uppenbrink, and V. Metreveli, eds. *Conservation of Biological Diversity as a Prerequisite for Sustainable Development in the Black Sea Region.* Dordrecht: Kluwer, 1998. Contains the proceedings of an international workshop held in the Georgian port city of Batumi in 1996. Emphasizes the prevention of further deterioration of plant and animal life in the Black Sea and along its coasts.

Mavrodiev, Strachimir Chterev. *Applied Ecology of the Black Sea.* Commack, N.Y.: Nova Science, 1999. A technical report covering the geological history of the Black Sea as a route to understanding the origins of some of the natural resources of the region. Emphasizes the need to educate populations living around the Black Sea and to regulate activities that can lead to depletion of productive potential contained in its ecology.

Ryann, Amy L., and Nathan J. Perkins. *The Black Sea: Dynamics, Ecology and Conservation.* Hauppauge, N.Y.: NOVA Science Publishers, 2011. Presents topical research data in the study of the Black Sea. Aimed primarily at general readers.

Sorokin, I. O. U. *The Black Sea: Ecology and Oceanography.* Kerkwerve: Backhuys Publishing, 2002. Discusses the unique features of the Black Sea that have caused it to have the largest amount of anoxic waters of any of Earth's seas, using information from Russian and Romanian publications that have been unavailable in the West.

Sosson, M, et al., eds. S*edimentary Basin Tectonics from the Black Sea and Caucasus to the Arabian Platform.* Special Publication 340. London: Geological Society of London, 2010. A compilation of articles discussing the geology, geomorphology, natural history, and seismology of the Black Sea region. Includes an abstract and references following each paper. Useful to graduate students and researchers.

Yanko-Hombach, Valentina, et al., eds. *The Black Sea Flood Question: Changes in Coastline, Climate, and Human Settlement.* Dordrecht: Springer, 2007. A compilation of articles discussing aspects on both sides of the Noah's Flood Hypothesis. Provides information without bias to allow the reader to draw their own conclusions regarding the hypothesis. Previously available in Cyrillic only.

Zaitsev, Y. P., and B. G. Alexandrov, eds. *Black Sea Biological Diversity: Ukraine.* Black Seas Environmental Series 7. New York: United Nations, 1998. A collaborative effort by United Nations-sponsored scientists to study local flora and fauna in key zones around the Black Sea. Focuses on the coast of Ukraine, with particular attention to the effects of the major urban port area around Odessa.

**See also:** Aral Sea; Arctic Ocean; Atlantic Ocean; Caspian Sea; Deep Ocean Currents; Gulf of California; Gulf of Mexico; Gulf Stream; Hudson Bay; Indian Ocean; Mediterranean Sea; North Sea; Ocean Tides; Ocean Waves; Pacific Ocean; Persian Gulf; Red Sea; Storms; Surface Ocean Currents; Surface Water

# BLEACHING OF CORAL REEFS

*Coral bleaching is a phenomenon in which normally brightly colored, hard, tropical corals lose their pigments and appear bright white. Bleached corals are starved and weakened, and will eventually die if they do not recover. Coral bleaching occurs when zooxanthellae, symbiotic algae that normally live in the tissues of the transparent corals, are expelled by the corals. Coral bleaching has many causes, both natural and anthropogenic, but large-scale mass bleaching events are primarily linked with rising ocean temperatures.*

## PRINCIPAL TERMS

- **anthropogenic:** caused by or resulting from human activities; mainly used to describe environmental pollution and pollutants
- **cnidarian:** any organism, including corals, hydras, jellyfish, and sea anemones, which belongs to the phylum Cnidaria; all cnidarians are radially symmetrical and have stinging cells known as nematocysts
- **coral bleaching:** a pale or whitened appearance that arises from the loss of zooxanthellae, the symbiotic algae that are present in healthy reef-building corals
- **coral polyp:** the basic unit of a coral reef; a single invertebrate marine animal with a protective skeleton made up of limestone, or calcium carbonate, around a soft inner body
- **coral reef:** a living ecosystem in the shape of a ridge or mound, composed of colonies of coral polyps and the organisms that live in, on, and around them
- **fluorescent pigment:** a coloring matter that absorbs light at a particular wavelength and emits it at a longer wavelength; these pigments, found in zooxanthellae, give hard corals their bright colors
- **hermatypic:** of or relating to corals with stony skeletons that are formed from limestone and that grow in colonies to form coral reefs
- **resilience:** in ecological terms, the ability of an ecosystem as a whole to recover from natural or anthropogenic disturbances and stresses and to continue to thrive and reproduce; the opposite of resilience is vulnerability
- **SST:** commonly used abbreviation for sea surface temperature, or the temperature of water measured within the top few meters of the surface; a critical factor affecting ocean ecosystems, including coral reefs
- **zooxanthellae:** single-celled organisms that inhabit the living cells of hard corals and provide

them with energy in the form of oxygen and nutrients; a type of tiny algae

## CORAL BIOLOGY AND CORAL-ALGAL SYMBIOSIS

Corals, like hydras, jellyfish, and sea anemones, are cnidarians: living organisms that belong to the phylum Cnidaria. All cnidarians have a single body cavity, known as a coelenteron, through which they ingest food and release waste.

Members of this phylum also are characterized by radial symmetry and the presence of stinging cells on their surfaces known as nematocysts. These nematocysts are used to poison and capture prey. The term *coral* is most often applied to hermatypic corals, which are corals that build hard skeletons around their soft bodies by secreting calcium carbonate (limestone); they also grow and reproduce in large colonies, forming impressively intricate branching structures called reefs. Reefs typically comprise thousands of individual coral polyps, each of which is a single invertebrate marine animal.

Most hermatypic corals share another critical attribute: Some of their cells contain vast numbers of a particular type of single-celled organism. These organisms are tiny algae known as zooxanthellae. A single square centimeter of tissue from the gastrodermis (the lining of the digestive cavity) of a coral may contain millions of zooxanthellae. The relationship between the corals and the zooxanthellae is generally described as being symbiotic, or mutually beneficial. The coral provides the algae that inhabit its cells with a protected environment, and the waste products of its digestion also provide the algae with beneficial inorganic nutrients, such as nitrates and phosphates.

In return, the zooxanthellae remove waste products from the coral's tissues, serve as a crucial source of oxygen, and produce carbohydrates that the coral can use to manufacture fats and its protective calcium carbonate skeleton. As a result of their photosynthesis, the zooxanthellae also provide the coral with many useful organic products, including nitrogen compounds, vitamins, amino acids, and nutrients that are essential for supporting and speeding the coral's growth.

Color is another by-product of the symbiotic relationship between corals and zooxanthellae. Although the skeletons of hard tropical corals appear to be beautifully and brightly colored, they are actually white. Their apparent colors come from vast quantities of fluorescent pigments produced by the algae; these striking pigments can be seen through the clear tissues that make up each coral polyp's soft body.

## CAUSES OF CORAL BLEACHING

Coral bleaching is the phenomenon in which a coral expels most or all of the zooxanthellae that normally inhabit its tissue. When this happens, the coral is left appearing bright white. This is a generalized reaction that occurs when corals, and therefore the algae they contain, are subject to stress.

Stressed zooxanthellae produce a form of oxygen that is toxic to the corals. Rather than suffer the cellular damage that would result from this toxic compound, corals are forced to expel the zooxanthellae. Most coral colonies go through natural, temporary periods of mild bleaching, often in response to seasonal alterations in water temperature, but they recover when conditions return to normal.

Other natural environmental stresses that can lead to bleaching include increased predation, changes in the salinity or oxygen levels of ocean water, or an outbreak of a coral disease, such as a bacterial infection. Two pathogens that have been shown to cause bleaching in corals are the bacteria *Vibro shiloi* and *V. patagonica.*

Normally, coral reefs display a high level of resilience to such localized environmental stresses. Mass bleaching events (when bleaching occurs to vast swathes of reef at the same time) rarely occur in nature, and in time corals that have expelled their zooxanthellae will typically regain them and recover their health. Since about 1980, however, human activities have reduced the resilience of coral reefs and have introduced new and much more powerful causes of bleaching. These harmful activities include pollution, oil spills, overfishing, and stress caused by reef tourism.

The most significant cause of mass coral bleaching in recent years, however, has been increased SST (sea surface temperatures). Elevations in SST of as little as 1 to 2 degrees Celsius (33.8 to 35.6 degrees Fahrenheit) above the long-term average, if they persist for between one to two months, can lead to bleaching. Widespread increases in SST are occurring as a result of global climate change.

According to the 2007 Intergovernmental Panel on Climate Change, unequivocal evidence now exists that both the atmosphere and oceans are warming, primarily as a result of anthropogenic greenhouse gas emissions, especially carbon dioxide. Temperature records indicate that the past twenty years with the warmest sea temperatures on record have all occurred since about 1980.

Rising global temperatures also are being linked with an increase in the frequency of El Niño Southern Oscillation (ENSO) events. ENSO events are fluctuations in the surface temperature and pressure of the Indian and Pacific oceans; the events are named for a warm-water current that flows along the coast of Ecuador and Peru every few years. ENSO events can lead to such extreme weather events as floods and droughts and also may cause mass coral bleaching.

## EFFECTS OF CORAL BLEACHING ON OCEAN ECOSYSTEMS AND HUMANS

In addition to losing their color, corals that lack zooxanthellae are deprived of up to 90 percent of the nutrients and oxygen they require for their survival, reproduction, and growth. The stress caused by zooxanthellae expulsion also leaves corals susceptible to disease and predation. Corals that have lost all their algae and do not recover them can typically survive a few months only before starving.

Research has shown that branching corals and plate corals, which grow quickly, are more susceptible to bleaching and less likely to survive it. Massive corals, which grow more slowly, are more resilient. Thus, even when coral reefs do eventually recover from bleaching events, their overall genetic and species diversity is likely to be reduced.

Bleaching also has a profound negative impact on the numerous other organisms that live in and around the coral ecosystem. For example, coral reef degradation has been shown to cause declines in fish communities. A variety of studies of coral reefs around the world indicate that bleaching events result in marked declines in the numbers of species that depend heavily on coral reefs for food and shelter. Some coral species also have been shown to disappear completely from particular areas, a phenomenon known as a local extinction, when reefs undergo mass bleaching. These types of negative effects often persist for many years after the bleaching event itself.

Finally, bleaching has a negative impact on human communities. Colorless reefs are less likely to attract dive tourism, which may make it more difficult for local communities to sustain their economic activities. In addition, weakened reefs provide less protection from waves, which can lead to faster and more damaging erosion of the shorelines near coral reefs.

*M. Lee*

## FURTHER READING

Crabbe, M. James C., Emma L. L. Walker, and David B. Stephenson. "The Impact of Weather and Climate Extremes on Coral Growth." In *Climate Extremes and Society*, edited by Henry F. Diaz and Richard J. Murnane. New York: Cambridge University Press, 2011. Synthesizes scientific research linking extremes in ocean temperatures to bleaching events. Technical, and best suited to students with some background in marine ecology.

Grimsditch, Gabriel D., and Rodney V. Salm. *Coral Reef Resilience and Resistance to Bleaching*. Gland, Switzerland: International Union for Conservation of Nature, 2006. Prepared by a diverse international union of conservation agencies and nongovernmental organizations, this work is an extremely accessible and well-organized introduction to the problem of coral bleaching that includes photographs and a useful glossary of terms. Available at http://www.iucn.org/dbtw-wpd/edocs/2006-042.pdf.

Hutchings, P. A., Michael Kingsford, and Ove Hoegh-Guldberg, eds. *The Great Barrier Reef: Biology, Environment, and Management*. New York: Springer, 2008. A lavishly illustrated and beginner-friendly collection of articles about coral bleaching in Australia's Great Barrier Reef. Clearly organized. Suitable for high school students and undergraduates.

Lesser, Michael P. "Coral Bleaching: Causes and Mechanisms." In *Coral Reefs: An Ecosystem in Transition*, edited by Zvy Dubinsky and Noga Stambler. New York: Springer, 2011. A logically organized introduction to coral bleaching, written by a marine microbiologist and illustrated with figures and several black-and-white photographs.

_____. "Coral Reef Bleaching and Global Climate Change: Can Corals Survive the Next Century?" *PNAS* 104, no. 13 (2007): 5259-5260. A brief commentary on a study examining the role of anthropogenic factors in the mass bleaching of coral reefs in the Caribbean in 2005. Considers the study's implications for the future of coral reefs.

Rosenberg, Eugene, and Yossi Loya, eds. *Coral Health and Disease*. New York: Springer, 2004. Dozens of color photographs accompany this pointed, comprehensive look at coral physiology. Case studies of reef health in five different regions are followed by a section on the relationships between corals and microbes and one examining coral diseases. Covers multiple possible mechanisms for coral bleaching.

Van Oppen, Madeleine J. H., and J. M. Lough, eds. *Coral Bleaching: Patterns, Processes, Causes, and Consequences*. New York: Springer, 2009. A technical but relatively accessible volume, suitable for college students. Each chapter addresses a broad topic such as the evolution of coral-algal symbiosis and methods for detecting and monitoring bleaching events, and begins with its own introduction.

**See also:** Climate Change Theories; Geochemical Cycles; Greenhouse Effect; IPCC; Long-Term Weather Patterns; Observational Data of the Atmosphere and Oceans; Recent Climate Change Research; Reefs; Remote Sensing of the Oceans; Tropical Weather

# C

## CARBONATE COMPENSATION DEPTHS

*Carbonate compensation depths are the depths in ocean basins that separate calcium carbonate-containing seafloor sediments from carbonate-free sediments. The carbonate compensation depth (CCD) may be different in different ocean basins, and it may rise or fall at different times in the same ocean basin as a result of the balance between surface production of carbonate and deep-water carbonate dissolution. Carbonate solubility increases with increasing pressure and decreasing temperature.*

### PRINCIPAL TERMS

- **deposition:** the process by which loose sediment grains fall out of seawater to accumulate as layers of sediment on the sea floor
- **paleodepth:** an estimate of the water depth at which ancient seafloor sediments were originally deposited
- **plankton:** microscopic marine plants and animals that live in the surface waters of the oceans; these floating organisms precipitate the particles that sink to form biogenic marine sediments
- **precipitation:** the formation of solid mineral crystals from their chemical components that are dissolved in water
- **productivity:** the rate at which plankton reproduce in surface waters, which in turn controls the rate of precipitation of calcareous or siliceous shells or tests by these organisms
- **red clays:** fine-grained, carbonate-free sediments that accumulate at depths below the CCD in all ocean basins; their red color is caused by the presence of oxidized fine-grained iron particles
- **test:** an internal skeleton or shell precipitated by a one-celled planktonic plant or animal

### MARINE SEDIMENTS

Marine sediments are composed of a variety of materials of biological or mineral origin. Individual sediment grains in oceanic deposits may be either clastic or biogenic particles. Clastic sediments are materials derived from the weathering, erosion, and transportation of exposed continental rocks, and these grains are classified by particle diameter into gravels, sands, silts, and clays. Clastic sediment particles become less common with increasing distance from the continental landmasses and are nearly absent from deep-water sediments on the abyssal plains. Biogenic sediment particles are composed of skeletons and tests precipitated by planktonic plants and animals living in the shallowest waters of the ocean. Biogenic particles composed of calcium carbonate or of opaline silica make up the majority of oceanic sediments, which are deposited at great distances from land. Deposition of biogenic sediments is controlled by two factors: the biological productivity of surface waters and the dissolution of biogenic particles by corrosive bottom waters. Increased levels of atmospheric carbon dioxide lead to an increase in the corrosiveness of bottom waters.

Biogenic sediment particles are produced in shallow, well-lighted surface waters as a result of the biochemical activity of microscopic plants and animals, which precipitate solid shells and tests formed from minerals dissolved in seawater. Biological productivity is a measure of the number of the organisms present and their rate of reproduction. The productivity of planktonic organisms is directly related to chemical and physical conditions in the surface waters. High-productivity waters have abundant supplies of oxygen, with dissolved nutrients and chemicals needed for the formation of shells and tests, and enough light for photosynthesis by plants. Generally, high productivity is found in warm-water areas with abundant dissolved oxygen and nutrients.

When planktonic plants and animals die, their shells and tests fall through the water column to be deposited on the bottom of the ocean. This "planktonic rain" causes deposition of seafloor sediments as the biogenic particles produced in the surface waters sink and accumulate on the sea floor, and it moves chemicals from surface waters to deep waters. The higher the productivity values in surface waters, the greater the supply of biogenic sediment particles to the sea floor. As biogenic particles sink through the water column to the sea floor, they may be dissolved by corrosive seawater. Surface waters are saturated with dissolved calcium carbonate, so most dissolution takes place in deep waters, which are undersaturated with carbonate.

Calcareous sediments are produced by the accumulation of biogenic particles of calcium carbonate that survive the fall through the water and are deposited and buried on the sea floor. Accumulation rates of biogenic sediments are controlled by the balance between surface productivity and deep-water dissolution: The higher the biological productivity in the surface, the greater the number of shells and tests that will sink to the ocean bottom. In certain high-productivity areas associated with upwelling of cold, nutrient-loaded water masses to the surface, seafloor sediment accumulation rates may be as high as 3 to 5 centimeters per thousand years. In areas with higher dissolution rates, fewer calcareous particles will survive to be deposited on the ocean floor. In these low-productivity areas, all the carbonate produced at the surface may be dissolved, and sediment accumulation rates may be as low as 1 millimeter per million years.

## LEVEL OF CARBONATE COMPENSATION DEPTH

The carbonate compensation depth (CCD) marks the boundary between carbonate-rich sediments (calcareous oozes and chalks) and carbonate-free sediments (red clays) in the oceans. It is the result of deep-water dissolution rates exceeding the rate of supply of calcium carbonate to the deep sea by surface productivity. The depth of the CCD marks the zone in which the supply of carbonate sinking in the planktonic rain from surface waters is exactly balanced by the rate of removal of carbonate dissolution in deep waters. Calcareous sediments will be deposited on the sea floor in water depths shallower than the CCD, because in these areas, the rate of supply of calcium carbonate is higher than the dissolution rate. In these areas, individual particles of calcium carbonate will survive the trip through the water column and be deposited as biogenic sediments on the sea floor.

Below the compensation depth, dissolution exceeds the rate of supply, so all carbonate particles supplied from the surface become dissolved, and seafloor sediments are carbonate-free. Surface

sediments in water depths below the CCD tend to be red clays, or combinations of fine-grained materials derived from continental sources and carried to the deep sea by wind, mixed with micrometeorites and other particles from extraterrestrial sources. Red clays generally lack fossils, as a result of complete dissolution of carbonate and opaline silica, so only a few dissolution-resistant fossils, such as phosphatic fish teeth and whale ear bones, are found in these sediments. The reddish-brown color of deep-sea clay deposits is a result of the presence of iron particles, which have reacted with oxygen in seawater to form the rust-brown color.

Much information on the past history of the compensation depth in different ocean basins has been provided by ocean-floor drilling programs. The level of the CCD has changed dramatically throughout geologic history, with fluctuations of up to 2,000 meters being recorded in deep-sea sediments. Changes in the CCD are believed to be caused by changes in either the rate of supply of carbonate to the oceans or the rate of carbonate dissolution in the deep sea, which may be caused by changes in the shape of ocean basins, by changes in the location of carbonate deposition within different ocean basins, or by changes in the concentration of carbon dioxide in the atmosphere. During the last 100,000 years, carbonate sediments formed at greater depths during glacial intervals than during interglacials, indicating a less corrosive deep-water environment resulting from lower concentrations of atmospheric carbon dioxide.

## CALCIUM CARBONATE DEPOSITION

Calcium carbonate is delivered to the oceans by rivers carrying particles eroded from continental rocks. Ocean sediments are the primary geochemical reservoir for calcium carbonate, so most of the calcium carbonate on Earth remains dissolved in ocean water or in the form of calcareous sediments on the sea floor. The amount of carbonate deposition on the sea floor depends on the input of calcium carbonate derived from continental weathering and delivered to the oceans by rivers. Because oceanic plankton can precipitate solid calcium carbonate at a much faster rate than the rate of input of dissolved carbonate from rivers, most of the calcium carbonate that is deposited in the oceans must dissolve in order to maintain the chemical balance between dissolved carbonate and solid carbonate in oceanic sediments.

Any changes in the locations of carbonate deposition may cause corresponding changes in the level of the CCD as the compensation depths in different oceans change as a result of bathymetric fractionation or basin-basin fractionation. In bathymetric fractionation, a balance is established between the rates of carbonate deposition in shallow-water and deep-water sedimentary basins. Greater deposition of calcium carbonate in shallow waters atop the continental shelves will cause a shallowing of the CCD as more deep-water carbonate deposits are dissolved so as to balance the shallow-water deposition. Similarly, the level of the compensation depth may vary by basin-basin fractionation of carbonate, which establishes a balance between the compensation depths in different ocean basins. For example, greater deposition of calcium carbonate in the Pacific Ocean will cause the Pacific CCD to become deeper, while at the same time the Atlantic compensation depth must become shallower, because greater deposition of carbonate in Pacific sediments will leave less dissolved carbonate for precipitation in the Atlantic Ocean.

Even within an ocean basin, the level of the CCD may vary, depending on the balance between carbonate productivity and dissolution in a local area. For example, in the equatorial Pacific Ocean, the compensation depth is 500 to 800 meters deeper than in areas immediately to the north and south of this high-productivity area as a result of the greater supply of carbonate in the planktonic rain below these high-productivity surface waters. Also, the compensation depth tends to shoal near the edges of ocean basins, because higher biological productivity in shallow water near the continents causes rapid sinking of large amounts of organic carbon produced by planktonic plants and animals. Breakdown of this organic carbon by seafloor bacteria produces increased amounts of dissolved carbon dioxide gas, which reacts with water molecules to form carbonic acid; carbonic acid is corrosive to solid calcium carbonate. Greater carbonic acid concentrations lead to increased carbonate dissolution in bottom waters and cause upward migration of the CCD into shallower waters.

## STUDY OF CARBONATE COMPENSATION DEPTHS

Carbonate compensation depths may be studied by obtaining a series of deep-sea sediment samples from different depths within an ocean basin to determine the relationship between sediment type and water depth. In 1891, a study was published describing the global patterns of seafloor sediment type in each ocean basin, based on sediment samples obtained on the HMS *Challenger* oceanographic expedition. It was discovered that virtually no calcium carbonate was present below depths of 4,500 meters as a result of dissolution of carbonate.

The CCD in the modern ocean was first described in detail in 1935 based on sediment core transects taken across the South Atlantic Ocean by the 1925-1927 German *Meteor* expedition. Similar studies of sediment cores from different depths in the Pacific Ocean revealed that calcareous oozes are common seafloor sediments in water depths to 4,400 meters, with noncalcareous red clays being found in surface sediments deeper than 4,400 meters. Cores with calcareous oozes underlying red clays were obtained, however, in depths well below the present CCD, indicating that this geochemical boundary has migrated vertically throughout geologic history.

One innovative experiment to measure the rate of calcium carbonate dissolution with increasing water depth involved the placement of a stationary mooring for a period of months in deep water in the Pacific Ocean. Calcite spheres and calcareous microfossils were hung in permeable nylon bags at different water depths on the mooring; the nylon bags allowed seawater to come in contact with the calcium carbonate and thus permitted carbonate dissolution to occur. By measuring the weight loss of spheres suspended for a few weeks to months on the mooring, the rate of dissolution was determined for different water depths. In this experiment, little carbonate dissolution was observed at water depths shallower than 3,700 meters, while a rapid transition from minimal dissolution to extreme dissolution was seen. Rapid loss of carbonate by dissolution occurred in carbonate spheres suspended between the lysocline (the depth at which carbonate dissolution first begins to occur) and the compensation depth. Below 4,500 meters, the carbonate compensation depth, all carbonate was removed within a matter of weeks, demonstrating the ability of bottom waters to dissolve calcium carbonate.

Information on compensation depths may also be provided from microfossils preserved in seafloor sediments. Calcite dissolution can be measured by enrichment of dissolution-resistant forms of planktonic organisms, by benthic-planktonic foraminiferal ratios (foraminifera are one-celled animals that secrete a calcium carbonate internal test), by fragmentation indices (the percentage of broken planktonic tests compared to whole tests), or by the coarse-fraction ratios of seafloor sediments. Different microfossils will have differing susceptibilities to dissolution, depending on the thickness of the walls of the microfossil tests. Thin-walled plankton will be more dissolution-susceptible, while thicker-walled tests will resist dissolution. Thin-walled planktonic tests in seafloor sediments deposited in water depths between the lysocline and the CCD will be removed by dissolution more rapidly than thicker-walled tests, leading to greater enrichment of dissolution-resistant fossils with greater carbonate dissolution. Relative dissolution rates may be measured by the ratio between dissolution-susceptible and dissolution-resistant planktonic shells in seafloor sediments.

Similarly, the deeper-water bottom-dwelling benthic foraminifera tend to have thicker walls than the tests of planktonic foraminifera, which live floating in the shallow surface waters. Seafloor sediments may contain both benthic and planktonic varieties of foraminifera. Dissolution of calcium carbonate will preferentially remove the thinner-walled planktonic foraminifera, thus leading to enrichment in the proportion of benthic forms remaining in the sediment. The relative amount of carbonate dissolution in sediments may be determined by measuring the proportion between benthic and planktonic foraminifera in sediment samples.

Finally, fragmentation of planktonic tests may provide an indication of dissolution of carbonate from marine sediments. The percentage of broken tests to whole tests will increase with greater dissolution of carbonate. Also, coarse-fraction percentages of sediments provide information on carbonate dissolution, because unbroken foraminiferal tests are sand-sized, with particle diameters greater than 63 microns in size (1 millimeter is equal to 1,000 microns). As foraminifera are dissolved, they tend to break into smaller fragments, so dissolution tends to break down sand-sized particles into smaller silt-sized fragments. By measuring the proportion between sand-sized and silt-sized particles (the coarse-fraction percentage) in calcareous sediments, it is possible to obtain an indicator of the relative amount of dissolution that has affected those sediment deposits. The

more dissolution that has occurred, the smaller will be the percentage of coarse (sand-sized) sediment particles.

## DEPTH ESTIMATES

All these methods provide similar depth estimates for the CCD at approximately 4,500 meters below the sea surface, about halfway between the crests of the midocean ridges and the abyssal plains. Individual compensation depths may vary in the different ocean basins of the world, however, as a result of basin-basin fractionation of carbonate. For example, in the Pacific Ocean, the CCD is typically between 4,200 and 4,500 meters, while in the Atlantic and Indian Oceans the CCD is deeper, being found near a depth of 5,000 meters. Even within an ocean basin, the level of the CCD may vary, depending on the balance between carbonate productivity and dissolution in a local area. In the centers of the North and South Pacific Oceans, the average compensation depth is between 4,200 and 4,500 meters, while near the equator it is found near 5,000 meters because of the higher biological productivity of equatorial surface waters.

Analyses of deep-sea cores drilled by the *Glomar Challenger* have revealed that the level of the carbonate compensation depth has changed by up to 2,000 meters in the South Atlantic, Indian, and Pacific Ocean basins. For example, one of the results of Deep Sea Drilling Project Leg 2 was the discovery of significant vertical excursions in the compensation depth of the Atlantic Ocean. In seafloor boreholes drilled in the North Atlantic, 2- to 5-million-year-old calcareous ooze sediments were found atop older (5- to 23-million-year-old) red clays, which in turn were deposited atop carbonate deposits older than 23 million years. In order for these sediments to have accumulated in that order, large vertical changes must have occurred in the compensation depth, starting when the sea floor was shallower than the CCD more than 23 million years ago. Red clays were deposited between 23 and 5 million years ago, when the CCD became shallower. After these red clays accumulated, deepening of the compensation depth allowed the deposition of younger calcareous sediments atop the carbonate-free red clays.

In order to study the past history of the compensation depth within an ocean basin, it is necessary to obtain a series of cores that were deposited at different paleodepths at the same time in the past. (The paleodepth is the depth at which ancient seafloor sediments were deposited.) Paleodepth estimates for sea floor sediments are calculated by studying the cooling history of seafloor basement rocks. After new seafloor is produced by volcanic activity at the midocean ridge system, these rocks cool and contract, thus sinking to greater water depths as they move away from the ridge system. The older the sea floor, the greater its water depth will be, so sediments deposited atop volcanic sea floor will accumulate in progressively deeper water. Once a series of sediment cores deposited at the same time in the past has been obtained, it is possible for the oceanographer to determine the paleodepth of the compensation depth by finding the paleodepth below which no calcium carbonate is present in ancient sediment deposits.

*Dean A. Dunn*

## FURTHER READING

Ahr, Wayne M. *Geology of Carbonate Reservoirs.* Hoboken, N.J.: John Wiley & Sons, 2008. Covers many principles of mineralogy with a focus on carbonates. Covers rock properties, petrophysical properties, stratigraphy, deposition, and diagenesis of carbonate reservoirs. A summary of the geology of carbonate reservoirs ties together fundamental topics with specific field examples. Offers references and an extensive index.

Allen, Philip A. *Earth Surface Processes.* Oxford, England: Blackwell Science, 1997. Serves as a clear introduction to oceanography and Earth sciences. Details the processes, properties, and composition of the ocean. Color illustrations, maps, index, and bibliography.

Berger, Wolfgang H. "Sedimentation of Deep-Sea Carbonate: Maps and Models of Variations and Fluctuations." *Journal of Foraminiferal Research* 8 (October, 1978): 286-302. A summary of the oceanographic influences on carbonate dissolution and carbonate sediment deposition patterns, with abundant illustrations.

Berger, Wolfgang H., and E. L. Winterer. "Plate Stratigraphy and the Fluctuating Carbonate Line." In *Pelagic Sediments: On Land and Under the Sea*, edited by Kenneth J. Hsü and Hugh C. Jenkyns. Oxford, England: Blackwell Scientific, 1974. A thorough review of the factors that may cause vertical migrations of the compensation depth through time.

The text is suitable for college-level readers and contains many explanatory figures.

Emerson, Steven, and John Hedges. *Chemical Oceanography and Marine Carbon Cycle.* New York: Cambridge University Press, 2008. Provides a good overview of geochemistry topics in oceanography. Discusses chemical composition, thermodynamics, carbonate chemistry, the carbon cycle, and calcium carbonate sedimentation. Includes appendices following the chapters and an excellent index.

Garrison, Tom S. *Oceanography: An Invitation to Marine Science.* Belmont, Calif.: Brooks/Cole, Cengage Learning, 2010. Discusses sediments, including calcium carbonate precipitates. Includes abundant diagrams to aid readers from the layperson to advanced undergraduates.

Hay, William A. "Paleoceanography: A Review for the GSA Centennial." *Geological Society of America Bulletin* 100 (December, 1988): 1934-1956. Reviews all aspects of the study of ancient oceans, from the work of the first geologists to current research. Covers the history of oceanographic examination of seafloor sediments, the development of deep-sea sediment sampling, and the history of research on the lysocline and the CCD.

Kennett, James P. *Marine Geology.* Englewood Cliffs, N.J.: Prentice-Hall, 1982. A college-level textbook on all aspects of marine geology and geological oceanography. Describes the deposition of seafloor sediments and the factors influencing the positions of the lysocline and the CCD, as well as changes in these depths through time.

Rigo, M., et al. "A Rise in the Carbonate Compensation Depth of Western Tethys in the Carnian (Late Triassic): Deep-Water Evidence for the Carnian Pluvial Event." *Palaeogeography, Palaeoclimatology, Palaeoecology* 246 (2007):188-205. Examines the carbonate deposition in rock of the upper Triassic. Interprets the carbonate compensation depth from these carbonate deposition levels.

Segar, Douglas. *An Introduction to Ocean Sciences.* 2d ed. New York: W. W. Norton, 2007. Comprehensive coverage of all aspects of the oceans and their chemical makeup. Readable and well illustrated. Suitable for high school students and above.

Sliter, William V., Allan W. H. Be, and Wolfgang H. Berger, eds. *Dissolution of Deep-Sea Carbonates.* Washington, D.C.: Cushman Foundation for

Foraminiferal Research, 1975. Contains a number of research papers analyzing the dissolution of carbonates in the water column and on the sea floor, along with studies of the factors influencing the position of the lysocline and the compensation depth in ocean basins. Suitable for college-level readers.

Sverdrup, Keith A., Alyn C. Duxbury, and Alison B. Duxbury. *An Introduction to the World's Oceans.* 8th ed. Boston, Mass.: McGraw-Hill, 2004. A freshman-level review of the marine environment. Describes the physical, chemical, biological, and meteorological structure of the continental margins. Color plates review satellite and submarine technology.

Swart, Peter K., Gregor Eberli, and Judith A. McKenzie, eds. *Perspectives in Carbonate Geology.* Hoboken, N.J.: Wiley-Blackwell, 2009. A collection of papers presented at the 2005 meeting of the Geological Society of America. Presents studies on carbonate sediments and the comparison of modern and ancient sediments. Suited for the professional geologist or graduate student.

Talley, Lynne D., George L. Pickard, William J. Emery, and James H. Swift. *Descriptive Physical Oceanography: An Introduction.* 6th ed. London: Elsevier, 2011. Provides an overview of the science of physical oceanography with emphasis on the chemical processes occurring in ocean water that determine such phenomena as the carbon compensation depth. College-level.

Williams, Richard G., and Michael J. Follows. *Ocean Dynamics and the Carbon Cycle: Principles and Mechanisms.* New York: Cambridge University Press, 2011. Uses a multidisciplinary observational approach to discuss the basic principles controlling ocean biogeochemistry. Combines basic physical theory with schematic illustrations, numerical models, and observations to examine how marine chemistry and biological productivity are controlled by ocean circulation. Deals with carbonate chemistry and the carbon cycle in oceans.

**See also:** Carbon-Oxygen Cycle; Dams and Flood Control; Deep Ocean Currents; Drainage Basins; Floodplains; Floods; Gulf Stream; Hydrothermal Vents; Ocean-Atmosphere Interactions; Ocean Pollution and Oil Spills; Oceans' Origin; Oceans' Structure; Ocean Tides; Ocean Waves; River Flow; Sea Level; Seamounts; Seawater Composition; Sediment Transport and Deposition; Surface Ocean Currents; Surface Water; Tsunamis; Turbidity Currents and Submarine Fans; Water Quality; World Ocean Circulation Experiment

# CARBON-OXYGEN CYCLE

*The carbon-oxygen cycle is the process by which oxygen and carbon are cycled through Earth's environment. The cycle includes phenomena such as photosynthesis, which makes oxygen by producing sugar from carbon dioxide, and respiration, which uses oxygen to break sugar down into carbon dioxide. The cycle is critical for the homeostasis of the environment.*

## PRINCIPAL TERMS

- **biosphere:** life on Earth, and the area it inhabits
- **carbon sequestration:** the process of storing carbon in a stable state to negate carbon's effects on climate
- **combustion:** reactions by which oxygen and organic materials become carbon dioxide and water
- **decomposition:** process by which organic matter is broken down into its most basic components by microorganisms
- **greenhouse effect:** process by which heat is trapped on Earth
- **homeostasis:** property of a system to maintain a certain internal state, such as temperature
- **hydrosphere:** water on Earth, and its area
- **pedosphere:** the soil
- **photosynthesis:** set of chemical reactions that use solar power to make sugars and oxygen from carbon dioxide and water
- **respiration:** process by which organisms break down organic material for energy

### THE CARBON-OXYGEN CYCLE

The carbon-oxygen cycle involves the flow of carbon and oxygen through the environment. Carbon is recycled between carbon dioxide and carbon compounds in life-forms. The cycle includes the transfer mechanisms of the carbon compounds and oxygen in the biosphere and in the atmosphere, hydrosphere, and pedosphere. Climatological research is focusing on the rates of processing and the homeostasis of the system to determine carbon dioxide release and capture.

The carbon-oxygen cycle is the result of the storing of carbon from atmospheric (or in the case of aquatic organisms, dissolved) carbon dioxide as organic material through photosynthesis, and its release through respiration, combustion, or decomposition, all of which combine the carbon in the organic material with atmospheric or dissolved oxygen as $CO_2$. This $CO_2$ is then released into the atmosphere, where it is reabsorbed by photosynthetic organisms such as

cyanobacteria, some protists, and plants, starting the process anew.

The cycle shows how different parts of Earth's environment and climate are interrelated. For example, analysis of the flow of material from rivers into seas allows scientists to consider the impact of biologic materials, and the release of the gases into the atmosphere and the capturing of carbon by plants allow scientists to see the effect of land use on climate.

By examining the environmental factors in a given area—factors such as land use or propensity for fire—one can get an idea of how a given region will impact other areas and can estimate the total amount of gas it might contribute to the atmosphere. Additionally, environmental factors provide hints as to effective methods of managing or sequestering carbon. An example of carbon sequestration would be a forest.

Based on the carbon-oxygen cycle, it is known that trees take in a certain amount of carbon in time, that the trees will produce living space for some number of animals, and that those animals will produce a certain amount of carbon dioxide throughout their lives and when they die and decompose. Also factored is whether a forest is prone to fires. By looking at such factors, one can get an idea of how much carbon dioxide a forest can remove from the atmosphere, for how long, and if there is anything humans can do to make that forest more efficient.

Most of the actual oxygen released into Earth's atmosphere occurs because of oceanic plankton, however, but rainforests absorb huge amounts of carbon. Northern forests are another major site of oxygen turnover.

### PROCESSES

Photosynthesis is a set of chemical reactions that occur in plants and in some bacteria and some protists. Photosynthesis uses energy from the sun to turn carbon dioxide and water into sugar and oxygen.

The process itself is the interaction of several sets of reactions: the light-dependent reactions and the Calvin or dark cycle. The light-dependent reactions separate the hydrogen from the oxygen in water and are thus the oxygen-producing portion of the cycle; they prepare reactants needed for the other cycles. The Calvin or dark cycle turns the carbon dioxide and light cycle products into sugar. The light cycle runs directly on captured photons, whereas the Calvin cycle does not depend directly upon sunlight.

Along with rates of release from combustion, which is the conversion of organic material to carbon dioxide and water vapor, also needing to be factored are rates of respiration and decomposition. Respiration is how organisms get energy from organic compounds, such as sugars. Plants and animals use aerobic respiration, which evolved after oxygen became prevalent in the atmosphere. Aerobic respiration occurs in special organelles of eukaryotic cells called ribosomes. Here sugars are broken down into carbon dioxide and water in a process known as the Krebs or citric acid cycle.

Decomposition releases various greenhouse gases, as decomposing organisms are often excellent habitats for anaerobic life-forms; anaerobic respiration releases such gases as carbon dioxide, hydrogen sulfide, and methane. Thus the total impact of the disruption of an environment, such as the burning of a forest, is not only in the combustion of the trees but also in the drop in photosynthesis and decomposition of any detritus.

## HISTORY OF THE CYCLE

The understanding of the carbon-oxygen cycle has existed since the eighteenth century, after the discoveries of Antoine Lavoisier, who found that respiration was a process similar to burning, and of Joseph Priestley, who outlined the idea of the relationship between blood and air and who also discovered oxygen. The full mechanics of the carbon-oxygen cycle and its implications, however, were discovered by scientists in subsequent centuries.

The workings of the chemical reactions of photosynthesis were discovered in the 1940's, and research continues into the role of quantum physics in the light-harvesting stage of photosynthesis. With the growing awareness of climate change, more work is being done to examine the cycle.

Current research often takes samples from the area in question, be they soil samples to estimate bacterial carbon production or air samples to measure carbon levels at a given site. These samples are then analyzed together to get a picture of how each part

is affecting the collective carbon-oxygen balance for that area. The carbon data are often mixed with data on other climatological factors, such as albedo, humidity, and temperature, to build a picture of the effects on climate.

## ATMOSPHERIC COMPOSITION AND THE GREENHOUSE EFFECT

The carbon-oxygen cycle affects climate because it affects the earth's atmosphere. Ignoring water vapor, which is variable, it is known that the standard dry atmosphere is 78.09 percent nitrogen, 20.95 percent oxygen, 0.93 percent argon, and 0.039 percent carbon dioxide; the rest of the atmosphere comprises various trace gases. Water vapor composes, on average, about 1 percent of the atmosphere.

The carbon-oxygen cycle plays an important role in maintaining homeostasis in the atmosphere. The cycle "cycles" air and carbon to maintain equilibrium. The gases in the atmosphere and their relative amounts have an important impact upon climate. Some gases absorb infrared rays reflected or emitted from the earth's surface and send them back to Earth, thus trapping the energy and making Earth hotter. This process is known as the greenhouse effect.

Gases that are best at emitting in the infrared range are known as greenhouse gases and include water vapor, carbon dioxide, methane, nitrous oxide, and ozone. Water vapor and carbon dioxide make up the largest contribution to the effect, with methane and ozone playing a smaller role. Many of these gases are produced in industrial reactions such as combustion. Combustion from cars and factories and from other human activities (namely since the start of the Industrial Revolution) have been major factors in anthropogenic climate change. Here the carbon-oxygen cycle comes into play. There is a maximum rate at which a particular environment can cycle the carbon dioxide into oxygen, meaning that excess carbon dioxide builds up in the atmosphere, increasing the greenhouse effect.

By studying the components of the cycle (that is, the plants and animals that are part of the process), scientists can find the effects of increased temperature on plants and animals. The environmental impact (increased environmental stress) leads to less capacity to handle the increased carbon dioxide. This creates a positive feedback loop that further increases the temperature. Increased global temperature could cause increased cloud cover, which would increase albedo. Increased temperature also could spawn algal blooms in the oceans, which would absorb carbon dioxide and thus reach a new equilibrium. Much environmental damage will occur before such an event happens.

The greenhouse effect is not wholly negative, however. It is needed to some degree for life to exist on this planet. In the absence of greenhouse gases, the planet would be significantly colder than it is now. Additionally, Earth's atmospheric composition has varied throughout its history. The earth's early atmosphere was different from today's atmosphere. It comprised mostly water vapor, carbon dioxide, hydrogen sulfide, and ammonia. It did not take long, in geologic time, for the oceans to form from rainwater. All of the greenhouse gases of the time kept the average Earth temperature fairly high.

Once photosynthetic life evolved about 3.5 billion years ago, that life began to break down the methane and carbon dioxide in the atmosphere, and the level of oxygen began to increase. This resulted in what is called the oxygen catastrophe, in which many of the organisms at the time became extinct, leaving those that adapted to the new situation by developing aerobic respiration. This development had a great impact upon later life, as aerobic respiration became far more efficient.

The oxygen catastrophe also may have caused the Huronian glaciation event, which occurred most likely as a result of greenhouse gas depletion by photosynthetic life-forms. The event was one of the longest glaciation periods in Earth history (300 million years). It and the later Snowball Earth periods that likely caused the evolution of multicellular life and the Cambrian explosion were ended only by the buildup of greenhouse gases from volcanic eruptions and by the weathering of rocks.

Equally, after this period, greenhouse gas levels have oscillated by period; for example, during the Devonian period, carbon dioxide was higher and oxygen was lower than present levels. In the Cretaceous period, both carbon dioxide and oxygen levels were higher. During both periods, the average temperature was higher than it is today.

A dramatic example of how the carbon-oxygen cycle figures into climate would be the Permian-Triassic mass extinction event (250 million years ago), in which a shutdown of the carbon dioxide cycle may have been responsible for the greatest mass extinction in Earth's history. While the exact causes remain unknown, the mass extinction is thought to have begun with the eruption of the Siberian Traps, a massive set of volcanoes that released carbon dioxide and debris into the atmosphere.

The largest eruptions in known history probably caused a small period of global cooling, which stressed many environments. As a result, many trees died before they could process the carbon dioxide that was added to the atmosphere. Evidence seems to show that global warming set in around the peak of the extinction, and it is thought that the oceans became saturated with carbon dioxide, to the point in which there was a degassing event. The seas released their carbon in huge clouds that swept over the land, suffocating all nearby creatures. Massive erosion suggests that the climate became arid and the interior of continents were more or less lifeless. While another event on this scale is unlikely, it provides a strong example of the kind of phenomena that can arise from the disruption of the carbon-oxygen cycle.

## GREENHOUSE GAS EMISSIONS

More than 3 million tons of carbon dioxide is released each year into the atmosphere. Compounding that are ongoing releases of other gases such as methane, which has twenty-five times the effect of carbon dioxide (as a greenhouse gas) in a one-hundred-year period.

The largest sources of anthropogenic emissions are deforestation and the impact of industrial technology, including factories, cars, and cement making (which releases carbon dioxide). However, agriculture also releases greenhouse gases; for example, cows release 16 percent of anthropogenic methane.

Humans have always released carbon dioxide into the environment. Various indigenous North American cultures burned forests annually to encourage habitats for large game animals. Also, much of Europe has been deforested for agricultural space and shipbuilding. However, the emissions from industrial and postindustrial technology are on a wholly greater scale. Equally important is that the carbon being released now has not been part of the carbon-oxygen cycle since it was stored as coal or oil: in many cases, carbon dioxide released now was stored more than 250 million years ago in ancient swamps. Thus, with the combination of environmental disruption and greenhouse gas emission, recent activities are more akin to a sustained volcanic eruption than to any other Earth event.

As the system moves further from homeostasis, the compensatory mechanisms, such as increased plant growth, are overloaded or are not given the chance to function (as is the case with heavy deforestation). This forms positive feedback, pushing the environment further from equilibrium. Something as simple as planting additional trees can stall the harmful trend, however. An important point is that the carbon in a living tree has not been expelled; it remains stored as long as the tree lives. Once the tree dies the carbon dioxide is released.

## CARBON SEQUESTRATION

Carbon sequestration involves storing carbon in a stable state to negate carbon's harmful effects on the environment. As discussed, trees serve as reservoirs for carbon.

Many efforts focus on manipulating the carbon-oxygen cycle to increase the amount of carbon stored. Means of sequestering carbon range from genetically modifying plants to employ a more efficient form of photosynthesis to timing forest harvesting and replanting to ensure peak carbon storage, as in the Canadian lumber industry. Other forms of carbon sequestration under consideration include injecting carbon into coal seams; sinking carbon to the bottom of the ocean through algae growth; producing certain carbon-absorbing compounds; and conducting controlled burns that increase the health of an ecosystem, as in the Australian savanna. With proper manipulation of the carbon-oxygen cycle through methods such as carbon sequestration, many of the worst effects of climate change can be mitigated.

*Gina Hagler*

## FURTHER READING

Bashkin, V. N., and Robert W. Howarth. *Modern Biogeochemistry*. New York: Kluwer Academic, 2002. An excellent detailed description of the carbon-oxygen cycle, its precise mechanics, means for quantitative analysis, the development of the system, and its relationship to other cycles in the environment. Most suitable for college students and other readers with some background in chemistry and biology.

Kirby, Richard R. *Ocean Drifters: A Secret World Beneath the Waves*. Winchester, England: Studio Cactus, 2010. Focuses on plankton, the life-forms responsible for most of the oxygen release on Earth. Includes enlarged photographs. Recommended for general readers.

Kondratyev, K. Y., V. F. Krapivin, and Costas A. Varotsos. *Global Carbon Cycle and Climate Change*. New York: Springer, 2003. Detailed analysis of the relationship of the carbon-oxygen cycle to climate change. Includes the effects of other greenhouse gases and how it relates with Earth's environment. Recommended for those with some background in the subject.

Lévêque, C. *Ecology from Ecosystem to Biosphere*. Enfield, N.H.: Science, 2003. Contextualizes the carbon-oxygen cycle in its ecological role, which allows readers to see how the cycle fits with the larger functioning of the ecosystem and biosphere. Suitable for general readers.

Vallero, Daniel A. *Environmental Biotechnology: A Biosystems Approach*. Amsterdam: Academic, 2010. Examines the carbon-oxygen cycle in the context of environmental biotechnology. Systems based, it deals with the larger scale and interactions of the cycle.

Walker, Sharon, and David McMahon. *Biochemistry Demystified*. New York: McGraw-Hill, 2008. An accessible introduction to the processes behind respiration and photosynthesis that underlie the carbon-oxygen cycle. Written for beginners. Also covers biochemistry.

**See also:** Climate Change Theories; Geochemical Cycles; Greenhouse Effect; Ice Ages and Glaciations; Nitrogen Cycle; Observational Data of the Atmosphere and Oceans

# CASPIAN SEA

*The Caspian Sea, the largest inland body of water on Earth, holds petroleum and wildlife resources that are crucial for Eurasia's economic and political stability.*

## PRINCIPAL TERMS

- **littoral:** adjacent to or related to a sea
- **petrochemical:** a chemical substance obtained from or derived from natural gas or petroleum
- **petroleum:** a naturally occurring liquid composed of hydrocarbons found in Earth's subterranean strata

### CASPIAN SEA GEOGRAPHY

The Caspian Sea, the world's largest landlocked body of water, is located in central Eurasia. During the Miocene era 26 million years ago, the Caspian Sea was part of a larger sea that included the Black and Aral Seas before tectonic uplift eventually transformed them into distinct basins. Actually a saltwater lake, the Caspian Sea is bordered by five countries. Kazakhstan lies to the north and east of the sea, while Turkmenistan is also on the sea's east coast. Azerbaijan and Russia (containing the two autonomous republics of Daghestan and Kalmykia) are on the Caspian Sea's western border, and Iran is on the southern coast. Known to humans for centuries, the Caspian Sea was called by the ancient Roman name of Caspium Mare. Its proximity to the trade route known as the Silk Road assured its importance as a center of trade and commerce until ships came to dominate trade routes.

Unsymmetrically shaped and somewhat resembling the silhouette of a seahorse, the Caspian Sea's size varies because it undergoes cycles of contraction and expansion. On average, the sea stretches 1,210 kilometers from north to south and from 210 to 436 kilometers from east to west, encompassing an approximate area of 371,000 square kilometers. Six rivers feed freshwater into the Caspian Sea, diluting its brine. The Volga and Emba Rivers enter from the north. The Ural River, considered the geographical division line between Europe and Asia, also ends at the sea's northern edge. The Gorgan and Atrek Rivers flow from the east, and the Kura River is the sea's western tributary. The Elburz and Greater Caucasus mountain ranges are located south and southwest of the Caspian Sea.

Deepest in its southern part, the Caspian Sea has an average depth of 170 meters. There are two gulfs along the eastern coast, the Krasnovodsk Gulf and the shallow Kara-Bogaz-Gol, which evaporates and leaves salt deposits. Although the Caspian Sea's level fluctuates every year, it measures an average of 28 meters below mean sea level. During the 1960's and 1970's, the Caspian Sea's level was substantially lowered because water that usually flowed into the sea from its tributaries was diverted for agricultural irrigation. In an attempt to stop water loss, engineers built a dike across the mouth of the Kara-Bogaz-Gol in 1980. Instead of creating a long-lasting lake, the dammed-up water evaporated within three years. Subsequently, an aqueduct was necessary to deliver water to the Kara-Bogaz-Gol. Other river and sea dams block both water and fish from moving naturally while protecting oil refineries on land.

Lacking a natural outlet to other large bodies of water, the Caspian Sea is connected by tributaries to adjacent waterways (such as the Black Sea) that can be navigated to reach distant markets. The Caspian Sea's primary ports are Krasnovodsk in Turkmenistan, Makhachkala in Russia, and Baku on the Apsheron Peninsula in Azerbaijan. The frigid winter climate causes ice to form in the northern parts of the Caspian Sea, hindering travel and fishing. Weather patterns produce dangerous storms, mostly moving southeast across the sea, damaging vessels and oil-drilling platforms. These storms are also hazardous to humans, plants and animals living in the water, and reefs and other geological structures.

### PETROLEUM AND GEOPOLITICS

Humans have known about the Caspian Sea's valuable natural resources for hundreds of years. Thirteenth-century explorer Marco Polo commented that Caspian oil was so plentiful that it gushed like fountains. By the 1870's, speculators equipped with drilling technology became wealthy selling oil, and the Caspian Sea region enjoyed prosperity until the 1917 Russian Revolution, when oil barons were arrested and their wells seized. Caspian Sea oil was considered a strategic raw material during both world wars. Soviet leaders abused Caspian oil production, negligently removing oil with inadequate tools and polluting adjacent areas in the process. Because the Caspian Sea contains some of the world's greatest untapped petroleum deposits, international competition to secure this valuable fuel resource was the catalyst for an oil rush in the late twentieth century that filled the sea with oil derricks.

The largest known oil deposits in the Caucasus region and Central Asia are located in the Caspian states of Kazakhstan and Azerbaijan. Geologists surveyed three major areas of petroleum lying in deposits that began about 3,600 meters beneath the Caspian Sea's surface. Although the size of these deposits was not determined, their potential seemed to rival the measured deposits of other oil-producing countries. One supply is underneath the Caspian Sea. The others are onshore. The second deposit reaches eastward from Baku to Turkmenistan. The third area spreads west of Kazakhstan underneath the sea's northern tip. The Caspian Sea's oil offered an alternative source of petroleum to countries that had been relying on Persian Gulf reserves.

Some experts suggested that almost 200 billion barrels of oil remained in the Caspian Sea area. On average, 1.1 million barrels was removed daily, in addition to by-products of sulfur. The area was also rich in natural gas, with an estimated 7.89 million cubic meters. During the 1990's, European and American oil companies such as Amoco and Chevron secured agreements with the governments of several of the countries bordering the Caspian Sea to invest money, equipment, and labor to build pipelines and extract petroleum. The Azerbaijan State Oil Academy prepared workers for the technical demands of oil extraction. Geologists rely on seismic imaging to analyze the Caspian Sea for petroleum. Drillers encountered problems with currents and mud volcanoes on the sea floor, and scientists developed refined technology and geophysical extraction methods to locate and remove oil from deep reservoirs of hydrocarbons existing at high temperatures and pressures.

Investors were aware of tensions in the Caucasus region between the independent Russian states and their neighbors and carefully planned to place pipelines so as to avoid territorial claims on oil. Competing commercially for the oil, foreign nations strove to achieve strategic alliances and to consider

diplomatic consequences. They were aware of the vulnerability and instability of the Caspian Sea countries. Several treaties were signed by the littoral Caspian states to legally divide the sea into zones for each state to claim resources. These nations encouraged competition to boost prices because access to Caspian Sea energy resources became an urgent post-Cold War geopolitical objective.

## CASPIAN SEA INHABITANTS

The Caspian Sea and its surrounding lands provide habitats for an eclectic assortment of wildlife and plants that enhance the area's environment and economy. The sea's primary water source, the Volga River, forms a swampy delta near the confluence with the sea. This marsh is home to waterfowl, including ducks, swans, flamingos, and herons, as well as eagles that rely on the wetlands for shelter and food. Along parts of the uninhabited eastern shore, camels thrive and wildflowers grow among reeds on the dunes. Fish living in the Caspian Sea include salmon, perch, herring, and carp. The sea is also home to seals, tortoises, and crustaceans unique to the region. Most significantly, three species of sturgeon have flourished in the Caspian's salty water. The Beluga sturgeon stretches as long as 4 meters and weighs 1,000 kilograms, the Asetra sturgeon is about 2 meters long and weighs 200 kilograms, and the Sevruga sturgeon is just over 1 meter in length and weighs approximately 25 kilograms.

Approximately 90 percent of the global sturgeon population lives in the Caspian Sea, producing what gourmets consider the world's tastiest caviar. Eggs compose as much as 15 percent of a sturgeon's weight. Russian czars generously served Caspian caviar, and Soviet and Iranian entrepreneurs became wealthy exporting the processed sturgeon eggs for hundreds of dollars per ounce. In an effort to maintain a monopoly and keep prices high, these countries carefully controlled fishing and cultivated sturgeon in fisheries to replenish the population. After the collapse of the Soviet Union in 1991, sturgeon poaching became a serious problem that threatened the Caspian Sea. Nineteenth-century Russians reported yearly catches of 50,000 tons of sturgeon. In the 1980's, the Soviet Union caught as much as 26,000 tons of sturgeon annually according to statistics maintained by the Caspian Fishery Research Institute. By the late 1990's, only 3,000 tons of sturgeon was reported to be harvested by commercial fishermen from Caspian countries each year.

Motivated by potential profits and desperate economic conditions plaguing the independent Russian states, Caspian fishermen poached sturgeon to survive financially by selling fish eggs and meat. One ounce of caviar often sold for more than ten times the amount of an individual's monthly salary. Poachers also saved portions of fish to feed their families. Placing nylon line traps beneath the surface of the Caspian's tributaries, poachers waited for spawning sturgeon to swim upstream. Such uncontrolled snaring reduced the sturgeon population to the point that it was eventually considered an endangered species. Experts stressed that overfishing resulted in the deaths of approximately 90 percent of sturgeons too immature to spawn (most sturgeons require an estimated eighteen years to develop reproductive capabilities), which detrimentally reduced the number of fish returning to the Caspian's rivers every year. Fewer Beluga sturgeon were captured annually, and not many reached their potential to be as heavy as 1 ton and survive for a natural one-hundred-year life span.

## ENVIRONMENTAL ISSUES

Pollution also posed substantial dangers for sturgeon. The Caspian Sea was contaminated by oils and petrochemicals from drilling platforms, as well as by pollutants and industrial residues dumped into rivers by factories. Careless drilling methods caused oil films as thick as 6 millimeters to float on the Caspian Sea, creating an unhealthy environment for fish. Raw sewage was also discharged from drains into the sea's tributaries. Scientists found that the waters near the cities of Baku and Sumqayit were so saturated with poisonous wastes and were so depleted of oxygen and nutrients that they could not harbor aquatic life.

The Caspian Sea benefits from the marshes in the Volga River's delta, which strain out some of the debris and contaminants before the water reaches the sea. Near the river's mouth, which consists of thousands of slender streams moving through the stalks of indigenous water plants, the Caspian Sea's shallow pools of sparkling blue water seem clean. Moving toward the sea's interior, however, the water becomes murky and appears gray in color.

Experts warn that the Caspian Sea is experiencing an environmental crisis that will only worsen as extraction of natural resources increases and industrialization expands in bordering countries. Authorities blame these pollutants for contributing to the growing rates of miscarriages and stillbirths in Caspian nations, and industries are now encouraged to reduce pollution and to cease using rivers for effluent disposal. Petroleum companies operating onshore oil fields and platforms in the Caspian Sea have been advised to adapt their operations to rigid standards regarding drilling to prevent tainting the water with contaminating fluids and detritus.

Ecologists want extractors of natural resources to become more aware of their impact and accountability to maintain the Caspian Sea's environmental quality. Scientists demand that bans on poaching and illegal caviar processors be enforced more diligently, with police monitoring the sea and arresting offenders. They also seek to prohibit open-sea fishing, which kills young sturgeon, and recommend that Caspian Sea nations agree on caviar quotas so that fair prices are set to discourage black-market profiteering and poaching. Public relations campaigns now work to promote public knowledge of the Caspian Sea's environmental crises and provoke more responsible treatment of its resources.

## RISING OF THE CASPIAN SEA

Biologists monitoring the Caspian Sea noted that after a lengthy period of contraction, its level rose 2 meters from the late 1970's to the 1990's. They optimistically believed that the increased amount of water might weaken the effect of pollutants on the sea and on its inhabitants unless the amount of toxic material and oil spills entering the sea drastically increased.

Many scientists speculated that global warming and the region's climate might have contributed to the rise in sea level. Other scholars hypothesized that the water level increases might have been caused by natural processes, possibly initiated by climatic deviations that affect the atmosphere and sea over periods of years, decades, or greater periods of time. The erratic changes in the sea's composition and contents that were documented in the past and that periodically recur will crucially affect future development of the Caspian Sea's resources. The quantity and quality of the Caspian Sea's water are vital to ensure Eurasian economic and political stability and to maintain international technological and business interests in the region.

*Elizabeth D. Schafer*

**FURTHER READING**

Amineh, Mehdi Parvizi. *Towards the Control of Oil Resources in the Caspian Region.* New York: St. Martin's Press, 1999. A technical text that analyzes the political aspects of the Caspian Sea petroleum industry. Provides bibliographical sources and index.

Amirahmadi, Hooshang, ed. *The Caspian Region at a Crossroad: Challenges of a New Frontier of Energy and Development.* New York: St. Martin's Press, 2000. Presents scholarly essays discussing the geopolitical aspects of future oil extraction from the Caspian Sea. Suggests additional sources to consult. Indexed.

Brunet, M.-F., M. Wilmsen, and J. W. Granath, eds. *South Caspian to Central Iran Basins.* Special Publication 312. London: The Geological Society of London, 2009. Discusses the geology, geomorphology, natural history, and seismology of the Caspian Sea. Each of these technical papers includes an abstract and references. Useful to graduate students and researchers.

Croissant, Cynthia. *Azerbaijan, Oil, and Geopolitics.* Commack, N.Y.: Nova Science, 1998. Focuses on historic, current, and future Caspian Sea oil developments, outlining political concerns and potential economic benefits. Illustrated with maps, tables, and figures and includes a bibliography and an index.

Cullen, Robert. "The Rise and Fall of the Caspian Sea." *National Geographic* 195 (May, 1999): 2-35. A well-illustrated article offering insight into the experiences and motivations of people who live in the Caspian Sea region and seek profits from the oil and caviar industries. Useful for high school students to supplement geophysical descriptions.

Dumont, Henri, Tamara A. Shiganova, and Ulrich Niermann, eds. *Aquatic Invasions in the Black, Caspian, and Mediterranean Seas.* Boston: Kluwer Academic Publishers, 2004. Provides an overview of the links between the Black Sea, Caspian Sea, and Mediterranean Sea. Discusses the physical connections and the resulting movement of invasive species due to shipping. Examines the invasion of jellies into the Caspian Sea. Provides a comparison to other invasive species, including invertebrates, plankton, and ctenophores.

Kostianoy, Andrey G., and A. Kosarev. *The Caspian Sea Environment.* Berlin: Springer-Verlag, 2005. Presents a systematic description of the physical oceanography, marine chemistry and pollution, marine biology and geology, meteorology, and hydrology of the Caspian Sea based on observational data. Primarily aimed at specialists in physical oceanography, marine chemistry, pollution studies, and biology.

Marvin, Charles Thomas. *The Region of the Eternal Fire: An Account of a Journey to the Petroleum Region of the Caspian in 1883.* London: W. H. Allen, 1891. Republished by Elibron Classics, Adamand Media Corporation, 2005. Details the nineteenth-century Caspian Sea oil boom, providing historical perspective to modern studies. Illustrated and suitable for high school readers.

Nihoul, Jacques C. J., Peter O. Zavialov, and Philip P. Micklin, NATO Scientific Affairs Division. *Dying and Dead Seas: Climatic versus Anthropic Causes.* Dordrecht: Kluwer Academic Publishing, 2004. The first comprehensive study of dead and dying seas. Presents an analytical look at past, present, and future roles of climatic and anthropic effects relevant to the Caspian Sea.

Rodionov, Sergei N. *Global and Regional Climate Interaction: The Caspian Sea Experience.* Dordrecht: Kluwer, 1994. Describes how the cyclical contraction and expansion of the Caspian Sea affects the environment and the petroleum industry. Suggests possible scientific reasons for rising levels. Contains maps, bibliographical notes, and index.

Stolberg, F., United Nations Environment Program. *Caspian Sea.* New York: United Nations, 2006. One of a series of strategic impact assessments carried out as part of the Global International Waters Assessment Project. Presents the current and historical condition of the Caspian Sea and surrounding region, and suggests various policy options for its ecological reclamation and maintenance.

Zonn, Igor S., et al. *The Caspian Sea Encyclopedia.* New York: Springer-Verlag, 2010. Contains about 1,500 articles that discuss the natural resources, history, and geographical features of the Caspian Sea.

**See also:** Aral Sea; Arctic Ocean; Atlantic Ocean; Beaches and Coastal Processes; Black Sea; Coal; Gulf of California; Gulf of Mexico; Hudson Bay; Indian Ocean; Mediterranean Sea; North Sea; Pacific Ocean; Persian Gulf; Red Sea

# CLIMATE

*Climate is the long-term combined effects of atmospheric variations. Over shorter periods of time, climate is referred to as weather. Climate always refers to a specific geographical location or region and is determined by many factors, including wind belts, topography, elevation, barometric pressure, the movement of air masses, the amount of solar radiation available, proximity to oceanic influences, and planetary and solar cycles.*

## PRINCIPAL TERMS

- **air mass:** a mass of air in the lower atmosphere that has generally uniform properties of temperature and moisture
- **atmosphere:** the envelope of gases surrounding Earth, consisting of five clearly defined stratigraphic regions
- **general circulation models (GCMs):** comprehensive, mathematical-numerical formulas in climate studies that attempt to express in equations the basic dynamics thought to govern the large-scale behavior of the atmosphere
- **greenhouse effect:** the enhanced surface heating effect of solar radiation due to the absorption and redirection of infrared radiation by various gases in the atmosphere
- **parameterization:** the arbitrary assignment of a value to physical processes that occur on scales too small to be resolved by a general circulation model
- **precipitation:** phenomena such as rain, snow, and hail that form through condensation of atmospheric water vapor into liquid and solid forms that subsequently fall toward the surface due to gravity

## SCIENCE OF CLIMATOLOGY

From early times, humans have attempted to classify climates. The ancient Greeks in the sixth century B.C.E. visualized Earth as having three temperature zones based on the sun's elevation above the horizon. (The Greek word *klima*, meaning "inclination," is the origin of the word "climate.") They called these three zones torrid, temperate, and frigid. This system did not consider the differences between climates over land and over water, the effects of topography, and other atmospheric elements such as precipitation.

People tend to confuse weather with climate. The two phenomena are related, but weather is concerned with day-to-day atmospheric conditions, such as air temperature, precipitation, and wind. Climatology is concerned with the mean (average) physical state of the atmosphere, along with its statistical variations in both time and location over a period of many years. In addition to the description of climate, climatology includes the study of a wide range of practical matters determined by climate and the effects and consequences of climatic change. As a result, it has become an interdisciplinary science, relevant to and affected by a wide range of other fields, such as geophysics, biology, oceanography, geography, geology, engineering, economics, statistics, solar system astronomy, and political and social sciences.

For centuries, the content of the atmosphere and the amount of solar radiation reaching Earth's surface have been fairly constant. Until the middle of the 1950's, it was assumed that the climates of Earth were also relatively unchanging, but climate scientists now understand that the climate is never constant. The global climate consists of effects generated by the oceans, the atmosphere, the cryosphere (areas of permanent ice), the land surface, and biomass. The various components of climate are coupled to one another in a nonlinear feedback process by which a change in any one of these factors produces a change or feedback effect in all of the other factors. Therefore, all or any of these processes can change the statistical state of the system called climate.

The science of climatology has developed along two main lines. Regional climatology studies the discrete and characteristic qualities of a particular region of the globe. The second approach of climatology is a physical analysis of the basic relationships existing among various atmospheric elements of temperature, precipitation, air pressure, and wind speed. A third branch now widely used, which originated in the 1960's, is called dynamical climatology. This branch of climatology uses models to simulate climate and climatic change based on the averaged forms of the basic equations of dynamic meteorology.

## CLASSIFICATION OF CLIMATES

The classification of climates is of two types: genetic and empirical. A genetic system is based on air masses and global wind belts, which control climates. The empirical classification system utilizes the observed elements of climate, such as temperature.

An empirical classification system called the Köppen system uses five designated general categories to classify the different climatic regions of the world by their characteristics: tropical forest climates; dry climates; warm, temperate rainy climates with mild winters; cold forest climates with severe winters; and polar climates.

Another empirical classification system, which may be more useful in describing world climates, is based on decreasing temperature and increasing precipitation. This system of classification is convenient for use in many other sciences, as well. The first category is the desert climate, which has the highest temperatures and the lowest precipitation. The next category is the savanna, which may be either temperate or tropical. In either case, this climate is characterized by nearly treeless grassland. The steppes of Eastern Europe and the prairie regions of Canada and the United States are examples of a temperate savanna, with plants adapted to very hot conditions and extremely limited water supply. The grassy plains of western Africa are typical tropical savanna. The next category or climate type is temperate and tropical Mediterranean climate. An example of temperate Mediterranean climate is the coast of southern California. Tropical Mediterranean climate is found typically in the northern region of Africa. The next climate type is the temperate and tropical rainforests. An example of a temperate rainforest is found on the Olympia Peninsula in Washington State, while the Amazon Basin in Brazil is the standard example of a tropical rainforest. The two last climate types are the coldest and have a great amount of precipitation in the form of snow. The first of these two, the taiga, is characterized by great forests of evergreen coniferous trees. Taigas occur only in the temperate zone or at great elevation. Large portions of Canada and Russia are typical of this climate type. The last region is the tundra. This coldest climate type may have small amounts of precipitation. Again, this region is found in either the temperate zone or the highest elevations on Earth. The best examples are the northern Territories of Canada and much of eastern Alaska.

## FACTORS OF CLIMATE CHANGE

Climate is not a steady, unvarying cycle of weather; it fluctuates and varies over a period of time. One of the major factors causing these variations is the role of atmospheric circulation. The atmosphere extends outward from Earth's surface for hundreds of kilometers and consists of five clearly defined regions. The troposphere, the region closest to Earth, extends about 11 kilometers above the surface. It is in this region that most weather phenomena affecting climate take place.

In the tropics there is intense solar heating, whereas in the polar regions there is little solar heating. These differences of heating and cooling of the atmosphere result in a large-scale global circulation of air that carries excess heat and moisture from the tropical areas of intense heating into the higher latitudes, where there is little excess heat and moisture. The great movements of air masses that are produced create and alter various local climates.

Global cloud patterns are also extremely important factors in regulating worldwide climate. Heavy cloud cover reduces the amount of solar radiation reaching Earth's surface, producing a cooling effect. Yet cloud cover also acts as a kind of blanket to reduce the amount of heat normally lost from the ground, by absorbing surface-emitted infrared radiation and directing about half of it back toward the surface. A clear sky in winter always results in colder temperatures at the surface than when there is cloud cover. Even so, only the solar heat that already reached Earth's surface is able to be kept under this protective cover. Extended cloud cover will result in less solar heat reaching the surface, leading to a cooler climatic condition.

The activity of the climate on various parts of the globe is also strongly influenced by the movement of ocean currents and global wind belts. An oceanic effect termed El Niño is periodically responsible for dramatic climate changes, possibly on a global scale. Scientists are not completely certain of the initial cause of this phenomenon, but it appears to be dependent upon the accumulation of warmer surface water driven by prevailing surface winds.

It is important to remember that climate has never been stable and unchanging. Many factors can cause changes in local and even global climate; when one factor changes in the level of its activity, it inevitably modifies or influences other factors affecting climate.

## FACTORS RESULTING FROM HUMAN ACTIVITY

Some factors causing climate change are of natural origin, while many others result from human activity. The globe's tropical rainforests are vital to maintaining the equilibrium of the carbon dioxide balance in the global atmosphere. Heavy deforestation in South America and Asia has endangered this natural means for maintaining the balance of atmospheric carbon dioxide by removing great numbers of the trees that remove vast quantities of carbon dioxide from the atmosphere. At the same time, industrial activity and automobile emissions release massive quantities of carbon dioxide into the atmosphere, contributing to the buildup of carbon dioxide in the atmosphere and measurable increases in global temperatures. Carbon dioxide absorbs heat emitted from Earth's surface and reradiates much of it back toward the surface, creating a greenhouse effect that normally raises the average global temperature to a level that supports the existence of life. Even a small increase of 2 degrees Celsius in the average global temperature, however, would have a potentially disastrous effect on climatic conditions. Global warming would mean that as the oceans warmed, they would expand appreciably. In addition, ocean temperature and the direction of upper-level winds are the main factors in the development of hurricanes. Some scientists argue that the warming of the oceans has already increased the frequency as well as the power and destructive force of hurricanes.

One of the main dangers of a warming climate is flooding from rising sea levels. Scientists have calculated that if the average temperature increases by about 3.7 degrees Celsius, the sea levels could increase by approximately 81 centimeters, enough to flood huge areas of unprotected coastal land. Nearly 30 percent of all human beings live within about 60 kilometers of a coastline. A rise in sea level of only 50 centimeters would have a profound effect, as it would result in the flooding of many of the world's most important cities and ports.

Large urban areas alter the local climate through the "heat island effect," by which the average temperature of an urban area is raised significantly, forming rising air columns that would not otherwise exist to influence the activity of the larger air mass. Various forms of pollution over large urban areas such as Los Angeles, Tokyo, and London have also caused local climate changes in these areas. Concentrations of ozone that build up at ground level from industrial and automobile exhausts cause extreme eye and lung irritation. Oxides of nitrogen, carbon dioxide, sulfur dioxide, and particle-laden air, which can cause a depletion in the amount of total solar radiation, also accumulate and alter climate conditions over local urban areas. Changes in surface characteristics, such as converting forests and prairies to agricultural fields, building cities, damming rivers to create lakes, and spilling oil at sea, all affect local climate conditions. Usually, these surface changes appear to have only a minor global effect, though they do alter local conditions significantly, but the cumulative effects of these minor changes are not well understood and should, therefore, not be overlooked.

There is now widespread national and international interest in how human activity is fundamentally altering the global climate. Organizations are seeking to inform the general public regarding society's role in maintaining a healthy and healthful climate. The Climate Institute in Washington, D.C., sponsors world-class conferences where scientists and government leaders from many nations gather, providing a forum for information interchange among climate researchers and analysts, policymakers, planners, and opinion makers.

**DEPARTURES FROM THE "NORM"**

For centuries, the content of the atmosphere and the amount of solar radiation reaching Earth's surface have been fairly constant. Scientists now accept that climate is never constant, and that it is the departures from the expected "norm" that provide the greatest insight into climatic processes and have the greatest human impact.

The ice ages were a decided departure from the "norm." Drought in central North America during the 1920's and 1930's resulted in the Dust Bowl disaster, which disrupted human life and crop production in the fertile middle region of the continent. The impact of the devastating drought of the Sahel region in Africa led, in the early 1970's, to increased desertification. Loss of the anchovy fisheries along the Peruvian coast in the 1970's (probably the result of the oceanic fluctuations known as El Niño), severe droughts in the United States' farmbelt in the 1980's, and large-scale flooding in many parts of the world all attest to the fact that dramatic climatic fluctuations and changes are the true norm. It is now understood by climatologists that climate modification may even result in microclimates as small as portions of a backyard garden.

## DATA-GATHERING TOOLS

Because the science of climatology is the study of weather patterns for a geographic area over a long span of time, records of scientific data on the weather are valuable for interpreting climate trends. Because of the international agreements and standardized procedures concerning climate that were adopted in 1853, about 100 million observations have been taken from ships since then. The quantities recorded were sea-surface temperature, wind direction and speed, atmospheric pressure and temperature, the state of the sea, and cloudiness.

A great variety of instruments are used by meteorologists in gathering weather information. The net radiometer, pyranometer, and Campbell-Stokes sunshine recorder are used for measuring radiation and sunshine. The sling psychrometer is used to determine relative humidity. For remote sensing, a humidity gauge is used in which the variations of electrical transmission by the device depend on the passage of an electrical current across a chemically coated strip of plastic that is proportional to the amount of moisture absorbed at its surface. This type of gauge is used in radiosondes for upper-air observations.

Measurement of wind velocity, its speed and direction, involves a simple device called an anemometer. Anemometers have four-bladed propellers that are driven by the wind to record wind speed. When the propellers are mounted at right angles, an anemometer responds to air movement in three dimensions, the sum of which reveals its true direction. Weather pilot balloons, or "pibals," are released and then tracked visually using a theodolite, or a right-angled telescopic transit that is mounted in a way to make possible the reading of both azimuthal (horizontal) and vertical angles. The progress of the balloon is then plotted minute by minute, and the direction and speed are computed for each altitude. This technique does not work, however, for upper-air wind observations when there is a low cloud cover. Under these conditions, "rawin" observation is used. Here, a metal radar target is attached to the balloon and tracked by a radar transceiver or radio theodolite until the balloon breaks or is out of range.

Precipitation is measured in several ways. Rain measurements are made using a rain gauge. An improved version is the tipping-bucket gauge, which has a small divided metal bucket mounted so that it will automatically tip and empty measured quantities of rainfall. This tipping closes electrical contacts that activate a recording pen. Another type of improved rain gauge is a weighing-type gauge. Here, as precipitation falls on a spring scale, a pen arm attached to the scale makes a continuous record on a clock chart. Snow depth is determined by averaging three or more typical measured depths. Where snow depths become very great, graduated rods may be installed in representative places so that the depth can be read directly from the snow surface. Heat and air pressure are measured by thermometers and barometers, most of which record and transmit data automatically using digital electronics.

## AERIAL PHOTOGRAPHS AND CORE SAMPLES

Specially equipped airplanes and satellites can obtain previously inaccessible data of immeasurable value in determining climate changes. The National Aeronautics and Space Administration (NASA) has supplied a modified U-2 plane to carry instruments to an altitude of 20 kilometers for seven-hour flights up to 80 degrees north latitude. A DC-8 is also used at the same time because of its extended range. The Nimbus 7 satellite is also employed. The U.S. satellite Landsat continually photographs Earth's surface. These photographs are radioed back to Earth stations and are computer-enhanced to provide visual information about such things as the amount and type of vegetation, precipitation, and underground waterways. The enhanced photographs also provide information about changing conditions that may cause changes in local and global climates. More recent satellites using much more sensitive instrumentation now provide researchers with real-time observations of essentially all points of Earth's atmosphere and surface, and have rendered much of the earlier technology obsolete. One has only to access satellite images of Earth over the Internet to appreciate the observational capabilities that are available.

In order to gain information about the relationship between carbon dioxide and global temperature, scientists have drilled deep into Arctic and Antarctic ice to obtain core samples of the polar ice pack. By studying the amount of carbon dioxide trapped in air bubbles in very old layers of ice pack, scientists have been able to chart the ebb and flow of carbon dioxide in the atmosphere. Fossilized plant tissues indicate how warm the air was during the same period as the bubbles in the core samples. This clue helps provide a picture of the warming and cooling trends that have occurred in the past, and makes comparisons with current climate conditions possible.

## GENERAL CIRCULATION MODELS

Climatologists have found mathematical or computer-modeling methods very useful, primarily with regard to general circulation models (GCMs). The most important part of a GCM is the parameterization of a wide range of physical processes too small to be resolved by the model. These may include all of the turbulent fluxes that occur in the surface boundary layer as well as the occurrence of convection, cloudiness, and precipitation. These relatively small processes must be parameterized in terms of the variables that are capable of being resolved by GCMs. It is primarily this parameterization that makes GCMs different from one another and leads to variations between the results of different GCM models. The value of the models, however, depends mostly on the accuracy with which the model actually simulates natural conditions. Assumptions sometimes must be made in modeling, and these assumptions can give a false picture under certain conditions. Models, therefore, are intensely tested against actual weather data as newer methods of observation and more data become available, to ensure the accuracy of the models.

As newer methods for measurement, comparison, and prediction become available, greater understanding of global climate change, and more accurate predictions, will lead to better understanding of the factors affecting the climate.

*George K. Attwood*

## FURTHER READING

Ackerman, Steven, and John Knox. *Meteorology.* 3d ed. Sudbury, Mass.: Jones & Bartlett Learning, 2011. An introductory-level college textbook intended for undergraduates not majoring in this field. Presents an overview of weather and climate phenomena.

Ahrens, C. Donald *Meteorology Today: An Introduction to Weather, Climate and the Environment.* 8th ed. Belmont, Calif.: Thomson Brooks-Cole, 2007. One of the most widely used and authoritative introductory textbooks for the study of meteorology and climatology. Explains complex concepts in a clear, precise manner, supported by numerous images and diagrams.

Aquado, Edward, and James E. Burt. *Understanding Weather and Climate.* 5th ed. Upper Saddle River, N.J.: Prentice Hall, 2009. Discusses meteorology and climatology concepts with reference to common, everyday events. Presents conclusions from the IPCC as well as many other scientific studies on climate change. Examines weather events; the structure and dynamics of the atmosphere; and the past, present, and future climate of Earth.

Bolius, David. *Paleoclimate Reconstructions Based on Ice Cores: Results from the Andes and the Alps.* Berlin: SVH-Verlag, 2010. Presents a study of climate change using samples taken from ice cores in the Andes and Alps. Includes sampling methodology and, chemical analysis, and documents the history of anthropogenic air pollution. Best suited for graduate students and professional geologists and paleoclimatologists.

Curry, Judith A., and Peter Webster. *Thermodynamics of Atmospheres and Oceans.* San Diego, Calif.: Academic Press, 1999. Offers a look at the effects of the interaction between oceans and the atmosphere on weather patterns and climatic changes. Provides good insight into the role that atmospheric thermodynamics play in meteorology. Illustrations, maps, and index.

Fagan, Brian M., ed. *The Complete Ice Age: How Climate Change Shapes the World.* New York: Thames & Hudson, 2009. Discusses the development and cycle of the ice age. Examines the effects of the ice age on human evolution and includes information on global warming and interglacial periods. Written in a manner accessible to high school students.

Graham, N. E., and W. B. White. "The El Niño Cycle: A Natural Oscillator of the Pacific Ocean Atmosphere System." *Science* 240 (June 3, 1988): 1293-1302. Discusses the oceanic phenomenon known as El Niño. The journal *Science* provides excellent scientific materials on the cutting edge of discovery.

Hopler, Paul. *Atmosphere: Weather, Climate, and Pollution.* New York: Cambridge University Press, 1994. Offers a wonderful introduction to the study of Earth's atmosphere and its components. Explains the causes and effects of global warming, ozone depletion, acid rain, and climatic change. Illustrations, color maps, and index.

Lutgens, Frederick K., Edward J. Tarbuck, and Dennis Tasa. *The Atmosphere: An Introduction to Meteorology.* 11th ed. Upper Saddle River, N.J.: Prentice Hall, 2010. An excellent introduction to and description of the atmosphere, meteorology, and weather patterns. A perfect textbook for the reader new to the study of these subjects. Color illustrations and maps.

Marshal, John, and R. Alan Plumb. *Atmosphere, Ocean and Climate Dynamics: An Introductory Text.* Burlington, Mass.: Elsevier Academic Press, 2008. Offers an excellent introduction to atmospheres and oceans. Discusses the greenhouse effect, convection and atmospheric structure, oceanic and atmospheric circulation, and climate change. Best suited for advanced undergraduates and graduate students with some background in advanced mathematics.

Pierrehumbert, Raymond T. *Principles of Planetary Climate.* New York: Cambridge University Press, 2011. Written for the undergraduate or graduate student, as it guides the reader toward original research and primary source material. Covers the fundamental principles of Earth's climate, climates around the solar system, and climates of past and present. Provides examples of current research practices on planetary climate beyond the study of paleoclimatology.

Schneider, Stephen H., and Terry L. Root. *Wildlife Responses to Climate Change: North American Case Studies.* Washington, D.C.: Island Press, 2002. Provides a number of case studies presenting information on the repercussions of climate change on the biological world. Species discussed include sachem skipper butterflies, grizzly bears, and the two-lobe larkspur. Intended to draw the public into the global warming discussion.

Stevens, William Kenneth. *The Change in the Weather: People, Weather, and the Science of Climate.* New York: Random House, 2001. Describes various natural and human-induced causes of changes in the climate. Includes a twenty-page bibliography and an index.

**See also:** Air Pollution; *AR4 Synthesis Report;* Atmosphere's Global Circulation; Atmosphere's Structure and Thermodynamics; Atmospheric and Oceanic Oscillations; Atmospheric Properties; Barometric Pressure; Climate Change Theories; Climate Modeling; Clouds; Drought; Earth-Sun Relations; Floods; Global Energy Transfer; Greenhouse Effect; Hurricanes; Hydrologic Cycle; Ice Ages and Glaciations; Lightning and Thunder; Long-Term Weather Patterns; Monsoons; Ozone Depletion and Ozone Holes; Precipitation; Remote Sensing of the Atmosphere; Satellite Meteorology; Seasons; Severe Storms; Tornadoes; Tropical Weather; Van Allen Radiation Belts; Volcanoes: Climatic Effects; Weather Forecasting; Weather Forecasting: Numerical Weather Prediction; Weather Modification; Wind

# CLIMATE CHANGE THEORIES

*Earth's climate is a complex system in constant flux. An understanding of how those changes occur has emerged in recent decades. This insight has allowed for a better understanding of Earth's history and has helped to mitigate the dangers of modern climate change.*

## PRINCIPAL TERMS

- **albedo:** the reflecting power of a substance
- **carbon-oxygen cycle:** the process by which oxygen and carbon are cycled through Earth's environment
- **cosmic ray:** high-energy subatomic particles that are produced by phenomena in space, such as supernovae
- **eccentricity:** the departure of an ellipse from circularity; less circularity means greater eccentricity
- **El Niño:** an eleven-year weather cycle in the Western Hemisphere that creates alternating wet and dry periods
- **greenhouse effect:** the process by which some gases trap heat on Earth
- **obliquity:** the angle of tilt between the earth's rotational axis and an axis perpendicular to the plane of its orbit
- **precession:** a change of the axis of rotation in a rotating body or system
- **proxies:** traces of ancient environments that reveal details, such as climatic data, about those environments
- **sedimentary rock:** rock formed by the repeated deposition of sediment in a body of water or by the layering of material on land

## CLIMATE CHANGE THEORIES

An understanding of Earth's climate as a dynamic system originated in the eighteenth century from discoveries in geology and paleontology. Continued research in the twentieth century provided an increasingly complex image of how Earth systems are interrelated. An understanding of future anthropogenic (human-caused) climactic shifts is thus based on an understanding of Earth's climate mechanisms, which are derived from matching historical data with the geologic record.

## METHODOLOGY

Climatology employs many methods, including tree ring analysis, ice core sampling, sediment core sampling, and satellite observation. Tree ring analysis, or dendroclimatology, works by examining the rings of trees, which depict the growth of a tree during its growing season. Rings are wider under favorable growing conditions and are narrower under less favorable growing conditions. Dendroclimatology works only to the point at which preserved trees can be found; these trees have provided high-resolution data as far back as 11,000 years ago.

Another common method in climatology is to examine ice cores from glaciers. Ice cores are shafts of ice pulled from an ice cap by drilling into that cap with a core drill. Cores are commonly taken from Greenland and Antarctica, but cores also can be taken from mountaintop glaciers. Additionally, ice cores trap air; however, that air is not always the same age as the ice because rates of compression of snow to ice can be low. The oldest data obtained are less than 800,000 years old, but speculation exists that samples could be obtained from 1.5 million years ago. Given the age of ice caps, the oldest records are in Antarctica. Ice cores also can catch useful proxies, such as iridium, pollen, and volcanic ash.

Another common method for climate data collection is sediment drilling from lakes and sea floors. Like their icy counterparts, these sediment cores retain similar data, but they also can record fossil data. Seafloor cores can theoretically go back 200 million years.

Terrestrial rocks go back to about 3.5 billion years, with some isolated instances 1 billion years older. Analysis of their constituent materials and minerals can give an idea of the environment in which they were formed. Ancient sedimentary rocks can give ideas of water levels and temperature, while other stones record the shores and flows of lakes and rivers. Along with these markers, the fossil record also often gives an idea of ancient Earth conditions.

To study the current climate, networks such as the Earth Observation System (EOS) of the National Aeronautic and Space Administration gather various sorts of data. The EOS includes more than fifteen satellites, all designed to monitor different aspects of Earth's environment. The satellite *Aqua* observes the water cycle; *CloudSat* measures cloud altitude and properties.

## MILANKOVITCH THEORY

The Milankovitch theory is one of the most comprehensive long-term climate change theories available. The theory considers how Earth's movement affects climate. The theory, discovered by a Serbian engineer during World War I, considers the processes of orbital eccentricity, obliquity, and precession, allowing for scientists to explore the impact of these processes' individually and in combination.

Earth's orbit is not a perfect circle; it is an ellipse that varies in eccentricity in time. Changes in the eccentricity of the earth from gravitational interactions with other planets mean that, on occasion, the earth is narrower, which results in greater difference in seasons, and at other times, the earth is more circular, which leads to a lesser difference in seasons. These changes occur because of Earth's proximity to the sun.

The effect of the variance of eccentricity can be great; at its most elliptical, the earth receives about 23 percent more radiation at its perihelion (the point closest to the sun) than it does at its aphelion (the point farthest from the sun). The eccentricity ranges between about 0.0034 and 0.058. This process takes about 413,000 years. By interaction with other cycles the process combines to create a cycle of about 100,000 years.

The obliquity of the ecliptic (the tilt of Earth's axis) varies between about 22.1 to 24.5 degrees in about 41,000 years. When the obliquity is greater, the difference in solar radiation received is also greater, meaning a greater difference in the seasons. When the obliquity is lower, less difference occurs between the seasons. It is thought that low obliquity favors ice ages. The cycle now is heading toward its lowest obliquity and, therefore, toward an ice age. It is expected to reach lowest obliquity in about 8,000 years; however, anthropogenic climate change has mitigated this trend.

The axis of Earth's orbit also precesses in a cycle of about 26,000 years. When the earth's axis tilts toward the sun at perihelion, the hemisphere pointing toward the sun has a greater difference in seasons that year. When the axis is not pointing toward the sun at a solstice, but instead does so at an equinox, less difference in the seasons occurs between the hemispheres.

The orbital ellipse itself also precesses and operates with changes in eccentricity to alter the length of the seasons. Orbital inclination was not an original part of Milankovitch's work but has since been added. Earth's orbit is pulled up and down relative to the plane of the solar system by the gravitational effects of Jupiter. Orbital inclination works on a time scale of about 100,000 years and is thought to have an effect because of the dust clouds and other debris altering the amount of light reaching the earth.

The Milankovitch theory does have its weaknesses: Some trends have been observed but not predicted, and some trends have been predicted but not observed. On the whole, however, the theory fits the observed periodicities of climate.

## COSMIC RAYS

Another space-based phenomenon that could affect terrestrial climate is cosmic rays. Though disputed, there seems to exist evidence for such a proposition.

Cosmic rays ionize the atmosphere and cause increased cloud cover, resulting in a lower temperature. This would be most noticeable on a long-time scale as the earth moves through the spiral arms of the galaxy and their supernovae density. Earth moves through a spiral arm about once every 135 ± 25,000,000 years, causing flux in cosmic rays with a period of 143 ± 10,000,000 years, matching the cycle of 145 ± 8,000,000 years. An absence in ice ages on earth between 2 billion and 1 billion years ago coincides with a drop in star formation, giving the idea further credence.

## GREENHOUSE GASES AND THE CARBON-OXYGEN CYCLE

Another theoretical framework for climate shifts is examination of the carbon-oxygen cycle, which works in tandem with the Milankovitch cycle and is used to explain some points in which the climatic reaction far exceeds the change in the cycle. Because carbon is a greenhouse gas, it has great influence on the climate. Disruption of this mechanism is one of the main causes of anthropogenic climate change.

A good example of how the carbon-oxygen cycle figures into climate is the *Azolla* event of 49 million years ago. At that time, the Arctic Ocean was warm and closed off, such that it started to collect a layer of freshwater. It is thought that the fresh-water plant *Azolla* grew and covered the Arctic Ocean. After the *Azolla* started to die, it sank to the ocean bottom; the freshwater and seawaters did not mix, which sequestered the carbon. It is estimated that this single phenomenon could have caused the 80 percent estimated drop in atmospheric carbon dioxide, dramatically reducing the greenhouse effect and turning the climate from a much warmer one to a far cooler one.

## GREENHOUSE AND ICEHOUSE EARTH

Greenhouse Earth is characterized by high temperatures, which can sometimes reach the polar regions. Eighty percent of Earth's history has featured this climate. The rest of the climate has been Icehouse Earth. (Today's Earth, more specifically, is in an interglacial era. Interglacials make up 20 percent of the time in the average glacial period.)

A third state, called Snowball Earth, occurred when the entire planet was believed to have frozen over, though there is some debate about whether or not there was an open or seasonally open ocean near the equator. The state was most likely ended by buildup of volcanic greenhouse gases. It is thought that the transition of icehouse to greenhouse was caused by plate tectonics.

## CONTINENTAL DRIFT

An important factor in determining climate is continental placement because continents shift the workings of various cycles. It is thought that the closing of the Isthmus of Panama, the opening of the Drake Passage, and the opening of the Tasmanian Gateway helped Earth's climate transition from "greenhouse" to "icehouse" by changing the flow of ocean currents. The surrounding seas were open to circulation around Antarctica and less able to circulate at the equator. In the south, this led to the creation of the Antarctic Circumpolar Current, which kept warm water from Antarctica, allowing it to freeze. This led to increased albedo, which led to more freezing. A positive feedback loop like this also is thought to have led to Snowball Earth.

## IMPACT

Increased greenhouse gases has caused a drop in the amount of ozone, a gas that shields the planet from ultraviolet radiation. When the temperature drops, as in an ice age, ozone is depleted. When depleted, the stratosphere gets still colder, further depleting the ozone. Ozone also cools the troposphere. Additionally, low-altitude ozone is a greenhouse gas.

An understanding of climate change not only illuminates historical processes, but also gives an idea of what occurs with anthropogenic climate change. Understanding Earth's climate provides warning of climate change and also aids in finding solutions to related climate concerns.

*Gina Hagler*

## FURTHER READING

Battarbee, R. W., and H. A. Binney. *Natural Climate Variability and Global Warming: A Holocene Perspective.* Malden, Mass.: Blackwell, 2008. An overview of climate change and modern anthropogenic change in the context of historical shifts. Makes clear overall patterns and recent changes to the system. Recommended for those with a strong background in the material.

Bradley, Raymond S. *Paleoclimatology: Reconstructing Climates of the Quaternary.* San Diego, Calif.: Academic Press, 1999. Details the process of paleoclimatology through the example of reconstructing the Quaternary period. While advanced, it is a general overview and is suitable for all readers.

Cowie, Jonathan. *Climate Change: Biological and Human Aspects.* New York: Cambridge University Press, 2007. A study of climate change that covers anthropogenic processes. Explains the phenomena and covers Earth's history. Highly readable and recommended for all levels.

Cronin, Thomas M. *Principles of Paleoclimatology.* New York: Columbia University Press, 1999. A detailed explanation of the methods and means of the study of paleoclimatology. Also deals with issues in the modern era. As a guide to methodology, it is most suitable for college students.

Gornitz, Vivien. *Encyclopedia of Paleoclimatology and Ancient Environments.* New York: Springer, 2009. Covers all aspects of climate change across the history of the planet. A comprehensive work covering methodology, effect, periods, and theory.

Wilson, Elizabeth J., and David Gerard. *Carbon Capture and Sequestration: Integrating Technology, Monitoring, and Regulation.* Ames, Iowa: Blackwell, 2007. An introduction to carbon sequestration as a system. Covers technological, political, and theoretical considerations. An accessible text.

**See also:** *AR4 Synthesis Report*; Atmospheric Properties; Bleaching of Coral Reefs; Carbon-Oxygen Cycle; Climate Modeling; Earth-Sun Relations; Greenhouse Effect; Ice Ages and Glaciations; Icebergs and Antarctica; Impacts, Adaptation, and Vulnerability; IPCC; Long-Term Weather Patterns; Recent Climate Change Research; Severe and Anomalous Weather in Recent Decades; Tropical Weather

# CLIMATE MODELING

*Climate models are computer programs utilizing the basic laws of physics and data from atmospheric measurement to predict the development and evolution of regional or global climate. One of the primary aims of modern climate modeling is to investigate the phenomenon of global warming and to determine to what extent human activity is contributing to global warming trends.*

## PRINCIPAL TERMS

- **carbon dioxide:** gaseous combination of carbon and oxygen that functions as one of the most important greenhouse gases
- **climatology:** branch of science that studies climate and climate change and chemical and physical forces within the atmosphere
- **convection:** patterns of molecular movement within liquids generally related to heat
- **deforestation:** removal of forest areas or stands of trees, including for commercial and agricultural development
- **fluid dynamics:** branch of physics that deals with the movement of fluids, including gases and liquids
- **general circulation model:** climate model that measures variations across three dimensions
- **ozone:** gaseous compound composed of molecules containing three atoms of oxygen bonded together
- **parameterization:** process of setting parameters for a measurement or other analysis system
- **precipitation:** process by which atmospheric water vapor condenses and returns to Earth

### DATA FOR CLIMATE MODELS

Climate models are computer programs that use atmospheric and temperature data to calculate the state and development of the climate. Climate models are designed to simulate the fundamental forces of nature, which are based on the laws of physics. These physical laws are translated into differential equations, which can be used to simulate climate variation when supplied with relevant data.

Climate scientists use equations from the study of fluid dynamics to describe the movement and formation of atmospheric fluids and combine that study with input from the chemical reactions involving energetic exchange between the fluids of the atmosphere and the hydrosphere. One of the most basic factors considered when creating a climate model is the exchange of energy between Earth and space.

Solar radiation impacting the earth is the primary source of heat and the main force that drives climate variation across the globe. To create models of the climate, atmospheric scientists need to assess the amount of electromagnetic radiation distributed across the earth from the sun.

Solar radiation is concentrated in areas where the sun is directly above the earth, such as along the equator. High levels of solar radiation heat the land, water, and atmospheric gases and cause low-density pockets to form within the atmosphere. Circulating currents are then formed when high-density pockets of air meet low-density pockets of air. These currents develop into winds and weather patterns and help to distribute solar energy across the earth. As pockets of air gain and lose density, they also gain and release water vapor in the form of precipitation; this leads to rainfall and snowfall over the terrestrial environment. Climate models also simulate the development of precipitation patterns by creating equations based on relative levels of pressure, solar radiation, and the chemical relationships among atmospheric gases.

Climate models also must take into account levels of atmospheric gases and their relationship to solar radiation, the formation of precipitation, and the sources of atmospheric gases in the marine and terrestrial environments. The ozone layer is a thin membrane of gases in the upper atmosphere largely consisting of ozone, which is a molecule that contains three atoms of oxygen. The ozone layer blocks harmful solar radiation, making ozone concentration an important factor to consider when attempting to model any climate.

Greenhouse gases, such as carbon dioxide and methane, also are important factors in climate models. Concentrations of these gases help to determine how much of the heat that enters the environment from solar radiation remains within the atmosphere and how much of this energy radiates back into space.

## Types of Climate Models

One of the most basic types of climate models is the energy balance model (EBM), which simulates the global distribution of radiation and the movement of heat energy from the equator to the poles. This basic type of model is one dimensional, measuring the distribution of energy across the latitude of the earth by measuring the temperature of the earth at the surface.

Another type of basic one-dimensional model is the radiative-convective model (RCM), which measures the distribution of energy through convection, which is measured by the heat distributed through a vertical column of air. RCMs ignore horizontal distribution of heat across the earth's surface and focus on the heat gradient within the vertical dimension of the atmosphere. RCM models are useful for measuring the effect of increased greenhouse gases on vertical heat distribution; they are most effective when measuring climate variation in regional areas.

RCM and EBM models are useful for calculating climate change in regional areas, but they are not sufficient for effectively predicting global climate change. By combining the data from RCM and EBM models, atmospheric scientists create statistical dynamical models (SDMs), which are useful for modeling the development of weather patterns and storm systems. For instance, SDMs, which are used to predict the development of tropical cyclones, take into account changes in pressure, humidity, and wind currents and often use data from one-dimensional RCM and EBM systems.

General circulation models (GCMs) are three-dimensional models that track climate change in the three basic dimensions and also through time. Generally, GCMs are designed to measure climate in either marine or terrestrial ecosystems. These two models can be combined to form an atmospheric and oceanic general circulation model (AOCGM), which is the basis of most modern climate modeling programs.

AOGCMs are important for modeling global changes in temperature, precipitation patterns, and other general patterns but lack the accuracy needed for modeling specific changes in regional climate. For this reason, atmospheric scientists still use limited models, such as RCMs and EBMs, to study localized climate variation and then use these data to create AOGCM systems.

## Applications of Climate Models

Many researchers studying climate change are involved in the ongoing debate over global warming, specifically the degree to which human activity and environmental modification contribute to the warming trends. GCMs and AOGCMs are helping scientists to understand the significance of current data regarding climate change.

Many of the causal and contributing factors involved in climate change are poorly understood, and it is hoped that GCMs and other climate models can help scientists to understand how factors including increases in greenhouse gas emissions, atmospheric ozone concentrations, and deforestation contribute to climate change. By refining climate models, researchers also hope to discover unknown elements that might influence climate models.

Climate models also predict weather patterns. Regional climate modeling, for instance, can allow scientists to predict the development of tropical storms and other weather systems that may pose a threat to human settlements. Computer-generated climate models have been used to refine warning and alert systems that help protect communities from potential storms and other weather-related threats.

Climate models are used primarily to study the interaction between humans and climate change, but they also are useful in ecological research, helping ecologists and biologists understand the relationship between nonhuman animals and climate change. These data help researchers create systems that can be used to preserve and protect wildlife and ecosystems from climate variation.

## Climate Modeling Organizations and Groups

The National Aeronautics and Space Administration conducts climate modeling research through its Goddard Institute for Space Studies (GISS). Researchers at GISS use RCM, EBM, and SDM systems for regional and localized analyses and also develop detailed GCM calculations for studying individual variations within the oceanic and terrestrial systems. Data from these localized systems are integrated to create AOGCM models using the most advanced technology available from international participants.

The primary interest for the GISS climate modeling system is investigating the impact of human activity on climate change, including the emission of

123

greenhouse gases and deforestation. GISS research focuses on climate sensitivity, which attempts to measure the climatological response to various perturbations, such as local or widespread increases in gas emissions or other similar variables. GISS researchers also help develop GCM models that are used by other organizations to conduct climate research. In this capacity, GISS works on parameterization, which is the process of setting parameters for the various inputs used in calculating climate variation and development.

The Atmosphere-Ocean Dynamics Group (AODG) of Cambridge University's Department of Applied Mathematics and Theoretical Physics conducts many climate modeling programs utilizing GCM and AOGCM models and uses many regional climate models. Cambridge researchers focus on fluid dynamics, including the movements of atmospheric gases and water in the oceans and in precipitation systems.

One of the AODG's primary ongoing programs deals with the calculation of ocean turbulence and its effect on climatic patterns. Other research programs at the university study global fluid dynamics of the atmosphere as a whole, using AOGCM data and regional data to study the mathematical principles behind fluid exchange in planetary atmospheres.

The Met Office Hadley Centre is one of the most prominent organizations for climate change research in the United Kingdom and is funded by the U.K. Department of Energy and Climate Change and the Department of Environment, Food, and Rural Affairs. The Hadley Centre focuses on creating programs aimed at modeling the climate of the earth from the twentieth century and utilizing this information to create detailed models predicting climate development in the twenty-first century. Researchers utilize a variety of GCM and AOGCM programs to complete their climate models and have been leading contributors of data used in government policy meetings regarding climate change and global warming.

The World Climate Research Programme (WCRP), under the leadership of the International Council for Science, is one of the largest global organizations contributing to research on climate modeling systems. The WCRP funds and coordinates research programs with meteorological organizations, universities, and professional research organizations studying climate change and creating new systems

and methods for climate modeling technology. The WCRP is interested both in investigating the fundamental processes governing climate development and in studying human interaction with the climate and the effects of human activity on climate change. Many research programs sponsored or conducted by WCRP members focus on global warming and human activity.

*Micah L. Issitt*

**FURTHER READING**

Dodson, John. *Changing Climates, Earth Systems, and Society.* New York: Springer, 2010. Advanced student text covers climate change and Earth systems as related to human culture. Includes a detailed discussion of climate modeling as it applies to researching climate change.

Henderson-Sellers, Ann, and Kendal McGuffie. *The Future of the World's Climate.* Boston: Elsevier, 2012. Complex overview of climate science, including reviews of global warming research. Includes a discussion of GCMs and other climate modeling techniques and their role in investigating climate change.

Kiehl, J. T., and V. Ramanathan. *Frontiers of Climate Modeling.* New York: Cambridge University Press, 2006. Complex text on climate modeling written for working professionals and students. Covers the basic elements of climate modeling techniques, including one-, two-, and three-dimensional models.

Neelin, David J. *Climate Change and Climate Modeling.* New York: Cambridge University Press, 2011. Comprehensive text covering many aspects of climate modeling and climate change research. Covers the history of GCM climate modeling and a variety of other issues.

Phalen, Robert N. *Air Pollution Science.* Burlington, Mass.: Jones and Bartlett, 2011. Complex text covering issues in air pollution. Includes a discussion of climate modeling in terms of measuring and studying air pollution in the atmosphere.

Trenberth, Kevin E. *Climate System Modeling.* New York: Cambridge University Press, 2005. Text covering computer-aided climate modeling including reviews of a variety of data management systems and climate modeling techniques. Includes description of how to use climate modeling to investigate global warming.

Washington, Warren M., and Claire L. Parkinson. *An Introduction to Three Dimensional Climate Modeling*. 2d ed. South Orange, N.J.: University Science Books, 2005. Detailed text covering modern methods used to make three-dimensional models of climate, including GCM methods. Discusses factors used to measure and utilize atmospheric agents.

**See also:** *AR4 Synthesis Report*; Atmospheric Properties; Carbon-Oxygen Cycle; Climate Change Theories; Earth-Sun Relations; Greenhouse Effect; Impacts, Adaptation, and Vulnerability; IPCC; Long-Term Weather Patterns; Observational Data of the Atmosphere and Oceans; Recent Climate Change Research; Remote Sensing of the Atmosphere; Remote Sensing of the Oceans; Severe and Anomalous Weather in Recent Decades; Tropical Weather

# CLOUDS

*Clouds provide an indication of the current weather and a forecast of weather to come as well as information regarding climate and other aspects of the atmosphere. They are also a resource for the investigation of the dynamic interactions of solid, liquid, and gaseous substances.*

## PRINCIPAL TERMS

- **cirrus:** trailing or streaky clouds, at altitudes ranging from 5 to 13 kilometers, that are feathery or fibrous in appearance
- **condensation:** the transformation of a substance from the vapor state to the liquid state; atmospheric condensation occurs when droplets of liquid form (or condense) around small particles in the atmosphere
- **convection:** the transmission of heat by cyclic mass transport within a fluid substance; the movement of warmer, less dense material that rises as cooler, denser material sinks
- **cumulus:** clouds with vertical development rising from a seemingly flat base, often appearing as fluffy masses, at altitudes ranging from ground level to 6 kilometers above the ground; sometimes called heap clouds
- **radiation:** the transfer of energy emitted from one body through a transparent medium to another body, as occurs when light and heat energy from the sun impinge on Earth
- **stratus:** sheet or layer clouds, at altitudes ranging from 2 to 6 kilometers above the ground (altostratus, or middle) or from 0 to 2 kilometers above the ground (stratocumulus, or low)
- **supersaturation:** a state in which the air's relative humidity exceeds 100 percent, the condition necessary for vapor to begin transformation to a liquid state

## CLOUD FORMATION AND LEVEL

Clouds are the single most obvious feature of the atmosphere. The formation of clouds is essentially a two-part process. The heat provided by the radiation from the sun warms liquid water on Earth's surface. This moisture evaporates and the vapor rises, cooling as it reaches higher altitudes. Numerous particles in the atmosphere provide surfaces to which the water molecules adhere and become nuclei for the formation of water droplets or ice crystals, depending on the temperature of the atmosphere. The particles are usually composed of combustion products, meteoritic dust, volcanic material, soil, or salt. When a sufficient number of droplets or crystals have formed, a visible cloud has come into existence. The size, shape, and growth patterns of the cloud will depend upon the moisture and the atmospheric energy available during its formation.

Although clouds appear to be relatively stable, they are always in motion—rising or falling, expanding or shrinking—depending on the temperature and humidity of the surrounding air and the direction of the winds. Eventually, clouds either precipitate as the water droplets in the form of rain, snow, or some typoe of ice fall back to the surface, or evaporate as the water droplets return to vapor that remains in the atmosphere. Clouds have been divided into three stratigraphic levels: high, middle, and low. The upper limits of the high stage, at the tropopause, range from 8 kilometers in the polar regions to 18 kilometers in the tropics. In the temperate zones, which include most of the world's landmasses, the high stage ranges up to approximately 13 kilometers.

## CLOUD TYPES

Cirrus clouds, the highest clouds, are composed of ice crystals and some supercooled liquid water. The air temperature of these clouds is usually below −25 degrees Celsius. Cirrus clouds form when air rises slowly and steadily over a wide area. They are usually delicate in appearance, white in color, and tend to occur in narrow bands. They are thin enough to permit stars or blue sky to be seen through them, and are responsible for the ring or halo effect that often appears to surround the sun. They typically signal large, slow-moving warm fronts and rain. Cirrus clouds sometimes seem to be gathered in branching plumes or arcs with bristling ends. Because the color of a cloud depends on the location of the light source that illuminates it, the position of the sun at sunset or sunrise may cast these clouds in bright reds or yellow-oranges. A variant of this classification is the cirrocumulus, a cloud composed of ice crystals gathered in columns or prismatic shapes. The cirrocumulus usually is arranged in ripples or waves; it is the cloud that sailors describe when they speak of a "mackerel sky" because of its similarity to the coloration of the mackerel fish. A close relation is the cirrostratus, which resembles a transparent white veil. Suffusing the blue of the sky with a milky tone, it is composed of ice crystals shaped like cubes. Because cirrus clouds often form in long, wispy streamers under the influence of strong winds, herdsmen described such clouds as "mares' tails."

On the middle atmospheric level are found the altocumulus clouds. They signal an approaching cyclonic front. The altocumulus clouds reach an altitude of 6 kilometers, where they intersect the lower level of the cirrocumulus, and tend to resemble the cirrocumulus at their uppermost reaches. Below this point, they tend to be larger and to have dark shadings on their lower boundaries. They are composed primarily of water drops and are formed by a slow lifting of an unstable layer of air. Typically, they look like large, somewhat flattened globules and are often arranged evenly in rows or waves. In the summer, they may appear as a group of small, turretlike shapes, and in that form they often precede thunderstorms. The other basic middle-level cloud is the altostratus. These clouds are formed in air that is ascending slowly over a wide area, and they often occur in complex systems between the higher cirrostratus and the low-level nimbostratus. They are composed of both ice crystals and water droplets, with raindrops or snowflakes in the lower levels of the cloud. They are usually gray or bluish-gray, resembling a thick veil that is uniform in appearance, and often cover a substantial portion of the sky. There is a reciprocal relationship between altocumulus and altostratus clouds, in that a sheet of altocumulus clouds—particularly high, scaly ones—may be forced down and become transformed into altostratus. Conversely, as weather improves, altostratus may be altered toward altocumulus.

Clouds on the lowest levels generally are associated with the onset of precipitation. They often produce what is frequently called a leaden sky—one that is flat and "dirty" in appearance. These clouds range in altitude from 2 kilometers to the air just above the surface. Stratocumulus clouds may appear in sheets, patches, or layers close together, in which case the undersurface has a wavelike appearance. These clouds are formed by an irregular mixing of air currents over a broad area and suggest instability in the atmosphere. They are composed primarily of water droplets and, because they often have a marked vertical development, they can be confused with small cumulus clouds; however, they have a softer, less regular shape than cumulus clouds. Stratus clouds tend to be the lowest level of cloud formation. Sometimes resembling fog, they are usually an amorphous, gray layer, and while they do not produce steady rain, drizzle is not uncommon. They are formed by the lifting of a shallow, moist layer of air close to the ground and are composed almost exclusively of water droplets. They produce the dullest of visible sky conditions.

## CUMULUS AND NIMBUS CLOUDS

The cumulus cloud is probably the cloud form most frequently depicted in artistic illustrations of the sky. In fair weather, cumulus clouds are separated into cottonlike puffs with distinctive outlines against a deep blue background. When they are few in number and almost pure white in color, they are a prominent feature of the most pleasant weather during the spring and summer; however, these clouds also have the potential to change into storm-producing systems. They are initially formed by the ascent of warm air in separate masses or bubbles, the air rising with increasing velocity as the cloud takes shape. Their development is driven by convection currents, and while their base is usually about 1 kilometer above the land surface (or somewhat less over the ocean), they may grow vertically to an altitude of more than 7 kilometers; in the case of the cumulonimbus, the classic anvil-shaped thunderhead, they may reach an altitude of more than 15 kilometers. In the transition to cumulonimbus, the warm air rises rapidly into a cooler layer, and the convergence of radically varying temperatures produces a "boiling" motion at the tops of the clouds. When the rising air encounters strong winds at its upper reaches, it becomes flattened against the stable, colder layer of air that resists the convection from below. This shearing effect is what creates the classic anvil shape of the thunderhead.

The cumulonimbus is the great thundercloud of the summer sky. It tends to be white or gray, dense, and massive, with a dark base. Generally growing out of cumulus clouds, these are the tallest of conventional cloud forms, created by strong convection currents with updrafts of high velocity within the clouds.

127

The thunderstorm is a product of considerable turmoil within a cumulonimbus cloud. Violent updrafts in the center of the cloud result in an increasing size of water droplets that alternately descend and rise within the cloud until they fall as some form of precipitation. The size of the drops that eventually fall is an indication of the altitude of the cloud as well as the severity of the storm; in their lower regions, cumulonimbus are composed of water droplets, but as they rise, ice crystals, snowflakes, and hail may form. The development of the cumulonimbus is generally cellular (or compartmentalized) so that there is usually some space between clouds of this type, although they may occur in a long "squall line" that is often the boundary between warm and cold airflows. Cumulonimbus clouds, because of the great energy involved in their production, often generate subsidiary clouds as well. On lower levels, the sky seems to take on a convoluted, disorganized appearance, while upper levels often feature extensions that are like cirrus clouds. The forward edge of a cumulonimbus cloud often produces what looks like the front of a wave or rolling scud just under the base of the main cloud.

Although original attempts at cloud classification included the word "nimbus" to describe a separate order of low, rain-producing clouds, the term is now applied as a modifier. In addition to its application to the storm-producing version of the cumulus cloud, there are nimbostratus clouds, or low, gray, dark clouds with a base close to the ground, which generally carry rain. They are formed by the steady ascent of air over a wide area, usually associated with the arrival of a frontal system. They differ from stratus clouds in that they are much darker and are composed of a mixture of ice crystals and water droplets. These clouds have a ragged base, highly variable in shape, that is often hard to discern in steady rain. Depending on the temperature, the base will be primarily snowflakes or raindrops.

## OTHER CLOUD-RELATED PHENOMENA

In addition to the standard forms, other cloud-related phenomena may occur in the atmosphere. A cloud is essentially composed of tiny droplets of liquid water or of ice particles suspended in the air, or both; however, other substances that may not technically be considered the components of clouds may be suspended in the atmosphere and gathered in cloudlike masses. Haze, for example, is a suspension of extremely fine particles invisible to the naked eye but sufficiently numerous to produce an opalescent effect. Clouds of smoke are usually composed of the products of combustion gathered in a dense, swirling mass that rises from its source. Particles of sand may be gathered by strong, turbulent winds into clouds that move near the ground surface. One other type of cloud, the noctilucent, is a high-altitude form found in the mesosphere at 8 to 90 kilometers above the surface in northern latitudes. Its formation is somewhat mysterious, but one theory maintains that it is created by turbulence that carries water vapor to unusual heights during the arctic summer. The nuclei for the formation of these clouds may be deposited in the atmosphere by the residue of meteors. Noctilucent clouds are either pure white or blue-white and have pronounced band or wave structures. One theory maintains that this wave configuration is a result of gravity waves in the atmosphere, but the source of these gravity waves is unknown. Noctilucent clouds resemble high cirrus clouds more than any other form.

On rare occasions, clouds have also been observed in the stratosphere. Their composition has not been determined, but at a range of 20 to 30 kilometers above Earth's surface, nacreous or "mother-of-pearl" clouds, so designated because they may display a full spectrum of colors, have been seen in the sky over Scandinavia and Alaska during the winter.

## STUDY OF CLOUDS

Recorded descriptions of the sky and weather-related phenomena date at least as far back as Aristotle's *Meteorologica* (c. 350 B.C.E.). The first formal system of cloud classification owes its development to the work of Luke Howard in London in the early nineteenth century. Howard proposed the use of the Latin names currently used to identify clouds. In 1897, C. T. R. Wilson's cloud chamber experiments simulated the cloud formation process in the laboratory. By 1925, radio wave observations were beginning to augment free balloon ascensions as a method of examining the atmosphere. In 1928, the first radiosonde apparatus—a kind of electrical thermometer that transmits temperature, pressure, and humidity information to a receiver—was placed into operation. Immediately after World War II, high levels of atmospheric exploration became possible with rockets.

In 1960, the first meteorological satellite, Tiros 1, was launched by the United States. It transmitted pictures of cloud patterns back to Earth, and whereas previously the motion of cloud systems could be developed only from separate ground sightings, the satellite provided an overview of areas more than 1,000 kilometers wide. The use of time-lapse imaging permitted an observer to follow the birth and decay of cloud systems and, in conjunction with computer data storage, made it possible to develop a history of the entire range of cloud formation in the atmosphere.

There are seven basic factors that must be considered to determine the fundamental properties of cloud systems. They are, according to Horace Byers, the amount of sky covered, the direction from which the clouds are moving, their speed, the height of the cloud base, the height of the cloud top, the form of the cloud, and the constitution of the cloud. Each of these categories depends upon the location of the observing system. The amount of sky covered will vary considerably, depending on whether a ground observer is reporting or satellite data are being analyzed. The direction of cloud motion also depends on the point of observation because relatively specific motion with regard to a fixed point is much easier to chart than the complex series of measurements necessary to describe the motion of a large-scale cloud formation. The upper and lower reaches of a cloud may be measured from balloons, from airplanes, by satellites, and in ground stations. The refinement of cloud investigation by satellite imaging has led to the consideration of what the eminent meteorologist Richard Scorer calls "messages," which consider the fact that cyclonic patterns are more varied than previously realized, requiring the assimilation of much more data into an explanation of how large-scale cloud systems are formed and why they exist for a particular duration in time.

In addition to measurements that are essentially external, the inner mechanics of a cloud must be studied to determine its basic properties and potential behavior. The inner cloud physics that influence its development and its potential for precipitation involve the supercooling and freezing of water, the origin and specific form of the ice nuclei that lead to the growth of a cloud and that depend on the rise and descent of particles within the cloud, the rate of collisions between particles in a cloud that influence the growth of droplets and crystals, and the reflecting and refracting properties of the components of a cloud that determine the color of a cloud in terms of both visual observation and radiophotometric measurement. Aside from spectographic observations, most cloud physics depend on an understanding of thermodynamics, or the effects of heat changes on the motion and elemental properties of particles in the atmosphere.

*Leon Lewis*

## FURTHER READING

Ahrens, C. Donald. *Essentials of Meteorology: An Invitation to the Atmosphere.* 6th ed. Belmont, Calif.: Brooks/Cole Cengage Learning, 2012. An updated and enhanced edition written in an authoritative style that is easy to read and that explains complex concepts clearly. Contains more than 250 new images and figures, and focuses on the latest research and coverage of meteorology.

_____. *Meteorology Today: An Introduction to Weather, Climate and the Environment.* 9th ed. Belmont, Calif.: Brooks/Cole Cengage Learning, 2009. An up-to-date, clearly written examination of a number of meteorological phenomena that might be of interest to the nonspecialist. Discusses cloud systems within the context of atmospheric science. Extensively illustrated. Suitable for college-level students.

Day, John A., and Vincent J. Schaefer. *Peterson First Guides: Clouds and Weather.* New York: Houghton Mifflin, 1991. Provides basic information on cloud formations, atmospheric dynamics, and storm events. Offers many images to assist the reader with identification.

Dunlop, Storm. *The Weather Identification Handbook.* Guilford, Conn.: Lyons Press, 2003. Provides a simple and useful guide to various atmospheric objects, patterns, dynamics, and phenomena. Begins with identification of cloud formations, followed by optical phenomena such as rainbows. Discusses various weather patterns and events. Best used as a reference for the weather enthusiast as an introduction to the study of meteorology.

Moran, Joseph M., and Michael Morgan. *Meteorology: The Atmosphere and the Science of Weather.* 5th ed. Upper Saddle River, N.J.: Prentice Hall, 1996. One of the best current texts available for college-level students. Written by two longtime scholars in the field. Includes charts, photographs, and illustrations in addition to clear and informative discussions of weather phenomena.

Parker, Sybil P., ed. *Meteorology Source Book.* New York: McGraw-Hill, 1988. Clear definitions and comprehensive illustrations make this a good source for basic information about meteorology. Different sections have been written by specialists in the field.

Pretor-Pinney, Gavin. *The Cloud Collector's Handbook.* San Francisco, Calif.: Chronicle Books, 2011. Provides information for those interested in spotting and identifying clouds, using full-color images and descriptions of clouds of all types.

_____. *The Cloudspotter's Guide: The Science, History and Culture of Clouds.* New York: Berkeley Publishing Group, 2006. A very useful book, discussing the history of cloud classification and explaining what cloud formations indicate in terms of local weather patterns. Also offers amusing anecdotes about clouds in popular culture, music, and art.

Rogers, R. R., and M. K. Yau. *A Short Course in Cloud Physics.* 3d ed. Woburn, Mass.: Butterworth-Heinemann, 1989. For the student who wishes to go beyond preliminary material. Some knowledge of mathematics and physics is required.

Shaw, Glenn E. *Clouds and Climatic Change.* Sausalito, Calif.: University Science Books, 1996. A good look at the physics of clouds, their evolutionary processes, and the effects those changes have on the climate. Illustrated, index, and bibliography.

Straka, Jerry M. *Cloud and Precipitation Microphysics: Principles and Parameterizations.* New York: Cambridge University Press, 2009. Covers cloud formation principles, vaporization and saturation dynamics, and analysis of vapor collection in the process of cloud formation. Contains highly technical writing well suited for graduate students, researchers, and professional meteorologists. References and indexing are substantial.

Trefil, James. *Meditations at Sunset: A Scientist Looks at the Sky.* New York: Charles Scribner's Sons, 1988. Tone is clear and enthusiastic, and the chapter "When Clouds Go Bad" is a superb description of a storm; vivid and energetic.

**See also:** Atmospheric and Oceanic Oscillations; Atmospheric Properties; Barometric Pressure; Climate; Climate Modeling; Cyclones and Anticyclones; Deep Ocean Currents; Drought; Global Energy Transfer; Global Warming; Gulf Stream; Hurricanes; Lightning and Thunder; Monsoons; Ocean-Atmosphere Interactions; Satellite Meteorology; Seasons; Severe Storms; Surface Ocean Currents; Tornadoes; Tropical Weather; Van Allen Radiation Belts; Volcanoes: Climatic Effects; Weather Forecasting; Weather Forecasting: Numerical Weather Prediction; Weather Modification; Wind; World Ocean Circulation Experiment

# COSMIC RAYS AND BACKGROUND RADIATION

*Cosmic rays are highly energetic protons and heavier atomic nuclei that continually rain down on Earth from space. Ranging in energies up to $10^{20}$ electron volts (about 20 joules), cosmic rays are not well understood. Primary cosmic rays can produce a secondary "cosmic-ray air shower," an abundance of secondary particles that reach Earth's surface as a result of collisions of primary rays with atoms in the atmosphere.*

## PRINCIPAL TERMS

- **alpha particle:** the nucleus of a helium atom, consisting of two protons and two neutrons
- **electronvolt (eV):** a unit of energy used for atomic and subatomic measurements; 1 eV is the kinetic energy acquired by an electron accelerated through a potential difference of 1 volt
- **flux:** the number of particles striking a unit of surface area per unit of time
- **isotopes:** atoms of a given element with the same number of protons but different numbers of neutrons
- **neutrinos:** massless (or nearly massless) particles given off in certain types of nuclear reactions
- **photon:** the smallest energy packet of light for a given frequency; X rays and gamma rays are examples of high-energy photons
- **positron:** the antiparticle to the electron
- **proton:** the nucleus of the primary hydrogen atom; a proton carries one unit of positive electrical charge, and the number of protons in the nucleus of an atom determines the identity of the particular element

## COSMIC-RAY PARTICLES

Cosmic-ray particles are continually striking Earth's atmosphere. They are highly energetic, electrically charged subatomic particles traveling at high velocity. About 90 percent of primary cosmic rays are bare protons, each carrying a single positive charge, 9 percent are alpha particles carrying two positive charges, and the rest are positively charged nuclei of heavier elements and negatively charged electrons (beta rays). The velocities of most cosmic rays are so high that they must be described in relativistic terms; that is, their kinetic energies are comparable to or much greater than their rest mass energies. Cosmic rays are believed to come from various sources that include the sun; from other stellar sources in this galaxy, such as supernovas; and from other galaxies, particularly those with active galactic centers.

The flux of cosmic-ray particles decreases exponentially with increasing energy. The highest-energy cosmic rays are in the range of $10^{20}$ electron volts or approximately 20 joules—energy that is about eleven orders of magnitude greater than the mass-energy of the proton at rest.

Cosmic rays are generally divided into two categories: primary and secondary. Primary rays strike Earth's atmosphere; secondary rays are the resulting cascades of particles that are produced by collisions of the primary rays with atoms in the atmosphere. Primary rays have the following characteristics: Their intensity seems to be essentially constant; their flux and spectrum appear to be isotropic in space (the same in all directions); they are anomalous in composition; and their spectrum includes very energetic particles.

An important feature of cosmic rays is that their chemical abundance is significantly different both from that of the sun and from that of the universe in general. Cosmic-ray composition is particularly rich in heavier elements. However, even with the lighter elements, cosmic rays contain about 1 million times as much lithium, beryllium, and boron relative to hydrogen as does the sun.

Not only are cosmic rays individually energetic but, given their high spatial particle density, they also represent a large fraction of the total energy associated with astrophysical phenomena. The energy density of cosmic rays is comparable to that of photons, interstellar magnetic fields, and the turbulent motion of interstellar material, each of which is approximately 1 electron volt per cubic centimeter.

## SOURCE AND EFFECT OF COSMIC RAYS

Since their discovery in 1911 by V. F. Hess, cosmic rays have presented scientists with an enigma. The fundamental questions with regard to cosmic rays—where do they come from, and how are they accelerated to such high energies—have yet to be fully answered. Although the sun is a source of some of the lower-energy cosmic rays, it is clear that it cannot be responsible for the higher energy of the cosmic-ray spectrum. As yet, there has not been a satisfactory mechanism developed that can realistically explain the high acceleration necessary to account for the higher-energy rays. However, some promising models have recently been developed whereby cosmic rays may be repeatedly accelerated by shock waves from violent astrophysical phenomena. Sources for these shock waves would include exploding galaxies and supernovas (exploding stars). It is well known that supernovas are rich in heavier elements. Indeed, this is presently thought to be the primary mechanism whereby heavier elements are synthesized.

The flux of primary cosmic rays incident upon Earth's atmosphere is about 1 particle per square centimeter per second. However, Earth's magnetic field effectively prohibits cosmic rays of less than $10^8$ eV (100 million electron volts) from reaching the surface, the radius of curvature of the path of the charged particles being sufficiently short that they are turned away by Earth's magnetic field.

There is a measurable latitude effect in the flux of cosmic rays reaching Earth. This effect is caused by Earth's magnetic field, which manifests a higher cutoff energy (the minimum kinetic energy required for the cosmic rays to reach the atmosphere) at the geomagnetic equator than at higher latitudes. For example, the energy cutoff for vertically arriving protons at the geomagnetic equator is about 15 gigaelectron volts (GeV), whereas at geomagnetic latitude 50 degrees, the cutoff energy is 2.7 GeV.

Furthermore, there is a significant east-west effect in the flux of cosmic rays incident upon Earth, which is also produced by Earth's magnetic field. The cutoff energy for cosmic rays reaching Earth's surface is less from the west (about 10 GeV) than from the east (about 60 GeV) for positive charges interacting with Earth's magnetic field.

High-energy gamma-ray photons have also been observed in cosmic rays. Gamma rays and X rays produced by the interaction of primary cosmic rays with interstellar material yield information on the distribution and composition of matter in the galaxy.

Several point sources of X rays and gamma rays have been observed, including the Crab Nebula, Hercules X-1, and Cygnus X-3. These sources have

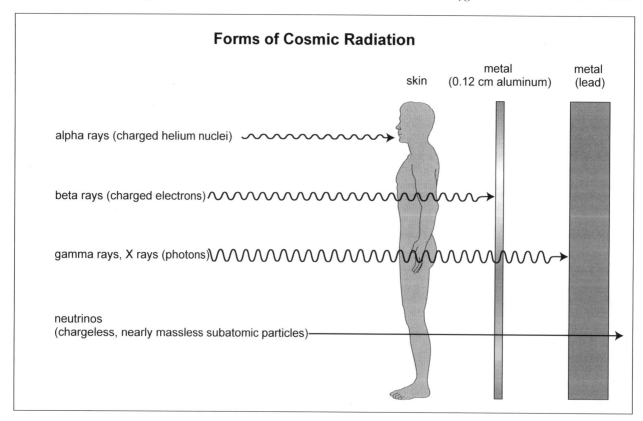

**Forms of Cosmic Radiation**

skin    metal (0.12 cm aluminum)    metal (lead)

alpha rays (charged helium nuclei)

beta rays (charged electrons)

gamma rays, X rays (photons)

neutrinos (chargeless, nearly massless subatomic particles)

been observed to emit gamma rays in the range of $10^{12}$ eV.

Intense gamma-ray burst events (about five per year) have also been observed. The bursts are characterized by an initial intense pulse of 0.1 second to 4 seconds duration followed by one or more pulses, the entire event taking place within 1 minute.

## SHOCK WAVES AND REMOTE MATTER

Researchers have also begun to investigate the ability of shock waves to accelerate cosmic rays to some of the higher energies observed in the cosmic-ray spectrum. Shocks are common in the solar system. For example, "bow" shock waves are produced when the solar wind is deflected by planetary magnetic fields, from solar flares, and when fast solar-wind streams overtake slower streams and impart additional kinetic energy to the leading particles. Some solar-wind particles attain high energies through this mechanism, which is generally referred to as "diffusive shock acceleration." Additionally, mathematical models have shown that the shock waves generated by supernova explosions can dramatically accelerate charged particles.

The resulting acceleration from shock waves may also account for the anomalous component of cosmic rays. Thought to originate from neutral particles in the interstellar medium that become ionized via collisions with the solar wind or by solar photons, anomalous cosmic rays may then be further accelerated at their termination by the shock of the solar wind. However, this is still a very speculative model, and spacecraft have yet to detect the termination shock of the solar wind. Hence, its properties and effects are currently somewhat conjectural; the ultrahigh energies observed in the range of $10^{20}$ eV still defy a clear understanding as to their production and acceleration mechanisms.

The nuclear composition of cosmic rays enables scientists to sample matter that comes to Earth from remote parts of the galaxy. Indeed, some cosmic rays are thought to be of extragalactic origin. Data on the composition of cosmic rays suggest they may well have experienced nuclear conditions similar to those of the interstellar medium in the distant past. The cosmic-ray spectrum contains abundances of heavier nuclei in approximate proportion to their nuclear charge. For example, iron-56 (atomic number 26) is about twenty-six times more abundant than would

otherwise be expected. Likewise, this observation extends to elements as heavy as lead and uranium. The lighter elements lithium, beryllium, and boron are present in cosmic rays by a factor of about 100,000 times greater than that found in typical universal abundance. In contrast, the relative proportions of nuclei of odd-even charge are similar to those of universal abundance for those nuclei from carbon to iron.

The extremely high proportions of lithium, beryllium, and boron can be accounted for by collisions of the more abundant heavier nuclei, carbon through iron, with protons—overwhelmingly the most populous nuclei present in the interstellar medium. If most cosmic rays are confined to the galactic disk, then this process, known as "spallation," can be used to calculate a mean galactic age of the cosmic rays of several million years.

One radioactive species, beryllium-10, is particularly suitable as a radioactive clock when compared with its stable neighbor, beryllium-9. The results of this comparison yield a mean age for galactic cosmic rays of greater than 20 million years.

The "leaky box" model of galactic cosmic rays assumes that spallation is not complete by the time cosmic rays reach Earth and, hence, leads to the prediction that most cosmic rays leave the galactic disk at a relatively young age. The galactic magnetic field has an energy density slightly less than that of cosmic rays. Hence, it is thought that galactic gravity also plays a role in confining cosmic rays within the galactic halo.

Given the energy density of cosmic rays and the requisite confinement energies, it appears that the principal source of cosmic-ray energy is supernovas—events that are also the source for the synthesis of nuclei heavier than helium and lithium.

## STUDY OF COSMIC RAYS

There are a variety of methods for observing cosmic rays: ground-based arrays of detectors both at low altitudes and on mountain tops, balloons carrying a detector payload, and rockets and satellites carrying payloads designed for cosmic-ray research.

Ground-based instruments can employ a variety of detection techniques. Some of the more traditional techniques use cloud chambers, scintillation detectors, and photomultiplier tubes. Usually, these instruments are networked into an extensive two-dimensional array (or, in some cases, with limited three-dimensional information). Computer modeling of the arrival times at the various detectors then can be used to reconstruct a shower front, yielding the approximate direction of origin of the shower. A more recent technique detects the Cherenkov radiation given off by fast-moving electrons through Earth's atmosphere.

Ground-based detectors rarely observe a primary cosmic ray. Rather, they detect the shower of secondary particles produced by the interaction of the high-energy primary ray with atoms in Earth's atmosphere. Primary particles with energies in excess of a few hundred million electron volts (MeV) per nucleon will readily interact with atmospheric nuclei, mainly producing neutral pions, charged pions, protons, and neutrons. The neutral pions decay into gamma-ray photons that in turn produce electron-positron pairs in a cascading process known as an "extensive air shower." The charged pions decay into muons, some of which can be very penetrating. A case in point is the muon detector situated 1.6 kilometers underground at the Homestake Detector in South Dakota. The nucleons (neutrons and protons) produced by primary cosmic rays are sufficiently energetic to produce additional nucleons, resulting in neutron-proton cascades, which are also detected at ground-based array detectors. Production of these cascades is very sensitive to atmospheric conditions, especially temperature and pressure.

The use of rockets and satellites beginning in the 1960's enabled researchers to establish crucial characteristics of cosmic rays not directly discernible from ground-based detectors. In particular, scientists were able to determine the energy spectra and particle abundances of primary cosmic rays. For example, gamma-ray emissions from the plane of the galaxy were first detected by the OSO III satellite in 1967. The SAS-2 satellite then detected the diffuse gamma-ray background, and the COS-B satellite provided a detailed map of gamma-ray emissions in the galaxy. In addition, COS-B discovered twenty-five discrete gamma-ray sources, including 3C273 (a quasar) and the pulsars in the Crab and Vela supernova remnants. The Einstein X-Ray Observatory was launched in 1978, making high-resolution surveys of selected areas of the sky. The Einstein Observatory provided detailed images of many X-ray sources.

*Stephen Huber*

134

**FURTHER READING**

Balbi, Amedeo. *The Music of the Big Bang: The Cosmic Microwave Background and the New Cosmology.* Berlin: Springer-Verlag, 2010. Discusses details of the cosmic microwave background. Chapters are broad, but provide information from specific studies and developments in cosmic radiation research. A good introduction to cosmology.

Bychkov, Vladimir L., Gennady V. Golubkov, and Anatoly I. Nikitin, eds. *The Atmosphere and Ionosphere. Dynamics, Processes and Monitoring.* New York: Springer Verlag, 2010. Geared toward the advanced student of atmospheric physics. Discusses many of the features of and phenomena occurring in the ionosphere, including electromagnetic and optical phenomena.

Darrow, K. K. "Cosmic Radiation: Discoveries Reported in *The Physical Review.*" In *The Physical Review: The First 100 Years.* New York: American Institute of Physics, 1999. A lengthy introduction provides an overview of the development of cosmic-ray physics. Includes an excellent bibliography for further investigation.

Davies, P., ed. *The New Physics.* New York: Cambridge University Press, 1990. Several chapters include discussion of cosmic rays. The chapter "The New Astrophysics," by Malcolm Longair, is strongly recommended.

Dormin, Lev I. *Cosmic Rays in the Earth's Atmosphere and Underground.* Dordrecht: Kluwer Academic Publishers, 2004. Presents a full overview of the theoretical and experimental basis of atmospheric and subterranean cosmic-ray research, these cosmic rays' influence on meteorological effects, and these cosmic rays' effects on atmospheric properties. Discusses realized and potential applications of cosmic-ray research.

Durrer, Ruth. *The Cosmic Microwave Background.* New York: Cambridge University Press, 2008. Covers many topics in cosmic background radiation and the accompanying research methods. Includes many appendices, references, and indexing. Provides mathematical equations with the text; knowledge of advanced math is assumed. Well suited for graduate students or professionals.

Gaisser, Thomas K. *Cosmic Rays and Particle Physics.* New York: Cambridge University Press, 1990. Although this text can be somewhat technical at times, it is an excellent overview of the field of cosmic-ray astrophysics.

Giacomelli, Giorgio, Maurizio Spurio, and Jamal Eddine Derkaoui. *Cosmic Radiations: From Astronomy to Particle Physics.* Dordrecht: Kluwer Academic Publishers, 2001. A highly technical publication that discusses results and interpretations of cosmic radiation studies of neutrino oscillations and astronomy, high-energy cosmic rays, gravitational waves, magnetic monopoles, and dark matter.

Gombosi, Tamas I. *Physics of the Space Environment.* New York: Cambridge University Press, 2004. Explores the field of atmospheric physics to determine the makeup of the upper atmosphere and the space environment and how these forces affect Earth's atmosphere and weather cycles. Focuses on solar winds and the magnetosphere. Intended for the advanced reader. Illustrations, index, and bibliography.

Murthy, R., P. V. Wolfendale, and A. W. Wolfendale. *Gamma-Ray Astronomy.* 2d ed. New York: Cambridge University Press, 1993. A good overview of gamma-ray (high-energy photon) astronomy.

Partridge, R. B. *3K: The Cosmic Microwave Background Radiation.* New York: Cambridge University Press, 2007. Discusses the big bang theory, inflationary cosmology, and cosmic background radiation. Thirty years of study are incorporated into the well-organized pages of this book. Technical at times; well suited for advanced undergraduates or graduate students studying cosmology.

Silberberg, R., C. H. Tsao, and J. R. Letaw. "Composition, Propagation, and Reacceleration of Cosmic Rays." In M. M. Shapiro et al., eds., *Particle Astrophysics and Cosmology.* Boston: Kluwer Academic Publishers, 1993. An excellent, though somewhat technical, discussion of cosmic-ray astrophysics.

**See also:** Acid Rain and Acid Deposition; Air Pollution; Atmosphere's Evolution; Atmosphere's Global Circulation; Atmosphere's Structure and Thermodynamics; Atmospheric Properties; Auroras; Earth-Sun Relations; Global Energy Transfer; Global Warming; Greenhouse Effect; Nuclear Winter; Ozone Depletion and Ozone Holes; Radon Gas; Rainforests and the Atmosphere

# CYCLONES AND ANTICYCLONES

*Cyclones and anticyclones are large-scale weather systems with opposite properties. A cyclone is characterized by a central region of low atmospheric pressure and an anticyclone is characterized by a central region of high atmospheric pressure. Because cyclones are a major cause of stormy weather and anticyclones typically bring good weather, accurate meteorological predictions are greatly informed by an understanding of how these weather systems originate and develop.*

## PRINCIPAL TERMS

- **convergence:** a tendency of air masses to accumulate in a region where more air is flowing in than is flowing out
- **Coriolis effect:** the illusion of deflection observed when a body moves through the atmosphere with regard to an individual situated on the moving surface of the earth
- **cyclogenesis:** the series of atmospheric events that occur during the formation of a cyclone weather system
- **divergence:** a tendency of air masses to spread in a region where more air is flowing out than is flowing in
- **front:** the boundary between two masses of air with different densities and temperatures; usually named for the mass that is advancing (for example, in a cold front, the mass that is colder is moving toward a warmer mass)
- **hurricane:** a cyclone that is found in the tropics (between 23.5 degrees north and south of the equator) and that has winds that are equal to or exceed 64 knots, or 74 miles per hour
- **isobar:** on a map, a line connecting two or more points that share the same atmospheric pressure, either at a particular time or, on average, in a particular period
- **mid-latitude cyclone:** a synoptic-scale cyclone found in the mid-latitudes (between 30 and 60 degrees north and south of the equator)
- **synoptic scale:** a scale used to describe high- and low-pressure atmospheric systems that have a horizontal span of 1,000 kilometers (621 miles) or more

## ATMOSPHERIC PRESSURE AND AIR CIRCULATION

The term *atmospheric pressure* describes the physical force exerted by the weight of the air above a given area on the earth's surface. Meteorologists map pressure distributions with isobars, which appear as a series of curved lines connecting points that share the same atmospheric pressure. This pressure is measured in units called millibars.

Under average conditions, the atmospheric pressure at sea level is approximately 1,013.2 millibars. Values that are greater than the average sea-level pressure are considered high and those that are lower than average sea-level pressure are considered low. Pressure gradients are horizontal or vertical differences in atmospheric pressure. Pressure gradients create winds because air is constantly moving from areas of high pressure to areas of low pressure, seeking to create equilibrium. In other words, pressure gradients cause air to move perpendicular to isobars.

Pressure gradients, however, are not the only force at work. A phenomenon known as the Coriolis effect also affects the way air circulates in the atmosphere. The Coriolis effect is an apparent force that acts on moving objects, such as masses of air, in a rotating system, such as the rotating Earth. The result is that the moving object shifts perpendicular to the axis of this rotation.

To understand the Coriolis effect, one can imagine trying to throw a ball in a straight line from the North Pole to the equator. Because the earth is wider at the equator than it is at the poles, points at the equator must travel a greater distance than points at the poles in the same period of time. A ball thrown from the North Pole to the equator would thus appear to bend to the right. Similarly, when a mass of air is moving in the Northern Hemisphere, the Coriolis force appears to deflect that mass toward the right. When a mass of air is moving in the Southern Hemisphere, the Coriolis force appears to deflect that mass toward the left.

The force arising from pressure gradients and the force associated with the Coriolis effect are roughly equal in magnitude, and at the upper levels of the atmosphere they balance each other to create winds that travel more or less parallel to isobars. Friction, or air resistance, reduces the effects of the Coriolis force at the earth's surface.

## CYCLONES AND THEIR FORMATION

Cyclones and anticyclones are both large-scale weather systems that are shaped by atmospheric pressure gradients, the Coriolis effect, and surface friction. A cyclone has a central region of low atmospheric pressure with winds circulating around that center. On a weather chart, a cyclone appears as a series of roughly circular or oval isobars; the area inside the innermost isobar is the region of lowest pressure. Isobars that take this particular configuration are known as troughs.

The direction in which a cyclone's winds circulate depends on the hemisphere in which the weather system forms. In the Northern Hemisphere, a cyclone has winds that move in a counterclockwise direction. The reverse is true in the Southern Hemisphere. Because cyclones are a major cause of severe stormy weather, including blizzards and floods, meteorologists creating detailed and accurate weather forecasts pay close attention to how these atmospheric systems originate and develop.

The atmospheric events that take place as a cyclone forms are known collectively as cyclogenesis. Mid-latitude cyclones are cyclones that occur between 30 and 60 degrees north and south of the equator and are about 1,000 kilometers (621 miles) or more in diameter. These typically form at fronts, or the boundaries or transition zones between two masses of air that have different temperatures and densities. At this first stage of cyclogenesis, the heavy cold air and lighter warm air are simply pushing against each other. Because the air masses are not moving, the place where they meet is known as a stationary front.

At the next stage, the cold air—because it is denser and heavier—begins to sink below the light warm air. In turn, the light warm air is forced upward and then over the cold and heavy air mass. Instead of a single stationary front, two fronts are formed: one consisting of the advancing edge of the cold air and one consisting of the advancing edge of the warm air. Because of the Coriolis force, these masses of cold and warm air do not simply exchange places vertically but begin to revolve around each other, turning inward toward the area of low pressure in the center of the rotation. This pattern of winds causes warm air to "pile up" in the center of the cyclone, near the surface. This phenomenon is known as convergence.

When air converges low to the ground, it has nowhere to move but up and out. As the warm-air front rises and expands (diverges), it carries water vapor that cools and condenses into clouds and rain. Different characteristic weather patterns are seen along each front. Brief, intense thunderstorms tend to form along the cold front, and slow, steady rains tend to fall along the warm front. When the cyclone nears its end, the cold front pushes on the warm front so much so that the mass of warm air is entirely separated from the low pressure center. This is known as an occluded front, and it is usually associated with more rainy weather.

## ANTICYCLONES

As its name suggests, an anticyclone has properties opposite of those of a cyclone. Whereas a cyclone consists of winds circulating around a center of low atmospheric pressure, an anticyclone has a center of high atmospheric pressure with winds circulating around that center. In the Northern Hemisphere, an anticyclone has winds that move in a clockwise direction; the reverse is true in the Southern Hemisphere.

Like a cyclone, an anticyclone appears on a weather chart as a series of roughly circular or oval isobars; however, in an anticyclone, the area bounded by the innermost isobar is the region of highest pressure. When isobars take this configuration, they are known as ridges.

An anticyclone forms when dense cold-air masses in the upper atmosphere converge, or pile up. When this convergence reaches a high enough level, the air begins to sink to the earth's surface. As it descends, the air is compressed by increasing pressure and becomes warmer and drier. Anticyclones are generally associated with clear weather.

## TROPICAL CYCLONES (HURRICANES)

Tropical cyclones, which are known as typhoons in the western Pacific Ocean and as hurricanes in the Atlantic and eastern Pacific Oceans, are cyclones that form in the tropics (between 23.5 degrees north and south of the equator). They are typically smaller than mid-latitude cyclones but are characterized by extremely high winds, usually exceeding speeds of 119 km (74 mi) per hour.

Because of these winds and the intense thunderstorms, occasional flash floods, and storm surges with which they are associated, tropical cyclones can cause great damage to life and property in coastal areas. Storm surges refer to seawater pushed inland by strong winds.

Tropical cyclones differ from mid-latitude cyclones in not being associated with a front. Instead, this type of cyclone arises when each of a set of specific environmental conditions is present. The ocean waters above which a potential cyclone would form must be at least 27 degrees Celsius (80 degrees Fahrenheit) up to a depth of 46 meters (about 150 feet) or more. This condition makes it natural that tropical cyclones tend to originate relatively near the equator.

The air near the middle of the troposphere, the lowest region of the earth's atmosphere, must be moist. There must be relatively low vertical wind-shear between the ocean's surface and the upper levels of the troposphere, meaning that wind speeds must not be changing radically as they ascend. Air must be able to cool relatively quickly as it ascends. Finally, there must exist some kind of atmospheric disturbance, or "seedling," such as a trough or an elongated area of low pressure.

If all these conditions are present, a tropical cyclone may begin to form, first with the establishment of a pattern of convection over the ocean. Warm and moist air rises into the atmosphere, cooling and condensing. As the air releases heat, it becomes even lighter, thus powering its own ascent. The water vapor that is released forms the clouds and thunderstorms commonly associated with tropical cyclones. As air moves up and away from the surface, more air rushes in to take its place, creating high winds at the surface. This process creates a self-sustaining cycle that can cause the cyclone to grow and intensify as long as it remains over the water. Usually, however, tropical cyclones begin to dissipate as soon as they move inland because the cycle is broken when the storm system no longer has access to the warm, moist air that moves over the ocean.

*M. Lee*

**FURTHER READING**

Ackerman, Steven, and John Knox. "Extratropical Cyclones and Anticyclones." In *Meteorology: Understanding the Atmosphere*. Sudbury, Mass.: Jones and Bartlett, 2012. This chapter, richly illustrated with color photographs, diagrams, and charts, explains each stage of both weather systems' formations.

Includes an outline, summary, key terms, and review questions.

Ahren, C. Donald. "Air Masses, Fronts, and Middle-Latitude Cyclones." In *Essentials of Meteorology: An Invitation to the Atmosphere*. Belmont, Calif.: Brooks/Cole, 2011. Notable for being especially clear and well organized, this chapter walks readers through each fundamental concept in a logical order, from air masses and fronts to convergence, divergence, and storm formation. Suitable for beginning students.

Chan, Johnny C. L., and Jeffrey D. Kepert, eds. *Global Perspectives on Tropical Cyclones: From Science to Mitigation*. London: World Scientific, 2010. Includes chapters on forecasting and modeling tropical cyclones, the effects of climate change on cyclone activity, and approaches to disaster response. Highly technical; best suited to college students and above with some background in meteorology.

De Villiers, Marc. *Windswept: The Story of Wind and Weather*. New York: Walker, 2006. An accessible, scientifically accurate book of popular science that explores the history of human attempts to understand the weather. Contains black-and-white figures and several useful appendices, including two covering tropical cyclone statistics.

Fahy, Frank. *Air: The Excellent Canopy*. Chichester, UK: Horwood, 2009. A well-written introduction to the properties of air and the forces that govern atmospheric circulation, designed for nonspecialists. Explains physics concepts using detailed descriptions and analogies rather than equations.

Longshore, David. *Encyclopedia of Hurricanes, Typhoons, and Cyclones*. New York: Facts on File, 2008. A comprehensive reference book containing about four hundred cross-referenced entries covering the science, history, and cultural significance of severe weather phenomena. Contains black-and-white photographs and other images. Suitable for high school readers and older.

**See also:** Atlantic Ocean; Atmospheric Properties; Barometric Pressure; Beaches and Coastal Processes; Climate; Climate Modeling; Clouds; Deltas; Earth-Sun Relations; El Niño/Southern Oscillations (ENSO); Floodplains; Floods; Ganges River; Global Energy Transfer; Gulf of Mexico; Hurricane Katrina; Hurricanes; Indian Ocean; Long-Term Weather Patterns;

Monsoons; Remote Sensing of the Oceans; Saltwater Intrusion; Satellite Meteorology; Sea Level; Seasons; Severe and Anomalous Weather in Recent Decades; Severe Storms; Tornadoes; Tropical Weather; Weather Forecasting; Wind

# D

## DAMS AND FLOOD CONTROL

*Dams have provided a means of substantially reducing the risk of catastrophic floods, as well as saving lives. As an added benefit that helps to offset environmental costs, dams generate pollution-free hydroelectric power and provide a reliable water supply for consumption, irrigation, industrial use, and recreation.*

### PRINCIPAL TERMS

- **channelization:** the practice of deliberately re-routing a stream or artificially modifying its channel by straightening, deepening, widening, clearing, or lining it
- **floodplain:** a wide, flat, low-lying area, adjacent to a river, that is generally inundated by floodwaters
- **flood zoning:** passing laws that restrict the development and land use of flood-prone areas
- **levee:** a dike-like structure, usually made of compacted earth and reinforced with other materials, that is designed to constrain a stream flow to its natural channel
- **outlet works:** gates or conduits in a dam that are generally kept open so as to discharge the normal stream flow
- **spillway:** generally, a broad reinforced channel near the top of a dam, designed to allow rising waters to escape the reservoir without overtopping the dam

### FLOODPLAIN RISKS

Human beings and rivers have competed for the use of floodplains for millennia. The floodplain, a wide, flat, low-lying area adjacent to the river, is built by natural sedimentation as the river carries its load of sand, silt, and clay to the sea. During times of excess flow, the river spills over its banks and covers the floodplain with muddy water. When the water retreats, a new layer of fertile soil is left behind. Humans have long exploited floodplains for these rich soils in order to grow crops. Floodplains were also attractive places to settle because the nearby river provided an accessible source of usable water, a transportation route, and sewage disposal. In some rugged terrains, the development of communities in any area other than floodplains is almost impossible.

Unfortunately, floodplain development has often proceeded despite the risks involved, with disastrous consequences. In the United States, a few flood-related deaths occur almost every year, and sometimes one flood kills many. In other parts of the world, flooding is a major annual concern, made even more pressing as precipitation patterns are affected by global changes in the atmosphere, particularly due to global warming. For every death that occurs, many more are left homeless or experience property damage, hardship, or suffering. Heavy rains in the winter of 1926-1927 caused disastrous floods on the Mississippi River that killed 313 people and left up to 650,000 homeless. In 1993, flooding of the Red River and the Upper Mississippi drainage basin killed 50 people and caused some $12 billion dollars worth of damage. In early 2012, major flooding of river floodplains due to heavy rains occurred in Southeast Asia, Australia, Venezuela, and several other locations around the world.

### DAM CONSTRUCTION

Human attempts to control floods by the use of dams goes at least as far back as ancient Egypt, where about 2700 B.C.E., a dam was built at Sadd-el-Karfara. The basic principle of flood control by the use of dams is to store floodwaters in the dams' reservoirs, from which they can be released in a controlled manner over a period of time instead of allowing them to spread suddenly and disastrously over the natural floodplain and the valuable human-made structures therein. After the smaller tributary streams below the dam have passed their floodwaters safely, the main reservoir is drained in a controlled fashion. The overall effect is to lengthen the time of passage of the flood, while drastically reducing the peak flow.

Dams are constructed of three basic materials: earth, rocks, and concrete. Eighty percent of all dams in the United States and Canada are earthen dams. In these, sand and soil are compacted into a broad triangular embankment surrounding a watertight clay core. The upstream face must be reinforced with rock or concrete to prevent wave erosion. Earthen dams are the most practical in broad valleys and are relatively inexpensive to construct. Rockfill dams are similar to earthen dams, but the heavy weight of the rock requires a more solid natural foundation. The upstream side must be sealed with a watertight material to prevent water from leaking through the dam.

Concrete dams require narrow valleys with hard bedrock floors to anchor and support them. A concrete gravity dam uses its great bulk and weight to resist the water pressure. These dams can generally remain stable when floodwaters overtop them, but they are costly. Washington State's Grand Coulee Dam, which is of this type, required 8.1 million cubic meters of concrete. A concrete buttress dam relies both on its weight and structural elements to support it: The watertight upstream face slopes underneath the reservoir, which helps to distribute the water pressure to the foundation and, in effect, uses the great weight of the water as an anchoring force for the dam. Buttresses on the downstream face of the dam both counteract the force of the water and help the dam to withstand minor foundation movements, a distinct advantage in earthquake-prone areas. Concrete arch dams have a convex upstream face that spans the steep valley walls. The water pressure transmits the force along the arch to the side abutments and foundation, bonding the dam to the canyon. These dams are less expensive to construct than gravity dams, but also are more likely to fail in the event of a small rupture.

## OUTLET WORKS AND SPILLWAYS

All dams must contain properly designed outlet works, which are gates or conduits near the base of the dam kept open to discharge the normal low-water flow of the stream. These gates can be operated manually or automatically, but they must be carefully regulated to control the reservoir flood storage in an optimal manner and to prevent overwhelming the spillway or overtopping the dam. This means that the engineers of the outlet works must have an intimate knowledge of the design flood (the statistical probability and approximate return period of the maximum flood for which the dam was designed), the reservoir capacity, characteristics of past flood behavior, downstream flood hazards, accurate meteorological forecasts, and a good dose of intuition.

Finally, all dams must be constructed with a spillway. A spillway is generally a broad reinforced channel near the top or around the side of the dam that acts as a safety valve because it allows rising waters, which might otherwise overtop the dam and cause its collapse, to escape harmlessly. Outlet works cannot be depended on to relieve the rising waters because they may become either blocked with debris or inoperative during a flood. The spillway must be large enough to convey the maximum probable flood.

## MULTIPLE-PURPOSE RESERVOIRS

An additional benefit of dams is that they not only provide flood control, but the reservoir can also be used as a water supply for irrigation, industrial, and municipal purposes. When the water falls from the top of the reservoir to the dam base through turbines, it generates inexpensive and pollution-free electricity. The reservoir also can be used for fishing, boating, and swimming and can become a haven for certain kinds of wildlife.

Unfortunately, a multiple-purpose reservoir contains built-in conflicts of interest. Because a reservoir used to control floods must have storage space for the floodwaters, ideally the water level should be kept low, or the reservoir kept nearly empty. Conserving water for irrigation or domestic use, however, requires holding floodwaters in storage, sometimes for years. For hydropower generation, the reservoir must be kept as full as possible and certainly never emptied. The area's fish and wildlife are best served by maintaining a stable reservoir level, as are recreational uses. Thus, the management of a multipurpose reservoir for flood control is a very complicated enterprise. The goal is to derive the maximum value from the water while keeping the threat of flood damage to a minimum. All forecasts of possible and probable floods must be carefully weighed against the need for keeping the reservoir full for other purposes.

## INTEGRATED FLOOD-CONTROL PROGRAMS

Flood protection is rarely implemented by the construction of a single dam. An integrated flood-control program also involves the construction of levees, floodways, and channel modifications. Levees are dikes or structures that attempt to confine the stream flow to its natural channel and prevent it from spreading over the floodplain. They have the advantage of increasing the flow velocity in the channel, which diminishes the deposition of sediment in it. The increased velocity, however, also increases the tendency to undercut and erode the levee. Levees block off the floodplain, but the increased volume of water in the channel raises the level to which the waters will flood. When levees are breached, the floodwaters often spill out over the floodplain suddenly, catching residents by surprise. A breached levee may also trap floodwaters downstream by preventing their return to the channel, thereby increasing the damage. A way to prevent the breaching of levees is to install emergency outlets (like dam spillways) to specially constructed floodways, or flood-diversion channels. These are a means for safely returning the river to its natural floodplain.

Channelization, or modifications to the stream channel, is also used in conjunction with flood-control dams. This generally involves straightening, deepening, widening, clearing, or lining the channel in such a way that the stream flows faster. In this way, potential floodwaters are removed from the area more quickly. The lower Mississippi River was shortened by 13 percent between 1933 and 1936 (by short-cutting meander bends), which reduced the flood levels 61 to 366 centimeters for equal rates of flow.

## DISADVANTAGES OF FLOOD-CONTROL SYSTEMS

Channelization may have deleterious effects. Erosion of the channel from faster flows may drain adjacent wildlife habitats and may undermine levees and bridges. The rapid passage of floodwaters also increases the hazard to locations downstream of the channelized area.

Flood protection by means of dams and their attendant structures has some other disadvantages. The term "flood control" is often misinterpreted by the general public to mean absolute and permanent protection from all floods under all conditions. This leads to a false sense of security and promotes further economic development of the floodplain. Every levee and every dam has a limit to its effectiveness. To compound the danger, there is always the possibility that the dam could fail entirely. The main

causes of dam failure are overflowing because of inadequate spillway capacity, internal structural failure of earthen dams, and failure of the dam foundation material. During the twentieth century, more than 8,000 people perished in more than 200 dam breaks. An earthen dam across the Little Conemaugh River above Johnstown, Pennsylvania, was overtopped and failed on May 31, 1889, killing 2,209 people. To prevent such failures, geologists and engineers must cooperate closely to design the safest possible structure matching the geology of the dam's foundation.

Dams can also cause both undesirable sedimentation and erosion of a riverbed. As the river enters the reservoir, it is forced to slow and drop its sediment load, which decreases the water-holding capacity of the reservoir and limits its usable lifetime. The resulting delta can grow upstream and engulf adjacent properties and structures. The opposite effect results from the discharge of the reservoir water, now deprived of its sediment load, back into the natural channel below the dam. Here, the water can rapidly erode the bed. At Yuma, Arizona, for example, 560 kilometers downstream of the Hoover Dam, the riverbed has been lowered by 2.7 meters. If a river formerly supplied sediments to beaches, the deprivation of this material results in coastal erosion. The loss of topsoil deposition as a result of flood control on the Nile by the Aswan High Dam has required farmers to add expensive fertilizers to their crops.

Other harmful effects of the construction of dams and flood-control projects include an increase in the water's temperature, salinity, and nutrient content (from the strong solar heating and evaporation in reservoirs and the return flow of irrigation waters). This decrease in water quality can result in undesirable weed growth in reservoirs and in fish kills. Fish that migrate up rivers to spawn, such as trout and salmon, are physically prevented from doing so by dams. Fish ladders around dams are expensive and have proved to be only partially effective. The loss of wet floodplain habitat to reservoir inundation has been detrimental to many water birds, some of which are already endangered species. Unfortunately, the reservoir itself does not usually substitute for this loss because its elevation must fluctuate. Finally, dams are expensive to build and maintain, and they cause the loss of valuable farmland because of reservoir inundation and sometimes force the relocation of entire communities. Construction of the Three Gorges

Dam, in China, a project costing some $24 billion, has required the relocation of more than 1 million people from their homes and farms, and of at least 116 towns and villages. It is now also thought to be responsible for affecting local weather patterns, increasing seismic activity, and destabilizing the surrounding land and causing landslides and other catastrophic failures. Such failures have already claimed many lives in the Three Gorges area of China.

## PREVENTION OF FLOODPLAIN MISUSE

One of the best methods to avoid the economic, social, and environmental disadvantages of dams is to prevent floodplain misuse. This is often effectively done through the use of flood zoning laws, which restrict or prohibit certain types of development in flood hazard areas. Flood insurance laws, building codes, and tax incentives have a similar effect. Appropriate uses of flood hazard zones include parkland, pastureland, forest, or farmland.

If necessary, existing buildings can be flood-proofed or relocated. Flood-proofing involves reinforcing the building against flood damage, physically raising the elevation of the building above flood levels, or both. Land-treatment procedures can improve the ability of the natural ground surface to retain water and release it slowly to streams. These techniques include reforestation, terracing, and building contour ditches and small check dams. In urban areas, rooftop or underground water retention tanks and porous pavements can be installed to reduce the risk of flooding from excessive precipitation.

## MISSISSIPPI RIVER PROJECT

The Mississippi River Project is an example of a massive flood-control project. Levees were first constructed along the banks of the Mississippi River in 1727 to protect the city of New Orleans, and they continued to be constructed (privately and haphazardly) into the 1830's. Major funds were granted in 1849 and 1850 by Congress to the U.S. Army Corps of Engineers for a flood-control study. After the Civil War, the entire 1,130-kilometer lower Mississippi Valley was organized into levee districts. In spite of the levees, major floods struck the Mississippi River in 1881, 1882, 1883, 1884, 1886, 1890, 1903, 1912, 1913, and 1927. The worst of these came in the spring of 1927 after a long winter of heavy rains, when the levees failed

in more than 120 places. Many people were killed or left homeless. This inspired Congress to pass the Flood Control Act of May 15, 1928, which provided for levee expansion and improvement, dams and reservoirs, bank stabilization coverings (revetments), floodway diversion channels, artificial meander cutoffs, and emplacement and operation of river gauging stations (to continuously monitor the water level). The first test of the system came in 1937. Although the potential for another disastrous flood was just as great, the damages caused in 1937 were far less severe than a decade prior.

Work has continued on the Mississippi. A major contribution was the 1944 project on the Missouri River, a tributary of the Mississippi River system. On the Missouri, six huge dams and a 1,500-kilometer chain of reservoirs and levees have been built. From 1950 to 1973, the combined effect of natural events and human-made structures produced a lull in the floods. The calm was broken when a massive flood in 1973 moved down the Mississippi, a significant test of the control projects. The maximum flow was approximately 56,800 cubic meters per second, enough to supply New Orleans with its daily water needs in less than 10 seconds.

Despite the flood-control facilities, losses were great. Thirty-nine levees were breached or overtopped. Property losses were estimated at $1 billion. In spite of this, the river crested 21 centimeters lower than it had in 1927 in Cairo, Illinois, and 207 centimeters lower than it had the same year in Vicksburg, Mississippi. Engineers estimated that flood-control works had reduced damages from a possible $15 billion. Still, the flood left 23 dead and another 69,000 homeless.

Changing global weather patterns, due primarily to the effects of global warming, have resulted in the occurrence of increasing amounts of rainfall in many flood-prone parts of the world, including the Mississippi River drainage basin. This effectively guarantees that higher and more frequent floods will occur in those regions. A major flood in 1993 extending from central Manitoba in Canada, along the Red River, and south throughout the Mississippi River drainage basin resulted in numerous deaths and more than $12 billion in damages. The event made obvious many weaknesses and shortcomings in the flood control system and underscored the need for continuous monitoring and maintenance of the system. It is a certainty that the potential for even larger floods exists.

## IMPORTANCE OF WISE RIVER-MANAGEMENT POLICIES

Where humans have been short-sighted in settling and building cities on floodplains, wisely built and well-managed dams and their attendant structures are an excellent way to reduce the flood hazard, though they cannot eliminate the risk entirely. The expenses that have been saved have been great, and no price can be put on the lives saved by flood control programs. At the same time, however, the environmental losses, which are harder to measure, also have been great. Wild, natural rivers are rapidly becoming one of nature's rarest possessions in the United States. As the demands for electric power generation and water resources continue to increase, managing the rivers wisely is in the nation's best long-term interest.

*Sara A. Heller*

## FURTHER READING

Aldrete, Gregory S. *Floods of the Tiber in Ancient Rome.* Baltimore: Johns Hopkins University Press, 2006. Offers a study of the floods in ancient Rome. Examines primary sources and supplements them with current knowledge of hydrology and geology to hypothesize about the physical and social effects of flooding in ancient Rome. Discusses flood prevention methods and safety precautions that were used or could have been used to protect the city. An interesting melding of history and science, accessible to all readers.

ASCE Hurricane Katrina External Review Panel. *The New Orleans Hurricane Protection System: What Went Wrong and Why.* Reston, Va.: American Society of Civil Engineers, 2007. A technical report of the events during and following Hurricane Katrina. Discusses the hurricane protection system, levees, and the results of levee failure. Reviews what can be learned from the events and what can be done to prevent future disasters.

Billington, David P., and Donald Conrad Jackson. *Big Dams of the New Deal Era: A Confluence of Engineering and Politics.* Norman: University of Oklahoma Press, 2006. Delivers a thorough analysis of the social and political forces behind the construction of large dams during the 1930's. Reviews the engineering concepts that resulted in the preferred construction of massive gravity dams over other designs. Includes a discussion of the negative ramifications of the dams after the fact.

Bowker, Michael, and Valerie Holcomb. *Layperson's Guide to Flood Management.* Sacramento, Calif.: The Foundation, 1995. Serves as an excellent handbook and introduction to flood control, floodplain management, and water drainage for persons without prior knowledge in the field. Illustrations.

Chadwick, Wallace L., ed. *Environmental Effects of Large Dams.* New York: American Society of Civil Engineers, 1978. Collects short articles concerning various environmental effects of large dams and reservoirs. Topics include harmful temperature effects; fish, algae, and aquatic weed problems; loss of wildlife habitat; erosion; and seismic activity. Contains an extensive list of dams in the United States that have experienced some earthquake difficulties. Suitable for the college-level reader. Index.

Committee on the Safety of Existing Dams, Water Science and Technology Board, Commission on Engineering and Technical Systems, and the National Research Council. *Safety of Existing Dams: Evaluation and Improvement.* Washington, D.C.: National Academy Press, 1983. A practical and not overly technical consideration of dam safety problems intended for the college-level reader. Considers hydrologic, geologic, and seismologic factors, foundation, outlet works, reservoir problems, and available instrumentation. For each problem additional technical references are included. Index.

Freitag, Bob, et al. *Floodplain Management: A New Approach for a New Era.* Washington, D.C.: Island Press, 2009. Each chapter presents a different case study that focuses on a new topic in flood control. Strategies of floodplain management revolve around the natural processes and dynamics of rivers. Discusses a multiple approaches, depending on the locations in which they are used. Best suited for engineers and hydrologists taking part in floodplain management.

Ghosh, Some Nath. *Flood Control and Drainage Engineering.* 3d ed. Boca Raton, FL: Taylor & Francis, 2006. Looks at the engineering methods and practices used in the construction and maintenance of dams and other flood-control structures. Appropriate for college-level readers.

Orsi, Jared. *Hazardous Metropolis: Flooding and Urban*

*Ecology in Los Angeles.* Berkeley: University of California Press, 2004. A history of flood control efforts in Los Angeles from 1870 to 2004. Demonstrates clearly how engineering efforts have consistently failed to meet the demands of nature; presents a view of cities as ecosystems.

Palmer, Tim. *Endangered Rivers and the Conservation Movement.* Lanham, Md.: Rowan and Littlefield Publishers, 2004. Looks at dam construction from the point of view of damaged and lost habitats, as well as the ecological and environmental effects that dams have. Focuses on the history of the conservation movement and its drive to preserve river habitats rather than sacrifice them for economic gain.

Schneider, Bonnie. *Extreme Weather: A Guide to Surviving Flash Floods, Tornadoes, Hurricanes, Heat Waves, Snowstorms, Tsunamis and Other Natural Disasters.* Palgrave Macmillan, 2012. Presents vivid explanations of how, when, and why major natural disasters occur. Discusses floods, hurricanes, thunderstorms, mudslides, wildfires, tsunamis, and earthquakes. Offers background information on weather patterns and natural disasters and presents a guide of how to prepare for and what to do during an extreme weather event.

Sene, Kevin. *Flood Warning, Forecasting and Emergency Response.* Dordrecht: Springer Science+Business Media B.V., 2008. Provides a comprehensive summary of various aspects of the title topics, based on up-to-date research.

Singh, V. P. *Dam Breach Modeling Technology.* Boston: Kluwer Academic Publishers, 2010. Discusses dam failure from the perspective of multiple science disciplines. Covers disaster mitigation, hydraulics, mathematical and analytical modeling, and data from dam breaches around the world. Intended for researchers, environmental managers, and graduate students studying disaster prevention and mitigation related to hydrology.

Smith, Norman. *A History of Dams.* London: P. Davies, 1971. Well-written and interesting historical discussion of dam building through the ages, beginning with the Egyptians and continuing to the end of the nineteenth century, but not including India or the Far East. Suitable for the high school-level reader. Photographic plates, glossary, and index.

**See also:** Artificial Recharge; Floodplains; Floods; Groundwater Movement; Groundwater Pollution and Remediation; Hydrologic Cycle; Precipitation; Salinity and Desalination; Saltwater Intrusion; Surface Water; Waterfalls; Water Quality; Watersheds; Water Table; Water Wells

# DEEP OCEAN CURRENTS

*Deep ocean currents, the dynamics of which are not yet well understood, involve significant vertical and horizontal movements of seawater. They distribute oxygen- and nutrient-rich waters throughout the world's oceans, thereby enhancing biological productivity.*

## PRINCIPAL TERMS

- **bathymetric contour:** a line on a map of the ocean floor that connects points of equal depth
- **bottom current:** a deep-sea current that flows parallel to bathymetric contours
- **bottom-water mass:** a body of water at the deepest part of the ocean identified by similar patterns of salinity and temperature
- **continental margin:** the part of Earth's surface that separates the emergent continents from the deep-sea floor
- **Coriolis effect:** an apparent force, acting on a body in motion, caused by the rotation of Earth
- **salinity:** a measure of the quantity of dissolved salts in water
- **surface water:** relatively warm seawater between the ocean surface and that depth marked by a rapid reduction in temperature
- **thermohaline circulation:** vertical circulation of seawater caused by density variations related to changes in salinity and temperature
- **turbidity current:** a turbid, relatively dense mixture of seawater and sediment that flows downslope under the influence of gravity through less dense water
- **upwelling:** the process by which bottom water rich in nutrients rises to the surface of the ocean

## EVIDENCE FOR EXISTENCE

Deep-sea currents—ocean currents that involve vertical as well as horizontal movements of seawater—are generated by density differences in water masses that result in the sinking of colder, denser water to the bottom of the ocean. For many years, however, most oceanographers refused to accept the presence of these currents. Even when the Deep Sea Drilling Project, an international effort to drill numerous holes into the ocean floor, was initiated, most researchers envisioned the deep sea as a tranquil environment characterized by sluggish, even stationary, water. More recently, however, oceanographers and marine geologists have accumulated abundant evidence to suggest the opposite: that the deep sea can be a very active area in which currents sweep parts of the ocean floor to the extent that they affect the indigenous marine life and even physically modify the sea floor.

In the 1930's, Georg Wust argued for the likelihood that the ocean floor is swept by currents. Furthermore, he suggested that these currents play an important role in the transport of deep-sea sediment. Wust's ideas were not widely accepted; in the 1960's, however, strong evidence for the existence of deep-sea currents began to accumulate. In 1961, for example, oceanographers detected deep-sea currents moving from 5 to 10 centimeters per second in the western North Atlantic Ocean. These researchers also determined that the currents changed direction over a period of one month.

In 1962, Charles Hollister, while examining cores of deep-sea sediment drilled from the continental margin off Greenland and Labrador, noted numerous sand beds that showed evidence of transport by currents. The nature of these deposits suggested to Hollister that they did not accumulate from turbidity currents, dense sediment-water clouds that periodically flow downslope from nearshore areas. Moreover, it appeared to Hollister that the sand was transported parallel to the continental margin rather than perpendicular to it, as might be expected of sediment transported by a turbidity current. He argued that the sand beds in the cores were transported by, and deposited from, deep-sea currents moving along the bottom of the ocean parallel to the continental margin. Since then, extensive photography of the ocean floor has provided direct evidence for the existence of deep-sea currents. Such evidence includes smoothing of the sea floor; gentle deflection, or bending, of marine organisms attached to the sea floor, as though they were standing in the wind; sediment piled into small ripples by saltation; and local scouring of the sea floor.

## THERMOHALINE CIRCULATION

Essentially all earth scientists now agree that the deep-sea floor is swept by rather slow-moving (less than 2 centimeters per second) currents. The driving force behind these currents, and all oceanic currents for that matter, is energy derived from the sun. Differential heating of the air drives global wind circulation, which ultimately induces surface ocean currents. The vertical circulation of seawater, and thus the generation of deep-sea currents, is controlled by the amount of solar radiation received at a point on Earth's surface. This value is greatest in equatorial regions; there, the radiation heats the surface water, the seawater that lies within the upper 300 to 1,000 meters of the ocean. As this water is heated, it begins to move toward the poles along paths of wind-generated surface circulation, such as the Gulf Stream current of the northwestern Atlantic Ocean.

The cold waters that compose the deep-sea currents originate in polar regions. There, minimal solar radiation levels produce cold, dense surface waters. The density of this water may also be increased by the seasonal formation of sea ice, ice formed by the freezing of surface water in polar regions. When sea ice forms, only about 30 percent of the salt in the freezing water becomes incorporated into the ice. The salinity and density of the nearly freezing water beneath the ice are therefore elevated. This cold, saline seawater eventually sinks under the influence of gravity to the bottom of the ocean, where it moves slowly toward the equator. Deep-sea circulation driven by temperature and salinity variations in seawater is termed "thermohaline circulation" and is much slower than surface circulation; the cold, dense water generated at the poles moves only a few kilometers per year. After moving along the bottom of the ocean for anywhere from 750 to 1,500 years, the cold seawater rises to the surface in low-latitude regions to replace the warm surface water, which, as noted above, moves as part of the global surface circulation system back to the polar regions.

Thermohaline circulation and related deep-sea currents are commonly affected by the shape of the ocean floor. Although sinking cold seawater seeks the deepest route along the sea floor, deep-sea currents may be blocked by barriers. The Mid-Atlantic Ridge, the large volcanic ridge effectively bisecting the Atlantic Ocean basin, may prevent the movement of water from the bottom of the western Atlantic to the eastern Atlantic. Conversely, the funneling of deep-sea currents through narrow passages or gaps in seafloor barriers will lead to an increase in the velocity of the current. Once beyond the passage, however, the current spreads and velocity is reduced. Because both air and water act like a fluid, these effects and behaviors are entirely analogous to those of winds produced in the atmosphere by air density differences resulting from the uneven distribution of solar energy over Earth's surface.

## CORIOLIS EFFECT

The circulation pattern of deep-sea currents is controlled to a large extent by Earth's rotation. The Coriolis effect—the frictional force achieved by Earth's rotation that causes particles in motion to be deflected to the right in the Northern Hemisphere and to the left in the Southern Hemisphere—induces deep-sea currents to trend along the western margins of the major oceans. Thus, water sinking from sources in the North Atlantic Ocean and moving south toward the equator will be deflected to the right, causing it to run along the western side of the North Atlantic. Similarly, north-directed deep-sea currents generated by the sinking of cold water from the Antarctic region will also be deflected to the western margin of the Atlantic.

The Coriolis effect guides deep-sea currents along bathymetric contour lines, lines on a map of the ocean floor that connect points of equal depth. Deep-sea currents that have a tendency to move parallel to the bathymetric contours are known as bottom currents. Barriers to flow may locally deflect deep-sea currents from the bathymetric contours; nevertheless, bottom currents are most conspicuous along the western margins of the major oceans.

## SHORT- AND LONG-TERM CONTROLS

The formation of the cold seawater required to set deep-sea currents in motion can itself be considered in terms of short- and long-term controls. Seasonal sea ice formation is probably the most important process in the production of the north-flowing water generated at the south polar region, or the Antarctic bottom water (AABW). Velocities of the AABW are highest in March and April, that period of the year when sea ice production in the ocean surrounding Antarctica is greatest. During Southern Hemisphere summers, however, the sea ice melts and there is an increase in the freshwater flux to the ocean from the continent, both of which reduce the salinity and therefore the density of the seawater, thereby decreasing AABW production.

Many oceanographers and marine geologists have argued that long-term variations in the production of the cold, dense bottom water required to generate deep-sea currents may be related to global climatic changes. More specifically, deep-sea currents appear to be most vigorous during glacial periods, when sea ice production is enhanced and the sea ice remains on the ocean surface for a greater proportion of the year. Nevertheless, there is also evidence to suggest that the velocities of deep-sea currents in the North Atlantic Ocean were much lower during the most recent glacial periods than they were during the times between glacial phases. Much more work is required to gain a more complete understanding of long-term controls on deep-sea currents.

## MEASUREMENT TOOLS AND TECHNIQUES

The most common methods for the study of deep-sea currents include direct measurement of current velocities, bottom photography, echo sounding, and the sampling of ocean-floor sediment. The speed and direction of deep-sea currents have been determined by the use of free-fall instruments, such as the free-instrument Savonics rotor current meter. This device, dropped unattached into the ocean, is capable of recording current velocities and directions over a period of several days. It returns automatically to the surface of the ocean, at which time a radio transmitter directs a ship to its position. Other current-measuring devices can be suspended at various depths in the ocean from fixed objects, such as buoys or light ships, to monitor currents for long periods. One such anchored meter measures the flow of water past a fixed point. Flowing water causes impeller blades, similar to the blades of a fan, to rotate at a rate proportional to the current's speed. In addition, the blades cause the meter to align with the current's direction. Electrical signals indicating the direction and speed of the current are transmitted by radio or cable to a recording vessel. Current velocities of less than 1 centimeter per second can be detected by this meter.

To get the most complete picture of the variability of the ocean, a combination of various measurement techniques with remote sensing may be employed. Such a multidimensional approach may involve the measurement of current velocity, pressure (a measure of depth), water temperature, and water conductivity (a measure of salinity). These data can be transmitted via satellite to a land station or even directly to a computer.

## ADDITIONAL STUDY METHODS

Perhaps the most persuasive evidence for the existence of deep-sea currents and their influence on the ocean bottom has been gained through bottom photography. Sediment waves, or ripples, apparently formed by the saltation of sediment carried by deep-sea currents, along with evidence of current-induced scour of the ocean floor, were first photographed in the Atlantic Ocean in the late 1940's. Since then, the technology of bottom photography has advanced greatly. Bottom photography permits detailed study of some of the smaller features on the ocean floor apparently formed by deep-sea currents. Benthonic, or benthic, organisms, marine organisms that live attached to the ocean floor, bending in the flow of the current, are a particularly intriguing example of the phenomena recorded by this technique.

Echo-sounding studies of the sea floor have yielded abundant information on ocean-floor features that are either formed or modified by deep-sea currents. Notable among these are very long ridges in the North Atlantic evidently constructed from sediment carried by deep-sea currents. In echo sounding, a narrow sound beam is directed from a ship vertically to the sea bottom, where it is reflected back to a recorder on the ship. The depth to the sea floor is determined by multiplying the velocity of the sound pulse by one-half the amount of time it takes for the sound to return to the ship. The depths to the ocean floor are recorded on a chart by a precision depth recorder, which produces a continuous profile of the shape of the sea floor as the ship moves across the ocean.

Sediment transported by and deposited from deep-sea currents can be studied directly by actually sampling the ocean floor. Sampling of these deposits is best accomplished by the use of various coring devices capable of recovering long vertical sections, or cores, of seafloor sediment. Sediment recovery is achieved by forcing the corer, a long pipe usually with an inner plastic liner, vertically into the sediment. The simplest coring device, the gravity corer, consists of a pipe with a heavy weight at one end. This type of corer will penetrate only 2 to 3 meters into the sea floor. The piston corer, used to obtain longer cores, is fitted with a piston inside the core tube that reduces friction during coring, thereby permitting the recovery of 18-meter or longer cores. Analysis of the sediment recovered from the ocean floor by these and other coring devices reveals much information about small-scale features formed by deep-sea currents.

## IMPORTANCE TO LIFE ON EARTH

Because cold bottom-water masses often are nutrient-rich and contain elevated abundances of dissolved oxygen, deep-sea currents are extremely important to biological productivity. There are areas of Earth's surface where nutrient-rich cold bottom waters rise to the ocean surface. These locations, known as areas of upwelling, are generally biologically productive and therefore are important food sources. Especially pronounced upwelling occurs around Antarctica. Bottom waters from the North Atlantic upwell near Antarctica and replace the cold, dense, sinking waters of the Antarctic.

The great amount of time required for seawater to circulate from the surface of the ocean to the bottom and back again to the surface has become an important practical matter. If pollutants are introduced into high-latitude surface waters, they will not resurface in the low latitudes for hundreds of years. This delay is particularly important if the material is rapidly decaying radioactive waste that may lose much of its dangerous radiation by the time it resurfaces with the current. The introduction of toxic pollutants into a system as sluggish as the deep-sea circulation system, however, means that they will remain in that system for prolonged periods. Nations must, therefore, be concerned with the rate at which material is added to this system relative to that at which it might be redistributed at the surface of the ocean by wind-induced surface circulation. The multinational Geochemical Ocean Sections (GEOSECS) program, introduced as part of the International Decade of Ocean Exploration, attempted to better assess the problem of how natural and synthetic chemical substances are distributed throughout the world's oceans. The GEOSECS program, carried out from 1970 to 1980, yielded abundant information regarding the movement of various water masses and, among other things, the distribution of radioactive material in the oceans. For example, GEOSECS demonstrated that tritium produced in the late 1950's and early 1960's by atmospheric testing of nuclear

weapons had been carried to depths approaching 5 kilometers in the North Atlantic Ocean by 1973.

*Gary G. Lash*

## FURTHER READING

Baker, D. J. "Models of Oceanic Circulation." *Scientific American* 222 (January, 1970): 114. A somewhat complex discussion of surface circulation in the world's oceans. Generally suitable for college-level readers.

Broecker, Wally. *The Great Ocean Conveyor.* Princeton, N.J.: Princeton University Press, 2010. Discusses ocean currents, focusing specifically on the great conveyor belt. Written by the great ocean conveyor's discoverer; explains the conception of this theory and the resulting impact on oceanography. Written in an easy-to-follow manner, yet still relevant to graduate students and scientists.

Colling, Angela. *Ocean Circulation.* 2d ed. Oxford: Butterworth-Heinemann, 2001. Discusses the effects of ocean circulation as reflected in various phenomena.

Garrison, Tom S. *Oceanography: An Invitation to Marine Science.* Belmont, Calif.: Brooks/Cole, Cengage Learning, 2010. Discusses the circulation of the oceans, including deep-water and surface currents. Describes various aspects of waves and the physics of tides. Provides abundant diagrams to aid readers from the layperson to advanced undergraduates.

Goni, G. J., and Paola Malanotte-Rizzoli. *Interhemispheric Water Exchange in the Atlantic Ocean.* Amsterdam: Elsevier B.V., 2003. While the focus of this book is the dynamics of the tropical Atlantic Ocean, the overall goal is to relate those dynamics to the ocean globally through the operation of deep ocean currents. Written for ocean science specialists.

Hollister, C. D., A. Nowell, and P. A. Jumar. "The Dynamic Abyss." *Scientific American* 250 (March, 1984): 42. Addresses the formation of bottom waters that flow away from the polar regions toward the equator. Suitable for high school students.

Ittekko, Venugopalan, et al., eds. *Particle Flux in the Ocean.* New York: John Wiley and Sons, 1996. Contains descriptions of the chemical and geobiochemical cycles of the ocean, as well as the ocean currents and movement. Suitable for the high school reader and beyond. Illustrations, index, bibliography.

Jeleff, Sophie, ed. *Oceans.* Strasburg: Council of Europe, 1999. This collection of debates and lectures is a detailed account of oceanography and ocean ecology. Discusses ocean circulation, marine chemistry, and the ocean's structure. Illustrations and bibliography.

Kennett, James P. *Marine Geology.* Englewood Cliffs, N.J.: Prentice-Hall, 1982. Contains an excellent discussion of deep-sea currents and thermohaline circulation. Discusses the major methods of study of deep-sea currents, including bottom photography (there are two pages of black-and-white bottom photographs). Best suited to the college student.

Oceanography Course Team. *Ocean Circulation.* 2d ed. Oxford: Butterworth-Heinemann, 2001. Discusses surface currents and deep water currents, with a focus on the North Atlantic Gyre, Gulf Stream, and equatorial currents. Describes the El Niño phenomenon and the great salinity anomaly. Offers a good introduction to oceanography.

Ross, David A. *Introduction to Oceanography.* 5th ed. New York: HarperCollins College Publishers, 1995. A fine introductory oceanography textbook with an informative discussion of deep-sea currents and their mechanisms of generation. Includes a section on oceanographic instrumentation. Suitable for high school students.

Siedler, Gerold, John Church, and John Gould. *Ocean Circulation and Climate: Observing and Modelling the Global Ocean.* London: Academic Press, 2001. Written to guide the reader consistently through the broad range of world ocean circulation experiment science, with cross-referenced contributors' list, a comprehensive index, and unified reference list. Easily readable at the undergraduate level.

Smith, F. G. "Measuring Ocean Currents." *Sea Frontiers* 18 (May, 1972): 166. Discusses methods used to determine the speed and direction of ocean currents.

Teramoto, Toshihiko. *Deep Ocean Circulation: Physical and Chemical Aspects.* New York: Elsevier, 1993. Provides a detailed look at ocean circulation and currents. Offers much information on the chemical processes that occur in the deep ocean. Illustrations, maps, and bibliographical references.

Trujillo, Alan P., and Harold V. Thurman. *Essentials of Oceanography.* 10th ed. Upper Saddle River, N.J.: Prentice-Hall, 2010. Describes deep ocean currents in the broader context of the whole ocean and oceanography. Uses a systems approach that is useful to all earth science students.

Vallis, Geoffrey K. *Atmospheric and Oceanic Fluid Dynamics: Fundamentals and Large-Scale Circulation.* New York: Cambridge University Press, 2006. Begins with an overview of the physics of fluid dynamics to provide foundational material on stratification, vorticity, oceanic and atmospheric models. Discusses topics such as turbulence, baroclinic instabilities, wave-mean flow interactions, and large-scale atmospheric and oceanic circulation. Best suited for graduate students studying meteorology or oceanography.

**See also:** Carbonate Compensation Depths; Deep-Sea Sedimentation; Gulf Stream; Hydrothermal Vents; Observational Data of the Atmosphere and Oceans; Ocean-Atmosphere Interactions; Ocean Pollution and Oil Spills; Oceans' Origin; Oceans' Structure; Ocean Tides; Ocean Waves; Remote Sensing of the Oceans; Sea Level; Seamounts; Seawater Composition; Surface Ocean Currents; Tsunamis; Turbidity Currents and Submarine Fans; World Ocean Circulation Experiment

# DEEP-SEA SEDIMENTATION

*Deep-sea sedimentation occurs by the settling of particles to the ocean floor and by the transport of material from shallow to deep water. Knowledge of the distribution of sediments and the processes of sedimentation will help to evaluate the potential of the world's oceans for the mining of natural resources and the storage of waste.*

## PRINCIPAL TERMS

- **calcareous ooze:** sediment in which more than 30 percent of the particles are the remains of plants and animals composed of calcium carbonate
- **carbonate compensation depth (CCD):** the depth in the oceans at which the rate of supply of calcium carbonate equals the rate of dissolution of calcium carbonate
- **clay:** a sediment particle less than 0.1 millimeter in diameter
- **clay minerals:** a group of hydrous silicate minerals characterized by a structure of layers of thin sheets that are held together loosely
- **pelagic:** "of the deep sea"; refers to sediments that are fine-grained and are deposited very slowly at great distances from continents
- **siliceous ooze:** fine-grained sediment in which more than 30 percent of the particles are organic remains of plants and animals composed of silica
- **turbidity current:** a mass of water and sediment that flows downhill along the bottom of a body of water because it is denser than the surrounding water; common on continental slopes
- **upwelling:** an ocean phenomenon in which warm surface waters are pushed away from the coast and are replaced by cold waters that carry more nutrients up from depth

## INVESTIGATION OF DEEP-SEA SEDIMENTS

Sedimentation in the deep sea differs substantially from that in any other environment. Deep-sea sedimentation is pelagic. It typically occurs at great distances from continents and at depths in excess of 1,000 meters, primarily by the settling of particles through the overlying water. Every 100,000 liters of seawater contains about 1 gram of extremely small particles. Their settling is extremely slow, and sediments accumulate at rates on the order of a few millimeters per thousand years. Transport of material by turbidity currents along the bottom of the ocean from shallow to deep regions is also a mechanism by which deep-sea sediments are deposited, but this is far less important than the settling of particles from the surface. In contrast, current transport dominates sedimentation in all other environments, where sedimentation rates typically exceed several centimeters per year.

The oceans cover 70 percent of Earth, and, as a result, deep-sea sedimentation may be the most common sedimentary process. Early investigations of the seas were limited, however, to coastal processes and oceanic circulation because of the lack of techniques available to study the deep ocean bottom. Knowledge of deep-sea sedimentation was profoundly increased by the voyage of HMS *Challenger* from 1872 to 1876, which made the first systematic study of the ocean floor. Prior to the *Challenger* expedition, practical investigations of the oceans concentrated on winds, tides, and water depths in harbors to provide information for commercial navigation and for the laying of submarine telegraph cables. The *Challenger* crew collected data on the temperature and depth of the oceans, marine plants and animals, and sediments on the ocean bottom. Deep-sea sediments were described by John Murray, naturalist to the expedition. He noted that sediments in the deepest parts of the ocean are fine-grained and are composed of particles that settled from the surface. Comparison of rocks exposed on land with those dredged from the ocean led to the assertion that deep-sea sediments could not be found on land. This conclusion caused great debate, and the interpretation of some rocks as deep-sea deposits by a few scientists persisted despite opposition by most of the scientific community.

Investigation of deep-sea sedimentation flourished again during and immediately after World War II. Despite the voluminous amount of data available by 1950, the prevailing doctrine in the first half of the twentieth century was similar to that in the previous century: The continents and oceans were generally believed to be permanent features of Earth's surface that had developed to their present form near the beginning of geologic time. Thus, the ocean basins were billions of years old, ocean-bottom sediments were tens of kilometers thick, having accumulated from early in the geological history of Earth to the present, and deep-sea sediments were not exposed on land, a point that was still severely contested. Then, during the 1950's two discoveries shocked the oceanographic community. First, fossils taken from seamounts (submarine mountains) in the Pacific Ocean at depths of 2 kilometers were only 100 million years old, suggesting that subsidence of the ocean floor was recent. Second, deep-sea sediments were found to be less than 200 meters thick. The advent of the theory of plate tectonics in the 1960's profoundly changed ideas about the ocean basins by providing both an explanation for the two observations of the previous decade and a mechanism for the emplacement of deep-sea sediments onto continents. In addition, the motion and cooling of the oceanic plates predicted by plate tectonics explain the global distribution of deep-sea deposits.

## PELAGIC SEDIMENTS

Pelagic sediments are characterized by their fine grain size. Few particles are larger than 0.025 millimeter, and most are smaller than 0.001 millimeter. Particles include terrigenous debris (derived from the continents), such as clay minerals, quartz, and feldspar; volcanogenic grains (derived from volcanoes), such as volcanic glass, pumice, and ash; biogenic material (derived from living organisms), such as fecal pellets (waste produced by organisms inhabiting the surface waters) and skeletons of planktonic plants and animals; and cosmogenic matter (derived from space or the cosmos), such as "space dust" and pieces of meteorites. Pelagic sediments that contain more than 30 percent biogenic material are called oozes. If the biogenic debris is composed of calcium carbonate, the sediment is a calcareous ooze. If the biogenic debris is composed of silica, the sediment is a siliceous ooze. Most calcareous oozes consist of foraminiferans, which are single-celled animals that secrete calcium carbonate and live in the surface waters. In contrast, siliceous oozes contain diatoms (green, unicellular algae) in cold water and radiolarians (silica-secreting single-celled animals) in warm water. Pelagic clays are sediments that contain less than 30 percent biogenic debris.

The two major controls on pelagic sedimentation are the calcium compensation depth (CCD) and the fertility of the surface waters. The CCD reflects the interplay between the release of carbon dioxide during the decay of surface organisms and the dissolution of the calcium carbonate skeletons as they descend through the ocean. Above the CCD, calcareous oozes are dominant, whereas below the CCD, siliceous oozes and pelagic clays are common. For example, in one region of the ocean, sediments decreased from 90 percent calcium carbonate at a depth of 1,000 meters to 20 percent calcium carbonate at a depth of 5,000 meters. Oozes accumulate below areas of high fertility where upwelling brings nutrient-rich bottom water to the surface. Because upwelling creates increased productivity of both calcareous and siliceous organisms, siliceous oozes form only where the water depth exceeds the CCD. Calcareous oozes, the most abundant biogenic sediments in the ocean, cover only 15 percent of the Pacific Ocean bottom but blanket 60 percent of the Atlantic Ocean floor. This difference reflects the combination of a shallower CCD and greater water depth in the Pacific Ocean than in the Atlantic Ocean. Siliceous oozes are in areas of elevated CCDs such as the equatorial regions, the subarctic and subantarctic zones, and the continental margins of northwestern South America and eastern Africa.

Sedimentation of pelagic clays is restricted to the central portions of the oceans where the fertility is low. This reflects the difference in settling rates between nonbiogenic particles and fecal pellets. Because of its small size, a particle may take many years to sink through the ocean. The concentration of many particles into fecal pellets by organisms feeding at the surface, therefore, is thought to be an important mechanism by which debris settles to the bottom. The greater size of the pellets allows them to fall much faster than unconsolidated particles. It is only in areas of low productivity, therefore, where biogenic material does not overwhelm the sediment. Wind is the primary means by which nonbiogenic particles reach the sea. Dust is carried in the atmosphere at great distances from continents and constitutes as much as 10 percent of pelagic sediment.

## OTHER PROCESSES

Additional processes in deep-sea sedimentation include the rafting of glacial debris by ice, the transport of terrigenous and shallow-water material by turbidity currents along the ocean floor, and the precipitation of nodules on the ocean bottom. Pieces of ice calve off glaciers in the polar regions and float toward warmer waters where they eventually melt, releasing glacial sediment into the ocean. Although this process is most common in the south polar region today, glacial sediments greater than ten thousand years old elsewhere on the ocean floor suggest the process extended over a much greater area in the past. Turbidity currents generated along continental slopes may traverse hundreds of kilometers of ocean basin where ridges and depressions do not occur. Far away from continents, the currents commonly contain only silt but may deposit the grains over great distances. These currents also may locally erode the sea floor, creating breaks in the continuity of the sediment pile by removing the most recent sediments and exposing older sediments at the abraded surface. Chemical reactions of seawater with the surface of the sediment cause the precipitation of ferromanganese (composed of iron and manganese) nodules in areas where sediments accumulate slowly. The nodules form by nucleating around a manganese source in the sediment and by accreting minerals from the seawater. Such nodules may cover more than 40 percent of the sea floor.

Sedimentation in the deep sea is a complex interaction between water depth, fertility of the surface waters, and current transport. To illustrate how the depth dependence of sedimentation affects the global distribution of sediments, one must understand the relationship between plate tectonics and the topography of the ocean floor. At a midocean ridge, which rises above the surrounding ocean basin, molten rock is extruded to form new sea floor. Calcareous oozes accumulate on the new sea floor because of its depth above the CCD. As more sea floor forms, the older sea floor moves away from the midocean ridge toward the deep ocean basin (seafloor spreading), sinking to lower depths as it does so. The old sea floor gradually descends below the CCD, where siliceous oozes accumulate above the calcareous oozes. Eventually the motion of the sea floor carries it below a region of low surface productivity such that pelagic clays are deposited above the siliceous oozes. This vertical sequence has been documented in several localities throughout the world's oceans.

## STUDY OF DEEP-SEA SEDIMENTATION

Techniques used to study sedimentation in the deep ocean are diverse and include piston coring,

seismic reflection profiling, the use of deep-diving submarines, and experiments performed both at sea and in the laboratory. Piston coring is the most important technique because it enables direct sampling of the ocean floor. It uses the same kind of rotary-drilling equipment that is standard in petroleum exploration to drill into the top few hundred meters of the sea floor. The cores are 7 centimeters in diameter and provide samples that are large enough to preserve features that yield insights into the processes by which the sediment accumulated. For example, current-produced features in some samples were the first direct evidence of currents in the deep seas. In addition, piston coring confirmed the vertical sequence of sediments predicted by plate tectonics.

To map the ocean floor using a 7-centimeter drill core is time-consuming and inefficient. Seismic reflection profiling has the advantage of providing information about the thickness and distribution of layers in the sediments on the sea floor relatively quickly. The two major disadvantages are that it cannot determine the composition of the sediment and that it identifies only layers that are tens of meters thick. In seismic reflection profiling, a low-frequency sound source is towed over the bottom. The sound waves penetrate and are reflected by the various layers within the sea floor to several kilometers below the boundary between the sediment and the water. The returning signals are displayed on strip charts that reveal the water depth and the thickness and pattern of layers in the sediments. One of the most significant early contributions of seismic reflection profiling was the evidence that sediments on the ocean floor were too thin to have been accumulating since early in Earth's history, an observation explained by plate tectonics.

The use of deep-diving submarines has allowed scientists to observe the ocean floor directly. Previously the ocean floor was only photographed. Deep-diving submarines allowed scientists to confirm or reject hypotheses based on other indirect techniques. The ranges and endurance times of the submersible vessels, however, are limited. In addition, they frequently cannot be used in areas where current velocities are greater than 1 meter per second.

Simple experiments either at sea or in the laboratory also can be illuminating. For example, the existence of the CCD was verified by lowering spheres of calcium carbonate to various water depths in the

Pacific Ocean. When the spheres were examined, it was discovered that the spheres in the deepest water had dissolved the most. Scientists in the laboratory conduct experiments on settling rates by dropping particles of various sizes in large beakers and measuring the time necessary for the particles to reach the bottom. Thus the amount of time represented by the thickness of sediment on the ocean floor was roughly estimated at a few hundred million years.

One of the most important developments in deep-sea sedimentation was the creation in 1965 of the Deep Sea Drilling Project (DSDP), now known as the Ocean Drilling Project (ODP), to drill and investigate the ocean floor. The project began in earnest in 1968 with the maiden voyage of the American ship the *Glomar Challenger*, which was outfitted with state-of-the-art navigational, positioning, and drilling equipment. International teams of scientists boarded the ship in ports all over the world and remained at sea for two months. Since its inception, ODP has undertaken more than sixty-five voyages and contributed significantly to scientists' understanding of the ocean.

## SIGNIFICANCE

The oceans cover nearly three-fourths of the surface of Earth. To study deep-sea sedimentation, therefore, is to understand one of the most globally prevalent processes. Understanding deep-sea sedimentation is important for several other reasons, however. First, sediments on the ocean floor yield insight into bottom currents and water depths, which affect the navigation of submarines and the installation of underwater communications cables. Second, the sediments preserve the climatic record of the past few hundred thousands of years. For example, glacial deposits approximately ten thousand years old are found over a much greater area of the bottom of the ocean than recent glacial deposits, suggesting that transport of glacial debris by ice was very common ten thousand years ago. This implies that the temperature of Earth's environment was lower in the past, allowing the formation of significant quantities of ice. Scientists can use this information to gauge present-day climate patterns and predict future fluctuations in the global climate. Third, the seas receive a significant amount of the garbage produced by modern society. Knowledge of sedimentation rates and currents in the deep sea helps to establish the length of

time necessary to bury the waste, the direction the waste will travel prior to its burial, and the effect of dumping waste on the overall health of the ocean.

Fourth, and most important, understanding deep-sea sedimentation and the global distribution of deep-sea sediments provides a framework within which one can estimate the economic potential of the deep sea and minimize the damage from ocean exploitation. As the reserve of natural resources on the continents dwindles because of expanding demand from high-technology industries, exploding population growth, and increasing consumer appetite, attention focuses on the oceans as a possible source of raw materials. Important resources that may be found locally in the deep sea include oil and gas. The decay and accumulation of plants and animals are essential to the generation of hydrocarbons and are processes that occur over a large area of the sea floor. Emphasis also is placed on mining the deep sea for important materials such as manganese, gold, cadmium, copper, and nickel. Similar to most new industries, large-scale exploitation of the natural resources of the deep sea is limited at present by the current state of technology. The incentive to mine the oceans, however, will grow as cheaper and more sophisticated technologies are created.

*Pamela Jansma*

## FURTHER READING

Blatt, Harvey, Robert Tracy, and Brent Owens. *Petrology: Igneous, Sedimentary, and Metamorphic.* 3d ed. New York: W. H. Freeman, 2005. Describes sedimentary rocks and processes. Provides an excellent introduction to mechanics of sedimentation, classification of sediments, features of sediments that form under different conditions of water flow, environments of sedimentation, and transformation of sediments into rocks. Some chapters assume quantitative knowledge. Suitable for the college-level student and the layperson who is technically inclined.

Coe, Angela L., ed. *The Sedimentary Record of Sea-Level Change.* New York: Cambridge University Press, 2003. Discusses changes in the sea level throughout time. Discusses the factors influencing the sea level, including ice ages and sedimentation. Includes multiple case studies, chapter summaries, references, and an index. Suited for undergraduates and graduate students.

Grasshoff, Klaus, Klaus Kremling, and Manfred Ehrhardt, eds. *Methods of Seawater Analysis*. 3d ed. New York: Wiley-VCH, 1999. Describes the techniques and instrumentation used in determining the chemical makeup of seawater. Illustrations, bibliography, and index.

Grotzinger, John, et al. *Understanding Earth*. 5th ed. New York: W. H. Freeman, 2006. An introductory text on the evolution of Earth, designed to accompany a beginning class on geology. Discusses sedimentary rocks, the oceans, plate tectonics, and deep-sea sediments. Suitable for high school students.

Gurvich, Evgeny G. *Metalliferous Sediments of the World Ocean*. Berlin: Springer-Verlag, 2010. Discusses sedimentation in the Pacific and Indian Oceans. Chapters discuss the metalliferous sediment and sedimentation in hydrothermal plumes.

Huneke, H., and T. Mulder. *Deep-Sea Sediments*. Amsterdam: Elsevier, 2011. Focuses on the sedimentary processes operating within the various modern and ancient deep-sea environments, using new information from recently developed sophisticated exploration techniques that have been driven by immense industrial interest in deep-sea sediments.

Joseph, P. *Deep Water Sedimentation in the Alpine Basin of SE France*. Special Publication 221. London: Geological Society of London, 2004. A compilation of articles discussing stratigraphy, depositional models, and basin-floor topography. Articles are very technical; best suited for researchers.

McKinney, Frank. *The Northern Adriatic Ecosystem: Deep Time in a Shallow Sea*. New York: Columbia University Press, 2007. Covers the paleogeography of the Adriatic Sea. Discusses the succession of the ecosystem as the sea's geography changed. Discusses such oceanography topics as circulation and sedimentation. Topics are well described and logically ordered, making this book accessible to undergraduates.

Murray, John, and Alphonse J. Renard. *Deep Sea Deposits*. Edinburgh: Neill, 1891. Written by two members of the crew of the *Challenger*, which made the pioneering voyage in deep-sea studies.

An excellent source for the historical framework of deep-sea investigations; dramatically emphasizes the importance of observation in science.

Reading, H. G., ed. *Sedimentary Environments: Processes, Facies, and Stratigraphy*. 3d ed. Cambridge, Mass.: Blackwell Science, 1996. Discusses sediments in terms of the environment (beach, pelagic) in which they occur instead of by type (sand, mud). An excellent text for the comparison of modern and ancient deep-sea sediments. Exceptionally well illustrated; references list is thorough. Suitable for college-level students.

Stein, Ruediger. *Arctic ocean Sediments: Processes, Proxies and Paleoenvironment*. Amsterdam: Elsevier, 2008. Written by a leading scholar and industry expert. Presents background research, recent developments, and future trends, aimed at specialists and graduates in this field.

Trujillo, Alan P., and Harold V. Thurman. *Essentials of Oceanography*. 10th ed. Upper Saddle River, N.J.: Prentice-Hall, 2010. An introductory college text designed to give the student a general overview of the oceans in the first few chapters. Provides a well-designed, in-depth study of the ocean involving chemistry of the ocean, currents, air-sea interactions, the water cycle, and marine biology. Includes a well-developed section on the practical problems resulting from human interaction with the ocean, such as pollution and economic exploitation.

Usdowski, Eberhard, and Martin Dietzel. *Atlas and Data of Solid-Solution Equilibria of Marine Evaporites*. New York: Springer, 1998. Offers the reader an illustrated guide to seawater composition, including phase diagrams of seawater processes. Accompanied by a CD-ROM that reinforces the concepts discussed in the chapters.

**See also:** Alluvial Systems; Beaches and Coastal Processes; Deltas; Desert Pavement; Drainage Basins; Floodplains; Lakes; Ocean-Atmosphere Interactions; Ocean Tides; Ocean Waves; Reefs; River Bed Forms; River Flow; Sand; Sea Level; Sediment Transport and Deposition; Surface Ocean Currents; Tsunamis; Weathering and Erosion

# DELTAS

*Deltas are dynamic sedimentary environments that undergo rapid changes over very short periods of time. Found in lakes and shallow ocean waters, they are rich in organic material and provide food and shelter for fish and wildlife.*

## PRINCIPAL TERMS

- **crevasse:** a break in the bank of a distributary channel causing a partial diversion of flow and sediment into an interdistributary bay
- **delta:** a deposit of sediment, often triangular, formed at a river mouth where the wave action is low and the river's current slows suddenly
- **distributary channel:** a river that is divided into several smaller channels, thus distributing its flow and sediment load
- **geoarchaeology:** the technique of using ancient human habitation sites to determine the age of landforms and when changes occurred
- **interdistributary bay:** a shallow, triangular bay between two distributary channels; over time, the bay becomes filled with sediment and colonized with marsh plants or trees
- **natural levee:** a low ridge deposited on the flanks of a river during a flood stage
- **prodelta:** a sedimentary layer composed of silt and clay deposited under water; it is the foundation on which a delta is deposited
- **sediment:** fragmented rock material such as gravel, sand, silt, or clay that is deposited by a river to form a delta
- **wave energy:** the capacity of a wave to erode and deposit; as wave energy increases, erosion increases

## FORMATION AND SHAPE

Deltas contain many valuable resources. Government agencies such as the U.S. Fish and Wildlife Service study the surface properties of deltas because of the enormous wetlands and abundant wildlife that occupy these landforms. Geologists study deltas because they are favored places for the accumulation of oil and gas resources. This low topographical feature serves society in many ways, and that is why it has been the object of intense study.

Deltas are deposits of sediments, such as sand or silt, that are carried by rivers and deposited at the shoreline of a lake, estuary, or sea. As the river meets the water body, its velocity is greatly decreased, which

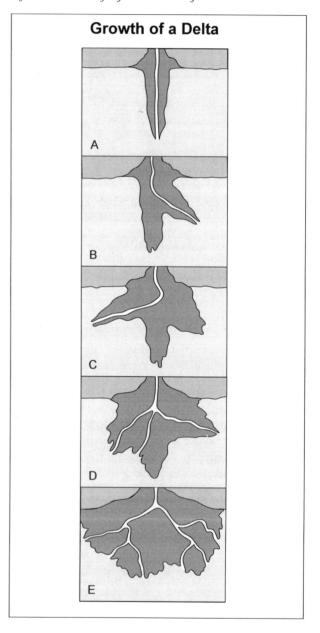

**Growth of a Delta**

reduces its ability to support the sediment load that it has been carrying. As a result, the river sediment becomes deposited. If the accumulated sediment is not removed by waves or currents, a delta will accumulate and continue to extend itself into the lake or ocean.

156

The term "delta" is used to describe this depositional landform because it is often triangular. It is believed that the Greek historian Herodotus coined the term with the shape of the capital Greek letter *delta* in mind. Herodotus visited Egypt, where the Nile Delta is located, and he correctly defined the shape of that delta. Not all deltas are triangular, however.

The Ganges Delta, the Colorado River Delta, the Mississippi River Delta, and many other deltas have different shapes. The shape postulated by Herodotus is, in fact, somewhat unusual, but it is applicable to the Nile River Delta. The Nile Delta has a smooth but curved shoreline –and so is designated an "arcuate" delta, whereas the Mississippi Delta has been extended into the Gulf of Mexico and has spreading channels resembling the digits of a bird's foot. Current and wave action at the shoreline can cause sediment to be distributed to the left and right of a river channel, forming smooth beaches on either side. Such a delta acquires a shape more like a cone, the point of which projects toward the sea, and is called a "cuspate" delta. The Tiber River, which empties into the Mediterranean Sea, is a classic example of this type of delta. Rivers such as the Seine, in France, may deposit sediment in elongated estuaries, forming shoals and tidal flats.

Earth scientists have noted that the shapes of deltas are associated with several conditions, such as the character of river flow, the magnitude of wave energy and tides, and the geologic setting. The bird-foot delta of the Mississippi River has extended itself well into the Gulf of Mexico because the river carries and deposits a high volume of sediment on a shallow continental shelf. The wave and tidal forces are low, and the delta deposit is not redistributed along the shoreline or swept away. Conversely, a cuspate delta, such as that of the Tiber, is a product of strong waves moving over a steep continental shelf. Persistent high wave energy redistributes the sediment often, forming beaches and sand dunes along the delta shoreline. The Nile Delta is an arcuate delta characterized by moderately high wave energy and a modest tide range. Occasional high wave conditions deposit beaches and sand dunes along the arc-shaped delta front at the Mediterranean Sea. Tides also play a direct role in the creation of deltas. Deltas in estuaries are formed because of a high tidal range coupled with low wave conditions. The Seine estuary, with its distinctive mud flats exposed at low tide, provides a good example.

## LANDFORMS

Although deltas have different shapes that reflect differences in the intensity of river, wave, current, and tidal processes, certain landforms may be identified as characteristic of delta formation. Submarine features are deposited below sea level, and subaerial features form at or just above sea level. As a river empties into the sea, the finest sediments, usually very fine silt or clay, are deposited offshore on the sea floor. This submarine deposit forms the foundation on which the delta sits and is appropriately referred to as a "prodelta deposit." The deposit can often be detected on navigation maps as a relatively shallow, semicircular deposit under the water.

As deposition continues, the prodelta deposit is covered with the extending subaerial delta, which is composed of coarser sediments. Deltaic extension occurs along the distributary channels. During higher river flow, the distributary channels overflow, depositing natural levees along their sides. The digitate distributary pattern of the Mississippi Delta illustrates this process well. As the distributaries extend to deeper water, the shallow areas between the distributaries are better developed. These areas, known as interdistributary bays, are shallow landforms colonized by aquatic plant life. Over time, deposition occurs in the interdistributary bays through breaches in the natural levees. As the river mouth distributaries enter a flood stage, the lower regions of a natural levee are broken, and fine suspended sediments are introduced into the interdistributary bay area. Such overbank splays, or crevasse splays, are primarily responsible for the infilling of a delta. The crevassing is usually a very rapid but short-lived process, occurring during a high river stage and operating over a ten- to fifteen-year period. With the passage of time, the open-water interdistributary bays become silt-filled and colonized. Eventually, however, the marshy bays may subside due to compaction of the sediment, creating water areas once again.

## GEOLOGIC HISTORY

Although the geologic history of large deltas such as the Mississippi is complex, the succession and behavior of shifting deltas have been determined in some detail. Over the past twenty thousand years, the large continental glaciers that occupied much of the upper midwestern United States began to melt. As Earth's climate continued to warm, the meltwater was returned to the oceans, and the sea level rose some 100 meters, inundating valleys that had previously been cut by streams. The oceans reached their present approximate level about five thousand years ago. The Mississippi and similar valleys were flooded and became elongated bays. Over time, the shallow water bays were choked with sediments that formed broad floodplains extending down the valleys. Once a depositing river extended beyond the confines of its valley, a delta was deposited in deeper water. Because the river was no longer confined, it was free to shift over greater distances. The Mississippi River Delta is actually composed of seven distinct delta lobes extending over an approximate distance of 315 kilometers. The oldest delta, Salé Cypremort, was deposited some 4,600 years ago; the most modern delta was deposited within the past 550 years. Older Mississippi Delta lobes, such as the Teche Delta, have subsided since they were deposited, giving an opportunity for a more recent delta (in this case, the Lafourche) to be deposited on top and more seaward of the older feature. The different delta lobes making up the enormous deltaic plain have resulted from a shifting of the Mississippi River well upstream in its valley. This process may be visualized as a hand movement occurring because of a shoulder movement.

Because of significant changes in the shoreline environments, many deltas in marine coastal zones are eroding. The Mississippi River is at the edge of the continental shelf and cannot build out into deep water. Also, subsidence and rising sea levels are causing the delta to erode. The Nile Delta is eroding as well. With the construction of the High Aswan Dam upstream, there has been a significant decrease in the sediment supplied to the Nile Delta. This lack of sediment, along with a slight rise in sea level, has led to erosion. Some earth scientists have suggested that the wave action in coastal Egypt is cutting back the Nile Delta at a rate of 15 to 30 meters per year in some areas.

## STUDY OF DELTAS

Deltas are difficult to study because many of the features are very flat, marshy, or under water. Because deltas change rapidly over time, however, maps are an important tool with which to determine changes. Navigation maps and maps that illustrate the topography of coastal areas around the world have been made for generations. By comparing the size and location of a delta on old maps and new maps, changes can be analyzed. Also, aerial photographs and pictures taken from satellites aid in identifying the erosion and deposition of delta landforms.

Often, older delta lobes were settled by ancient peoples. Through the science of geoarchaeology, it can be determined when changes occurred. As deltas such as the Mississippi shift from side to side over time, the human population follows the deltas from place to place. By examining the location of archaeological sites over the past fifteen hundred years, scientists have determined the minimum age of the several deltas forming the Mississippi deltaic plain. The Indian pottery found there reveals that the delta framework was deposited very recently. Cultural remains indicate other environmental changes—in salinity, subsidence, and delta deterioration.

By boring holes into the soft sediments of a delta, geologists can decipher its subsurface aspects. Because oil and natural gas are often associated with deltas, oil companies have bored holes in many delta landforms. Information derived from this method of study reveals the composition of the thick delta sediments and the rate of delta accumulation. In fact, boreholes in some deltas have indicated that older deltas once existed and are now buried beneath younger deltas.

Finally, deltas can be created in the laboratory. In nature, deltas are very large and complex. To make the study of deltas less difficult, scientists use tanks filled with water and sediments. Experiments can be performed that, for example, control the amount of sediment used to build deltas. Relationships between sediments and current velocities may be studied to gather information on such properties as the rate of delta growth. By controlling the phenomena that cause deltas to form, geologists can gain an overview of the behavior and processes of delta development.

## SIGNIFICANCE

Deltas, with their marshes and bogs, are not generally aesthetically pleasing; however, depositional landforms have been useful to prehistoric and historic populations in many ways. Deltas, along with estuaries, are among the most biologically productive areas on Earth. Most deltas are colonized with wetland swamps or marshes, which are breeding areas for wildlife. In many countries, for example, duck hunting is a popular sport associated with wetlands.

In marine deltas where there is tidal influence, fresh-water and saltwater mix. The river brings oxygen and nutritive substances into the delta, and the result is an enormous production of sea life. High biological productivity attracted humans to this land feature. Deltas have often been centers of civilization; the deltas of the Nile, in Egypt, and the Tigris-Euphrates, at the head of the Persian Gulf, have supported important societies. Soils in delta regions are nourished through seasonal flooding, and water tables are high, guaranteeing adequate water with which to irrigate crops, even in the dry season. Food and crop production from tropical deltas is significant because most tropical soils are not very productive. Deltas such as those of the Mekong and the Ganges are outstanding examples.

Deltas are transition zones between the land and the sea, and between river and marine processes. Their rivers are also links between ocean and continent. Cities such as New Orleans, Venice, Amsterdam, and Rotterdam owe their prosperity to their delta geography. Such cities, known as *entrepôts*, thrive on marine traffic entering a country or on overland traffic exiting the country.

Because deltas are areas of vast accumulations of sediments, they generate building material for future mountains. The young mountains of the world, such as the Alps and the Himalayas, parallel coastal areas and are composed of sedimentary rocks. Marine fossils frequently found in such rocks reveal that they not only are composed of sediment but also were once deposited under water, later to be thrust upward to great heights.

*C. Nicholas Raphael*

## FURTHER READING

Bird, Eric. *Beach Management.* New York: John Wiley, 1996. An introductory text on coastal zones and processes. Most examples are taken from Australia. The chapter on deltas is well illustrated with maps and diagrams. Nontechnical and comprehensible to readers with little scientific background.

Coleman, J. M. *Deltas: Process of Deposition and Models for Exploration.* Boston: International Human Resources Development Corporation, 1982. A detailed review of deltaic processes, including an overview of the Mississippi River Delta and discussions of other deltas and their variability. Numerous maps and diagrams illustrate specific points. For readers with some background in the subject.

Davis, R. A., Jr. *Oceanography: An Introduction to the Marine Environment.* 2d ed. Dubuque, Iowa: W. C. Brown Publishers, 1991. Covers deposition in coastal areas, including deltas, beaches, marshes, and estuaries. The treatment of deltas is generally narrative, and equations are sparingly used. A comparative presentation of deltas is instructive and not difficult to understand; some background in physical geology is useful but not necessary.

LaBlanc, R. J., ed. *Modern Deltas.* Tulsa, Okla.: American Association of Petroleum Geologists, 1976. A college-level text describing some of the world's deltas. The treatment is generally nonmathematical and descriptive, but will be most useful to those who have had a course in geology. Well illustrated with diagrams, maps, and pictures.

Levinton, Jeffery S., and John R. Waldman. *The Hudson River Estuary.* New York: Cambridge University Press, 2006. Discusses the geology, physics, and chemistry of the Hudson Bay, including sedimentary processes of the estuary. Covers microbial and nutrient dynamics and primary production of the estuary. Reviews ecology and conservation efforts. Well organized and well indexed.

Middleton, Gerard V., ed. *Encyclopedia of Sediments and Sedimentary Rocks.* Dordrecht: Springer, 2003. Cites a vast number of scientists, who are listed in the author index. Subjects include biogenic sedimentary structures, Milankovitch cycles, deltas and estuaries, and vermiculite. Includes an index of subjects. Designed to cover a broad scope and a degree of detail useful to students, faculty, and professionals in geology.

Morgan, J. P. "Deltas: A Resume." *Journal of Geologic Education* 18 (1970): 107-117. An excellent introductory article on deltas. The emphasis is on the Mississippi River Delta, which the author studied for many years. Covers different processes and their influence on delta development. An excellent and not overly technical paper. Includes maps and tables.

National Research Council. *Drawing Louisiana's New Map: Addressing Land Loss in Coastal Louisiana.* Washington, D.C.: National Academies Press, 2006. Produced by the U.S. National Research Council. Presents a critical assessment of the environmental status of the Mississippi River Delta in Louisiana with regard to anthropogenic modification, erosion, and potential future effects under rising sea levels.

Nett, Mary T., Martin A. Locke, and Dean A. Pennington. *Water Quality Assessments in the Mississippi Delta: Regional Solutions, National Scope.* Washington, D.C.: American Chemical Society, 2004. Discusses watershed management of the Mississippi Delta. Articles cover topics such as runoff, water quality, sedimentation, plankton, and herbicides.

Newson, Malcolm. *Land, Water and Development: Sustainable and Adaptive Management of Rivers.* 3d ed. London: Routledge, 2008. Presents land-water interactions. Discusses recent research, study tools and methods, and technical issues, such as soil erosion and damming. Suited for undergraduate students and professionals. Covers concepts in managing land and water resources in the developed world.

Oti, Michael N., and George Postma, eds. *Geology of Deltas.* Brookfield, Vt.: A. A. Balkema, 1995. A detailed yet easily understood account of the geology of deltas and their function in the environment and in river systems. Illustrations, maps, and bibliographical references.

Peterson, J. F. "Using Miniature Landforms in Teaching Geomorphology." *Journal of Geography* 85 (November/December, 1986): 256-258. Discusses small-scale landforms and their advantages in the classroom. Highlights deltas and related features, such as alluvial fans. This nontechnical text is supplemented with photographs. A good article for those interested in constructing a delta on a small scale. Suitable for high school students.

Schmidt, Paul E. *River Deltas: Types, Structures and Ecology.* New York: Nova Science Publishers, 2011. Presents current research on a broad variety of river deltas, discussing the delta's role in such areas as a carbon source. Describes changes to soil fertility, as well as political and ecological perspectives.

Raphael, C. N., and E. Jaworski. "The St. Clair River Delta: A Unique Lake Delta." *The Geographical Bulletin* 21 (April, 1982): 7-28. A delta in the Great Lakes is described and compared with the Mississippi River Delta. Suggests that although many deltas look the same, different processes are at work in them. Provides good aerial photographs, maps, and cross-sections. Offers a nontechnical treatment of delta processes, forms, and vegetation.

Wohl, Ellen. *A World of Rivers.* Chicago: University of Chicago Press, 2011. Covers the Amazon, Ob, Nile, Danube, Ganges, Mississippi, Murray-Darling, Congo, Chang Jiang, and Mackenzie Rivers in individual chapters. Provides figures and diagrams full of straightforward and well-organized information. Discusses the natural history, anthropogenic impact, and future environment of these ten great rivers. Provides a bibliography organized by chapter.

**See also:** Alluvial Systems; Beaches and Coastal Processes; Carbonate Compensation Depths; Deep Ocean Currents; Deep-Sea Sedimentation; Desert Pavement; Drainage Basins; Floodplains; Lakes; Reefs; River Bed Forms; River Flow; Sand; Seamounts; Seawater Composition; Sediment Transport and Deposition; Surface Ocean Currents; Turbidity Currents and Submarine Fans; Weathering and Erosion

# DESERT PAVEMENT

*Desert pavements are concentrations of stones on the land surface of arid areas, produced by wind and running water erosion and upward movement of stones through the soil. Stone pavements may signal serious soil erosion that must be addressed before the land can be irrigated for agriculture.*

## PRINCIPAL TERMS

- **creep:** the slow, gradual downslope movement of soil materials under gravitational stress
- **deflation:** the sorting out, lifting, and removal of loose, dry, silt- and clay-sized soil particles by turbulent eddy action of the wind
- **dirt cracking:** a process in which clays accumulate in rock cracks, take on water, and expand to rupture the rock
- **expansion-contraction cycles:** processes of wetting-drying, heating-cooling, or freezing-thawing, which affect soil particles differently according to their size
- **salt weathering:** the granular disintegration or fragmentation of rock material affected by saline solutions or by salt-crystal growth
- **thermal fracture:** the formation of a fracture or crack in a rock as a result of temperature changes
- **ventifact:** any stone or pebble that is shaped, worn, faceted, cut, or polished by the abrasive action of windblown sand, generally under desert conditions

### OCCURRENCE OF DESERT PAVEMENT

Desert pavements are extensive stony surfaces in arid areas that occur not only on slopes but also on a range of lowland surfaces, including water-eroded and deposited areas. The stones tend to be closely packed on flat or moderately inclined plane surfaces. Stone pavements are rendered prominent by their lack of vegetation but, in any case, they form most readily by processes typical of arid regions where plant roots are not normally established to bind the soil to any extent. The pavements are most striking in areas of low relief, where it is the largely flat, stony surface itself, rather than the contours of the land, that impresses the observer.

Desert stone pavements range from rocky or boulder-strewn surfaces to smooth plains of fine gravel. Terminology for different types commonly derives from Arabic, as one might expect, given the prevailing aridity across North Africa and the Middle East. Thus the term "hamada" (Arabic for "unfruitful") describes a boulder-strewn terrain, and "reg" (meaning "becoming smaller") indicates a finer pavement of small stones. The term "serir" is a synonym for "reg" in the central Sahara.

Most hamada pavements are residual, consisting of stones derived from the bedrock beneath, or constitute boulders transported only short distances. Most regs consist of transported stones. These distinctions of size and transport distance become rather blurred over time, as the larger rock fragments are progressively weathered to finer sizes and transported farther by wind and water. Nevertheless, a residual origin of many hamada pavements is clearly evident in a number of places by the angularity and lack of sorting of the rock fragments and by the similarity to the bedrock beneath. The water-deposited nature of a reg pavement can be indicated by sorted and rounded gravels of mixed composition and distant origin and also by the pavement's occurrence close to dry stream channels. Residual regs, which are usually closely associated with hamadas, commonly consist of angular flakes of local bedrock or of a less weatherable residue. Reg pavements can be composed of two elements in varying proportions: a compact mosaic of small stones embedded in soil and a more uneven component of larger fragments lying loose on that surface or protruding through it.

### WIND AND WATER EROSION

Desert pavements are polygenetic, which is to say there is more than one process by which they can be produced. The most traditional explanation, however, is that the stones are concentrated by means of wind erosion or deflation of the fine particles. Deflation may eventually settle the pebbles into such stable positions that they fit together almost like the pieces of a mosaic or the blocks of a cobblestone street.

The effectiveness of deflation on desert pavements is less, however, than has been commonly believed. Under natural conditions, soils that have been tested by scientists can resist wind erosion because they are

161

silty and cohesive and tend to form crusts as a result of repeated wetting and drying. A larger grain size of the soil also makes it more resistant to wind erosion. Soil cohesiveness is an important factor. Even where particles are spherical, the magnitude of the cohesive forces between particles less than 0.1 millimeter in diameter is greater than the weight of the particle. Small particles, however, are usually more irregular in shape, or even plate-like, a factor that increases their cohesion. Since cohesion is so important between small particles, silts and clays should be quite resistant to wind erosion. That is observed to be true, but only where the surfaces are smooth, the clays and silts are uniformly fine, and no other material is blown onto them. Otherwise, large grains are dislodged from fine soils as aggregates of particles or knocked loose by the impact of other large sand grains, a process known as saltation. In any case, the amount of lowering and stone concentration by wind action is limited. Deflation diminishes markedly as the protective stone cover increases, and deflation becomes virtually ineffective when stones cover 50 percent of the surface. Undisturbed stone pavements are among the most windstable of desert surfaces.

Rain and water flow appear to be more effective than wind in eroding fine-textured soils on sloping desert pavements. In test plots cleared of stones in one experiment on 5-degree slopes, water wash accounted for most of the 5- to 50-centimeter surface lowering that occurred in a period of five years. During this time, the stone pavement was renewed with stones from below this differential erosion.

## UPWARD DISPLACEMENT

Where subsoils are so clay-rich that they do not erode easily by either wind or water and the subsoils are also largely stone-free, the formation of stone pavements by deflation is more difficult to understand. In these situations, the mechanism of formation seems to have been forces of expansion and contraction within the soil that cause upward displacement and concentration of stones on the surface. Soils that exhibit this phenomenon contain expansive clays in alkaline chemical conditions and are subject to swelling and heaving upon wetting and to shrinkage and cracking upon drying. Periods of heating followed by cooling and of freezing followed by thawing also contribute to the expansion-contraction cycles, which cause stones to move upward and concentrate on the surface.

The exact mechanism for upward stone movement in deserts is not precisely known, although by analogy to known stone movement in areas of intense freeze and thaw cycles, some details are understood. It is possible, for example, that the stones shift upward as the underlying soils swell in wet periods and that, as the soil shrinks in dry periods, fine particles fall into cracks beneath the stones and prevent the return of the rock fragments. Stones may also induce differential swelling in the soil by speeding the downward advance of a wetting front around and over them. The stones are then likely to be displaced upward, away from the dry zone beneath them, and held tightly by the wet and sticky soil above. Soil may squeeze into the space left and thus prevent a return movement of the stones. These processes also resemble those postulated for certain types of sorted, patterned ground, where stones become arranged in polygons or striped zones. In these cases, it is clear that no process other than upward movement along certain zones could produce such patterns.

Stone pavements of the upward-movement type occur on and in the topsoils of weakly salty soils on the stony tablelands of arid Australia, where the subsoils are almost stone-free. The pavement stones are silica-rich, originally precipitated within the soils by mobilization and concentration of silica in unusual chemical reactions. For those stone pavements to have originated as a residue from erosion would have required a stripping of more than 1.5 meters of erosion-resistant, clay-rich soil. Such a process is unlikely, so an upward movement is far more probable, given the swelling and cracking potential of salty clays. Similar pavements have been noted in deserts in Nevada and California as well as in the Atacama Desert of Peru.

An alternative form of displacement is possible where the soil material has been transported, particularly by the wind. For example, in South Australia, a stone pavement occurs on a layer of clay rich in gypsum (calcium sulfate), which is thought to have been blown into place by the wind. The stone pavement on the surface resembles a buried pavement. It is possible that the original pavement first trapped a small amount of windborne dust among the stones. Some of the rock fragments were then displaced upward, little by little, through wetting and swelling of the aeolian (wind-deposited) clays during the course of their accumulation, and thus the stones were never deeply buried.

Concentration of stone pavement through winnowing by wind or wash can be relatively rapid, but the contribution by movement through the soil may take longer. Once formed, a pavement is relatively stable. The closely spaced stones act as a drag on the surface wind, restrict the entrainment of finer intervening materials, and so limit deflation. On moderate slopes, runoff water is spread over the surface by the stone mantle and thus does not tend to cut a deep gully. Selective erosion is countered further as the stone concentration increases. Whether residual or transported, pavements naturally consist of materials resistant to weathering and therefore serve as "armor" for deserts.

## SURFACE WEATHERING

Another manner in which stone pavements can be generated is by relative concentration through surface weathering. Rock fragments in the relatively moist subsurface environment of a desert soil are more susceptible to weathering than those on the arid surface, particularly where the soil is impregnated with salt or gypsum. Consequently, a stone pavement may survive above a soil that itself has few stones because they disintegrate at depth over time. The phenomenon is pronounced in granitic gravels. Some terrace sediments, for example, are known to have larger-sized stone pavements above horizons of small granite fragments formed by the chemical weathering breakdown of boulders beneath the surface. These subsurface zones of fine particles are deepest and most free of subsurface stones on the highest and oldest terraces, where they may be more than 50 centimeters thick. In such cases, it is thought that stone pavements first formed from the abundant rock fragments in the area, but as subsurface weathering progressed all the buried stones were destroyed, while those on the surface remained relatively unweathered.

In spite of their status as a protective armor on desert surfaces, stone pavements are subject to further evolution over thousands of years as the stones weather and their secondary products are redistributed. This evolution is generally toward an increasingly even surface of particles of small grain size and characterized by greater compaction. Also common is progressive darkening by surface weathering and the formation of rock or desert varnishes of manganese and iron stains, precipitated on the stone surfaces by microbial and weathering action.

Weathering of pavement stones occurs not only beneath the stones, where they are in contact with the protected, mildly corrosive soil layers, but also on their exposed surfaces. Pavement stones can be wetted frequently by dew, which also contributes salts to both the surface and the subsurface in the weathering process. Because pavements are generally unchanneled and provide little runoff, the features are particularly subject to episodic or seasonal cycles of shallow wetting by rainfall and evaporative drying, through which weathering is activated. Bare and generally dark-colored pavements, among the most strongly heated surfaces of the desert, are areas of considerable evaporation, as (mostly saline) moisture is drawn upward through narrow cracks. As a result, thin salt crusts are widespread, the soil itself is impregnated with chlorides and sulfates of calcium and sodium, and salt weathering is significant. In this process, the growth of various salt crystals in the pores of the stones generates forces sufficient to disrupt the rocks.

Pavement stones also trap windborne dust in fissures and cracks, and dirt cracking can result from expansion of the dust particles when they are wetted. Lichens and algae also exploit the shaded and relatively moist environments under the stones, adding an organic element of chemical decomposition to the weathering process.

In the breakdown of pavement stones, there tends to be a further selective concentration of resistant fine-grained material. For example, siliceous flint or chert pebbles may accumulate on the surface as relatively soft limestones are weathered. Such stones can eventually break down by incorporation of water into their microcrystalline atomic structures, by spalling (chipping), by blocky fracturing (crazing), or by complete cleavage and radial splitting. Coarser-grained stones undergo granular disintegration and pitting. Fracturing of pavement stones also has been attributed to differential expansion and contraction caused by solar heating, but the idea is controversial. Such thermal fracturing may be only a partial cause, as much of the broken stone has been superficially altered chemically, and dirt-cracking expansion may be a more important factor. In some cases, stones below the surface, where solar heating is not a possibility, can also be seen to have been pried apart in this fashion.

Wind abrasion is a form of natural sandblasting. Its effectiveness is related to wind velocity, the hardness of the sand and dust carried by the wind, and the hardness of the rock fragments being eroded. It is probably most effective in certain polar areas where cold, dense air can carry large particles at very high velocities and where, at winter temperatures, even ice has the hardness of some minerals. The "dry valleys" of Antarctica, for example, have extensive stone pavements that have been affected in this way.

## BOULDER AND PEBBLE FORMS

Boulders and pebbles in stone pavements may be fluted, scalloped, and faceted by wind abrasion. Flute and scallop forms vary in length from a few millimeters to several meters, but large flutes are not common in hard rocks. It is thought that they are produced by turbulent helical (helix-shaped) flows carrying dust and sand and that the scallops grow downwind. Where they are cut on large, immobile boulders, they are clear indicators of the strongest (dominant) wind directions. For this reason, they are most commonly reported from places where wind directions have remained largely unchanged for thousands to hundreds of thousands of years.

Some of the best-known wind-eroded stones in pavements are ventifacts (a general term for wind-faceted pebbles and boulders). A rock face abraded by wind may be pitted if there is a range of hardness in the minerals of the face, or it may be smooth where the rock is fine-grained or composed of only one mineral. Thus, rocks such as coarse-grained granite have pitted surfaces, but fine-grained quartzite generally produces only smooth, polished surfaces. Ventifacts have a great range of surface shapes, with plane and curved faces and two or more facets. The German term "dreikanter" is used for ventifacts with three facets, and "einkanter" is used for two-faceted stones. Multiple facets indicate either that there was more than one wind direction or that the stones have been turned over through time. As the sizes of a stone pavement's fragments are progressively reduced, a matrix of increasingly fine particles is supplied to the pavement. The proportion of such secondary material reveals the maturity of development of a desert stone pavement, although in practice it may be difficult without sophisticated analysis to distinguish between

a new pavement in the process of being formed and an old one being degraded.

The fine particles in a degraded pavement are redistributed by wash and rain, a process that contributes to the smoothing and compaction of the pavement. Any exposed soil is puddled and sealed by heavy rainfall or runoff water so that a saturated layer flows into hollows between the stones. Bare soil interspersed through a pavement is generally crusted above a bubbly or vesicular horizon, 1 to 3 centimeters thick, that also extends around and beneath the pavement stones. The bubbles result from the escape of entrapped air. Equally important in compaction is the gravitational settling of the stones during the expansion and contraction of the surface upon heating and cooling and (especially in saline and expansive clay soils) upon wetting and drying. Pavements of this type are generally soft and puffy after rain, but, upon drying, the stones become ever more firmly embedded in a tight mosaic.

The slow downhill creep of water-saturated surface materials can also assist in either the smoothing or the roughening of a sloping stone pavement, particularly where dispersal of the matrix is accentuated by salinity. Microrelief on a stone pavement can be reduced by the differential flow of the fine sediments, and the pavement stones can thereby become more evenly distributed and further embedded. In some cases the stone pavements can even move into a series of rough steps, aligned along the contours of a slope.

*John F. Shroder, Jr.*

## FURTHER READING

Cooke, Ronald U. "Stone Pavements in Deserts." *Annals of the Association of American Geographers* 60 (1970): 560-577. One of the most comprehensive nontechnical articles available on desert pavements. Emphasizes polygenesis, with examples mainly from California and Chile. Explains the view that deflation may be a relatively unimportant process in pavement formation. Suggests that the relative importance of different processes may vary greatly between sites.

Cooke, Ronald U., Andrew Warren, and Andrew Goudie. *Desert Geomorphology.* London: UCL Press, 1993. One of the chief English-language sources on dry climate geomorphology. Offers a long section on the formation of desert pavements as well as the many other collateral processes and landforms that occur with or around such features.

Laity, Julie J. *Deserts and Desert Environments*. Hoboken, N.J.: John Wiley & Sons, 2008. Discusses the climate, ecology, plant and animal communities, soil, and weathering processes of deserts. Covers desert pavement and the hydrology of deserts.

Mabbutt, J. A. *Desert Landforms*. Cambridge, Mass.: MIT Press, 1977. Includes a chapter on stony deserts that is a superb exposition on desert pavements. Explains their polygenetic origin, with numerous examples and illustrations to aid the discussion. Replete with material on other arid landforms and processes that occur in the same conditions as desert pavements.

Mainguet, Monique. *Aridity: Drought and Human Development*. Berlin: Springer-Verlag, 2010. Discusses global and local aridity and drought, and related changes in vegetation and hydrology. Focuses on anthropogenic impact. Includes references and multiple indexes.

McGinnies, William G., B. J. Goldman, and P. Paylore, eds. *Deserts of the World*. Tucson: University of Arizona Press, 1968. Surveys research on the physical and biological environments of the world's deserts. One of the most comprehensive studies of arid regions ever attempted. Gives valuable information specific to many particular deserts and has an extensive references list. Contains many references not generally cited elsewhere.

Middleton, Nick. *Deserts: A Very Short Introduction*. Oxford: Oxford University Press, 2009. Intended for general readership. Delivers an introductory overview of desert environments around the world.

Nicholson, Sharon E. *Dryland Climatology*. New York: Cambridge University Press, 2011. Covers the geomorphology, hydrology, ecology, and climatology of dry land. Discusses adaptation, microhabitats, and desertification. Written for graduate students and professional researchers and scientists.

Parsons, A. J., and Athol D. Abrahams. *Geomorphology of Desert Environments*. Springer Science+Business Media B.V., 2009. Provides a research-level treatment of desert geomorphology, beginning with a global overview and progressing through various themes such as weathering processes and aeolian surfaces.

Schaetzl, Randall J., and Sharon Anderson. *Soils: Genesis and Geomorphology*. Cambridge, England: Cambridge University Press, 2005. A comprehensive and readable textbook of general soil studies. Provides excellent background for understanding the soil-affecting processes at work in arid regions. Suitable for readers at the senior undergraduate and graduate levels.

Thomas, David S. G. *Arid Zone Geomorphology: Process, Form and Change in Drylands*. 3d ed. Chichester: Wiley-Blackwell, 2011. Written by recognized experts in the field of geomorphology. Presents recent advances made in the investigation and explanation of arid zone landforms, with one of its five sections devoted to surface processes and characteristics.

**See also:** Alluvial Systems; Beaches and Coastal Processes; Dams and Flood Control; Deep-Sea Sedimentation; Deltas; Drainage Basins; Floodplains; Floods; Lakes; Reefs; River Bed Forms; River Flow; Sand; Sea Level; Sediment Transport and Deposition; Turbidity Currents and Submarine Fans; Weathering and Erosion

# DRAINAGE BASINS

*A drainage basin collects water from a large area and delivers it to a channel or lake. Drainage basins reflect the operation of physical laws affecting water flow over the ground surface and through rocks. These basins concentrate groundwater flow from a broad area into rivers, which, in turn, carry away both water and sediment.*

## PRINCIPAL TERMS

- **basin order:** an approximate measure of the size of a stream basin, based on a numbering scheme applied to river channels as they join together in their progress downstream
- **channel:** a horizontal depression in the ground surface caused and enlarged by the concentrated flow of water
- **erosion:** the displacement of sediment from one location to another on the ground surface
- **groundwater:** water that sinks below the ground surface to join the water table, slowly flowing toward river channels
- **hydrological:** relating to the systematic flow of water in accordance with physical laws
- **limestone:** a rock composed primarily of the mineral calcite or dolomite, which can be dissolved by water that is acidic

### CHARACTERISTICS OF DRAINAGE BASINS

A drainage basin is an area that collects water, which accumulates on the surface from rain or snow. Its slopes deliver the water to either a channel or a lake. Normally, the channel that collects the water leads to the ocean. In this case, the drainage basin is defined as the entire area upstream whose slopes deliver water to that channel or to other channels tributary to it. Thus, strictly speaking, drainage basins are defined as natural units only when streams enter bodies of water such as lakes or the ocean or when two streams join.

Less often there is no exit to the ocean: This type of basin is called a basin of inland or interior drainage. Notable examples are the basins containing the Great Salt Lake in Utah, the Dead Sea in Israel and Jordan, and the Caspian Sea in Asia. The Basin and Range province in the Rockies (an area of about 1 million square kilometers extending from southern Idaho and Oregon through most of Nevada, western Utah, eastern California, western and southern Arizona, southwestern New Mexico, and northern Mexico) has at least 141 basins of inland drainage. The center of these basins is usually marked by a "playa," a level area of fine-grained sediments, often rich in salts left behind as inflowing waters evaporate. At certain times in the geological past, when the annual rainfall was heavier, some of these basins completely filled with water to the point of overflowing, at which point the drainage system may have connected to another interior basin or connected to a river system that drained to the sea. The basin now containing the Great Salt Lake (known to geologists as Lake Bonneville) overflowed at Red Rock Pass about 15,000 years ago. The overflowing waters discharged into the Snake River system, and thus to the Columbia River and the Pacific. Drainage basins may change in character over relatively short periods of geological time. There is some evidence that the entire Mediterranean Sea was a basin of inland drainage for a period of time about 3 to 5 million years ago: Substantial salt deposits are found on its bed, and traces of meandering rivers have been seen in certain geological sections.

### GROUNDWATER AND EROSION

Although the term "drainage basin" is normally thought of as applying to the surface, an important component of the basin as a hydrological unit is the rock beneath the surface. Much of the water that arrives on the surface sinks into the soil and the underlying rocks, where it is stored as soil water in the "unsaturated zone" and as groundwater. Soil water either sinks farther to become groundwater, or flows through soil and back out onto the surface downslope when and where the soil is saturated. Groundwater moves very slowly through the rock (millimeters to centimeters per day), but it eventually seeps into stream channels and so provides the base flow of water in rivers long after rain has finished falling.

This characteristic of groundwater can lead to a circumstance that alters the definition of a drainage basin when the rocks are primarily composed of limestone or any rocks susceptible to solution. Because limestone is soluble in acidic water (natural rain is slightly acid and contact with various airborne pollutants increases the acidity), over thousands of years, percolating groundwater dissolves substantial volumes of rock and causes a system of underground channels to develop, which may eventually become enlarged into caverns. Yugoslavia is famous for its underground cave systems; the U.S. states of Kentucky, Florida, and New Mexico are well known for their limestone terrain and underground drainage systems. Essentially every karst landscape on Earth is home to a system of caverns that has been formed through chemical erosion over long periods of time. In these cases, the route of the water underground may bear little or no relation to the pattern of channels and slopes seen on the surface, some or all of which may have become completely inactive. The determination of the drainage basin is then very difficult as various types of tracers (colored dyes or other readily detectable chemicals) have to be placed in the water in order to identify the points of egress of the water so that an interpretation of the underground channels may be made. The pattern of water flow to any given site will also depend on the location of the storm waters causing the flow. The very slow solution of limestone to form underground river systems and caves emphasizes the fact that water moving through the basin removes solid rock. As the rivers dissolve their way downward, they leave some caves "high and dry" above the general level of the underground water (the water table), which points out the fact that rivers work down through the rock with the passage of time.

This aspect of drainage basins is harder to observe in areas of less soluble rock, even though water flowing out of the drainage basin carries sediment (small particles of soil and rock) and has been doing so for long periods of geological time. Thus, in the long run, the surface of Earth is gradually lowered and, even in the short run, enormous amounts of sediment may be removed from a basin every year. The Mississippi removes a total of about 296 million metric tons per year, or 91 tons per square kilometer, though this is small compared to the Ganges, which takes out 1,450 million metric tons per year, or 1,520 tons per square kilometer. Both of these are dwarfed by the average annual sediment load carried by the Yellow River (Huang He), in China, of 1.6 billion tons. If there were no corresponding uplift of the drainage basins or other interference, this removal would lead to the leveling of entire basins within 10

to 50 million years, depending on the lowering rate and the mean altitude of the basin.

The process of erosion proceeding at different rates in adjacent basins may cause the drainage divide (the line separating different flow directions for surface waters) to migrate toward the basin with the lower erosion rate. This is most common in geologically "new" terrain when stream systems are not deeply incised into the rocks. Drainage diversions may be simulated, as in the Snowy Mountain diversion in Australia, where waters are diverted across a divide by major engineering works in order to provide irrigation water for the Murray Darling river basin and also for hydroelectric power generation.

## FLOODS

When winter snow melts or severe storms bring heavy rain to large areas, the water that falls to the surface flows into channels and generates floods in the rivers. Floods are not abnormal; rather, they are an expectable occurrence in drainage basins. It is easy to understand that when basin relief is high and slopes are steep, as in the Rockies or the Appalachians, floods tend to generate higher flood peaks than when slopes are gentle. A basin that is round in shape tends to concentrate floodwater quickly because the streams tend to converge in the middle, whereas a long, narrow stream has the effect of attenuating the flow peak, even when the total amount of water falling on the basin may be the same. Similarly, a forest tends to attenuate flood peaks and to promote higher river flows between flood peaks than does open farmland. With the latter, there is a tendency for water to flow rapidly off the surface into channels, whereas in a forest much water is intercepted by the leaves of trees and the impact of rain on the surface is weaker, in part because leaf cover protects the soil. Because the soil is not so well protected, sediment loss from the surface into streams is greater from farmland than from forests and is higher again from land disturbed by major building projects.

## CLASSIFICATION AND MEASUREMENT OF
## DRAINAGE BASINS

Various methods have been devised to classify basins according to size. The most common method depends on a numbering system applied to the streams that drain them. All the "fingertip" streams are labeled with 1. When two of these tributaries meet the

167

channel, it is termed a second-order channel and is labeled with 2. Subsequently, the order of a stream increases by one only when two streams of equal order join. Otherwise, if two streams of unequal order join, the order given to the downstream segment is that of the larger of the two orders. The order of the drainage basin is then the order of the stream in the basin. In this type of numbering system (called Horton/Strahler ordering), the Mississippi drainage basin is an eleventh- or twelfth-order basin. The exact number depends on the detail (map scale) with which the fingertip streams are defined. The larger orders are rare because of the requirement for another river of roughly similar magnitude to join in to make the next higher-order basin.

Basin order may be used as a relatively natural basis for the collection of other data about the basin. The simplest measure is of the area in square kilometers. In addition, the basin relief, or the height difference between the lowest and the highest points, and the mean relief, or the average height of the basin above the outlet, may be recorded. The most precise method of recording basin relief is by computing the hypsometric (height) curve for the basin, which requires an accurate topographic map. When constructed, it shows, for any altitude, the proportion of the basin area above that particular altitude and, for comparative purposes, it may be produced in a dimensionless form by dividing both the height and the area measures by their maximum values or by the difference between the maximum and minimum heights if zero is not the minimum height.

Basin shape and basin dimensions (length and width) may also be recorded, although the notion of basin shape suffers from the problem that no completely unambiguous numerical measure exists that can be used to define the shape of an area in the plane (that is, on a map), and the problem is especially intractable if there are indentations in the edge of the basin. All measures are dependent to a considerable degree on the accuracy of source maps. In mountainous terrain, such maps may often be much less than perfect, if they exist at all. Even with automated drafting aids and digitizers (which automatically record positions on maps and save them as a data file), the measurement of basin properties is a tedious and time-consuming process. Unless there are pressing reasons for a new analysis, it is common to rely on data tabulations made by hydrological or environmental agencies whenever possible.

Measurements are made of drainage basin properties because they are often used in statistical analyses together with the known flow of the gauged rivers in order to predict flow characteristics for rivers that have not been metered. The direct measurement of stream flow, while straightforward in principle, is time-consuming, especially in the early stages, and a flow record is not very useful for predictive purposes until it has recorded at least twenty years of flow (preferably much more). Because of the high capital and maintenance costs involved in collecting river records, there has been an understandable emphasis on records for large rivers; the economic benefits from prediction (and eventual control) of the flow are more obvious, and measured flow records can sometimes be supplemented by anecdotal evidence of historic large floods, those flows that are often of most interest in land-use planning (for example, zoning of land for residential use). It has been acknowledged that the hydrological behavior of low-order basins is less well understood, and more information has been collected on them, especially for urban areas where the routing of the large quantities of water that run off from impermeable surfaces in the city (roofs and roadways) has been recognized as a serious planning problem, especially in regard to groundwater recharge and the flow systems that connect with urban sewage systems.

## SIGNIFICANCE

The control of water outflow from drainage basins is necessary in some regions in order to promote irrigation, to supply domestic and industrial water, to generate power, and to implement flood control. The Hoover Dam on the Colorado was originally conceived as a control dam, but hydroelectrical generators were also included in order to help defray its costs by selling power. There are nineteen major dams in the Colorado basin. Aside from the legal technicalities of water ownership and redistribution, difficulties arise from the fact that to control substantial amounts of water, large areas of the basin have to be regulated. In addition, there are economies of scale in large projects, particularly in the construction of large dams and reservoirs. A single control dam strategically placed may regulate flow downstream for hundreds of kilometers, whereas it would require hundreds of small dams on first- and second-order streams to achieve the same effect.

Large control dams do generate problems. The reservoirs trap sediment coming from upstream, which will eventually fill them, at which point they will become useless. Small reservoirs may fill within a few years. An original estimate for the Hoover Dam suggested that it would take four hundred years to fill Lake Mead; after only fourteen years, surveys revealed that the water capacity had been reduced by 5 percent and that sediment in the lake bottom reached a maximum of 82 meters where the upstream river entered the still waters of the lake. Downstream of a dam, the reduced sediment content and the regulated water flow often seriously affect riparian environments. There may be a variety of channel responses, often unpredictable, to the interference in the river regime caused by the dam. The stream may cut into its bed, it may change the dimensions of its channel, or it may even aggrade its bed. In the case of the Hoover Dam, the water downstream, deprived of its sediment by the dam, had an increased ability to remove fine sediment from the river bed but left coarser rocks behind because the flood peaks that would normally have removed them were now controlled and much reduced. The result is an "armoring" of the stream bed with coarse rocks, an effect that extends 100 kilometers downstream in the case of the Colorado River below the Hoover Dam. In the Colorado system as a whole, the net effect of controlling flood peaks has been for rapids to stabilize and to increase in size as sediment becomes trapped in them. A corollary of the "winnowing" of fine material has been the disappearance of river beaches and an increased propensity to pollution as sediment becomes much less mobile and more concentrated in space.

*Keith J. Tinkler*

## FURTHER READING

Dodds, Walter K., Matt R. Whiles. *Freshwater Ecology: Concepts and Environmental Applications of Limnology.* 2d ed. Burlington, Mass.: Academic Press, 2010. Covers physical and chemical properties of water, the hydrologic cycle, nutrient cycling in water, as well as biological aspects. Written by two of the leading scientists in freshwater ecology. An excellent resource for college students. Each chapter provides a short summary of main topics.

Foresman, Timothy, and Alan H. Strahler. *Visualizing Physical Geography.* New York: John Wiley & Sons, 2012. Uses a unique approach to present the concepts of physical geography by heavily integrating visuals from *National Geographic* and other sources with the narrative.

Graf, William L. *The Colorado River: Instability and Basin Management.* Washington, D.C.: Association of American Geographers, 1985. An excellent, well-written study of the particular management problems and practices associated with the Colorado River. Focuses on the way the river has adjusted to a variety of changes caused by climatic change, rangeland management, the building of large dams, and the extraction of water for irrigation. Easily understood by the layperson; sound bibliography.

Gregory, K. J., and D. E. Walling. *Paleohydrology and Environmental Change.* New York: John Wiley, 1998. A comprehensive academic textbook aimed at the serious undergraduate or a well-prepared, scientifically minded layperson. Provides examples from around the world; well illustrated with photographs, maps, and diagrams. Describes instrumentation and the implications for socioeconomic management. Extensive bibliography.

Henry, Georges. *Geophysics for Sedimentary Basins.* Translated by Derrick Painter. Paris: Editions Technip, 1997. Looks at the geological characteristics of sedimentary basins. Suitable for the college-level reader. Color illustrations, index, and bibliographical references.

McClain, Michael E., Reynaldo L. Victoria, and Jeffery E. Richey. *The Biogeochemistry of the Amazon Basin.* Examines the Amazon River basin from a number of angles. Articles are compiled to include biogeochemistry, nutrient cycling, land development, and conservation. Evaluates biomass, trace elements, organic matter, and nutrients levels.

Newson, Malcolm. *Land, Water and Development: Sustainable and Adaptive Management of Rivers.* 3d ed. London: Routledge, 2008. Presents land-water interactions. Discusses recent research, study tools and methods, and technical issues, such as soil erosion and damming. Suited for undergraduate students and professionals. Covers concepts in managing land and water resources in the developed world.

Parnell, John, ed. *Geofluids: Origin, Migration, and Evolution of Fluids in Sedimentary Basins.* London: Geological Society, 1994. Looks at the evolution of geofluids and their dynamics and migration in relation to sedimentary basins. Intended for the college student. Filled with helpful illustrations and maps. Index and bibliography.

Rodriguez-Iturbe, Ignacio, and Andrea Rinaldo. *Fractal River Basins: Chance and Self-organization.* Cambridge, England: Cambridge University Press, 2001. Describes the theoretical basis for the arrangement of river basins and networks in terms of fractal geometry. Assumes a background in mathematics.

Strahler, Alan H. *Modern Physical Geography.* 4th ed. New Delhi: Wiley India, 2008. Presents a comprehensive overview of physical geography, and discusses drainage basins as the basic unit of surface water flow.

Tarbuck, Edward J., and Frederick K. Lutgens. *Earth: An Introduction to Physical Geology.* 6th ed. Upper Saddle River, N.J.: Prentice Hall, 1999. An introductory textbook suitable for high school students. Discusses drainage basins in their broadest context as fundamental to surface water flow systems. Includes diagrams, photographs, review questions, and lists of key terms.

Wohl, Ellen. *A World of Rivers.* Chicago: University of Chicago Press, 2011. Discusses the Amazon, Ob, Nile, Danube, Ganges, Mississippi, Murray-Darling, Congo, Chang Jiang, and Mackenzie Rivers. Contains valuable figures with a lot of information. Discusses the natural history, anthropogenic impact, and the future environment of these ten great rivers. Bibliography is organized by chapter.

**See also:** Alluvial Systems; Beaches and Coastal Processes; Deep-Sea Sedimentation; Deltas; Desert Pavement; Floodplains; Floods; Freshwater and Groundwater Contamination Around the World; Lakes; Reefs; River Bed Forms; River Flow; Sand; Sediment Transport and Deposition; Weathering and Erosion

# DROUGHT

*Drought is an unusually long period of below-normal precipitation. It is a relative rather than an absolute condition, but the end result is a water shortage for plant growth, affecting the people who live in that region and beyond. Drought is particularly disastrous for farmers and the practice of agriculture.*

## PRINCIPAL TERMS

- **adiabatic:** a change of temperature within the atmosphere that is caused by compression or expansion without transfer of heat into or out of the system
- **desertification:** the relatively slow, natural conversion of fertile land into arid land or desert
- **evapotranspiration:** the combined water loss to the atmosphere from both evaporation and plant transpiration
- **Palmer Drought Index:** a widely adopted quantitative measure of drought severity that was developed by W. C. Palmer in 1965
- **potential evapotranspiration:** the water needed for growing plants, accounting for water loss by evaporation and transpiration
- **precipitation:** any form of liquid water or ice that falls from the atmosphere to the ground
- **Sahel:** the semiarid southern fringe of the Sahara in West Africa that extends from Mauritania on the Atlantic coast to Chad in the interior
- **soil moisture:** water that is held in the soil and that is therefore available to plant roots
- **subsidence:** in meteorology, the slow descent of air that becomes increasingly dry in the process, usually due to an area of high pressure

## IMPACT OF DROUGHT

Droughts have had enormous impacts on human societies since ancient times. The most obvious effect is crop and livestock failure, which have caused famine and death through thousands of years of human history. Drought has resulted in the demise of some ancient civilizations and, in some instances, the forced mass migration of large numbers of people. Water is so critical to all forms of life that a pronounced shortage could, and has, decimated whole populations.

The effects of drought remain profound. The dry conditions in the Great Plains of North America in the early 1930's in conjunction with extensive and improper farming activities resulted in the creation of the Dust Bowl, which at one point covered more than 200,000 square kilometers, or an area about the size of Nebraska. During the early 1960's, a severe drought affected the Mid-Atlantic states. Parts of New Jersey experienced sixty consecutive months of below-normal precipitation, so depleting local water supplies that plans were actively considered to bring rail cars of water into Newark and other cities in the northern part of the state, as the reservoirs that usually supplied the region were practically dry. The Sahel region south of the Sahara in West Africa had a severe drought beginning in the late 1960's and continuing into the early 1970's, creating an enormous negative impact on the local population, livestock, and vegetation. Hundreds of thousands of people starved, thousands of animals died, and many tribes were forced to migrate south to areas of more reliable precipitation.

## DROUGHT CHARACTERISTICS

Almost all droughts occur when slow-moving air masses that are characterized by subsiding air movements dominate an area. Often, the air comes from continental interiors where the amount of moisture that is available for evaporation into the atmosphere is very limited. When these conditions occur, the potential for precipitation is low for a number of reasons. First, the humidity in the air is already low, as the continental air mass is distant from maritime (moist) influences. Second, air that subsides undergoes adiabatic heating at the rate of 10 degrees Celsius per 1,000 meters. The term "adiabatic" refers to a change of temperature within a gas (such as the atmosphere) that occurs as a result of compression (descending air) or expansion (rising air), without any input or extraction of heat from external sources. For example, assume that air at a temperature of 0 degrees Celsius is passing over the Sierra Nevada in eastern California at an elevation of 3,500 meters. As the air descends and reaches Reno, Nevada, at an elevation of 1,200 meters, the higher atmospheric pressure found at lower elevations results in compression and heating at the dry adiabatic rate of 10 degrees Celsius per 1,000 meters, yielding a temperature in the Reno area of 23 degrees Celsius. Thus, adiabatic heating from subsiding air masses results in a decline in relative humidity and an increase in moisture-holding capacity. In addition, the movement of air under these conditions is usually unfavorable for vertical uplift and the beginning of the condensation process. The final factor that reduces precipitation potential is the decrease in cloudiness and corresponding increase in sunshine, which in turn leads to an increase in potential evapotranspiration demands, which favor soil moisture loss.

Another characteristic associated with droughts is that once they have become established within a particular location, they appear to persist and even expand into nearby regions, resulting in desertification in extreme cases. This tendency is apparently related to positive feedback mechanisms. For example, the drying out of the soil influences air circulation and the amount of moisture that is then available for precipitation farther downwind. At the same time, the atmospheric interactions that lead to unusual wind systems associated with droughts can induce surface-temperature variations that, in turn, lead to further development of the unusual circulation pattern. Thus, the process builds on itself, causing the drought to both last longer and intensify. The situation persists until a major change occurs in the circulation pattern in the atmosphere.

Many climatologists concur with the concept that precipitation is not the only factor associated with drought. Other factors that demand consideration include moisture supply, the amount of water in storage, and the demand generated by evapotranspiration. Although the scientific literature of climatology is replete with information about the intensity, length, and environmental impacts of drought events, the role of individual climatological factors that can increase or decrease the severity of a drought is not fully understood.

## DROUGHT IDENTIFICATION

Research in drought identification has been changing over the years. Drought was once considered solely in terms of precipitation deficit. Although that lack of precipitation is still a key atmospheric component of drought, sophisticated techniques are now used to assess the deviation from normal levels of the total environmental moisture status. These

techniques have enabled investigators to better understand the severity and length of drought events, as well as the extent of the affected area.

Drought has been defined in numerous ways. Some authorities consider it to be merely a period of below-normal precipitation, while others relate it to the likelihood of forest fires. Drought is also said to occur when the yield from a specific agricultural crop or pasture is significantly less than expected. It has also been defined as a period when soil moisture or groundwater decreases to a critical level.

Drought was identified early in the twentieth century by the U.S. Weather Bureau as any period of twenty-one or more days when precipitation was 30 percent or more below normal. Subsequent examination of drought events that were identified by this method revealed that soil moisture reserves were often elevated during these events to the extent that there was sufficient water to support vegetation. It was also determined that the amount of precipitation preceding the drought event was ample or even heavy. Thus, it became apparent that precipitation should not be used as the sole measure to identify drought. Subsequent research has shown that the moisture status of an area is affected by additional factors.

Further developments in drought identification during the middle decades of the twentieth century began to focus on the moisture demands that are associated with the evapotranspiration in an area. Evaporation is primarily the process by which water in liquid form is converted into water vapor at the surface and conveyed into the atmosphere. To a lesser extent, this also includes the conversion of "solid water," as ice and snow, ultimately to vapor either directly or through an intermediate liquid state. Transpiration refers to the loss of moisture by plants to the atmosphere. Although evaporation and transpiration can be studied and measured separately, it is convenient to consider them in applied climatological studies as the single process of evapotranspiration.

There are two ways to define evapotranspiration. The first is actual evapotranspiration, which is the actual or real rate of water-vapor return to the atmosphere from the earth and vegetation; this process could also be called "water use." The second is potential evapotranspiration, which is the theoretical rate of water loss to the atmosphere if one assumes continuous plant cover and an unlimited supply of water. This process could also be called "water need," as it indicates the amount of soil water needed if plant growth is to be maximized. Procedures have been developed that enable one to calculate the potential evapotranspiration for any area as long as one has monthly mean temperature and precipitation values.

Some drought-identification studies have focused on agricultural drought, looking at the adequacy of soil moisture in the root zone for plant growth. This procedure involved the evaluation of precipitation, evapotranspiration, available soil moisture, and the water needs of plants. The goal of this research was to determine drought probability based on the number of days when soil moisture storage is reduced to zero.

Evapotranspiration was also used by the Forest Service of the U.S. Department of Agriculture when it developed a drought index to be used by fire-control managers. The purpose of the index was to provide a measure of flammability that could create forest fires. This index has limited applicability to nonforestry users, as it is not effective for showing drought as an indication of total environmental stress.

## PALMER DROUGHT INDEX

One of the most widely adopted drought-identification techniques was developed by W. C. Palmer in 1965. The method, which became known as the Palmer Drought Index, defines drought as the period of time—usually measured in months or years—when the actual moisture supply at a given location is consistently less than the climatically anticipated or appropriate supply of moisture. The calculation of this index requires the determination of evapotranspiration, soil moisture loss, soil moisture recharge, surface runoff, and precipitation. The Palmer Drought Index values range from approximately +4.0 for an extremely wet moisture status class to −4.0 for extreme drought. Normal conditions have a value close to 0. Positive values indicate varying stages of abundant moisture, whereas negative values indicate varying stages of drought.

Although the Palmer Drought Index is recognized as an acceptable procedure for incorporating the role of potential evapotranspiration and soil moisture in magnifying or alleviating drought status, there have been some criticisms of its use. For example, the method produces a dimensionless parameter of drought status that cannot be directly compared

with other environmental moisture variables, such as precipitation, which are measured in units (centimeters, millimeters) that are immediately recognizable. In addition, the index is not especially sensitive to short drought periods, which can affect agricultural productivity.

In order to address these shortcomings, other researchers use water-budget analysis to identify deviations in environmental moisture status. The procedure is similar to the Palmer method inasmuch as it incorporates the environmental parameters of precipitation, potential evapotranspiration, and soil moisture. However, the moisture status departure values are expressed in the same units as precipitation and are therefore dimensional. Drought classification using this index method ranges from approximately 25 millimeters for an above-normal moisture status class to −100 millimeters for extreme drought. The index would be close to 0 for normal conditions.

## SIGNIFICANCE

Drought is invariably associated with some form of water shortage, yet many regions of the world have regularly occurring periods of dryness. Three different forms of dryness—perennial, seasonal, and intermittent—have a temporal dimension. Perennially dry areas include the major deserts of the world, such as the Sahara, Arabian, and Kalahari. Precipitation in these areas is not only very low but also very erratic. Seasonal dryness is associated with regions where the bulk of the annual precipitation comes during a few months of the year, leaving the rest of the year without rain or other precipitation. Intermittent dryness is associated with those instances where the overall precipitation is reduced in humid regions or where the rainy season in seasonally dry areas does not occur or is shortened.

The absence of precipitation when it is normally expected creates variable problems. For example, the absence of precipitation for one week in an area where daily precipitation is the norm would be considered a drought. In contrast, it would take two or more years without any rain in parts of Libya in North Africa for a drought to occur. In those areas that have one rainy season, a 50 percent reduction in precipitation would be considered a drought. In regions that have two rainy seasons, the failure of one could lead to drought conditions. Thus, the word "drought" is a relative term, as it has different meanings in different climatic regions.

User demands also influence drought definition. Distinctions are often made among climatological, agricultural, hydrologic, and socioeconomic drought. Climatological, or meteorological, drought occurs at irregular periods of time, usually lasting months or years, when the water supply in a region falls far below the levels that are typical for that particular climatic regime. The degree of dryness and the length of the dry period are used as the definition of drought. For example, drought in the United States has been defined as occurring when there is less than 2.5 millimeters of rain in a forty-eight hour period. In Great Britain, drought has been defined as occurring when there are fifteen consecutive days with less than 0.25 millimeter of rain for each day. In Bali, Indonesia, drought has been considered as occurring if there is no rain for six consecutive days.

Agricultural drought occurs when soil moisture becomes so low that plant growth is affected. Drought must be related to the water needs of the crops or animals in a particular place, since agricultural systems vary substantially. The degree of agricultural drought also depends on whether shallow-rooted or deep-rooted plants are affected. In addition, crops are more susceptible to the effects of drought at different stages of their development. For example, inadequate moisture in the subsoil in an early growth stage of a particular plant will have minimal impact on crop yield as long as there is adequate water available in the topsoil. However, if subsoil moisture deficits continue, then the yield loss could become substantial.

Hydrologic drought definitions are concerned with the effects of dry spells on surface flow and groundwater levels. The climatological factors associated with the drought are of lesser concern. Thus, a hydrological drought for a particular watershed is said to occur when the runoff falls below some arbitrary value. Hydrological droughts are often out of phase with climatological and agricultural droughts and are also basin-specific; that is, they pertain to the particular watersheds that they affect.

Socioeconomic drought includes features of climatological, agricultural, and hydrological drought and is generally associated with the supply and demand of some type of economic good. For example, the interaction between farming (demand) and naturally occurring events (supply) can result in inadequate water for both plant and animal needs. Human activities, such as poor land-use practices, can also create a drought or make an existing drought worse. The Dust Bowl in the Great Plains and the Sahelian drought in West Africa provide ready examples of the symbiotic relationship between drought and human activities.

In a sense, droughts differ from other major geophysical events such as volcanic eruptions, floods, and earthquakes because they are actually non-events—that is, they result from the absence of events (precipitation) that should normally occur. Droughts also differ from other geophysical events in that they often have no readily recognizable beginning and take some time to develop. In many instances, droughts are only recognized when plants start to wilt, wells and streams run dry, and reservoir shorelines recede.

There is wide variation in the duration and extent of droughts. The length of a drought cannot be predicted, as the irregular patterns of atmospheric circulation are not fully known and remain unpredictable. A drought ends when the area receives sufficient precipitation and water levels rise in the wells and streams. Because the severity and areal extent of a drought cannot be predicted, all that is really known is that they are an integral part of the overall natural system and that they will continue to occur.

*Robert M. Hordon*

## FURTHER READING

Ahrens, C. Donald. *Meteorology Today: An Introduction to Weather, Climate and the Environment.* 8th ed. Belmont, Calif.: Thomson Brooks-Cole, 2007. One of the most widely used and authoritative introductory textbooks for the study of meteorology and climatology. Explains complex concepts in a clear, precise manner. Provides numerous images and diagrams.

Bryson, Reid A., and Thomas J. Murray. *Climates of Hunger.* Madison: University of Wisconsin Press, 1977. An interesting account of the profound effect of climate on human societies going back to ancient times. Discusses major climate changes and droughts for various regions in the world in separate chapters. Treatment is nonmathematical and suitable for senior-level high school students and above.

*Climate, Drought, and Desertification.* Geneva, Switzerland: World Meteorological Organization, 1997.

Deals with the climatic factors, such as drought, that lead to desertification worldwide. Includes color illustrations.

Dixon, Lloyd S., Nancy Y. Moore, and Ellen M. Pint. *Drought Management Policies and Economic Effects in Urban Areas of California, 1987-92.* Santa Monica, Calif.: RAND, 1996. Examines the impacts of the 1987-1992 drought in California on urban and agricultural users. Assesses the effects of the drought on a broad range of residential, commercial, industrial, and agricultural water users.

Fisher, R. J. *If Rain Doesn't Come: An Anthropological Study of Drought and Human Ecology in Western Rajasthan.* New Delhi: Manohar, 1997. Focuses on the human ecology associated with India and other countries prone to drought. Examines the factors involved with drought and its durations, as well as drought relief tactics and evaluation of such programs. Illustrations, maps, and references.

Frederiksen, Harald D. *Drought Planning and Water Resources Implications in Water Resources Management.* Washington, D.C.: World Bank, 1992. Contains two papers on drought planning and water-use efficiency and effectiveness. Deals with policy and program issues in water-resources management from the perspective of the World Bank.

Hidore, John, John E. Oliver, Mary Snow, and Richard Snow. *Climatology.* 3d ed. Columbus, Ohio: Merrill, 1984. Discusses all aspects of climatology. Contains numerous black-and-white illustrations and maps. Suitable for college-level students. Although quantitative measures are included, no particular mathematical background is necessary.

Knight, Gregory, Ivan Raev, and Marieta Staneva, eds. *Drought in Bulgaria.* Hants, England: Ashgate Publishing Limited, 2004. Presents the occurrence of drought in Bulgaria as a case study to provide information on prospective future global conditions. Discusses the geological, economic, and ecological effects of drought in Bulgaria from 1982 to 1994. Concludes with policy and conservation recommendations for the future.

Mainguet, Monique. *Aridity: Drought and Human Development.* Berlin: Springer-Verlag, 2010. Discusses global and local aridity, drought, and associated changes in vegetation and hydrology. The second half of the text focuses on anthropogenic impact. Includes references and multiple indexes.

Lutgens, Frederick K., Edward J. Tarbuck, and Dennis Tasa. *The Atmosphere: An Introduction to Meteorology.* 11th ed. Upper Saddle River, N.J.: Prentice Hall, 2009. An excellent introduction and description of the atmosphere, meteorology, and weather patterns. Suitable for readers new to the study of these subjects. Color illustrations and maps.

U.S. Congress. House. *Effects of Drought on Agribusiness and Rural Economy.* 100th Congress, 2d session, 1988. Contains the testimony of a variety of witnesses at a congressional hearing held July 13, 1988. Includes text and tables documenting the impact of the drought of July, 1988, as well as a map of the United States that illustrates the use of the Palmer Drought Index. Offers the flavor of a congressional hearing on drought issues.

Wilhite, Donald A., ed. *Drought and Water Crises: Science, Technology, and Management Issues.* Boca Raton, Fla.: CRC Press, 2005. Covers water management practices currently used in areas around the world. Discusses monitoring, drought planning, and water conservation policies. Provides case studies that discuss many issues of and solutions to drought. Concludes with a chapter discussing the future of water conservation.

Wilhite, Donald A., and William E. Easterling, with Deborah A. Wood, eds. *Planning for Drought: Toward a Reduction of Societal Vulnerability.* Boulder, Colo.: Westview Press, 1987. An extensive collection of thirty-seven short chapters on drought covering a wide range of topics from the climatological to the institutional. Provides a good background to the many issues pertaining to drought, its social impacts, governmental response, and adaptation and adjustment. Suitable for college-level students.

Workman, James G. *Heart of Dryness: How the Last Bushmen Can Help Us Endure the Coming Age of Permanent Drought.* New York: Walker Publishing Company, 2009. Introduces the reader to the occurrence of drought and its impact on humans, animals, vegetation, and land. Describes in detail the issues faced and created by humans in desiccated regions through the eyes of the people facing it presently.

**See also:** Atmosphere's Structure and Thermodynamics; Atmospheric and Oceanic Oscillations; Barometric Pressure; Climate; Climate Change Theories; Climate Modeling; Clouds; Greenhouse Effect; Hurricanes; Lightning and Thunder;

Monsoons; Satellite Meteorology; Seasons; Severe Storms; Tornadoes; Van Allen Radiation Belts; Volcanoes: Climatic Effects; Weather Forecasting; Weather Forecasting: Numerical Weather Prediction; Weather Modification; Wind

# E

## EARTH-SUN RELATIONS

*The sun is the center of the solar system and the supporter of all life on Earth. It affects the weather, climate, and seasons; it drives the food chain through the process of photosynthesis; and it provides heat, light, and energy. Ultimately, the demise of the sun will bring about the end of Earth.*

### PRINCIPAL TERMS

- **aphelion:** the time at which the distance between the earth and the sun is smallest; generally occurs on one of the first days of January, two weeks after the December solstice
- **aurora:** a glowing light display resulting from charged particles from solar wind being pulled into Earth's atmosphere by Earth's magnetic field; most often visible near Earth's North and South Poles
- **equinox:** a twice-a-year occurrence during which the tilt of the earth's axis is such that the earth is not tilted toward or away from the sun; the center of the sun is directly aligned with Earth's equator
- **geomagnetic storm:** the effect of variations in solar wind's interactions with Earth's atmosphere; can result in communications disruptions and auroral displays in lower than usual latitudes
- **nuclear fusion:** atomic nuclei join together to form a heavier nucleus; in the sun and other main-sequence stars, hydrogen is fused to form helium
- **perihelion:** the time at which the distance between the earth and the sun is largest; generally occurs on one of the first days of July, two weeks after the June solstice
- **solar cycle:** an approximately eleven-year-long cycle of varying solar activity; solar cycles are tracked based on the visibility of sunspots
- **solar wind:** a stream of charged particles that the sun's atmosphere ejects into space, where it can interact with the magnetic fields of planets
- **solstice:** a twice-a-year occurrence during which the sun appears at its highest point in the sky (once a year as seen from the North Pole and once a year as seen from the South Pole)
- **sunspot:** a cooler area on the sun's surface that appears darker than the surrounding area; a zone of decreased temperature resulting from the complex shape of the sun's magnetic field

### OVERVIEW

The sun was born 4.57 billion years ago out of the collapse of a portion of a stellar nursery, a molecular cloud of enormous density and size. The giant, hot sphere of plasma and gas is the center of the solar system and the supporter of all life on Earth.

The sun accounts for approximately 99.86 percent of the mass of the solar system; its total mass is $2 \times 10^{30}$ kilograms (about 330,000 times the mass of Earth). Its diameter is roughly 1.4 million kilometers (more than one hundred times that of Earth), and its surface temperature is estimated to be 5,778 kelvin (5,505 degrees Celsius or 9,940 degrees Fahrenheit), while its core temperature is estimated at $1.571 \times 10^7$ kelvin. Meanwhile, on Earth, the maximum surface temperature is around 331 kelvins. Three-fourths of the sun is made of hydrogen and nearly one-fourth is made of helium. Trace amounts of heavier elements make up the rest.

The sun is the closest star to Earth and thus the brightest to appear in the sky. While the distance between the sun and the earth is constantly changing based on the earth's 365.25-day orbit around the sun and based on the earth's twenty-four-hour rotation around its own axis, the mean distance between the sun and the earth is 149.6 million km (93 million miles), a distance that defines one astronomical unit. It takes light from the sun eight minutes and nineteen seconds to cover this distance and reach Earth. For comparison, consider the 1.3 seconds that it takes light to reach the earth from Earth's moon, or, on the other end of the spectrum, the one hundred thousand years it takes for light to cross the Milky Way galaxy.

The sun affects the earth in a multitude of ways. Its complex magnetic field causes a range of effects known as solar activity, including sunspots and solar flares, and these phenomena in turn cause space weather in Earth's atmosphere, leading to beautiful visual effects (auroras) but troublesome disruptions of communications and electricity. Space weather also influences the earth's climate, weather, and seasons. All of this solar activity varies on a roughly eleven-year-long solar cycle. While these effects are mostly within Earth's upper atmosphere, the sun also directly affects life at the surface.

Sunlight drives photosynthesis in plants, algae, and some bacteria, allowing them to convert sunlight into food; without this process, the entire food chain would collapse. Humans also both enjoy and suffer direct effects from the sun, which can damage human eyes and skin while also providing sanitizing ultraviolet rays and driving vitamin D production. Solar power can influence the future of technology. Earth's ultimate demise will also be brought about by the sun.

## EFFECTS OF THE SUN'S MAGNETIC FIELD AND THE SOLAR CYCLE

The sun releases energy in two main forms: electromagnetic radiation and charged particle emission. Through the activities of the sun's magnetic field and the different ways in which these types of energy are released and moved, the earth experiences a variety of effects.

The sun's strong magnetic field has a complex spiral shape. The star's high temperature means that it is composed solely of gases and plasma, a composition that allows it to rotate faster around its equator than at the poles; this differential rotation twists the magnetic field into its distinctive form. Ultimately, it is the twists in the magnetic field that lead to such solar activity as sunspots and solar flares.

When solar activity is at its maximum in the eleven-year-long solar cycle, the magnetic field flips direction. The sun's magnetic field exerts its influence well beyond the sun itself because the solar wind, a plasma stream of charged particles, carries it through the solar system, altering the magnetic fields of the planets.

Varying solar activity, such as sporadic corona eruptions (coronal mass ejections), can alter the activity and intensity of the solar wind, resulting in geomagnetic storms on Earth. These storms cause beautiful glowing lights in Earth's atmosphere called auroras, particularly aurora borealis (the northern lights) and aurora australis (the southern lights). On the negative side, though, the storms also interrupt communications on Earth, disrupting radio signals and electric transmissions and interfering with compass-based navigation, among other effects.

The frequency of geomagnetic storms roughly aligns with the sunspot cycle, which in turn is determined by the solar cycle. Sunspots are visual representations of the sun's magnetic field; when the sun's magnetic field lines emerge from within the sun, they create magnetic loops in the photosphere (the sun's visible surface). The temperature is lower within these loops, appearing as the dark patches called sunspots. Other solar activity includes solar flares and solar plumes.

The solar cycle, which usually takes eleven years, is marked by a solar maximum (at which time solar activity is at its highest level) and, conversely, a solar minimum. During a solar maximum in 1859, for example, a geomagnetic storm was so intense that the northern lights could be seen as far south as Rome. Throughout history, several prolonged solar minimums have been observed, during which Earth temperatures were at an all-time low. One such period, known as the Maunder minimum, spanned the second half of the seventeenth century and into the beginning of the eighteenth century.

The Earth-sun relationship is partially driven by the solar cycle, during which solar activity is always varying. The Earth-sun relationship also is strongly influenced by the earth's orbit around the sun and by the earth's rotation around its own axis, both of which directly affect the distance between the earth and the sun. Perihelion (when the distance between Earth and sun is shortest) occurs once a year, in early January, and aphelion (when the distance is longest) occurs once a year, in early July. These generally occur two weeks after the December solstice and June solstice, respectively.

The solstices are the two points in the year when the sun appears highest in the sky, either in the north or in the south, because of the tilt of the earth's axis. The northern (or summer) solstice occurs in June, and it corresponds with summer in the Northern Hemisphere and winter in the Southern Hemisphere. Conversely, the southern (or winter) solstice occurs in December, during winter in the Northern Hemisphere and summer in the Southern Hemisphere. Similarly, two equinoxes occur each year, one in late March and one in late September, when the tilt of the earth's axis is such that the sun is aligned in the same plane as the equator. The equinoxes are hallmarks of spring and autumn.

The big picture is that variations in the sun's magnetic field cause different types of space weather, which affects Earth's upper atmosphere, ultimately influencing Earth's climate, weather, and seasons; the degree to which space weather occurs remains a topic of ongoing research. The whole relationship is also influenced by the ever-changing distance between the earth and the sun.

## PHOTOSYNTHESIS AND HEALTH EFFECTS
### OF THE SUN

The sun's effects on Earth are not limited to climate control; the sun is also entirely responsible for the food chain (through photosynthesis), and it has a number of other health effects, both positive and negative.

Photosynthesis is a process that occurs in plants, algae, and some bacteria. In basic terms, the process involves sunlight reacting with carbon dioxide and water to yield sugar (food) and oxygen. To elaborate somewhat, the first stage of photosynthesis, light-dependency, begins with an organism trapping and storing sunlight. The sunlight's energy is then converted into chemical energy, stored in the molecules ATP and NADPH.

The second stage does not require any more light; the already-created ATP and NADPH drive the reduction of carbon dioxide to glucose (sugar) and other useful organic molecules. Not only does photosynthesis serve as the basis for the food chain, but it also maintains the necessary oxygen levels in the atmosphere because oxygen is a waste product of the process.

The human health implications of sunlight, particularly the ultraviolet (UV) radiation in sunlight, are both positive and negative. On the positive side, human bodies synthesize vitamin D from sunlight; a deficiency in vitamin D can lead to the thinning and softening of bones, and some research has suggested a link between vitamin D deficiency and seasonal affective disorder, a seasonal depression.

On the negative side, too much UV radiation can lead to skin aging, skin cancer, and eye damage, such as cataracts and age-related macular degeneration. Looking at the sun through binoculars without a proper UV filter can cause instant retinal damage and even permanent blindness. It is particularly dangerous to look at the sun directly or with an optical aid during an eclipse.

## SOLAR POWER

Much as light energy can be converted to food by photosynthesis, it also can be converted to electricity. The harnessing of solar power began on a commercial scale in the 1970's and 1980's, and it continues to gain prevalence, showing promise for the future because of its numerous advantages over other power sources. Most importantly, light energy is renewable: Sunlight will not be depleted, at least not while the sun is still in roughly its current form for the next several billion years.

Solar power also is a clean energy; it does not release greenhouse gases or pollute the ocean, as does oil. Additionally, solar cells are long-lasting and require little maintenance. Although renewable, sunlight is not always available. Storage solutions are important because solar energy cannot be collected at night.

There are two major methods of solar power: first, the use of photovoltaics, which involves the use of the photoelectric effect of sunlight to convert light energy directly into an electric current; and second, the use of concentrated solar power, which involves the use of lenses or mirrors to focus sunlight into a small beam. This focused light is then converted to heat, which then powers a steam turbine or other heat engine to ultimately generate electricity. Both technologies are under development, and the hope is that costs will decrease and efficiency will increase in the coming decades.

## THE END OF EARTH

Just as the sun has supported more than 4 billion years of life on Earth, it is ultimately the sun that will bring about the end of Earth. The sun is a main-sequence (or dwarf) star, a type of star marked by fusion of hydrogen to helium within its core.

It is estimated that Earth's sun will exist about 10 billion years in its main-sequence form before it has fused all of its hydrogen; it has already covered nearly one-half of that time span. This means that in about 5 billion years it will enter its next phase of life, as a red giant.

The outer layers of the red giant sun will expand as its core contracts and heats up, and this overall decrease in mass will likely loosen its gravitational pull on the solar system, allowing the planets to move outward. The sun's radius will be greater than the current distance between the sun and the earth, and it is unclear whether the looser pull will allow Earth's orbit to move far enough away or if Earth will be swallowed up by the sun. Even if the earth does escape being enveloped by the sun, the heat will still likely be enough to boil off all of Earth's water and destroy much of its atmosphere. Even before any of this occurs, though (perhaps 1 billion years from now), it is quite possible that life on Earth will cease to exist because of rising surface temperatures of the sun (and thus the earth).

*Rachel Leah Blumenthal*

**FURTHER READING**

Alexander, David. *The Sun.* Santa Barbara, Calif.: Greenwood Press, 2009. Part of the Greenwood Guides to the Universe series, this easy-to-understand text gives readers the necessary background on the sun before delving into a comprehensive look at recent discoveries and future research goals.

Alley, Richard B. *Earth: The Operators' Manual.* New York: Norton, 2011. Through storytelling, this book illuminates the history of humankind's use of energy while also discussing the negative effects this has had on Earth. The book also delves into current and future alternative and renewable energy options, such as solar power.

Crosby, Alfred W. *Children of the Sun: A History of Humanity's Unappeasable Appetite for Energy.* New York: Norton, 2006. More historical than scientific, this book will nonetheless be interesting for science-minded readers who wish to put scientific knowledge of the sun in its broader historical and cultural context, particularly with regard to its energy potential for humanity.

Eddy, John A. *The Sun, the Earth, and Near-Earth Space: A Guide to the Sun-Earth System.* Washington, D.C.: National Aeronautics and Space Administration, 2009. Provides a comprehensive but easy-to-understand look at the sun's structure, function, and relationship with Earth.

Lang, Kenneth R. *Sun, Earth, and Sky.* New York: Springer, 2006. This beautiful text uses a multitude of illustrations and images to illuminate the workings of the sun and its relationship with Earth. It is easy to understand and perfectly suited for students new to the subject.

McFadden, Lucy-Ann Adams, Paul Robert Weissman, and T. V. Johnson. *Encyclopedia of the Solar System.* San Diego, Calif.: Academic Press, 2007. From the origins of the solar system to modern-day planetary exploration, this comprehensive text covers topics about the sun, solar wind, and the relationship between the sun and the earth.

**See also:** Atmospheric Properties; Auroras; Barometric Pressure; Carbon-Oxygen Cycle; Climate; Cosmic Rays and Background Radiation; Global Energy Transfer; Global Warming; Greenhouse Effect; Ozone Depletion and Ozone Holes; Seasons; Severe Storms; Van Allen Radiation Belts; Wind

# EL NIÑO/SOUTHERN OSCILLATIONS (ENSO)

*The El Niño Southern Oscillation (ENSO) is a series of linked atmospheric and oceanic events in which a reduction or reversal of the normal large-scale pressure systems acting in the equatorial regions of the Pacific Ocean results in dramatic changes in precipitation, currents, and wind patterns. Repercussions occur over wide areas from the interior of North America to India. Although not regularly periodic, El Niño phenomena generally occur every two to seven years.*

## PRINCIPAL TERMS

- **convection cell:** the cyclic path taken when warmer, less dense material in one part of a fluid or gas rises, cools, contracts, and becomes denser, then descends again to its original level
- **El Niño:** the warm extreme of the Southern Oscillation cycle, with unusually warm surface water in the equatorial Pacific Ocean and subdued trade winds
- **ENSO:** acronym for El Niño Southern Oscillation, used to denote the combined atmospheric/ocean phenomenon
- **La Niña:** the cold extreme of the Southern Oscillation cycle, with unusually cold surface water in the equatorial Pacific Ocean and enhanced trade winds
- **Southern Oscillation:** the reversal of atmospheric pressures that occurs between opposite sides of the tropical Pacific Ocean
- **thermocline:** the depth interval at which the temperature of ocean water changes abruptly, separating warm surface water from cold, deep water
- **trade winds:** winds blowing from the northeast in the Northern Hemisphere and from the southeast in the Southern Hemisphere and converging at the Intertropical Convergence Zone, which wanders back and forth across the equator during the course of a normal year

### GLOBAL WIND PATTERNS

El Niño is the term used for the oceanic aspects of one phase of a climatological phenomenon more generally referred to as an El Niño Southern Oscillation (ENSO) event. *El Niño* is a Spanish term meaning "the Christ child" that was used by Ecuadoran and Peruvian fishermen to refer to unusually warm surface currents that arrived around Christmastime at irregular intervals. ENSO events typically recur every two to seven years and involve dramatic changes in sea surface temperatures, precipitation, and wind patterns, accompanied by nutrient upwellings. Such environmental fluctuations influence, and in some cases may control, the climate and ecology of large areas of Earth where the effects of ENSO events are most pronounced.

The large-scale wind patterns of Earth result from the different amounts of solar energy received at the equator and at the poles. Generally, a low-pressure region of ascending air can be found at the equator. This air rises, cools, and releases its water vapor as rain, creating equatorial rainforests over land. High in the atmosphere, the dry air spreads out, eventually returning to Earth near the latitudes of the tropics, between 23 degrees and 30 degrees north and south. Most of Earth's large deserts are located in this latitude band. To finish its circuit, this air travels across the surface of Earth toward the equator. Because of Earth's rotation, however, its path is deflected, a phenomenon called the "Coriolis effect." In the Northern Hemisphere, the observed deflection is to the right, and in the Southern Hemisphere, it is to the left. Such a deflection causes the air to travel obliquely toward the west in both hemispheres, resulting in the system of winds that are known as the trade winds. (Most of the United States lies north of this area and is under the influence of a similar set of winds blowing in the opposite direction, known as the prevailing westerlies.)

In one pattern called the Walker circulation, warm, wet air rises over Indonesia, spreads out, and then falls as cool, dry air on the eastern Pacific Ocean and the coast of Peru. The existence of this phenomenon became apparent to British scientist Sir Gilbert Walker as he collected data from Tahiti and Darwin, Australia, during the early part of the twentieth century. He found that the atmospheric pressures in these two locations varied together but out of phase with each other. That is, if the pressure rose in Darwin, it fell in Tahiti. Sometimes conditions were normal, with high-pressure zones over Tahiti and lows over Darwin, while at other times they were reversed. Walker called this phenomenon the Southern Oscillation and showed that many other

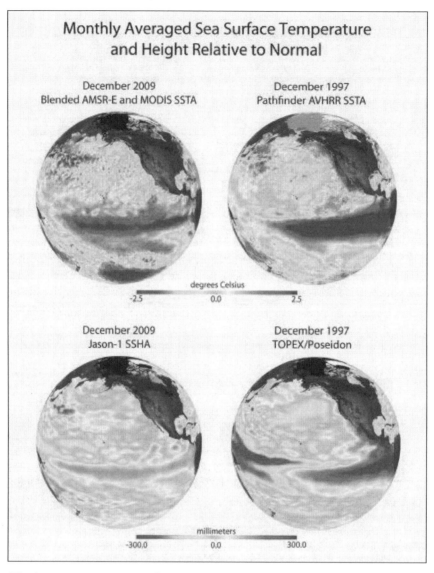

Monthly Averaged Sea Surface Temperature and Height Relative to Normal

*The images compare sea surface temperature (SST) and sea surface height (SSH) anomalies associated with the mature phase of the 1997-98 and 2009-10 El Niño events (in December 1997 and 2009, respectively). SST and SSH anomalies reflect the heat content in the mixed layer (approximately upper 50 m) and the upper ocean (approximatedly upper 150 m), respectively. They provide complementary views of the oceanic signature of climate variability known as El Niño. The comparisons indicate that the warming associated with the moderate 2009-10 El Niño actually extended farther west than the that associated with 1997-98 El Niño. Although the latter is considered the "event of the century" because of the large amplitude of the SST anomaly in the eastern equatorial Pacific, the warming in the central equatorial Pacific was not as pronounced as that for the 2009-10 El Niño. (NASA/JPL-Caltech)* meteorological phenomena from around the world seemed to be tied to it, though his particular goal was predicting when the monsoons would fall in India.

As the trade winds move across the Pacific Ocean, they drag along great quantities of water. This water is at the sea surface, where it is warmed by incoming solar radiation, and builds up in the western Pacific in the vicinity of Indonesia. The elevation of the sea there can be 40 centimeters higher than in the eastern equatorial Pacific. The temperature of the sea changes rather abruptly between the water that has been warmed by the sun and the water beneath it. The depth at which this change occurs is called the thermocline. Because the density of water varies with temperature, this abrupt temperature change corresponds to a density contrast that effectively acts as a barrier, and mixing of water across the thermocline does not readily occur. The buildup of warm surface water in the area east of Indonesia causes the thermocline to settle to as much as 200 meters deep. Off the coast of Peru, the thermocline is typically about 50 meters deep. The difference of 40 centimeters of water at the surface is balanced by a difference of 150 meters on the thermocline because the density contrast across the thermocline, between the cold water and warm water, is much less than the density contrast between the water and air at the surface.

## OCCURRENCE OF EL NIÑO AND LA NIÑA

Every few years, for reasons that are not well understood and are apparently the result of a complex atmospheric system, the strength of the trade winds decreases. Often this is accompanied by an eastward spreading of the surface pool of warm water. These two phenomena are linked: Each can and probably does, to some extent, cause the other. As the trade winds collapse, a major readjustment of the thermocline occurs. As the small surface slope dwindles, the much larger slope on the thermocline also disappears. As the thermocline rises, it generates Kelvin waves, which are large-scale gravity waves. These waves race across the Pacific Ocean, deepening the thermocline to the east as they go.

The deeper thermocline and warmer surface water reduce the effectiveness of upwelling in supplying nutrients to the surface water. This causes dramatic decreases in the primary productivity of the oceans off the coast of Peru and often spectacular die-offs in many species of fish and the seabirds and other animals that rely on them. During the 1972 El Niño, die-offs resulted in the collapse of the anchovy fishing industry, although its decline was exacerbated by questionable fishery management policies.

As the thermocline deepens and warmer surface waters move east, the area in the Pacific equatorial region where the ascent of warm, moisture-laden air produces intense precipitation shifts to the east. This area, called the Intertropical Convergence Zone (ITCZ), is generally found over the warmest water. During strong ENSO events, the Walker circulation may be completely reversed, with warm, moist air rising over coastal Peru, where cool, dry air normally falls. When this happens, regions that have not received a drop of rainfall for years are suddenly inundated with precipitation.

The paths and severity of many storms tracking across the continental United States are affected by the location of the ICTZ. During ENSO events, the jet stream is often moved or disrupted by the heavy storm activity above the zone. Sometimes this results in more precipitation across certain sections of the United States and sometimes less, but typically the result is abnormal weather.

Eventually, El Niño weakens, and the trade winds return to their usual strength. Cold water wells up from the depths to replace the warm, extra water as it dissipates, bringing nutrients to reestablish the high productivity of the waters. As warm surface water builds up in Indonesia, the thermocline there deepens again, and the ICTZ moves toward its earlier location. If, as sometimes happens, the return of the trade winds cools the surface waters beyond their normal temperatures, a phase called La Niña begins.

Apparently, the ocean/atmosphere system oscillates between the two states of La Niña and El Niño. What drives it from one state to the other remains a matter of considerable speculation. Although the complexities of the ocean/atmosphere system would seem more than adequate to produce almost any sequence of events, some researchers have suggested the involvement of yet another complex system, Earth's tectonic engine. ENSO events may be correlated with eruptions of large equatorial volcanoes, which inject dust into the air and can have dramatic impact on the atmospheric components of the cycle.

There is a great deal of interest, both scientific and general, in learning more about ENSO events. Computational capabilities and theoretical developments have progressed rapidly, and although scientists do not yet understand this complex interaction of air and water, they have begun to deploy water temperature sensing buoys across the Pacific to study it.

## STUDY OF EL NIÑO

El Niño Southern Oscillation events are huge. They last longer than one year, involve massive amounts of air and water, and affect a substantial percentage of Earth's surface. During the course of an ENSO event, the changes that have already occurred in the system influence the changes that will occur as it progresses. Winds affect sea surface temperatures, and sea surface temperatures affect the winds. Such feedback interactions make it difficult to develop models with predictive capability. In the past, computer models used for atmospheric studies often oversimplified the ocean, and those used for oceanic studies oversimplified the atmosphere. ENSO events involve a tight coupling of ocean and atmosphere, and any successful model will need to treat both with sophistication and detail.

Large computer models are used to study ENSO events. As they evolve, deficiencies in the available data set are identified, and additional data are acquired. Much of the necessary data are of the type normally gathered during almost any routine oceanographic study: temperatures, salinities, and surface wind velocities and directions. Researchers would like to have such data from before, during, and after an ENSO event. Because these events recur so frequently, almost any data must be collected within a few years of an event. The difficulty lies in figuring out which data are important. Although there is agreement on what typically happens during an ENSO event, the specifics vary widely from event to event. Just when scientists think they understand it well enough to venture forth with a predictive model, nature comes along and shows them that their model is inadequate.

Models, whether sophisticated computer models or simple physical models, can further the understanding of ENSO events. One simple physical model uses a clear Pyrex baking pan as the ocean basin; water as the deep, cool, ocean water; vegetable oil as the warmer, less dense, surface water; and a hair dryer (set to blow unheated air) as the trade winds. If the flow of air produced by the hair dryer is directed across the oil, the oil piles up at the far side of the baking pan. The surface of the oil will form a slight slope, with the higher elevation at the far side of the baking pan. A more pronounced, readily visible slope will form where the oil meets the water. Friction from the air produced by the hair dryer pushes surface oil toward the far side of the pan, but as it builds up, it increases the pressure on the water beneath it. This causes the water beneath the oil to move back toward the near side of the baking pan. Eventually a wedge of oil sits over the water, its upper slope maintained by the hair dryer "wind," and its lower slope positioned where the pressures at depth are the same. (The fraction of a centimeter of additional oil instead of air at the top surface is compensated by a considerably greater deflection of the oil-water boundary because the density contrast between oil and water is much less than that between oil and air.) This represents the normal state in the equatorial Pacific: a shallow thermocline off the coast of Peru; winds moving off the South American continent at Peru, preventing rainfall; a deep thermocline in Indonesia; and air moving from east to west across the ocean, becoming warm and moist, rising over Indonesia, and subsequently drenching it with rain.

In the model, the equivalent of an El Niño can be produced by turning off the hair dryer. The forces that maintained the slope on the upper surface of the oil are gone, so the oil sloshes down and its surface becomes horizontal. A wave forms where the oil meets the water, a boundary that corresponds to the thermocline. Such a wave is called an internal wave because it occurs between two layers of water with contrasting densities but in general is not visible at the surface. The wave races across the baking pan just as the waves race across the Pacific during an El Niño.

Although this physical model demonstrates many of the basics of an ENSO event, no physical model is capable of capturing the entire natural event because its scale is so enormous. The rotation of Earth plays a major role, and the nature of currents and waves as they move directly along the equator is also quite significant. Computer models can incorporate these effects, however, and developing such models is an active area of research.

Additional data regarding the ocean are also required. The Pacific Ocean between Indonesia and Peru has not been studied in enough detail to enable scientists to refine their models as much as they would like. The processes that occur during an ENSO event involve large areas and take place over a considerable period of time. Monitoring the sea to detect gradual changes in sea surface temperatures, depth of the thermocline, and surface wind velocity and direction is a complex, expensive project. As more data are acquired, understanding of ENSO events is certain to improve.

## SIGNIFICANCE

The El Niño Southern Oscillation cycle is as much a part of Earth's weather and climate as the cycle of seasons. Just as people spend about one-fourth of the year in winter, people spend about one-fourth of their lives in an El Niño phase. Therefore, El Niño events are not truly anomalous. Unlike winter, however, the timing of the ENSO cycle is not regular. For example, from 1991 through 1994, three El Niños developed. Analysis of historic records shows that El Niños have occurred at nearly the same frequency since at least 1525, when record-keeping began in Peru. In addition to historic records, evidence for prehistoric ENSO events is found in tree rings, flood deposits, ice cores, and other geological and biological record keepers. These, too, suggest that the frequency and severity of ENSO events has been fairly constant for at least several thousand years.

The severity of El Niños can vary remarkably. Substantial effort has been made to identify predictors that would permit an early estimate of the occurrence and severity of ENSO events. Several promising predictors of the timing or intensity have emerged but have been abandoned as additional events transpired in which the predictor was present but the expected development did not occur.

Sir Gilbert Walker's suggestion that local weather in widely separated parts of the planet might be closely related was ridiculed by some of his contemporaries. However, his belief has been borne out by the data and understanding acquired subsequent to his early work. ENSO-related changes can also be seen in the Indian and Atlantic Oceans. Some droughts in Africa and India and most droughts in Australia seem connected to ENSO events. For example, weak floods on the Nile River, records of which have been kept since 622 C.E., seem directly related to El Niño events.

Droughts and floods can probably never be prevented. However, many of the ensuing hardships could be avoided or reduced if they could be predicted. People and governments in affected areas could take preventive measures: Drought-resistant or flood-resistant varieties of crops could be planted, dikes could be reinforced, deforestation could be halted, and water could be temporarily impounded. Before these measures are taken, however, considerable confidence in any prediction is required. Preparing for a flood may worsen the effects of a drought. Someday, however, it may be possible to predict ENSO events with the same degree of accuracy with which meteorologists predict the arrival of a significant winter storm.

*Otto H. Muller*

## FURTHER READING

Ahrens, C. Donald. *Meteorology Today: An Introduction to Weather, Climate and the Environment.* 8th ed. Belmont, Calif.: Thomson Brooks-Cole, 2007. One of the most widely used and authoritative introductory textbooks for the study of meteorology and climatology. Explains complex concepts in a clear, precise manner, supported by numerous images and diagrams. Includes a detailed description of the El Niño/Southern Oscillation phenomena, their causes, and effects.

Clarke, Allan J. *An Introduction to the Dynamics of El Niño and the Southern Oscillation.* Burlington, Mass.: Academic Press, 2008. Presents the physics of ENSO, including currents, temperature, winds, and waves. Discusses ENSO forecasting models. Provides good coverage of the influence ENSO has on marine life from plankton to green turtles. Has references and a number of appendices and indexing. Best suited for environmental scientists, meteorologists, and similar academics studying ENSO.

Davidson, Keay. "What's Wrong with the Weather? El Niño Strikes Again." *Earth* (June, 1995): 24-33. Reviews the anomalous 1991-1994 period when El Niños of various strengths occurred in three out of the four years. Describes effects in the Indian Ocean. Includes some discussion of Rossby waves from earlier El Niños bouncing around the Pacific basin. No equations, several diagrams. Suitable for readers at the high school level.

Diaz, Henry F., and Vera Markgraf, eds. *El Niño: Historical and Paleoclimatic Aspects of the Southern Oscillation.* Cambridge, England: Cambridge University Press, 1992. A collection of twenty-two papers dealing with historical aspects of El Niño. Covers a variety of climate indicators from ice cores to tree rings to records of floods on the Nile River to reveal the long-term characteristics of El Niños. Has useful summaries and concluding sections. Suitable for college-level readers.

Hatfield, Jerry L., et al., eds. *Impacts of El Niño and Climate Variability on Agriculture.* ASA Special Publication Number 63. Madison, Wisc.: American Society of Agronomy, 2001. Presents the possibility of using knowledge of ENSO to forecast long-term weather patterns, which could aid farmers in crop planning. Discusses the topics of teleconnections and the global impact of ENSO.

Knox, Pamela Naber. "A Current Catastrophe: El Niño." *Earth,* (September, 1992): 31-37. Provides an easy-to-read overview of many aspects of El Niño events. Presents some historical perspectives along with various theories on causes, including possible tie-ins with volcanic eruptions. No equations, several good diagrams. Suitable for high school readers.

Nash, J. Madeleine. *El Niño: Unlocking the Secrets of the Master Weather-Maker.* New York: Warner Books, 2002. Discusses the extreme effects of El Niño events on the ecosystem and civilization throughout time. Describes the 1997-1998 flooding in California, both from the perspective of the people caught in the storm, and the scientists studying it. Knowledge of meteorology and climatology would aid the reader, but is not necessary to read this book.

Open University Oceanography Course Team. *Ocean Circulation.* 2d ed. Oxford, England: Butterworth Heinemann, 2001. Provides an introduction into the analysis and understanding of all sorts of circulation in the world's oceans as well as the mathematical background needed to understand many of the aspects of geophysical fluid dynamics. Intended for an undergraduate college-level reader. Builds on many well-presented discussions of topics such as Kelvin waves and Rossby waves, though treatment of El Niño is brief.

Philander, S. George. *El Niño, La Niña, and the Southern Oscillation.* San Diego, Calif.: Academic Press, 1990. Contains some quantitative material beyond the grasp of the average reader, but covers its subjects thoroughly and with enough qualitative discussions to make it worthwhile. Plenty of diagrams back up the material. Includes visual representations of much of the math. Suitable for most college-level readers, particularly those with some background in science.

Ramage, Colin S. "El Niño." *Scientific American* (June, 1986): 76-83. Presents several different theories on what causes and sustains ENSO events. Carefully distinguishes between the generally accepted model of El Niño and the actual course taken by any particular event. Several excellent diagrams and maps. Suitable for high school readers.

Sarachik, Edward S., and Mark A. Cane. *The El Niño-Southern Oscillation Phenomenon.* New York: Cambridge University Press, 2010. A comprehensive discussion of ENSO and other oceanic/atmospheric processes. Covers research measurements, models, and predictions of future occurrences. Includes many diagrams, appendices, a references list, and index.

Thompson, Russell D. *Atmospheric Processes and Systems.* London: Routledge, 1998. Discusses many different aspects of the mass, energy, and circulation systems that function in the atmosphere. Provides strong background and insight for understanding the El Niño phenomenon.

**See also:** Atmosphere's Global Circulation; Atmospheric and Oceanic Oscillations; Atmospheric Properties; Climate; Clouds; Cyclones and Anticyclones; Drought; Earth-Sun Relations; Global Energy Transfer; Global Warming; Greenhouse Effect; Hurricanes; Lightning and Thunder; Monsoons; Precipitation; Remote Sensing of the Oceans; Satellite Meteorology; Seasons; Severe Storms; Tornadoes; Tropical Weather; Van Allen Radiation Belts; Volcanoes: Climatic Effects; Weather Forecasting; Weather Forecasting: Numerical Weather Prediction; Weather Modification; Wind

# FLOODPLAINS

*Floodplains historically have been the locations of dense human population, providing fertile land and water that enabled early human settlements to attain some degree of permanence. They have also been the locations of significant recurring damage from floods.*

## PRINCIPAL TERMS

- **alluvium:** sediments that have been deposited by a stream
- **bluff:** the edge of the remnant higher-elevated land that marks the margin of the floodplain
- **discharge:** the volume of water that is transported by a stream, normally stated as cubic meters per second or cubic feet per second
- **fluvial:** pertaining to or resulting from the action of naturally flowing water in a streambed or watercourse
- **geomorphology:** the study of the origins of landforms and the processes of landform development
- **local base level:** that elevation below which a stream of equilibrium will not degrade
- **meander:** a large sinuous curve or bend in a stream of equilibrium on a floodplain
- **one-hundred-year-flood:** a hypothetical flood whose severity is such that it would occur on an average of only once in a period of one hundred years; equates to a 1 percent probability each year
- **oxbow:** a lake that is a remnant of a stream meander that has been isolated by sediment deposits
- **stream of equilibrium:** a stream that is carrying its maximum load of sediment; it will not erode its channel any deeper but instead will establish a floodplain

## CHARACTERISTICS

Floodplains are well named. They are some of the most level surfaces on Earth, and they are known for frequent floods. They also are probably the most agriculturally productive lands on Earth. The earliest concentrations of civilization are associated with the floodplains of historical rivers whose names are well known: the Nile, Ganges, Euphrates, Yangtze, and Huang He. These floodplains are still among the most populous areas on Earth.

A floodplain is a landform, a physical feature that is studied in the discipline of geomorphology. Its origin is linked with frequently recurring floods. One of the most studied floodplains, because of its importance to agriculture and its role in the history and the economy of the United States, is that of the Mississippi River. Extensive levees and flood walls have been built in an effort to restrict floodwaters to certain areas where they will do the least damage. A large flood, the so-called superflood or one-hundred-year-flood, will still do tremendous damage. A large river cannot be prevented from flooding; it is a part of the natural fluvial process. It can only be altered or controlled to some extent so that its floodwaters will do a minimum of damage.

The variables of climate and topographic relief will result in floodplain variation around the world, but all will possess certain identifiable characteristics. The foremost distinguishing characteristic of a floodplain is the very low relief, or its almost "flat" appearance. Of course, the land is not in fact level, or the river would not flow. The floodplain possesses nonetheless a gradient toward sea level. It is an expanse of sediment that was deposited by the river itself. The appearance is the result of the even distribution of sediment over the area. Thousands of years of intermittent flooding have spread layer upon layer of river-borne sands, silts, and clays, referred to as alluvium. As the waters receded after each flood, the river returned to its normal channel.

## FORMATION

In order to understand the floodplain, it is necessary to understand the processes that brought it into being. The originating river did not always possess the floodplain, and many rivers today do not have floodplains. A stream with a floodplain may have at one time, perhaps millions of years ago, originally flowed in a narrow channel with higher land bordering either side. There was no level land adjacent to the stream to form a flood-plain; its formation would require eons of time of erosion and deposition. As the stream eroded, its channel deepened and also eroded laterally, but the vertical downcutting was dominant and the lateral cutting ineffective. At some point in the fluvial process of downcutting, the stream reached an elevation, referred to as local base level, beyond which the stream could not downcut. The local base level is not some hard, resistant rock that stops downcutting, but rather is a particular elevation in respect to that stream, which, in relation to the stream's mouth at sea level, determines the gradient of the stream. At this point in any stream, downward erosion ceases until something changes the controlling condition.

When the stream downcuts its valley, the stream's gradient, or slope of flow, is lowered. As the gradient is lowered, so is the stream's velocity, and consequently the stream's ability to carry sediment load is lowered. At some gradient, the stream's ability to carry its sediment load equals the sediment load that has been brought to the master stream by its tributaries. When that happens, the stream is said to be in equilibrium. Its sediment-load carrying capacity is then equal to the sediment load. This "stream of equilibrium" condition is a prerequisite for the development of a floodplain. The Mississippi River is a classic example of such a stream. When the stream attains equilibrium and stops downcutting, lateral cutting becomes the dominant stream activity.

Thus, the processes for floodplain development are set in motion. The lateral cutting by the stream results in a widening of the valley, but without any deepening. A broader-level valley floor then begins to develop, as the stream swings side to side, without downcutting. Over many years, the floor of the valley becomes wider than the channel of the stream itself. This level land on the valley floor adjacent to the stream, small at first, is the floodplain. It will be enlarged over time through lateral movements of the stream. The stream's volume of water, or discharge, will naturally vary with seasonal precipitation. During a flood, the discharge exceeds the channel capacity, and the excess overflows onto the level adjacent land. The term "floodplain" is now appropriate.

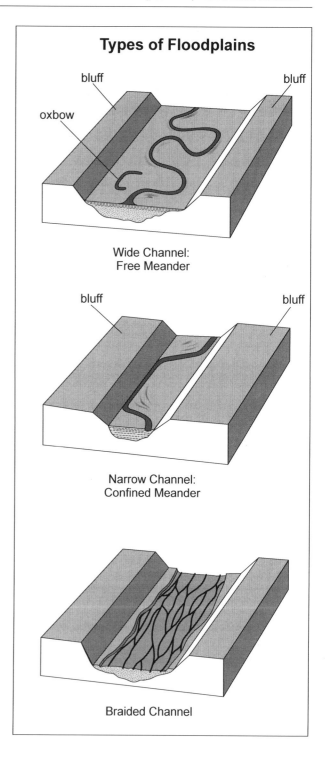

**Types of Floodplains**

Wide Channel:
Free Meander

Narrow Channel:
Confined Meander

Braided Channel

The floodplain is continually widened as the stream cuts laterally. The "bluffs" that mark the edge of the floodplain are driven farther back from the river as more of the adjacent higher land is eroded. When the floodplain is wide and horizontally level enough, the stream channel may develop great curving, sinuous "meanders." Meanders are perhaps the most prominent identifying characteristics of a stream of equilibrium, and true meanders are always on a floodplain. Meanders may subsequently be breached and cut off from the main channel by its persistent migrations or wanderings over the floodplain and the deposition of sediments across the breach. Such cutoff meanders may persist for hundreds of years as lakes, called oxbows. The oxbows will gradually become silt-filled and become oxbow swamps for a period of time. Even after thousands of years, the oxbow lakes and oxbow swamps may survive as meander "scars" on the floodplain, indicated by vegetation and soil differences.

These curving scars are even visible in fields of cropland. As the river continues to flood periodically, frequently on an annual basis, each time the floodwaters spread over the floodplain, an additional layer of silt is deposited. The entire floodplain is a compilation of layer upon layer of alluvial deposits from flood after flood, over thousands of years. The silts that are deposited on the plain have been eroded earlier from soils upstream. They tend to be fertile topsoils from upstream drainage basins, and thus the floodplain tends to receive almost annual increments of fertile topsoil. The recurring flooding and silting shortens the life span of the oxbows. Older oxbows are slowly obliterated, while new ones are formed by the meandering stream. Floodplains of large rivers can attain impressive dimensions. The Mississippi floodplain is 60 to 100 kilometers wide and some 800 kilometers long, extending from the confluence of the Ohio River in southern Illinois to the Gulf of Mexico, and is marked with old and new oxbows and oxbow remnants in many areas.

## FLOOD CONTROL

Agriculturists along the Nile River historically depended on the annual inundation to deposit a layer of fertile topsoil that made agriculture the foundation of Egyptian civilization. The Mississippi floodplain is also superb agricultural land. Great floods in 1927 and 1937, however, wrought such havoc that massive flood-control measures were initiated to prevent further such occurrences. The controls are not complete and probably never will

be, as it is a never-ending battle to stay abreast of the river. During the great Mississippi flood of 1973, floodwaters were largely confined, and a flood of catastrophic proportions was avoided. In 1993, the Mississippi River once again flooded. Big cities, such as St. Louis, were protected by floodwalls, either for aesthetic or economic reasons. However, Des Moines, which had no floodwalls, was severely affected: Its water-treatment plant flooded, leaving the community without a safe water supply for one month. The exceedingly costly flooding of 1993 once again threw into question many policies regarding levee and floodwall construction, which protect some farmland and communities but worsen flooding elsewhere. The U.S. Army Corps of Engineers breached at least one agricultural levee to "save" a town downstream during the 1993 flood. Also problematical is the federal flood insurance program, which "rewards" people who live in the floodplain by giving them benefits after each flood. The damage that occurred in 1993 prompted calls to change the program so that a recipient would receive benefits only for the first insured flood loss.

A major aspect of all floodplains is the relationship between the corresponding rivers' natural tendencies to flood and the human occupants' efforts to protect themselves. There are a number of flood-control measures and devices. Principal among them are levees. Natural levees are formed along the banks of a flowing river by silt deposition at the margins. Flood-control levees are human-made earthen ridges constructed parallel to the channel about 0.5 kilometer back from the riverbanks. The function of levees is to confine the overbank floodwaters to a narrow flood zone along the river and thus protect the greater portion of the floodplain. In essence, the levees form a secondary set of river banks, effectively making the river wider and deeper than its natural course. Without the levees, the entire floodplain could be flooded in a major flood all the way to the bluffs. Levee failure can be disastrous. Another major flood-control measure is the construction of reservoirs on tributary streams, where water can be held back when the master stream is in flood. These reservoirs are not on the floodplain itself and may be many miles upstream in the tributaries. For such broadly distributed measures, an area-wide, integrated plan must generally be adopted to ensure efficient flood control.

## STUDY OF FLOODPLAINS

Floodplains have been studied for almost as long as human civilization has existed. Most of the earliest permanent human settlements were in fact associated with floodplains, where fertile land and water were available. Early concerns were focused on two issues: distribution of irrigation water for agriculture and protection from excess water during floods. Modern concerns are similar. In the Mississippi floodplain, the concern is primarily flood control. The area's climate is humid, and river water has never been important for irrigation in the region. Research and study, therefore, have been directed toward developing effective measures and devices for control of overbank waters. An additional area for study, and for considerable expenditure as well, has been channel management for navigation purposes. The maintenance of an efficient channel is necessary for both navigation and flood control. Flood control, however, is concerned with maximum flow. Channel management for navigation purposes is primarily concerned with minimum flow, maintaining minimum shipping channel depths and widths during dry seasons and droughts.

Studies and data collection are directed toward the mechanics of stream flow. Stream-gauging stations are established in which measurements are taken to record discharge, velocity, and flow turbulence. Many of the gauging stations are automatic, producing data for long-term analysis and also alerting warning monitors to flash-flood conditions. Gauging networks are established for tributary streams as well as for the master stream on the floodplain. The morphology of the channel is studied. The width, depth, and cross-sectional areas of the streams are analyzed for their capabilities to pass the flood flows. Great effort and expense are applied in river-engineering works to improve channel flow and to increase capabilities for high water discharge. Some of the adjustments are channel straightening, bank clearing, dredging and sandbar removal, and the construction of levees and dikes. Models of the stream channels, floodplain, and levees are constructed to precise scale. Water is passed through the models in order to determine the best configurations for efficient flow. Dikes are placed in critical locations in the model, and the flow is studied to observe the effect of the dikes on bank erosion and sandbar formation. While the river conditions cannot be perfectly replicated in

a scale model, its use does aid in the selection of sites for river-engineering works.

## SIGNIFICANCE

Floodplains will continue to play an important role in human activities. They have always been areas of concentrated population and will be even more so in the future as the world population increases. Floodplains will acquire even more significance for agriculture. More people will live on the wide level surfaces that were once the undisputed domains of the rivers, yet the rivers can then no longer be allowed to flood the plains. It is natural for a stream to flood the plain periodically. It is not natural to confine the flood to a narrow strip along the channel as has been done with the Mississippi. More than 3,000 kilometers of levees have been constructed along the lower Mississippi River. The levees have been a success in preventing disastrous floods such as those that occurred in 1927 and 1937. They have brought protection to the valuable farmlands and to the towns and settlements on the plains. However, losses along the upper Mississippi were enormous during the 1993 flood, which caused fifty-four deaths and $12 billion in property damage.

The protection offered by the levees serves as an invitation for continued settlement and development, which proceed today on the world's floodplains as if there were no threats of flooding. A levee failure today could be more catastrophic than one in the past, even with the same amount of flooding, because so much more lies in the path of the flood. It is probably correct that the greatest flood is yet to come.

Adding to the threat of flooding on the floodplain is the fact that human habitation of floodplains increases the likelihood of floods. The floodplain can be thought of as the bottom of a funnel. Runoff from all basins upstream merges on the floodplain. Many flood-control measures that are applied to tributary streams, such as channel straightening, actually increase the likelihood of flooding downstream by speeding up the runoff. Additionally, changes in land use today exacerbate flooding. Deforestation and the drainage of wetlands increase runoff and intensify flooding. The increase in urbanization, with its growing expanse of asphalt paving and rooftops, accelerates runoff onto areas downstream by redirecting water that would normally enter the ground as part of the natural groundwater recharge function. As the drainage basins undergo change, so does the potential for flooding on the floodplain. The floodplain is a dynamic and changing environment, and it will require continued study and adjustment. Humans are now settled densely on the world's floodplains. The plains have become more significant as food-producing areas in a world of growing demands. Great engineering works have been constructed and have doubtless prevented many floods, yet the potential for disastrous floods has only increased. Continued vigilance is thus essential.

*John H. Corbet*

## FURTHER READING

Bloom, Arthur L. *Geomorphology: A Systematic Analysis of Late Cenozoic Landforms.* 3d ed. Upper Saddle River, N.J.: Prentice-Hall, 1998. Thoroughly covers geomorphology. Explains the processes and landforms of streams. Diagrams illustrate the stages of floodplain development. Assumes some knowledge on the part of the reader but is not difficult. Provides an extensive list of references with each chapter.

Changnon, Stanley A., ed. *The Great Flood of 1993: Causes, Impacts, and Responses.* Boulder, Colo.: Westview Press, 1996. Examines the Mississippi River basin flood damage of 1993 and its social and environmental effects. Also looks at the factors leading up to the flood and prevention methods put in place as a result. Illustrations and bibliographical references.

Chorley, Richard J., Stanley A. Schumm, and David E. Sugden. *Geomorphology.* New York: Methuen & Co., 1985. Written by three of the leading experts in the field of geomorphology. Provides a detailed description of the theory and physical principles of landform development, filled with numerous explanatory line drawings, working charts, and graphs.

Collier, Michael. *Over the Rivers: An Aerial View of Geology.* New York: Mikaya Press, Inc., 2008. Discusses the dynamic landscape of the rivers. Explains the processes that shape the landscape and its influence on humans. Written in the popular style, this book is easily accessible to the general public. Filled with bits of information and extraordinary photographs. Presents multiple examples drawn from the Mississippi River.

Darby, Stephen, and David Sear. *River Restoration: Managing the Uncertainty in Restoring Physical Habitat.* Hoboken, N.J.: John Wiley & Sons, Ltd., 2008. Begins with theoretical and philosophical issues with habitat restoration and provides a strong foundation for decision making. Addresses logistics, planning, mathematical modeling, and construction stages of restoration. References post-construction monitoring and long-term evaluations to provide a full picture of the habitat restoration process. Highly useful for anyone involved in the planning and implementing of habitat restoration.

Foresman, Timothy, and Alan H. Strahler. *Visualizing Physical Geography.* New York: John Wiley & Sons, 2012. Uses a unique approach to presenting the concepts of physical geography by heavily integrating visuals from *National Geographic* and other sources to vividly illustrate the manner in which physical geographic processes are interconnected.

Freitag, Bob, et al. *Floodplain Management: A New Approach for a New Era.* Washington, D.C.: Island Press, 2009. Presents different case studies that focus on a new topic in flood control in each chapter. Strategies of floodplain management revolve around the natural processes and dynamics of rivers. Discusses multiple approaches, as varying as the locations in which they are used. Best suited for engineers and hydrologists taking part in floodplain management.

Pavlopoulos, Kosmas, Niki Evelpidou, and Andrea Vassilopoulos. *Mapping Geomorphological Environments.* New York: Springer, 2009. Contains a chapter discussing fluvial environments and processes. Also discusses glacial formations and the waterfalls of glacial state park. Describes many other geologic formations and examines mapping methodologies.

Strahler, Alan H. *Modern Physical Geography.* 4th ed. New Delhi: Wiley India, 2008. Presents a comprehensive overview of physical geography. Discusses drainage basins as the basic unit of surface water flow.

Strahler, Arthur N., and Alan H. Strahler. *Introducing Physical Geography.* 5th ed. New York: John Wiley & Sons, 2010. A general introductory text on physical geography. Gives a good explanation of fluvial processes and the development of floodplains.

Well illustrated with maps and photographs. Recommended as a first source of information. Accompanied by a CD-ROM.

Tarbuck, Edward J., Frederick K. Lutgens, and Dennis Tasa. *Earth: An Introduction to Physical Geology.* 10th ed. Upper Saddle River, N.J.: Prentice Hall, 2010. A general beginner text for college Earth sciences, easily readable by the layperson. Employs quality color diagrams to explain the development of a floodplain, plus excellent illustrations throughout. Recommended as an initial source.

Wohl, Ellen. *A World of Rivers.* Chicago: University of Chicago Press, 2011. Describes the Amazon, Ob, Nile, Danube, Ganges, Mississippi, Murray-Darling, Congo, Chang Jiang, and Mackenzie Rivers. Contains valuable figures that provide straightforward and well-organized information. Discusses natural history, anthropogenic impact, and the future environment of these ten great rivers. Bibliography is organized by chapter.

**See also:** Alluvial Systems; Beaches and Coastal Processes; Deep-Sea Sedimentation; Deltas; Desert Pavement; Drainage Basins; Freshwater and Groundwater Contamination Around the World; Geochemical Cycles; Lakes; Reefs; River Bed Forms; River Flow; Sand; Sediment Transport and Deposition; Weathering and Erosion

# FLOODS

*Floods are extreme conditions of flowing water. They generally occur because of inordinate amounts of rainfall or snowmelt, but also result from other causes such as dam failures and volcanic eruptions. Floods exert a major role in shaping river systems, and their occurrence is critical to the human use of riparian lands.*

## PRINCIPAL TERMS

- **discharge:** the volume of water moving through a given flow cross-section in a given unit of time
- **flash floods:** rises in water level that occur unusually rapidly, generally because of especially intense rainfall
- **flood:** a rising body of water that overtops its usual confines and inundates land not usually covered by water
- **hydrology:** the branch of science dealing with water and its movement in the environment
- **jökulhlaup:** a flood produced by the release of water sequestered by a glacier, most often due to the failure of some type of glacial dam or to subglacial volcanic activity
- **monsoon:** a seasonal, reversing pattern of wind between warm ocean bodies and landmasses
- **recurrence interval:** the average time interval in years between occurrences of a flood of a given magnitude in a measured series of floods
- **runoff:** that part of precipitation that flows across the land and eventually gathers in surface streams

### CAUSES OF FLOODS

Floods involve extremely large flows of water in rivers and streams. Some technical hydrological definitions of floods involve stages, or heights, of water above some reference level, such as the banks of a river channel. In practice, however, floods can be thought of as any extreme flow of water that exceeds usual experience, often damaging or threatening life and property.

Floods may be caused by a variety of physical factors, including dam failures and the subsidence of land. The most common kind of flood occurs when excessive precipitation and physical factors of the land combine to produce maximum runoff of water. Precipitation can yield runoff directly, or water from melting snow can produce the flow. Very intense flash floods are usually associated with heavy, short-duration rainfall from thunderstorms. Such rainfall easily overwhelms the infiltration capacity of the ground, and

water rapidly runs off from the local drainage area or watershed into adjacent stream channels. If enough water concentrates in the channel to exceed the carrying capacity of the channel, it will constitute a flood.

Physical factors on the land surface also determine the rate of concentration of floodwater in stream channels. For example, less permeable soils will allow water to run off faster, as will a lack of vegetation. Artificial enhancement of runoff occurs when slopes are covered by impermeable materials. This situation commonly occurs in construction, when the natural ground surface is replaced by buildings and pavement. The result of such construction is extreme enhancement of runoff. In cities, the yield of floodwater from paved surfaces may be ten times greater than the same storm's yield under natural conditions. In this way, human construction tends to exacerbate flooding problems in urban areas.

### SEDIMENT AND BEDROCK

As the flow of water increases in a stream channel, it is usually associated with considerable sediment. If such water-borne sediment also composes extensive deposits adjacent to the stream channel, the river is termed "alluvial." An alluvial river commonly has a floodplain of deposited sediment adjacent to the channel. The floodplain is really an intimate part of the river system, as sediment is added to it every time the river rises above its banks. As the water rises, the river's depth rapidly increases, causing an increased ability to transport sediment that is eroded from the bed and banks. If the stream is appropriately loaded with sediment, it will be deposited when the banks are overtopped and the width of the flow greatly increases. If not enough sediment is supplied by erosion, however, the increasingly energetic flood flows will be erosive, attacking the banks, widening the channel, and restoring the appropriate sediment load to the stream.

When the bed and banks of a river are composed of bedrock, these adjustments of sediment load to flow energy cannot occur. Because the bedrock can be very resistant to erosion, the energy level in the flow can rise spectacularly without being damped by sediment. The excess energy of such sediment-impoverished floods goes into the development of turbulence. Turbulence at high energy levels takes on an organized structure of powerful vortices that produce immense pressure changes. These pressure effects may be sufficient to erode the bedrock boundary by a "plucking" action.

A famous geological controversy once surrounded the problem of bedrock erosion by great floods. In the 1920's, University of Chicago geologist J. Harlen Bretz proposed that immense tracts of eroded basalt bedrock in Washington State had been created by a catastrophic glacial flood. Bretz subsequently showed that the fascinating landforms of that region, known as the Channeled Scablands, were created when a great lake impounded by glacial ice burst. The lake, glacial Lake Missoula, had been more than 600 meters deep at its ice dam. It took Bretz nearly fifty years to convince his many critics that catastrophic flooding could explain all the bizarre features of the Channeled Scabland. Geologists now know that the physics of catastrophic flooding is completely consistent with Bretz's observations. Missoula flood flows moved at depths of 100 to 200 meters and velocities of 20 to 30 meters per second. The power (rate of energy expenditure) per unit area for such flows is thirty thousand times that for a normal river, such as the Mississippi, in flood. The reason for such immense flow power is that the Missoula flooding occurred on very steep slopes and that the potential energy of the water when the dam burst was great.

The Missoula floods occurred during the last Ice Age, more than twelve thousand years ago. There are, however, modern examples of glacial floods called jökulhlaups. In Iceland, jökulhlaups occur where glaciers overlie active volcanoes. Volcanic heat releases water that is stored in subsurface reservoirs by melting the overlying ice. Lakes may also form adjacent to the ice masses. Because ice is less dense than water, such juxtapositions are inherently unstable. When the pressure is high enough, the water may lift the ice dam and burst out from beneath the glacier. The jökulhlaups move house-sized boulders and transport immense quantities of sediment.

## DISCHARGE AND RECURRENCE INTERVAL

The volume of water released by a flood per unit of time is termed its discharge. This quantity is the magnitude of the flow that will potentially inundate

an area. The chances of experiencing floods of different magnitudes are expressed in terms of frequency. Large, catastrophic floods have a low frequency, or probability of occurrence; smaller floods occur more often. The probability of occurrence for a flood of a given magnitude can be expressed as the odds, or percent chance, of the recurrence of one or more similar or bigger floods in a certain number of years. Analyses of flood magnitude and frequency are achieved by measuring floods and statistically analyzing the data. Results are expressed in terms of the probability of a given discharge being equaled or exceeded in any one year. The reciprocal of this probability is the return period, or recurrence interval, of the flooding, expressed as a number of years.

The concept of a recurrence interval is sometimes confusing; some examples will illustrate the meaning. Suppose analysis of the flood frequency of a river flow indicates certain discharges for ten- and one-hundred-year recurrence intervals. The smaller, ten-year flood will have a probability of occurrence in one year of 0.1, or 10 percent. The larger one-hundred-year flood's probability of occurrence in one year is 0.01, or 1 percent. Note that these numbers do not preclude several such floods occurring in a given period; for example, two or more floods of the one-hundred-year magnitude could occur in the same year. Such an event, while unlikely, does have a small probability of occurring.

Flood magnitude-frequency relationships vary immensely with climatic regions. In the humid-temperate regions of the globe, such as the northern and eastern United States, stream flow is relatively continuous. Even rare floods are not appreciably larger than more common floods. A result of this relationship is that stream channels have developed in size to convey the relatively common, moderate-sized floods with maximum efficiency. In contrast, the more arid regions of the southwestern United States have immensely variable flood responses. Stream channels may be dry most of the time, filling with water only after rare thunderstorms. The flash floods that characterize these streams may also be highly charged with sediment. Indeed, the sediment may so dominate the flow that the phenomenon is of mud or debris flow rather than stream flow. Because extreme events dominate in these environments, the stream channels have developed in size according to these rare, great floods.

## MONSOONS AND ANCIENT FLOODS

In tropical areas, some of the greatest known rainfalls are produced by tropical storms and monsoons. Monsoons are seasonal wet-to-dry weather patterns driven by atmospheric pressure changes over the oceans and continents. They result in alternations between dry periods, which inhibit vegetation growth, and immensely wet periods, which facilitate runoff. Tropical rivers thus have a pronounced seasonal cycle of flooding, and some floods may be immense. These rivers also show channel size development in accordance with rare, great flows. Some of the most populous places on Earth are situated on seasonal tropical rivers, such as the Ganges and Brahmaputra Rivers in India and Bangladesh. Immense tragedies have occurred when a particularly severe monsoon or tropical storm has produced especially great floods. In 1974, monsoon-related flooding in Bangladesh killed 2,500 people. This pales in comparison to a flood in 1970 that took 500,000 lives and another in 1991 that killed more than 100,000 people.

Studies of ancient floods (paleofloods) through geological reconstruction of past discharges show that monsoons and other flood-generating systems have varied in the past. Between about ten thousand and five thousand years ago, floods were apparently much more intense in many world areas on the boundaries between the tropics and the midlatitude deserts. These intense floods may have been related to long-term glacial-to-interglacial cycles. Because of modern increases in carbon dioxide and other greenhouse gases, it appears quite likely that tropical floods may again become more intense. Such a situation could have grave consequences for flood-prone tropical countries.

## STUDY OF FLOODS

Hydrologists study floods by measuring flow in streams. Measurements are taken at stream gauges, where mechanical devices are used to record the water level, or stage of the river. To transform these stage measurements into discharge values, the hydrologist must perform a rating of the stream gauge. That is accomplished by measuring velocities in the stream channel during various flow events. A velocity meter with a rotor blade calibrated to the flow rate is used for this purpose. When the average measured velocity of the stream channel is multiplied by the cross-sectional area of the channel, the result is the discharge for that flow event. When several flows at different stages are measured, the data are used to generate a rating curve for the gauge. This curve shows discharges corresponding to any stage measured at the gauge.

The discharge values obtained at a gauge are collected over many years, constituting a record that can then be used in flood-frequency analysis. Several statistical procedures can be employed to plot the flood experience and to extrapolate to ideal values of the ten-year flood, the one-hundred-year flood, and so on. The discharges associated with these recurrence intervals are then used as design values for hazard assessment, dam construction, and other flood controls. "Flood control" is probably a misnomer, however, because flows larger than the design floods are always possible. Flood control really involves providing various degrees of flood protection.

Another approach to evaluating extreme flood magnitudes involves careful study of the precipitation values that generate the greatest known floods. By transposing the patterns of known extreme storms to other areas, scientists can, in theory, calculate what the runoff would be from hypothetical great storms. Such calculations involve the use of a rainfall-runoff model. These models are prevalent in hydrology because they can be easily programmed. The models give idealized predictions of how water from a given storm would concentrate. The flood discharge modeled from the assumed maximum rainfall is called a "probable maximum flood."

Unfortunately, there are problems inherent to both of these traditional hydrological approaches to flood studies. Both make assumptions in calculating potential flood flows. Another procedure is to study the natural records of ancient floods, or paleofloods, that are preserved in geological deposits or in erosional features on the land. It has been found for some sections of bedrock that nonalluvial rivers act as natural recorders of extremely large flood events. These natural "flood gauges" can be interpreted only by detailed studies that combine geological analysis with hydraulic calculations of the ancient discharges.

Paleoflood hydrology generates real data on the largest floods to occur in various drainages over several millennia. The data include the floods' ages, or the periods in which they occurred, and their discharges. The information can be used directly in a flood-frequency analysis, or it can be used to assess the probable validity of extrapolations from conventional data on smaller floods. Paleoflood data can also be compared with probable maximum flood estimates. In this way, the expense of overdesign and the danger of underdesign can be avoided in flood-related engineering projects.

## SIGNIFICANCE

People have lived with floods since the beginning of civilization. The first great civilizations on Earth developed along the fertile but flood-prone valleys of rivers such as the Nile, the Tigris and the Euphrates, the Indus, and the Yangtze and the Huang He. Various means of coping with floods have been documented since the biblical accounts of Noah. Early societies merely avoided zones that tradition told them were hazardous. It has only been in the modern era that large cities have systematically developed on immense tracts of flood-prone lands. Thus the natural process of flooding has become an unnatural hazard to humans.

There are immense consequences for the human insistence on occupying those areas near rivers that infrequently receive floodwater. A single tropical storm system, Hurricane Agnes, generated more than $3 billion in flood damage to the northeastern United States in 1972. Flooding of the Red River and Mississippi River basins generated more than $12 billion worth of damage, while the city of New Orleans may never fully recover from the flooding caused by Hurricane Katrina. Floods in Bangladesh have killed millions of people. In the 1930's, a national U.S. program began to respond to such problems by constructing large dams. Despite, or perhaps because of, this expensive effort, flood damage to life and property is much greater today than it was in the 1930's.

One trend has been to manage flood-hazard zones with multiple approaches that respond to the nature of the flood risk. The river is treated as a whole integrated system, rather than as individual segments, for engineering design. Management alternatives for this system are not limited solely to structural controls, such as dams and levees. Instead, options are considered for land-use adjustment. Flood-prone lands can be used for parks, greenbelts, and bikeways instead of industrial warehouses, stores, and housing. Even when construction must be done on floodplains, it may be possible to make provision for flood risks. Warehouses, for example, can be organized for

rapid transfer of materials to second stories or to temporary, safe storage sites. Such adjustments require accurate and timely warning systems that involve measuring rainfall in headwater areas and rapidly predicting the flood consequences to downstream sites at risk.

There is a general need to educate the public that floodplains are a natural part of rivers. Living on a floodplain is really choosing to play a game of "floodplain roulette," in that it is known with certainty that a flood will happen, but it is not known when that flood will happen. For this reason, the most accurate and reliable methods of evaluating flood magnitudes and frequencies are necessary. The choices made for land use on floodplains cannot be based merely on idealized theories of how floods behave.

*Victor R. Baker*

## FURTHER READING

Aldrete, Gregory S. *Floods of the Tiber in Ancient Rome.* Baltimore: Johns Hopkins University Press, 2006. Offers a study of the floods in ancient Rome through an examination of primary sources. Combines current knowledge of hydrology and geology to hypothesize about the physical and social effects of flooding in ancient Rome. Discusses flood prevention and safety precautions that were used or could have been used to protect the city. Provides an interesting melding of history and science, accessible to everyone.

ASCE Hurricane Katrina External Review Panel. *The New Orleans Hurricane Protection System: What Went Wrong and Why.* Reston, Va.: American Society of Civil Engineers, 2007. A technical report of the events during and following Hurricane Katrina. Discusses the hurricane protection system, levees, and the results of those levees failing. Also discusses what can be learned from the events and what can be done to prevent future disasters.

Baker, Victor R., ed. *Catastrophic Flooding: The Origin of the Channeled Scabland.* Stroudsburg, Pa.: Dowden, Hutchinson & Ross, 1981. Uses reprints of original papers to recount the history of the scientific controversy surrounding the origin of the Channeled Scabland. Covers the catastrophic flood hypothesis of J. Harlen Bretz in considerable detail. Follows the controversy through its resolution, and develops many modern concepts of erosion and deposition by examining the immense glacial floods that coursed the Scabland region. Well illustrated.

Baker, Victor R., R. C. Kochel, and P. C. Patton, eds. *Flood Geomorphology*. New York: John Wiley & Sons, 1988. A modern treatment of broad, interdisciplinary issues involving floods. Analyzes floods in terms of landscape, climate, and other geological factors. Documents their causes, effects, and dynamics, in addition to methods of management of flood-prone areas.

Comerio, Mary C. *Disaster Hits Home: New Policy for Urban Housing Recovery*. Berkeley, Calif.: University of California Press, 1998. Describes the destructive power of floods, hurricanes, and earthquakes, then explores the disaster relief measures and housing policies that have been implemented to handle natural disasters and their aftermaths. Appropriate for the layperson.

Freitag, Bob, et al. *Floodplain Management: A New Approach for a New Era*. Washington, D.C.: Island Press, 2009. Presents a different case study that focuses on a new topic in flood control in each chapter. Describes strategies of floodplain management and the natural processes and dynamics of rivers. Discusses multiple approaches, as varied as the locations in which they are used. Best suited for engineers and hydrologists taking part in floodplain management.

Knighton, David. *Fluvial Forms and Processes: A New Perspective*. 2d ed. London: Edward Arnold, 1998. Provides a succinct overview of water-related processes on landscapes. Treats floods as a part of that spectrum of processes. Reviews aspects of drainage basins, flow mechanics, sediment transport, channel adjustments, and changes in river channels. Illustrated with technical diagrams.

O'Loughlin, K. F., and James F. Lander. *Caribbean Tsunamis: A 500-Year History from 1498-1998*. Norwell, Mass.: Kluwer Academic Publishers, 2010. Presents characteristics of tsunamis and the history of their occurrences in the Caribbean Sea. Categorizes tsunamis by type and describes the effects of tsunamis in Chapter 4. Accessible to the general public and natural disaster specialists. Extensive bibliography and appendices as well as indexing.

Orsi, Jared. *Hazardous Metropolis: Flooding and Urban Ecology in Los Angeles*. Berkeley, Calif.: University of California Press, 2004. A history of flood control efforts in Los Angeles from 1870 to 2004. Demonstrates clearly how engineering efforts have consistently failed to meet the demands of nature. Presents a view of cities as ecosystems.

Philippi, Nancy S. *Floodplain Management: Ecology and Economic Perspectives*. San Diego, Calif.: Academic Press, 1996. Presents a thorough account of the destructive power of floods, emergency management policies, and floodplain management tactics. Easily understood by readers without prior knowledge of the field.

Prothero, Donald R. *Catastrophes!: Earthquakes, Tsunamis, Tornadoes, and Other Earth-Shattering Disasters*. Baltimore: Johns Hopkins University Press, 2011. Provides a detailed and clear explanation of the many natural and anthropogenic disasters facing our planet. Each chapter is devoted to a different catastrophe, including earthquakes, volcanoes, hurricanes, ice ages, and current climate changes.

Schneider, Bonnie. *Extreme Weather: A Guide to Surviving Flash Floods, Tornadoes, Hurricanes, Heat Waves, Snowstorms, Tsunamis and Other Natural Disasters*. Basingstoke, England: Palgrave Macmillan, 2012. Presents vivid explanations of how, when, and why major natural disasters occur. Discusses floods, hurricanes, thunderstorms, mudslides, wildfires, tsunamis, and earthquakes. Presents a guide of how to prepare for and what to do during an extreme weather event, along with background information on weather patterns and natural disasters.

Sene, Kevin. *Flood Warning, Forecasting and Emergency Response*. Dordrecht: Springer Science + Business Media, 2008. Provides a comprehensive summary of various aspects of the title topics, based on up-to-date research.

Singh, Vijay P., ed. *Hydrology of Disasters*. Boston: Kluwer Academic Publishers, 2010. Looks at the hydrological aspects of floods and other natural disasters, as part of the Water Science and Technology Library series. Also explores preventive measures and disaster relief programs that have been implemented to deal with floods. Illustrations, index, and bibliography.

**See also:** Aquifers; Artificial Recharge; Climate Change Theories; Cyclones and Anticyclones; Dams and Flood Control; Floodplains; Freshwater and Groundwater Contamination Around the World; Groundwater Movement; Groundwater Pollution and Remediation; Hydrologic Cycle; Precipitation; Salinity and Desalination; Saltwater Intrusion; Surface Water; Tsunami; Waterfalls; Water Quality; Watersheds; Water Table; Water Wells

# FRESHWATER AND GROUNDWATER CONTAMINATION AROUND THE WORLD

*The pollution of freshwater and groundwater is of critical environmental concern. The problem of pollution in drinking water is varied and diverse, with issues ranging from toxic metals and pesticides to bacteria growth and thermal pollution. Interconnected water systems allow for the free movement of pollution and for wide-ranging damage to drinking supplies and ecosystems.*

## PRINCIPAL TERMS

- **acute toxicity:** the effect of a toxic agent on an organism that manifests from a large dose in a short time
- **bioaccumulation:** the accumulation of substances, such as pesticides and metals, in an organism
- **chemical leaching:** the process of extracting a substance from a solid by dissolving it in a liquid
- **chronic toxicity:** the effect of a toxic agent on an organism that manifests from long-term, repeated exposure to a substance
- **epidemiology:** the evaluation and study of public health as it relates to the effects of environment, choice, and risk factors
- **eutrophication:** process of nutrient enrichment leading to dense algae growth
- **ground state metal:** metal in its elemental form with an oxidation state of zero
- **LD$_{50}$:** median lethal dose (LD) to 50 percent of subjects in a toxicity study
- **remediation:** the removal of pollution or contaminants from environmental systems
- **riparian:** transitional zones between terrestrial and aquatic systems that exhibit characteristics of both systems
- **silviculture:** the practice of controlling the establishment, growth, health, and quality of forests
- **water cycle:** the continuous movement of water above, below, and on the surface of the earth
- **watershed:** the region draining into a river, lake, stream, or other body of water

## INTRODUCTION

Earth's water cycle connects the water of the world through a vast and widespread system. Because of this interconnection, pollution in a single area can have far-reaching consequences. Just as water systems are connected in nature, so too are issues of water pollution.

## AGRICULTURAL RUNOFF

Agricultural runoff occurs when water, such as heavy rain, flows through farmland and washes fertilizers, soil, and pesticides into watersheds, those areas that drain into bodies of water, such as rivers, lakes, and streams. A major concern associated with agricultural runoff is the harmful effects of fertilizers.

Fertilizers, which are compounds used to replace nutrients in the soil for the growing of plants, are rich in the elements nitrogen, oxygen, and phosphorus; the most common combinations of these atoms in fertilizes are nitrates ($NO_3^-$) and phosphates ($PO_4^{-3}$). Nitrates and phosphates are both anions that are extremely effective at replenishing nutrients in the soil. The same properties that make these compounds effective fertilizers also make them potential environmental hazards when improperly managed.

Improper management of fertilizers includes overusage of fertilizers and the application of fertilizers in areas close to bodies of water. Both phosphates and nitrates are negatively charged ions, so they are highly water soluble; they easily dissolve in even small amounts of water runoff. The water runoff containing nitrates and phosphates can then be introduced into watersheds, where these anions undergo chemical reactions and impact ecosystems.

The introduction of excess phosphate into a watershed can significantly alter aquatic ecosystems and reduce the quality of surface waters. The presence of excess phosphate can lead to eutrophication and a significant increase in algae growth. A by-product of algae growth is an increase in the amount of organic matter and sediment in surface water. Excess sediment can cloud the water and reduce the amount of sunlight that reaches aquatic plants, causing the metabolic pathways of plants to slow, potentially leading to death. Sediment also can clog the gills of fish or smother fish larvae, further compromising ecosystems.

Additionally, the eruption of algae from eutrophication can cause further oxygen depletion in water systems where algae decrease the oxygen dissolved. Fish and other aquatic fauna are particularly sensitive to oxygen depletion in rivers and streams. Algae blooms also can ruin swimming and boating opportunities and can create a foul taste and odor in drinking water. Small increases in the level of phosphates can be detrimental to water systems and can potentially lead to eutrophication, which alters the life cycle of aquatic organisms and harms water quality.

Unlike phosphates, nitrates do not have the same immediate effect upon water quality and aquatic ecosystems. However, nitrates have a much further reach of environmental effects because of their chemical reactivity. One of the direct effects of nitrates in the water supply is the potential for methemoglobinemia, or blue baby syndrome. Methemoglobinemia is a potentially fatal disease in infants caused by the ingestion of large quantities of nitrates. The disease causes oxygen deprivation to vital tissues, such as the brain.

Well water and drinking water obtained from sources close to farming areas are the most common water sources affected. Pregnant women, adults with reduced stomach acidity, and people deficient in the enzyme that returns methemoglobin to normal hemoglobin are all susceptible to nitrite-induced methemoglobinemia.

Nitrates are in a category of chemicals called oxidizers, and nitrates are considered to be strong oxidizers. As a function of their reactivity, nitrates can have a wide range of environmental effects, particularly when interacting with toxic metals.

## HEAVY METALS

Metals in ores and minerals provide material for a wide array of products and industrial processes. However, when mines are abandoned or left unchecked, issues with chemical leaching frequently arise, as metals can find their way into, and contaminate, water systems.

Some of these metals are necessary for human metabolic processes, but humans need only a small amount of these metals, including cobalt, copper, chromium, manganese, and nickel. These and many other heavy metals are toxic when present in drinking water or food. The metals attack specific bodily systems and organs. For example, the central nervous system is impaired by manganese, mercury, lead, and arsenic, while mercury, lead, cadmium, and copper affect the kidneys and liver. Nickel, cadmium, copper, and chromium affect skin, bones, and teeth.

The damage to metabolic processes is not limited to animals; plants and ecosystems can also be harmed by constant exposure to heavy metals in freshwater. In both animals and plants, metals can build up through bioaccumulation and can be passed up the food chain. Unlike organic toxins such as pesticides, which can degrade over time, heavy metal toxicity remains present as long as the metal is present. Therefore, heavy metal remediation—involving the removal of heavy metal pollution and contaminants from a given environment—poses a daunting challenge because of the complex chemistry of the metals and because of the size of the water system in which they are dissolved.

A growing concern is that of the synergistic effects of different types of pollution. Chemical leaching of ground-state heavy metals (metals with an oxidation state of 0) into water systems is typically a slow process. Nitrates from agricultural runoff often can exacerbate and accelerate chemical leaching by oxidizing toxic metals and making the metals more water soluble. As a result, the oxidized metals are more readily dissolved into water systems and drinking water.

Heavy metal pollution also can be introduced to water systems in industrial processes, such as the smelting of copper and the preparation of metals as nuclear fuels; heavy metals also can be introduced into water systems through coal-burning power plants. Additional chemical leaching issues associated with heavy metals from mines include radioactive heavy metals in groundwater and arsenic poisoning of livestock from mine runoff in freshwater.

Mercury poses a unique hazard among the heavy metals in its ability to form organomercury compounds, such as methylmercury and dimethylmercury, which are more toxic than ground-state mercury. These more toxic forms of mercury are often found in fish and shellfish because of mercury in water systems. The organomercury compounds can work their way up the food chain through bioaccumulation and can ultimately arrive in larger fish and become part of the human food supply.

## ACID RAIN AND ACID RUNOFF

Just as toxic metals can be chemically leached into watersheds from both abandoned and active mines, so it is possible for ores and minerals containing sulfides to be leached into water systems. The sulfides that are leached into the water supply are then oxidized to form acidic compounds and acid runoff. The acid runoff produced from the sulfides causes a lowering of the pH in rivers, lakes, and streams, making water systems uninhabitable for native flora and fauna.

A prime example of how acid runoff can devastate a water system is that of the Ocoee River in the southern Appalachian Mountains of the United States. The Ocoee was polluted for one hundred years by acid runoff from a massive copper mine in Copperhill, Tennessee. Both copper metal and acidic runoff from the copper minerals would flow into the river from chemical leaching by rainwater and snowmelt.

A project administered by the U.S. Environmental Protection Agency initiated measures to treat the water runoff flowing into the Ocoee River by remediation of the copper and by neutralization of the acid runoff. This was achieved through the use of lime (a basic mineral containing calcium), wherein the water runoff was treated in holding areas. After being treated with lime, the metals settle out from the pH neutral water and then clean water is released into the river. After ten years of environmental cleanup, aquatic life began to return to the river near the mine.

A similar process gives rise to acid rain. Coal-burning power plants can produce oxidized sulfides from the coal they burn, resulting in these acidic compounds being driven into the atmosphere, where they become part of the water cycle. As part of the water cycle these compounds form acid rain. Acid rain, like acid runoff, increases rates of erosion and has toxic effects upon ecosystems and water quality. Acid rain, however, has a much larger area of impact, and because of its low pH, it can facilitate chemical leaching of heavy metals. As precipitation, acid rain readily makes its way into rivers, lakes, and streams, where it ultimately has the same negative impact as acid runoff on drinking water quality and on native flora and fauna.

## PESTICIDES

Pesticides are a class of chemical compounds that can repel, mitigate, or kill pests. Pesticides often are associated with agriculture or silviculture and are used to safeguard crops from such invading species as insects, weeds, and rodents.

In many developed nations, agencies regulate the sale and, in some cases, the use of pesticides. Specific pesticides, such as insecticides, fungicides, and rodenticides, require specific attention, as many of these compounds are toxic to humans. $LD_{50}$ values are used to describe the effectiveness of a particular pesticide so that a tangible measure can be used to assess toxicity.

There are two types of toxicity with regard to chemical compounds. The first type is acute toxicity, in which detrimental health occurs from a large dose of chemicals in a short time. The second type of toxicity is chronic toxicity, in which adverse health effects occur from continued exposure to smaller chemical doses over a prolonged period of time. The $LD_{50}$ values help to quantify toxicity in various organisms by considering the variables that affect toxicity.

With respect to pesticides in water and water systems, chronic toxicity is a much more common problem than is acute toxicity. Pesticides can be introduced to streams, lakes, and rivers through improper or excessive application. This can be exacerbated when heavy rains wash pesticides into watersheds. Pesticides and other chemicals that have been introduced to a stream are then deposited downstream and collect in reservoirs. Pollutants that are dissolved in reservoirs can be concentrated if evaporation rates are high, resulting in a higher level of exposure to toxic pesticides.

In developed nations, instances of poisoning from pesticides in drinking water are fairly rare; however, in developing nations, poisoning from pesticides in drinking water is more common because of the overuse of pesticides in those regions. Many developing countries do not regulate the sale or use of pesticides, and so have higher concentrations of toxic pesticides in their water supplies. The higher levels of pesticides also can cause problems with bioaccumulation, in which aquatic species, game animals, and livestock build up large amounts of pesticides in their tissue. This accumulation of pesticides not only has detrimental effects on the health of animals; it also has serious effects on animals that are higher up in the food chain.

## THERMAL POLLUTION

The temperature of freshwater is a physical property that directly influences the quality of water. Thermal pollution of water can be detrimental to agriculture and to the quality of drinking water. This type of water pollution occurs when water is introduced into another body of water at a temperature that is significantly above or below the body of water in question.

Of particular concern is when incoming water is significantly above the temperature of the accepting body of water, as is the case with some rivers, lakes, and streams. Serious issues then arise with land and water usage, leading to complications with naturally occurring flora and fauna near the affected body of water.

Coal-burning power plants and other industrial complexes utilize water as a cooling resource for various industrial processes. These facilities pull in large volumes of water from lakes, rivers, estuaries, and oceans to cool their machinery, and they then release water at an elevated temperature. Water at elevated temperatures also can arise from geothermal vents, wherein the energy from the earth's core warms subterranean water supplies. This heated water can eventually rise to the surface and can mix with freshwater and groundwater.

Issues associated with heated thermal pollution most often concern water usage. For example, water that is above 25 degrees Celsius (77 degrees Fahrenheit) is not suitable for human consumption because it is considered a heated resource. Additionally, water with a temperature of more than 35 degrees Celsius (95 degrees Fahrenheit) can be used only for irrigation (with caution). Water that has an elevated temperature also is likely to have a higher concentration of metal cations, such as boron and lithium, and anions, such as phosphates.

Additionally, there exists an inverse correlation between water temperature and dissolved oxygen. As the temperature of water rises, a smaller amount of oxygen is dissolved in the water; as a result, less oxygen is available for aquatic flora and fauna. Elevated water temperatures typically increase biological activity, resulting in greater oxygen needs for animals and causing a strain upon the ecosystems. This type of scenario can result in one species replacing another. For example, a species that favors warm water, such as the walleye fish, might replace a species that favors cold water, such as trout.

The natural ecosystem is affected in a variety of ways by thermal pollution in that native flora and fauna can become environmentally strained from increased biological activity and decreased oxygen supply. In extreme cases some native species are replaced by invading species that favor warmer temperatures. Furthermore, water intake from industrial complexes also traps and kills fish due to the intense rate at which water can be pulled from a body of water, thereby reducing native fish populations.

## Bacteriological Contamination

Bacteriologic factors are frequently considered in determining if water is of satisfactory quality for drinking or recreation. A number of dangerous bacteria, including coliform, fecal streptococcus, and fecal coliform, can arise in freshwater streams. Each of these bacteria carries its own risks to water quality, especially making water unsafe for drinking.

If bacterial levels exceed certain concentrations then drinking-water quality can be compromised and result in illness. Many of these bacteria are associated with the introduction of organic matter, including fecal matter and decomposing flesh, into rivers, lakes, and streams.

The principal sources of fecal coliform that can lead to high pollution levels include domestic livestock, game animals, and humans. The presence of these mammals in riparian areas serves as a vehicle for the introduction of bacteria; surface runoff from heavy rains, snowmelt, and irrigation outwash can deposit bacteria into groundwater. Studies have shown that livestock grazing in riparian systems are five to ten times as likely to have increased fecal coliform than are livestock near riparian systems with no grazing livestock. These increases in fecal coliform generally exceed drinking water standards.

Bacteria counts of this type also can be tied to seasonal trends, as heavy rains and snowmelt can play a large role in the introduction of bacteria to a water system. A general trend is seen in which spring and summer periods have a higher level of bacteria incidence, while winter periods show a time of low bacteria activity.

A unique issue of water quality arises in India. Millions of liters of untreated human sewage are introduced to the Ganges River each day. This leads to large concentrations of fecal coliform bacteria and results in a fecal coliform count that is ten times greater than the acceptable amount for water used for bathing, much less drinking. This gives rise to an unusual, specific issue: The Ganges is a holy river in Hinduism. It plays a number of different roles in Hindu culture, including the ritualistic cleansing and bathing that occur daily for many Indians. The constant pollution of the Ganges exposes many people to dangerous levels of bacteria, and does so as a function of their religious beliefs. This presents a unique situation in which drinking water, water pollution, epidemiology, and religion intersect.

## SUMMARY

Issues with heavy metals and pesticides can lead to the immediate contamination of drinking water, rendering a resource unusable because of the high toxicity of the pollutants. Other types of contamination, such as thermal pollution, acid rain, and acid runoff, have a chronic effect upon water quality, with repeated exposures harming ecosystems.

Agricultural runoff can have a synergistic effect upon other pollutants through chemical leaching and can degrade water quality through eutrophication. Finally, bacteriologic contamination arises from the abuse of riparian areas, leading to extreme health hazards from the consumption of contaminated water. All of these issues affect drinkable water throughout the globe and require constant monitoring and vigilance to protect Earth's freshwater supply.

*J. H. Shugart*

## FURTHER READING

Abraham, Wolf-Rainer. "Megacities as Sources for Pathogenic Bacteria in Rivers and Their Fate Downstream." *International Journal of Microbiology* (2011). Available at http://www.hindawi.com/journals/ijmb/2011/798292/abs. Discusses the impact big cities have on downstream water quality, including contamination and interactions of various pathogenic bacteria.

Brooks, Kenneth N. *Hydrology and the Management of Watersheds.* Ames: Iowa State University Press, 1997. A basic textbook that introduces the concepts of watershed management, erosion, and water movement and quality.

Krešić, Neven. *Groundwater Resources: Sustainability, Management, and Restoration.* New York: McGraw-Hill, 2009. A discussion of global groundwater management. Focuses on freshwater sources.

Machtinger, Erika T. *Riparian Systems.* Washington, D.C.: Natural Resources Conservation Service, 2007. Discusses the importance of the interface between water and land.

Mather, John R. *Groundwater Contaminants and Their Migrations.* London: Geological Society, 1998. A general overview of groundwater contamination, monitoring, and cleanup that focuses on Great Britain's role in this process. Also highlights a number of international sites that are especially significant.

Pauwels, H., M. Pettenati, and C. Greffie. "The Combined Effect of Abandoned Mines and Agriculture on Groundwater Chemistry." *Journal of Contaminant Hydrology* 115 (2010): 64-78. Results of a study that addresses the effects of both mining and agriculture on groundwater. Contains a brief overview of both topics.

U.S. Environmental Protection Agency. *Protecting Water Quality from Agricultural Runoff: Clean Water Is Everybody's Business.* Washington, D.C: Author, 2003. Brief overview of the major sources and the consequences of water pollution in the United States. Includes a list of further readings.

**See also:** Acid Rain and Deposition; Aquifers; Clouds; Deltas; Drainage Basins; Drought; Floodplains; Floods; Ganges River; Geochemical Cycles; Groundwater Movement; Groundwater Pollution and Remediation; Hydrologic Cycle; Lakes; Ocean Pollution and Oil Spills; Precipitation; River Bed Forms; River Flow; Sand; Sediment Transport and Deposition; Surface Water; Waterfalls; Water Quality; Watersheds; Water Table; Water Wells

# G

## GANGES RIVER

*The Ganges River flows from the Himalayas through the northern and eastern portions of India, then through Bangladesh to the Bay of Bengal. Although the Ganges is not a long river, it is the second muddiest river in the world, depositing sediment along the fertile floodplains and at the delta. The Ganges system annually floods and periodically causes great loss of life and damage.*

### PRINCIPAL TERMS

- **avulsion:** a natural change in a river channel, usually caused by flooding and excess deposition of sediment
- **base flow:** the natural flow of groundwater into a river, which commonly maintains the minimum flow of perennial rivers during the dry season
- **capacity:** the total amount of sediment a river or stream can transport at a given time
- **competency:** a measure of the largest particle that a river can carry at the time of measurement
- **delta:** a landform created when a river enters a relatively still body of water and deposits much of the sediment it was transporting; deltas may form in lakes and ponds, but the largest deltas form where a silt-laden river flows into the ocean
- **discharge:** the volume of water flowing past a measurement point in a river during a given time interval, usually measured in cubic meters per second
- **distributary:** a river that diverges from the main river and carries and distributes the flow of water across a delta
- **suspended load:** the total amount of sediment carried in suspension by moving water in a river, measurable over any time period but commonly determined on an annual basis
- **tributary:** a stream or river that flows into another stream that becomes the main stream; a tributary may be a significant river in its own right, but it loses its identity when it merges with the main river

### HISTORY

The Ganga River is known in the Western world by its anglicized name of the Ganges River. The Ganges River has been one of the most important Asian rivers in recorded human history. Indian civilizations dating back to the kingdom of Asoka in the third century B.C.E. have developed along the Ganges River or in the floodplains of the Gangetic Plain. Today, millions of people in India and Bangladesh rely on the waters of the Ganges River. The population of the Gangetic Plain is now more than 500 million people, making it the most heavily populated river basin in the world.

The Ganges River is the holiest of all rivers for Hindus. Many travel long distances just to bathe in its waters at Varanasi. Bathing in the river at India's oldest living city is considered an act of purification, and for many it is the culmination of one's lifetime. Hindus also believe that dying and having one's ashes scattered on the Ganges will ensure *moksha*, the release from the constant cycle of death and reincarnation. Because of these beliefs, there are numerous crematoria along the Ganges at Varanasi, and many ashes and bodies are given over to the river each year.

### GEOGRAPHY OF THE GANGES

The Ganges River arises from the runoff from the southern side of the Himalayas, the world's highest mountain range. The Ganges River flows about 2,500 kilometers across the Gangetic Plain and through a coastal delta in Bangladesh to empty into the Bay of Bengal. The river system drains a basin totaling more than 580,000 square kilometers, nearly one-fourth of the area of India. The average discharge of the Ganges is about 11,610 cubic meters per second, but floods from upland rains and snowmelt can increase the discharge by a factor of five or more.

Many believe that the source of the river is an ice cave at the base of a glacier at Gangatri, from which the Bhagairathi River flows. The true source of the Ganges is, however, considered to be some 21 kilometers to the southeast at Gaumukh. After flowing southward off the southern face of the Himalayas, the Ganges River flows predominantly eastward along the front of the Himalayan system, later turning southward as it makes its final way to the sea. The headwaters of the Ganges consist of five rivers that all arise in the northern Indian state of Uttarakhand. The main tributaries are the Bhagirathi, Alaknanda, Mandakini, Dhauliganga, and Pindar Rivers. The Bhagirathi and the Alaknanda Rivers merge at Devprayag to form the main stream of the Ganges River. The Ganges then cuts its way through the southern outer mountains of the Himalayas and emerges from the mountains at Rishikesh. From Rishikesh, the Ganges flows out onto the Gangetic Plain at Haridwar and continues eastward, still within the state of Uttarakhand. From this area eastward, the Ganges River is joined by the Yamuna River—which flows past the capital city of New Delhi—as well as the Tons, Ramganga, Gomati, and Ghaghara Rivers. The Ganges River then flows eastward into the Indian state of Bihar, where it is joined by the Gandak, Buhri Gandak, Ghugri, and Kosi Rivers, all of which arose in the mountains to the north.

These rivers provide more than 40 percent of the entire flow of the Ganges River. The surface waters of the area south of the Ganges are drained into the main river primarily by the Son River. After flowing northward around the Rajmahal Hills, the Ganges River turns southward and flows into the state of West Bengal. The river here is generally known as the Padma River and enters the delta region at Farakka. To the south, the Ganges River and the great Brahmaputra River merge and flow southward into the Bay of Bengal. The name of Padma is given to this part of the river, where many tributaries and many distributaries carry the water across the wide delta in Bangladesh.

## HYDROLOGY

The hydrologic regime in the Ganges River system is highly seasonal, with up to 80 percent of the rainfall occurring in the interval from July through October.

This heavy rainfall is caused by the southwesterly monsoons that cross the region during this time. Significant quantities of water are also added by melting snow in the Himalayas during the spring months. Rainfall, however, varies considerably over the entire river basin. In the western part of the basin, precipitation averages about 750 millimeters per year, but up to three times that much falls in the eastern part of the basin. In addition to the annual rainfall and snowmelt, water is added to the rivers by the base flow of the groundwater systems that are contiguous with the river.

The strong seasonal nature of precipitation is shown in the variations in the discharge of the Ganges system throughout the year. Discharge at the Hardinge Bridge in Bangladesh averages 39,400 cubic meters per second in August but may rise to a maximum of more than 53,000 cubic meters during flood stage. In contrast, the mean discharge at the same location during the height of the dry season from February through May is only slightly more than 21,000 cubic meters. Most of the flow during the dry season is derived from the northern tributaries of the Ganges River.

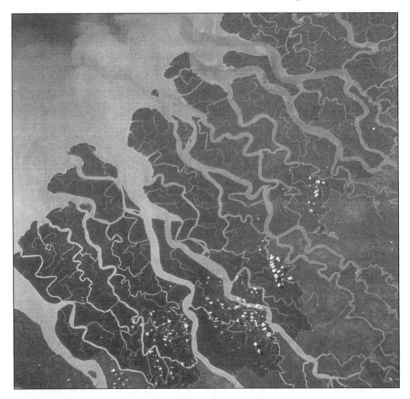

*The mouth of the Ganges River.* (PhotoDisc)

In addition, the delta region is commonly buffeted by severe cyclonic storms that generally just precede the onset of the monsoon season or occur at the end of the main rainy season. These storms have been responsible for many of the damaging and deadly floods in the delta region.

## SEDIMENTATION

The rivers that flow out of the highlands of the southern Himalaya generally have high velocity and therefore a very high competency. The land in many areas is highly erodible, particularly if it has been recently deforested. As a result, the rivers carry large amounts of sediment in suspension and as bed load. Based upon suspended load calculations, the Ganges River is second behind only the Yellow River in China in terms of sediment load. The average suspended load is about 1.4 billion metric tons per year. This leads to an average of approximately 1,500 metric tons of suspended sediment per square kilometer of the river basin. Although this is high, it is not as high as the Yellow, Ching, or Lo Rivers of China. The gentle gradient of the Ganges River after it leaves the Himalayas results in relatively little erosion and a low production of suspended load. In contrast, the Kosi River, a northern tributary to the Ganges, has an annual suspended load of only 172 million metric tons per year. However, because of its high gradient and relatively small drainage basin, the average sediment load per square kilometer of basin is almost twice that of the Ganges River as a whole. The high sediment load, coupled with large-scale flooding caused by rapid runoff from the Himalayas, has led to the frequent avulsion of the Kosi River. Its channel shifted more than 100 kilometers in the years between 1736 and 1964.

The gradient of the river across the Gangetic Plain (south of the Himalayas to the delta) is very shallow. The 1,600-kilometer stretch of river from the Yamuna River at New Delhi to the Bay of Bengal drops only about 1,100 meters in elevation. This shallow gradient leads to low water velocity and rapid deposition of sediment in the river channel and in the floodplains during flood stage. The combination of rapidly rising, very high mountains as a sediment source and a slowly moving major river leads to rapid aggradation of the floodplain surfaces along the river. The Gangetic Plain is composed of this rapidly deposited sediment and is, in some places, up to 2,000 meters

thick. Although the age of the bottom sediments in the basin is not definitely known, it is probable that the deposition of this thick accumulation of sediments was rapid, perhaps occurring within the last 10,000 years.

Sediment that is transported to the coastal area by the Ganges and Brahmaputra Rivers is deposited in the deltaic area. The delta built by these two great rivers is huge, totaling almost 60,000 square kilometers. Like most deltas, this landmass has been built up by the deposition of silts and sands along the main channel and in the many distributaries that channel water to the ocean. The areas between the channels are often heavily vegetated, which leads to the formation of peat when the plants die and are buried in the swampy areas. In some places, the peat has been mined and used as an energy source. The sediments that compose the delta include sand, silt, peat, and marl. Because of the heavy sediment load and the seasonally variable discharge of the river system, the delta has grown larger and extended farther seaward throughout historic times. New land areas, called *khadar*, are generally formed in the areas near the present-day channels where the river water meets the oceanic waters. As sediment accumulates, it blocks water flow and forces the river to change its course. In this way the sediment depositions continually change location, and the delta continues to prograde out into the Bay of Bengal. The current flow pattern of the river is toward the eastern side of the delta in Bangladesh, causing the buildup of land areas and some offshore islands. The western side of the delta has remained essentially unchanged since the eighteenth century. The delta will continue to grow as long as water and sediment are transported by the river to the coast. However, with increasing population growth and the resultant increase in agricultural and urban use of water, the total flow of the Ganges and Brahmaputra Rivers may be significantly reduced. This reduction will lead to the eventual erosion of the coastal delta if the sediment load is not sufficient to compensate for erosive mass wasting.

The seaward edge of the delta is characterized by a vast area of tidal forests and swamps that harbors an immense biological diversity. The tidal forests, known as the Sudarbans, have been declared protected areas by the governments of both India and Bangladesh, which have instituted conservation measures to preserve the biota, including the Bengal tiger.

## FLOODS

As in most areas, flooding has had both beneficial and detrimental effects on the land and the people in the floodplain of the Ganges River. The floods bring down sediment and nutrients into the lower reaches of the river and are responsible for building a fertile floodplain. Records from the sixteenth and seventeenth centuries indicate that the Ganges basin was once heavily forested and was home to a wide variety of animals and plants that are no longer found in the area. Deforestation of the area led to the loss of the animal life, and the floodplain today sustains a high level of agricultural productivity. The rapid increase in human population in the region has also led to increasing residential use of the floodplain and therefore a growing flood risk for the region. There are now more than 20 million people who are directly in the path of flooding along the Ganges.

Even though the area of the state of Bihar near the Ganges is swampy land, the population has grown in that region. Settlement in the floodplain has led to extensive loss of life and property damage during the frequent flooding of the area. Flooding in the upper part of the basin within Bihar is responsible for more than one-half of India's annual flood toll of 1,200 people. There is generally some damage each year from flooding, but major flood events have occurred in 1890, 1898, 1899, 1922, 1924, 1954, 1974, 1978, 1987, and 1988. In 1987, floods resulted in more than 400 deaths, the destruction of 70,000 homes, and damage to another 350,000 homes. The government of Bihar blamed the apparent recent increase in flooding in the region on Nepal, arguing that deforestation of upstream regions led to higher levels of runoff and flooding of the downstream basin. Although it is known that deforestation can lead to downstream flooding, scientific evidence does not clearly indicate that this has been the case for the more recent Bihar floods.

Flooding is also common in the lower reaches of the Ganges and Brahmaputra Rivers. The land in the delta area of Bangladesh is only a few meters above sea level, and when the rainy season arrives, the area is commonly inundated under 1 meter or more of water. To protect against these annual floods, many of the villages and homes are built upon hills that have been constructed by the local people. The delta area is also hit by cyclonic storms that come out of the Bay of Bengal. These storms wreak havoc on the low-lying areas. A cyclonic storm in November 1970 resulted in catastrophic flooding and caused more than 200,000 deaths.

## WATER RESOURCES

The Ganges River is a major water resource for the northern and eastern portions of India. It has been utilized for drinking water, transportation, and waste disposal for thousands of years. The waters of the great river have been diverted for irrigation purposes for hundreds of years. However, the great surge in population growth in India (which may become the most populous country on Earth by the end of the twenty-first century) and Bangladesh has placed huge demands upon the river system. India's heavy upstream use of freshwater for irrigation has dramatically reduced river flow for irrigation in Bangladesh and through the delta to the ocean. The loss of freshwater flow to the delta has resulted in a landward migration of seawater, leading to salinization of the coastal lands.

The Ganges River basin is shared by Nepal, India, and Bangladesh, and water rights questions have arisen concerning the share of water to which each country is entitled. The greatest conflict resulted when India built a barrage at Farakka in the early 1970's to divert water to Calcutta. The diversion facilitated transportation to the city, but, in turn, reduced river flow into Bangladesh. The Bangladesh government began a series of negotiations with the Indian government in order to obtain an equal share of the Ganges water. After many frustrating years, the two governments were finally able to craft a water-sharing agreement in 1996 that gave each country a 50 percent share of the water.

*Jay R. Yett*

## FURTHER READING

Barter, James. *Rivers of the World: The Ganges.* Farmington Hills, Mich.: Lucent Books, 2002. Presents information on the physical characteristics of the Ganges River, and dives deeper into the severe problems with pollution. Simply written, accessible to all ages.

Crow, Ben. *Sharing the Ganges: The Politics and Technology of River Development.* Thousand Oaks, Calif.:

Sage, 1995. Discusses the political tension that exists between India and Bangladesh regarding the distribution of the flow of the Ganges River. Bibliography and index.

Hollick, Julian Crandall. *Ganga: A Journey Down the Ganges River.* Washington, D.C.: Island Press, 2008. Recounts the experiences and observations of the author's journey from the source waters of the Ganges to its meeting with the Indian Ocean, describing the river as being devoutly worshipped and remorselessly exploited at the same time.

Mirza, M. Monirul Qader. *The Ganges Water Diversion: Environmental Effects and Implications.* Norwell, Mass.: Kluwer Academic Publishers, 2010. Points out the problems that arise in the long term when actions are taken for short-term gain. Discusses the diversion of water from the Ganges River at Farakka and how this has provided some immediate benefit to certain areas while endangering and adversely affecting other areas, both upstream and downstream.

Mirza, M. Monirul Qader Mirza, Ahsan Uddin Ahmed, and Qazi Kholiquzamman Ahmad. *Interlinking of Rivers in India: Issues and Concerns.* Boca Raton, Fla.: CRC Press, 2008. Renders a critical analysis of India's plan to interlink its rivers to mitigate irrigation, flooding, and drought concerns. Provides a strong rebuttal of arguments in favor of the plan and delivering a valid alternative approach based on international cooperation around the Ganges basin.

Pollard, Michael. *The Ganges.* London: Evans Brothers Publishing, 2003. Intended for juveniles. Traces the course of the Ganges River and describes its physical features, history, and importance. Bibliography and index.

Postel, Sandra. *Pillar of Sand.* New York: W. W. Norton, 1999. Details the water resource problems of world and gives good detailed information about the water rights issues of India and Bangladesh. Also provides additional information on the salinization of the delta area.

Shukla, Ashok Chandra, and A. Vandana. *Ganga: A Water Marvel.* New Delhi: Ashish, 1995. Focuses on the environmental problems faced by the Ganges River, as well as efforts to protect it from further degradation. Bibliography and index.

Wohl, Ellen. *A World of Rivers.* Chicago: University of Chicago Press, 2011. Describes the Amazon, Ob, Nile, Danube, Ganges, Mississippi, Murray-Darling, Congo, Chang Jiang, and Mackenzie Rivers. Offers valuable figures and graphs loaded with well-organized and straightforward information. Discusses the natural history, anthropogenic impact, and the future environment of these ten great rivers. Bibliography is organized by chapter.

**See also:** Amazon River Basin; Deltas; Floodplains; Great Lakes; Lake Baikal; Mississippi River; Nile River; River Bed Forms; River Flow; Yellow River

# GEOCHEMICAL CYCLES

*Geochemical cycles are processes that move environmental elements through the earth's systems, including the lithosphere, biosphere, hydrosphere, and atmosphere. The geochemical cycles are driven by solar energy and by thermal energy generated within the earth's core and are essential for the existence of life on the planet.*

## PRINCIPAL TERMS

- **atmosphere:** gaseous membrane surrounding the earth that is divided into layers according to chemical composition and temperature; protects the earth from cosmic and solar radiation
- **autotroph:** organism that produces complex organic molecules using materials and energy obtained through photosynthesis or chemiosynthesis
- **biosphere:** the collective relationship of living organisms on Earth and the other components of the environment
- **carbon fixation:** chemical process that incorporates carbon from its atmospheric state into complex organic molecules
- **heterotroph:** organism that cannot absorb atmospheric carbon for growth it must consume organic carbon in the bodies of other organisms to fuel growth and the development of complex organic molecules
- **hydrosphere:** the sum of the aqueous components of the earth, including the oceans, seas, rivers, lakes, and the water within the ground and lower levels of the atmosphere
- **lithosphere:** the outer solid layer of the earth consisting of the earth's crust and part of the upper mantle
- **photosynthesis:** chemical process that converts energy in solar radiation with carbon dioxide and water to form carbohydrates and oxygen
- **respiration:** physiological process that allows organisms to exchange carbon dioxide produced by biochemical processes within the organism's body for atmospheric oxygen
- **weathering:** breakdown of rocks, minerals, and other components of the earth's crust in contact with the atmosphere, biosphere, or hydrosphere

## EARTH AS A CHEMICAL SYSTEM

The earth can be thought of as a chemical system driven by light and heat energy from solar radiation and thermal energy generated within the earth's core. The system's raw material comprises the ninety-two naturally occurring chemical elements, most of which have been present in the chemical system since the earth solidified from cosmic dust and gas.

The chemical elements are stored and processed within Earth's four environmental spheres; the lithosphere, hydrosphere, atmosphere, and biosphere. The lithosphere consists of the solid rocks and sediment of the earth, which together constitute the earth's crust and part of the upper mantle. The hydrosphere consists of the collective waters of the earth, including the oceans, seas, rivers, lakes, and water vapor in the lower atmosphere. The atmosphere consists of a gaseous membrane, divided into layers of differing temperature and chemical composition, which surrounds the earth and differentiates the environment of the planet from that of outer space. The biosphere consists of all living matter on Earth.

The spheres are interconnected through a series of shared chemical reactions and exchanges. For instance, an organism (representing part of the biosphere) obtains chemical elements from the atmosphere through respiration, from the hydrosphere through drinking water, and from the lithosphere through eating plants and other materials that incorporate elements of the lithosphere. Organisms also contribute chemical elements to each of these systems through respiration, the production of waste, and their eventual death and decomposition. These patterns of relationships and chemical reactions are called geochemical cycles. Each of the spheres also serves as a reservoir for chemical elements, maintaining them in the form of molecular deposits until they are eventually filtered back into the other spheres.

Scientists have uncovered geochemical cycles associated with numerous elements, though the best known involve the elements that are considered most essential to the survival of life. Some of the major geochemical cycles include the hydrologic, nitrogen, phosphorus, carbon, and oxygen cycles. Other minor elemental cycles that have been discovered involve such elements as potassium, mercury, magnesium, aluminum, iron, and silicon. As elements cycle through the spheres, they are combined to form molecules and compounds that can be used to store or transfer energy within the environment. The geochemical cycles therefore not only result in the net movement in chemical elements but also function as a system for moving and distributing energy across the earth.

## THE NITROGEN CYCLE

Nitrogen is one of the most abundant elements on Earth, making up more than 78 percent of the atmosphere, about 50 percent of the earth's crust, and between 3 and 4 percent of the bodies of organisms. Atmospheric nitrogen ($N_2$) must be converted to another form before it can be utilized by organisms. Usually this is accomplished by nitrogen fixation, which is a set of processes that converts atmospheric nitrogen to ammonia ($NH_3$) or ammonium ($NH_4^+$), which can be utilized in a variety of biological processes.

Most nitrogen fixation takes place in the soil, where specialized types of bacteria, living either within soil or within the root tissues of plants, absorb atmospheric nitrogen and convert it to ammonia. Ammonia is then incorporated into the tissues of autotrophic microbes and plants, which are organisms able to manufacture complex molecules using only materials taken from the nonliving spheres of the earth. These organisms are later consumed by heterotrophs, which are organisms that cannot form their own complex molecules from environmental components and must obtain these molecules by consuming other organisms. Nitrogen within the tissues of organisms eventually returns to the lithosphere through animal waste or through organism decomposition.

Ammonium in the soil also may be converted to nitrate ($NO_3^-$) by a process called nitrification, which involves a chemical reaction with oxygen. Nitrate can be absorbed by autotrophs, too, thereby providing another route through which nitrogen enters the biosphere. Alternatively, when there is insufficient oxygen present, nitrate can be used in a process called denitrification, which involves a reaction between organic carbon and nitrate that releases both carbon dioxide ($CO_2$) and atmospheric nitrogen back into the system.

Some atmospheric nitrogen enters the lithosphere through the hydrologic cycle. Lightning causes combustion in the atmosphere, which leads to high-temperature oxidation of nitrogen to form nitric acid ($HNO_3$), which is water soluble and therefore becomes incorporated within raindrops as water vapor condenses in the atmosphere. Nitric acid is delivered to the lithosphere in rain, where it undergoes further chemical reactions that return some nitrogen to the atmosphere while the remaining nitrogen is incorporated into the bodies of organisms.

## THE OXYGEN CYCLE

Oxygen ($O_2$) is the second most abundant element in the atmosphere, making up more than 20 percent of atmospheric gases. The primary force in the oxygen cycle is photosynthesis, which is the process by which autotrophic organisms convert atmospheric carbon dioxide ($CO_2$), water, and energy from sunlight into complex organic compounds. The by-product of photosynthesis is atmospheric oxygen, which returns to the atmosphere.

Organisms also absorb atmospheric oxygen during respiration, which is a physiological process that exchanges carbon dioxide for oxygen. Oxygen taken in during respiration is utilized to break down carbohydrate molecules, thus generating energy. The by-product of respiration is carbon dioxide, which is returned to the atmosphere. Atmospheric oxygen is also absorbed during the decomposition process, which breaks down the tissues of organisms through oxidation into simpler molecules. Decomposition produces carbon dioxide as a by-product, which is returned to the atmosphere.

The lithosphere is a reservoir for vast amounts of oxygen contained within the physical structures of rocks and minerals. These deposits account for more than 99 percent of the oxygen on Earth. Weathering is a process by which the rocks and minerals of the lithosphere break down into their constituent molecules and minerals. This process can be the result of physical forces or chemical reactions on the surface of the rocks. As weathering occurs, some of the oxygen trapped in the lithosphere is returned to the hydrosphere and to the atmosphere.

## THE CARBON CYCLE

Life on Earth is often referred to as carbon-based life because carbon is the central molecule in the formation of living tissues. Carbon constitutes less than 0.05 percent of atmospheric gases and usually exists in the atmosphere as carbon dioxide.

Carbon dioxide is absorbed by autotrophs during photosynthesis, which releases oxygen into the atmosphere, and is then combined to form a variety of carbon-based molecules within the body, including proteins, lipids, and nucleic acids. Carbon then migrates through the food chain as heterotrophs consume autotrophs to fuel their metabolic needs.

Some of the carbon contained within the bodies of animals and other organisms is returned to the atmosphere through respiration. Respiring organisms inhale atmospheric oxygen and other gases and return (exhale) carbon dioxide to the atmosphere; carbon dioxide is a by-product of the processes used to convert oxygen and other molecules into energy. Carbon also returns to the atmosphere through decomposition of living tissues, which produce carbon dioxide and methane as by-products. Carbon from decomposing organisms also can be buried and compressed, thereby becoming part of the lithosphere. Periodically, lithospheric carbon will be returned to the atmosphere as a by-product of the chemical reactions that cause volcanic eruptions.

Carbon dioxide enters the hydrosphere through the decomposition of organisms, through the weathering of rocks, and by contact with carbon dioxide in the atmosphere. Carbon dioxide dissolves in ocean water to form carbonic acid ($CO_2H_2O$), which often dissociates to form carbonate ($CO_3^{2-}$). Carbonate added to the oceans tends to make ocean water more acidic, which is balanced by basic minerals introduced to the ocean from the weathering of the lithosphere. As oceanic acidity increases, carbon dioxide rises from the ocean into the atmosphere, thereby replenishing atmospheric supplies of carbon.

## THE PHOSPHORUS CYCLE

Phosphorus is an essential element for living processes and is also a common trace mineral in the lithosphere and hydrosphere. Phosphorus enters the hydrosphere through erosion and weathering. The phosphorus cycle is one of the few geochemical cycles that does not have a dominant atmospheric component.

Phosphorus exists in nature primarily as phosphate ($PO_4^{3-}$), which often binds oxygen within rocks and other parts of the lithosphere. Much of the phosphorus in the lithosphere consists of rocks that form from the compressed fossilized remains of animal and organismal tissues and waste. For instance, bird droppings, called guano, can form into a type of rock that has high levels of phosphate.

Both aquatic and terrestrial autotrophs utilize phosphate to build tissues. As phosphate moves through the food chain, it constitutes a major reservoir for phosphorus within the biosphere. Some of the phosphate that is absorbed into the hydrosphere undergoes further chemical reactions that form iron phosphate ($FEPO_4$), which becomes incorporated into the lithosphere, constituting a second lithospheric reservoir, until weathering and erosion dissolve iron phosphate and return dissolved phosphate to the hydrosphere.

## THE HYDROLOGIC CYCLE

The hydrologic or water cycle is the pathway that water ($H_2O$) takes through the various environmental spheres of the earth. The hydrosphere is the primary reservoir for water and exists in a complex and dynamic relationship with the atmosphere.

In response to solar radiation, water from the hydrosphere enters the atmosphere through evaporation, a chemical process that transforms water molecules on the surface of the water into atmospheric water vapor. Water vapor rises in the atmosphere as solar heating excites the gases of the atmosphere and reduces density. As columns of air rise, the molecules and water vapor contained within them gradually cool and condense to form clouds.

Solar radiation heats atmospheric gases differentially across the surface of the earth, leading to pockets of higher density and higher pressure and to pockets of lower density and lower pressure. The distribution of density causes the atmosphere to move in circular currents that distribute warm air across the planet. As this occurs, water vapor is transported above the lithosphere, where it descends to the earth as precipitation.

Some of the water that falls to the earth as rain returns to the hydrosphere as runoff into rivers that eventually flow back to the ocean. Some water is absorbed into the bodies of organisms through drinking and respiration. Living organisms are primarily composed of water, so the biosphere constitutes a major reservoir of liquid water. Water held in the biosphere is eventually returned to the environment through respiration and as a component of the waste produced by organisms during their metabolic processes.

In the lithosphere, water is stored as ice at high elevations and where temperatures support the development of surface and oceanic ice deposits. Additional water is stored in cavities beneath the earth's crust; a significant amount of water is incorporated into the crystal structures of rocks and minerals. Some of this lithospheric water is released through weathering, which often releases water vapor into the surrounding air.

Water contained under the surface of the earth tends to migrate toward the edges of the terrestrial environment, where it either rejoins the oceans or emerges in springs and rivers that eventually filter into the ocean. Cycles of heating and cooling associated with seasonal variations in sunlight drive the hydrologic cycle and thereby serve to distribute precipitation in various forms throughout Earth's ecosystems.

*Micah L. Issitt*

## FURTHER READING

Grotzinger, John, and Thomas H. Jordan. *Understanding Earth*. New York: W. H. Freeman, 2009. Detailed text covering aspects of Earth's systems, including geology, atmospheric sciences, and geochemical cycling. Covers the carbon, phosphorus, nitrogen, and water cycles.

Jacobson, Michael C., et al. *Earth System Science: From Biogeochemical Cycles to Global Change*. San Diego, Calif.: Academic Press, 2006. Introduction to Earth systems science and the role of geochemical cycles in Earth's processes. Contains detailed descriptions of the carbon, phosphorus, nitrogen, and water cycles.

National Research Council. *The Atmospheric Sciences: Entering the Twenty-First Century*. Washington, D.C.: National Academy Press, 2010. Modern treatment of atmospheric chemistry and physics written for advanced students of Earth systems science. Chapters detail the water and carbon cycles.

Stanley, Steven M. *Earth System History*. New York: W. H. Freeman, 2004. Historical overview of the development of the earth, written for general students. Contains information on geochemical cycling and the methods used to investigate them.

Stevenson, F. J., and M. A. Cole. *Cycles of Soil: Carbon, Nitrogen, Phosphorus, Sulfur, Micronutrients*. New York: John Wiley & Sons, 1999. Detailed coverage of the earth's geochemical cycles with individual chapters on each of the various elements and their paths through the environment.

Wallace, John M., and Peter V. Hobbs. *Atmospheric Science: An Introductory Survey*. 2d ed. Burlington, Mass.: Academic Press, 2006. Detailed introduction to the study of atmospheric processes, including discussions of the cycling of oxygen, nitrogen, and water in the atmosphere.

**See also:** Atmosphere's Evolution; Atmosphere's Global Circulation; Atmosphere's Structure and Thermodynamics; Atmospheric and Oceanic Oscillations; Atmospheric Properties; Carbon-Oxygen Cycle; Clouds; Earth-Sun Relations; Global Energy Transfer; Global Warming; Hydrologic Cycle; Hydrothermal Vents; Lightning and Thunder; Nitrogen Cycle; Observational Data of the Atmosphere and Oceans; Ocean-Atmosphere Interactions; Precipitation; Rainforests and the Atmosphere; Water and Ice; Weathering and Erosion

# GLOBAL ENERGY TRANSFER

*Earth is best thought of as a system with energy inputs and outputs. Energy, mostly heat, comes from the sun and from Earth's core. As this energy transfers around and through the planet, it drives major processes such as plate tectonics, climate, ocean and air currents, and the distribution of life across the globe.*

## PRINCIPAL TERMS

- **albedo:** the amount of radiation a surface reflects; higher albedo reflects more incoming radiation
- **atmospheric greenhouse effect:** the result of greenhouse gases in the atmosphere; trapped energy (heat) in the earth's system
- **electromagnetic radiation:** radiation that includes visible light, infrared radiation (heat), radio waves, gamma rays, and X rays
- **energy budget:** an accounting of all the incoming and outgoing energy for Earth as a system
- **greenhouse gas:** an atmospheric gas that contributes to the greenhouse effect by absorbing infrared radiation and reemitting that radiation
- **infrared radiation:** electromagnetic radiation with wavelengths longer than visible light but shorter than radio waves; the type of electromagnetic radiation perceived as heat
- **irradiance:** the power of electromagnetic radiation over a given unit of area, usually in watts per square meter; used to measure the influx of energy through an area such as the earth's surface
- **radiative equilibrium:** a state in which Earth's incoming radiation and outgoing radiation are equal; it results in a generally stable climate as there is no net gain or loss of energy from the planet's system
- **radiative forcing:** the total change in irradiance between different layers in the atmosphere; positive radiative forcing indicates a net increase in energy in the system (warming), whereas negative radiative forcing indicates a net release of energy (cooling)
- **solar flux:** the total energy entering Earth's atmosphere from the sun
- **specific heat:** the amount of heat (energy) it takes to raise the temperature of the unit mass of a given substance by a given amount, usually one degree; functionally, a substance's capacity to store heat

## EARTH AS A SYSTEM

Earth is governed by many integrated processes, which are together called a system. The study of global energy (heat) transfer aims to map the way these systems interact to move energy around the planet. The transfer of energy, in turn, drives everything from global weather patterns to the formation of mountains to the evolution of plants and animals.

The total amount of energy that moves through the earth's system amounts to hundreds of terawatts ($10^{12}$ watts). For perspective, a typical automobile engine is rated at 50,000 to 100,000 watts, and lightbulbs handle hundreds of watts; Earth deals in hundreds of billions of watts.

The rate at which Earth absorbs and radiates energy is essential to Earth's capacity to support life. The atmosphere, which slows the release of heat back into the universe, allows the planet to maintain a temperature conducive to liquid water and to the existence of life.

## EARTH'S SUBSYSTEMS

Earth's system can be broken down into four subsystems: atmosphere, lithosphere, hydrosphere, and biosphere. The atmosphere is most notable for the way it acts as a shell around the planet.

The atmosphere bounces back into space much of the incoming radiation that hits Earth, but what the atmosphere does let through, it also often traps on Earth; the atmosphere bounces energy radiating from the earth's surface back to the earth before that energy can leave the system.

The greenhouse gases, such as carbon dioxide ($CO_2$), get their name because they are especially good at trapping heat in the planetary system in a manner similar to a greenhouse, which traps heat. Earth's greenhouse gases trap heat by absorbing outgoing infrared radiation and by then reemitting it in all directions. Some radiation then returns toward Earth. The movement of air around the planet, as wind, also transfers some energy. However, the atmosphere is generally unable to hold much energy. Instead, it helps to drive some ocean currents, which move energy around.

The lithosphere is Earth's landmass. Land is better than the atmosphere at holding heat, but its most important role in global energy transfer comes from its relationship to the planet's core. At areas of high geothermal activity (volcanoes being perhaps the most dramatic example), energy from the earth's core escapes. The gases and chemicals produced by geothermal activity also can influence global energy transfer, such as when ash and gases spewed by volcanic eruptions lead to cooling because they reflect more incoming radiation than does the atmosphere.

The hydrosphere is arguably the most important element of Earth's energy system. The hydrosphere is the water that covers the planet's surface, largely the oceans, and the oceans serve as the major energy reservoir for the planet. This is true because of water's high specific heat (water can store a large amount of energy before its temperature increases). The oceans move this energy around the planet, shaping the climate.

As an example, one can consider London's comparatively mild winters, despite that city's northerly latitude. London benefits from the Gulf Stream, a major warm-water current that carries energy from near the equator, up the east coast of North America, and then northeast across the Atlantic Ocean to Europe. The interaction of the hydrosphere and atmosphere is largely responsible for Earth's weather, as heat moves between the oceans and the air above them.

Finally, the biosphere comprises all the living things on the planet, from microbes to plankton to humans to whales. Life uses the energy in Earth's system to power itself; photosynthetic organisms turn the sun's radiated energy into a form (that is, food) usable by other forms of life. Life-forms are most important to global energy transfer because of the roles they play in the way gases and chemicals cycle around the planet. Plants help trap carbon, keeping it from getting into the atmosphere as carbon dioxide and acting as a greenhouse gas. The processes of digestion in animals (cattle, for instance) can produce greenhouse gases such as methane. The reef-building of corals can help shape coastlines and the flow of energy in ocean currents.

Humans and industry alter the composition of the atmosphere and the way energy moves around the planet at an unprecedented rate. Burning fuel, for example, releases stored energy as heat, and the overall global effects of fuel-burning are significant.

## EARTH'S ENERGY SOURCES: SOLAR FLUX AND GEOTHERMAL ENERGY

Across all four subsystems, energy comes from two major sources: the sun and the earth's core. *Solar flux* is the term for the total amount of incoming solar radiation that enters the earth's system. Because of the tilt in Earth's axis, sunlight hits different parts of the planet with different intensities depending on the season. This uneven heating creates the three major climate zones: tropical, temperate, and polar. This uneven heating also creates weather systems.

Warmer fluids are less dense than colder fluids; warm air and water in the skies and oceans will be pushed away by heavier, colder air and water. This, coupled with the rotation of the earth, sets in motion the major ocean currents and wind patterns, which are important in moving energy around the planet.

The system's internal heat source, radiation from Earth's core, creates currents in the liquid rock that surrounds the core. These currents are what move the continents around on the surface of the earth. The motion of the crust sliding on the subsurface magma is what drives plate tectonics, which in turn gives the planet its mountains and basins and other physical features.

## CLIMATE CHANGE

All efforts to mitigate the effects of global warming are predicated on a thorough grasp of how energy moves around the planet. Global warming is, after all, essentially a global energy surplus: More energy remains in Earth's system than leaves it, thus raising global temperatures.

Through the use of observational satellites scientists can estimate how much total radiation enters and exits the earth's system. However, the details of what happens to this energy once it enters the system remain poorly understood. Scientists know that Earth currently has a large surplus of unaccounted-for energy that has "disappeared" somewhere on the planet. Earth is not in a state of radiative balance. For this reason, a detailed accounting of Earth's energy budget and a comprehensive model of the processes that drive global energy transfer are fundamental requirements for success in the face of the challenges presented by global climate change.

Consider once again London's mild climate. Knowing how ocean and air currents move warm water (energy) around supports predictions of some of the seemingly counterintuitive effects of global warming. London, for instance, could experience a drop in temperature if the Gulf Stream, a major warm water current, shifts due to climate change. Without the warm water mediating the temperature of the air around London, the city could see much more severe winters.

*Kenrick Vezina*

## Further Reading

Kandel, Robert. "Understanding and Measuring Earth's Energy Budget: From Fourier, Humboldt, and Tyndall to CERES and Beyond." *Surveys in Geophysics* (January, 2012): 1-12. An assessment of efforts to understand and quantify the planet's energy budget. Rather technical but extremely useful in providing a sense of the areas of weakness in accounting for the global energy budget—in particular, variations at the surface of the earth and in clouds. Suitable for undergraduates and advanced high school students.

Kiehl, J. T., and Kevin E. Trenberth. "Earth's Annual Global Mean Energy Budget." *Bulletin of the American Meteorological Society* 78 (February, 1997): 197-208. Dated estimate of the global energy budget that remains an excellent work for understanding how the process of constructing an estimate of the global energy budget occurs in the scientific literature. Suitable for advanced high school students and for undergraduates.

Smithson, Peter, Kenneth Addison, and Kenneth Atkinson. "Energy and Earth." In *Fundamentals of the Physical Environment*. 3d ed. New York: Routledge, 2002.

_____. "Energy Flows and Nutrient Cycles in Ecosystems." In *Fundamentals of the Physical Environment*. 3d ed. New York: Routledge, 2002.

_____. "Heat and Energy in the Atmosphere." In *Fundamentals of the Physical Environment*. 3d ed. New York: Routledge, 2002. These three chapters are part of a detailed textbook approach to the larger topic of understanding Earth as a physical system. They offer excellent depth without becoming too abstruse for general readers. The first chapter is an excellent overview of the earth as an energy system, and the two latter chapters explain how the atmosphere and biosphere relate to global energy transfer, respectively.

Trenberth, K. E. "An Imperative for Climate Change Planning: Tracking Earth's Global Energy." *Current Opinion in Environmental Sustainability* 1 (2009): 19-27. A detailed argument for the importance of tracking global energy transfer to understand and respond to climate change. Of particular note is figure 2, which presents a thorough but accessible schematic of global energy flows.

Trenberth, K. E., and John T. Fasullo. "Tracking Earth's Energy." *Science* 328 (April, 2010): 316-317. Brief and relatively accessible focus on the problem of "missing energy" in the earth's energy budget and its implications for climate change. Figure A offers an excellent overview of the global energy budget.

**See also:** Arctic Ocean; Atlantic Ocean; Atmosphere's Evolution; Atmosphere's Global Circulation; Atmosphere's Structure and Thermodynamics; Atmospheric and Oceanic Oscillations; Atmospheric Properties; Climate; Clouds; Earth-Sun Relations; El Niño/Southern Oscillations (ENSO); Geochemical Cycles; Global Warming; Greenhouse Effect; Gulf Stream; Hurricanes; Hydrologic Cycle; Hydrothermal Vents; Observational Data of the Atmosphere and Oceans; Ocean-Atmosphere Interactions; Pacific Ocean; Remote Sensing of the Oceans; Satellite Meteorology; Sea Level; Seasons; Surface Ocean Currents; Water and Ice; Wind; World Ocean Circulation Experiment

# GLOBAL WARMING

*Global warming is the term applied specifically to indicate rising average global air temperatures. This rise in temperature has the potential to cause drastic changes in climate and weather patterns worldwide by disrupting the equilibrium between incoming solar energy and the thermal energy that is reradiated away from the surface.*

## PRINCIPAL TERMS

- **climate:** the long-term weather patterns for a region, distinct from the day-to-day weather patterns
- **greenhouse effect:** the net accumulation of heat held in the atmosphere as thermal energy by certain gases
- **greenhouse gas:** any atmospheric gas that absorbs infrared radiation, preventing it from being radiated back into space; examples include carbon dioxide, methane, and water vapor
- **infrared radiation:** energy emitted as electromagnetic radiation having wavelengths between 0.75 micron and 1 millimeter, frequently confused with heat but actually a different phenomenon

## GREENHOUSE EFFECT

"Global warming" is the term for the rise in Earth's average atmospheric temperature. It is known that Earth's average global temperature has been rising slowly over time, a trend that has been particularly noted since the beginning of the Industrial Revolution in the second half of the eighteenth century. It is generally conceded that increases in air temperature could alter precipitation patterns, change growing seasons, result in coastal flooding, and turn some currently fertile areas into deserts. The exact cause or causes of this phenomenon are not known, but global warming could be part of a normal climate cycle, be caused by natural events, or arise through the activities of humankind.

When the ground surface is heated by sunlight, it emits much of the solar energy it has absorbed as infrared radiation. Gases in the atmosphere absorb the reradiated infrared radiation and prevent it from escaping into space. This is called the "greenhouse effect" because it was once believed that the glass panes of greenhouses captured the infrared radiation given off by the soil inside the greenhouse. Although it has since been shown that greenhouses work by trapping heated air and not allowing it to blow away, the name has stuck.

The atmosphere eventually releases its contained heat into space and the amount of heat stored in the atmosphere remains fairly constant as long as the composition of gases in the atmosphere does not change. Some gases, including carbon dioxide, water vapor, and methane, capture and store infrared radiation, or heat, more efficiently than others and are called "greenhouse gases." If the composition of the atmosphere changes to include more of these greenhouse gases, the atmosphere retains more heat and becomes warmer.

Global levels of greenhouse gases have been steadily increasing and in 1990 were more than 14 percent higher than they were in 1960. At the same time, the average global temperature has also been rising. Meteorological records show that from 1890 to the mid-1990's, the average global temperature rose by between 0.4 and 0.7 degree Celsius. About 0.2 degree Celsius of this temperature increase has occurred since 1950. In comparison, the difference between the average global temperature in the 1990's and in the last ice age is approximately 10 degrees Celsius, and it is estimated that a drop of as little as 4 to 5 degrees Celsius could trigger the formation of continental glaciers. Therefore, the rise in average temperatures is significant and already appears to be causing some changes in the global climate. Documented changes include the net loss of ice due to melting glaciers and slowly rising sea levels as the ocean gets warmer and its waters expand. Also, measurements of plant activity indicate that the annual growing season has become approximately two weeks longer in the middle latitude regions.

## EFFECTS OF GLOBAL WARMING

A common misunderstanding is that global warming simply means that winters will be less cold and summers will be hotter, while everything else will be basically the same. Actually, Earth's global weather system is very complex, and higher global temperatures will result in significant changes in weather patterns. Most changes will be observed in the middle and upper latitudes, particularly in North America, Europe, and most of Asia, with equatorial regions witnessing fewer changes. The Southern Hemisphere should experience less severe effects because it contains more water than the Northern Hemisphere does, and it takes more energy to heat water than land.

It is difficult to predict precisely what these changes will be, but observation of the changing climate and scientific studies allow researchers to make some estimates of the kinds of changes that will occur. Summers will be hotter, with more severe heat waves. Because hot air holds more moisture than cool air, rain will fall less frequently in the summer. Droughts can be expected to be more common and more severe. Through the late 1980's and into the early 1990's, annual temperatures climbed higher and higher, and summer heat waves became more frequent. This is a particularly troubling problem in areas where homes are typically built without air conditioning. The air in a closed-up house during a heat wave can reach temperatures well over 40 degrees Celsius. Because these temperatures exceed people's normal body temperature, which is about 36.5 degrees Celsius, it becomes very difficult for the body to cool down. For this reason, heat waves are particularly dangerous for the very young, the elderly, and the sick. More frequent heat waves will cause increases in the use of air conditioning, which requires more energy consumption that will in all likelihood lead to the release of additional greenhouse gases.

Global warming will also produce severe autumn rains. The overheated summer air will cool in the autumn and will no longer be able to hold all of the moisture it was storing. The moisture would be released as heavy rains, resulting in flooding. This phenomenon has already been observed, but not for a period of time that is scientifically significant as the difference between long-term changes and short-term fluctuations are not differentiable with only a few years of observations.

Changing rain patterns, as droughts and severe autumn storms, will have a strong effect on Earth's landscapes. Some fertile areas may become deserts, while other areas may be transformed from plains to forests.

One of the stranger aspects of global warming is that it is predicted to result in not only hotter periods during the summers but also colder periods during the winters. With more atmospheric energy available to move large cold-air masses from the polar regions in winter, it is possible that large winter storms will be colder, more violent, and more frequent. This

pattern has become more evident since about the mid-1970's. Winter storms have brought record low temperatures and enough snow to close cities for days. However, this time span is again too short to determine whether this is a temporary phenomenon or a trend. It is possible that some smaller, less permanent event than global warming is responsible for the more severe winter storms. Another consideration is that a shorter, though more severe, winter may produce a surge in pest populations and diseases that are normally controlled by longer winters.

Global warming will cause ocean levels to rise, in part because water expands as it becomes warmer, and because of the additional water that will enter the oceans due to melting glaciers on Greenland and Antarctica. These terrestrial icecaps are as much as 4 kilometers thick and lie atop land rather than ocean water. The melting of the ice in the northern polar areas should not contribute to rising ocean levels because that ice, unlike the ice on Greenland and Antarctica, is already in the ocean. Just as the melting of an ice cube in a drink does not cause the level of the drink to rise, the melting of the northern oceanic ice sheets will not affect sea levels.

## CAUSES OF GLOBAL WARMING

Although many scientists are convinced, based on the abundant evidence, that global warming is occurring, they are less sure of the causes. One significant factor, the rise in greenhouse gases, has been attributed to the activities of humankind. Burning forests to clear land and operating factories and automobiles produce carbon dioxide and water vapor. Livestock herds and rotting vegetation release methane, and fertilizers used on farms also release greenhouse gases. Power plants that consume fossil fuels such as coal, oil, or natural gas release massive amounts of carbon dioxide into the air. However, no one knows with certainty whether humankind's activities are the only reason for the increase in global temperature. Recent calculations demonstrate that human activity has released more than 1 trillion tons of carbon dioxide into the atmosphere over the past two hundred years. Yet this represents only a fraction of the amount of carbon dioxide naturally present, and an even tinier fraction, only some 0.03 percent, of the total gases in the atmosphere. The last ice age ended recently in geologic terms, and a number of changes are still taking place as the globe recovers from the presence of huge ice sheets on its surface. It is possible that the world's climate is still warming up from the last ice age. Volcanoes are another major source of greenhouse gases, and the level of volcanic activity has been increasing since approximately the beginning of the Industrial Revolution.

Global warming could also be part of a natural, cyclical change in the climate. Evidence indicates that Earth's climate varies between warmer and colder periods that last a few centuries. History provides many stories of dramatically changing climate, including one period from 1617 to 1650 that was so unusually cold that it is called the "Little Ice Age." Therefore, Earth may be merely experiencing another cyclical change in its climate. It has also been pointed out by some researchers that the global warming trend observed over the past thirty years coincides with an observed increase in solar energy emissions over that same period, suggesting that the cause is not to be found on this planet.

## STUDY OF GLOBAL WARMING

The problem with studying global warming is that no one can be sure of its extent, and many scientists continue to debate whether it actually exists. They argue that the observed changes may only be a warm phase of a climatic cycle that will make Earth warmer for a few years, then cooler for a few years. If this is true, then the coming decades may see average temperatures leveling off or even dropping.

A major source of the confusion involves the way global warming is studied. It would seem easy to record temperatures for a number of years and then compare them. However, detailed records on the weather have not been kept for more than a few decades in many areas. Scientists are forced to rely on interpretations of historical accounts and the clues left in fossil records. Tree rings, sedimentary deposits, and even old ice from deep within glaciers are examined to provide data about the global climate in the past.

In addition, the existing records must be reviewed carefully to identify local changes that may not reflect global ones. For example, as towns grow into cities, the temperature climbs simply because larger cities are warmer than smaller ones, a phenomenon known as the urban heat island effect. Measurements taken years ago in a more rural environment should be lower than those taken after the population around the measuring station

increased. This problem can be overcome with balloons. By sending instruments high in the atmosphere on weather balloons, air temperatures can be measured without being affected by urbanization. Although data recorded this way show a consistent rise in global temperature, such measurements go back only a few decades.

Measurements of the level of greenhouse gases in the atmosphere are also affected by urbanization. As a small town becomes a city, levels can be expected to rise. However, recording stations located in regions far removed from cities and factories also show an increase in the level of greenhouse gases. One station in Hawaii has shown a rise in the amount of carbon dioxide present in the air since early 1958, with similar reports coming from stations in Point Barrow, Alaska, and Antarctica.

Another important variable in looking at global warming is sea surface temperature. Measurements can be skewed by local effects that have no impact on the global climate. One method used to make detailed measurements of seawater temperature is to broadcast a particular frequency of sound through the water and measure it at distant locations. The speed and frequency of the sound are affected by the temperature of the water.

However, more accurate and more global data are becoming available through the Mission to Planet Earth program of the National Aeronautics and Space Administration (NASA). The program places satellites in orbit to study Earth and make a variety of detailed measurements, many of which are attributed to factors that contribute to global warming. These data will help immensely in enabling scientists to understand the global climate and the changes it is undergoing.

## SIGNIFICANCE

The global climatic environment is very complex, being the result of many contributing factors. All of these factors together produce the climate in which people live. For this reason, it is impossible to say which, if any, changes are attributable simply to a particular natural event or to the activities of humankind. However, at the same time, the activities of humankind are so large-scale and widespread that it is safe to conclude that they have definitely changed the world. Unfortunately, the effect of these changes is still not clear. People have generated massive amounts of greenhouse gases and released them into the air, but no one really knows the extent of their effect on the environment.

If scientists determine that humankind's activities are responsible for the observed changes in the environment, then people can identify what they must do to preserve the environment and take measures to change the way they live. People may eventually gain enough knowledge to plan the changes in the climate to occur in such a way as to provide maximum benefit to the environment and themselves. Even if researchers determine that global warming is caused by natural events beyond human control, people still must evaluate their role. Human activities are likely to contribute to the problem, so altering behavior and ways of life might lessen the negative effects of global warming. For example, although global warming might not be caused by emissions of greenhouse gases, those emissions might worsen droughts and induce more severe rainstorms. If so, then by reducing emissions, human activity might also be able to counteract some of the effects of the climatic changes.

If the changes in the climate approach what is expected based on scientific models of global warming, the impact on the environment and humanity will be staggering. The changes are such that virtually every person on Earth would be affected. If rain patterns change drastically, it is possible that the food supplies of the world will be sharply reduced. In some areas, if the average temperature increases, certain diseases that are endemic to areas with warmer climates will begin to affect people in other areas. An increase of only 3 meters in the mean ocean level will inundate many of the world's coastal areas and the large cities found in those areas. In addition to these large-scale changes, changes in the weather will have a great impact on the day-to-day lives of the majority of the world's people. Conversely, any solution to the problem of global warming will also probably have a dramatic effect on people's lives.

*Christopher Keating*

## FURTHER READING

Abrahamson, Dean Edwin, ed. *The Challenge of Global Warming*. Washington, D.C.: Island Press, 1989. Provides a detailed examination of what global warming is and how it occurs. Also discusses greenhouse gases and policymaking decisions regarding global warming. Suitable for all readers.

Fagan, Brian M., ed. *The Complete Ice Age: How Climate Change Shapes the World*. New York: Thames & Hudson, 2009. Discusses the development and cycle of the ice age. Discusses the effects of the ice age on human evolution and includes information on global warming and interglacial periods. Written in a manner accessible to high school students.

Gates, David M. *Climate Change and Its Biological Consequences*. Sunderland, Mass.: Sinauer Associates, 1993. Examines the effects that climatic changes would have on life on Earth, discussing many of the models used in making predictions. Suitable for high school students and up.

MacDougall, Doug. *Frozen Earth: The Once and Future Story of Ice Ages*. Berkeley, Calif.: University of California Press, 2006. Discusses current global warming as a snapshot within the current ice age cycle. Presents a history of the study of ice ages, including the work of many famous scientists.

Magill, Frank N., and Russell R. Tobias, eds. *USA in Space*. Pasadena, Calif.: Salem Press, 1996. Consists of an extremely well-written series of articles on U.S. space activities. Discusses data collection on the environment and detecting anthropogenic effects on the environment. Suitable for all readers.

Maslin, Mark. *Global Warming: Causes, Effects and the Future*. St. Paul, Minn.: MBI Publishing Company, 2007. Describes the evidence of global warming, discusses its causes and predicted impacts on the world, and suggests how the problem may be reduced.

Mooney, Chris. *Storm World. Hurricanes, Politics and the Battle Over Global Warming*. Orlando, Fla.: Harcourt Inc., 2007. Describes the conflicting purposes and understanding of scientists and politicians on the issue of global warming, its causes, and its effects.

Nance, John J. *What Goes Up: The Global Assault on Our Atmosphere*. New York: William Morrow, 1991. Clearly presented documentation of the investigation into the changes occurring in the atmosphere. Topics covered include global warming and the depletion of the ozone layer. Suitable for all readers.

Schneider, Stephen H., and Terry L. Root. *Wildlife Responses to Climate Change: North American Case*

*Studies.* Washington, D.C.: Island Press, 2002. Provides a number of case studies presenting information on the repercussions of climate change on the biological world. Species discussed include sachem skipper butterflies, grizzly bears, and the two-lobe larkspur. Intended to draw the public into the global warming discussion.

Sommerville, Richard C. J. *The Forgiving Air: Understanding Environmental Change.* 2d ed. Boston: American Meteorological Society, 2008. Focuses on the various consequences of air pollution, including the depletion of the ozone layer, climatic changes, the greenhouse effect, and acid rain. Intended as an introduction for the layperson. Bibliography and index.

Vaughn, Jacqueline. *Environmental Politics: Domestic and Global Dimensions.* 6th ed. New York: Wadsworth, Cengage Learning, 2011. Suited for undergraduate students studying environmental policy. Objectively written, provides the reader with information on current politics and environmental policy without persuasion. Discusses energy, water, air quality, land use, and waste.

Wallace, John M., and Peter V. Hobbs. *Atmospheric Science.* 2d ed. Burlington, Mass.: Academic Press, 2006. A complete study of the atmosphere that covers the fundamental physics and chemistry topics as well as specific topics in atmospheric science such as radiative transfer, weather forecasting, and global warming. Significant detail and technical writing, but still accessible to the undergraduate studying meteorology or thermodynamics.

Weiner, Jonathan. *Planet Earth.* Toronto: Bantam Books, 1986. Companion book to the *Planet Earth* series developed by the Public Broadcasting Service. Covers many aspects of the planet and its climate. Suitable for all audiences.

Wyman, Richard L., ed. *Global Climate Change and Life on Earth.* New York: Chapman and Hall, 1991. A collection of articles on various aspects of climate changes and the effects of these changes.

**See also:** Acid Rain and Acid Deposition; Air Pollution; *AR4 Synthesis Report*; Atmosphere's Evolution; Atmosphere's Global Circulation; Atmosphere's Structure and Thermodynamics; Atmospheric Properties; Auroras; Barometric Pressure; Bleaching of Coral Reefs; Climate Change Theories; Climate Modeling; Cosmic Rays and Background Radiation; Earth-Sun Relations; Global Energy Transfer; Greenhouse Effect; Ice Ages and Glaciations; Icebergs and Antarctica; Long-Term Weather Patterns; Nuclear Winter; Ozone Depletion and Ozone Holes; Radon Gas; Rainforests and the Atmosphere; Recent Climate Change Research; Remote Sensing of the Atmosphere; Remote Sensing of the Oceans

# GREAT LAKES

*The Great Lakes represent the largest freshwater lake complex on Earth. Created by continental glaciers over the past 18,000 years, these five major lakes (Ontario, Erie, Huron, Michigan, and Superior) and one minor lake (Saint Clair) provide significant resources for Canadians and Americans occupying the surrounding basin.*

## PRINCIPAL TERMS

- **epilimnion:** a warmer surface layer of water that occurs in a lake during summer stratification; during spring, warmer water rises from great depths, and it heats up through the summer season
- **greenhouse effect:** a natural process by which water vapor, carbon dioxide, and other gases in the atmosphere absorb heat and reradiate it back to Earth
- **isostatic rebound:** a tendency of Earth's continental surfaces to rise after being depressed by continental glaciers, without faulting
- **Pleistocene:** a geologic era spanning about 2 million years that ended about 10,000 years ago, often considered synonymous with the "Ice Age"
- **seiche:** rocking motion of lake level from one of the lake to the other following high winds and low barometric pressure; frequently, a seiche will follow a storm event
- **storm surge:** a rapid rise in lake level associated with low barometric pressure; the water level is frequently "pushed" above a shoreline on one end of the lake and depressed on the opposite end
- **thermocline:** a well-defined layer of water in a lake separating the warmer and shallower epilimnion from the cooler and deeper hypolimnion
- **wetlands:** areas along a coast where the water table is near or above the ground surface for at least part of the year; wetlands are characterized by wet soils, water-tolerant plants, and high biological production

## GEOLOGICAL DEVELOPMENT OF THE REGION

The Great Lakes are superlative features on the North American landscape. They make up the largest freshwater lake complex on Earth and represent about 18 percent of the world's water supply. Covering a total area of 245,000 square kilometers, the Great Lakes have a shoreline length of 17,000 kilometers. Lake Superior (82,100 square kilometers), Lake Huron (59,600 square kilometers), and Lake Michigan (57,800 square kilometers) are among the ten largest lakes on Earth.

The rocks forming the foundation of the Great Lakes date back some 600 million years. On the northern and northwestern shore of Lake Superior are remnants of the Canadian Shield, which is composed of igneous rocks of the Precambrian era, more than 1 billion years ago. Following volcanic activity and mountain building during the Precambrian era, the central region of North America was repeatedly covered by shallow tropical seas. At this time, during the Paleozoic era (600 million to 230 million years ago), sediments transported by rivers from adjacent eroding uplands were deposited in a shallow marine environment, and lime, salt, and gypsum precipitated from the seawater. All these soft materials were eventually hardened into sedimentary rock layers such as sandstone, shale, limestone, and halite. A multitude of fauna colonized the submarine environment, including corals, brachiopods, crinoids, and several species of mollusks.

As the layers of sediments accumulated over millions of years, the basin began to subside at its center. The Great Lakes basin structure may be compared to a series of bowls, one stacked on top of another. As viewed from above, only the top bowl is completely visible; however, the rims of the progressively deeper bowls are visible as a number of thin concentric rims along the perimeter of the basin.

The Paleozoic era was followed by the Mesozoic era (230 to 63 million years ago), a time of little deposition. In spite of the great age of the rocks making up the foundation of the Great Lakes, the lakes themselves were created in the relatively recent Pleistocene epoch. Between the 220 million years when the basin's bedrock was deposited and the onset of Pleistocene glaciers, the landscape now occupied by the lakes was occupied by streams. The streams eroded the softer bedrock to form channels and valleys. The divides between and parallel to the eroded valleys were more resistant to erosion and are represented by higher elevations.

The streams excavated shales, and weaker limestones now occupied the Lake Michigan and Lake Huron basins. An arc, composed of hard dolomitic

221

rock and known as the Niagara Escarpment, extends in a northwesterly direction from Niagara, forming the Bruce Peninsula that separates the east side of Lake Huron from Georgian Bay. The same structure continues across Michigan's upper peninsula, separating Lake Michigan from Green Bay as the Door Peninsula. The ancient stream channels were favored by the glaciers because they were at lower elevations and composed of more erodible bedrock. The linear shape of the lower Great Lakes is clearly related to the initial erosion by streams followed by the continental ice.

Lake Superior is partly located on the Canadian Shield, and its geologic origin is less obvious. East-west faults underlie Lake Superior, and the rocks form a structural sag, or syncline, oriented along the long axis of the lake in an approximately east-west direction.

The glacial origin and development of the Great Lakes is complex for several reasons. Each lake has a unique history and time of formation, making generalizations difficult. For example, the glaciers repeatedly advanced and retreated from many directions, covering and exposing each lake basin. There is abundant evidence regarding the size, elevation, and precise geographical distribution of each ancestral lake; this historical information is documented by coastal landforms such as higher ancient shorelines and relict wave-cut features. Yet the changing of the lakes' outlets to the ocean and reversals of drainage patterns complicate the sequence of events. Furthermore, because of the weight of the ice, the Paleozoic bedrock subsided to a lower level as the continental mass sought to "float" more deeply in the underlying mantle layer. As the glaciers receded, exposing different segments of the basin, the land began to recover its previous level and rise. This process, called isostatic rebound, is active today, causing elevation changes of many fossil shorelines. Although uplift has slowed since the ice exposed the newly created Great Lakes, the process is continuing.

As the ice began to retreat from the region, glacial landforms were deposited. Along many shorelines, moraines—composed of fragments of rock, sand, and silt—form spectacular bluffs. Along Lake Superior, the ice scraped and removed much of the soil, exposing the bedrock, which now forms high cliffs. Sand eroded from glacial sediment was transported by rivers to the lakes and deposited as beaches. The exposed beach sand was then transported inland by the wind to form coastal dunes.

## WEATHER AND CLIMATE

Weather and climate influence several processes occurring in the Great Lakes, including changing lake levels, storm surges and related seiches, and lake stratification and turnover. Through the hydrological cycle, moisture is evaporated from the lake surfaces and is then returned as precipitation over the water and as runoff from the land. During cooler and wetter years, evaporation is retarded, and more water is contributed to the lakes by excess precipitation, causing lake levels to rise. In warmer and drier years, evaporation increases, and precipitation is retarded, causing lower lake levels. Such changes in water levels are not cyclic and occur over several years. In 1988, Lake Huron and Lake Michigan had record levels of 177.4 meters above sea level. In 1995, the

*Lake Michigan.* (PhotoDisc)

average water level was 176.3 meters above sea level, a difference of 1.1 meters.

Unstable weather conditions generate storms that pass over the region, generally from west to east. When strong winds persist for several hours from a constant direction over a lake, the water level is "pushed" from one side of the lake to the other. This storm surge, accompanied by low atmospheric pressure, may elevate the water level as much as 2 meters along a shoreline in a matter of a few hours. Gale-force winds on October 30 and 31, 1996, over Lake Erie raised the lake level 1.25 meters at Buffalo, New York. Concurrently, as the water rose at Buffalo, it was lowered in Toledo, Ohio, at the opposite end of the lake, by 2.25 meters. The total difference of water level was 3.50 meters. Following a storm, the level of a lake rocks back and forth as a "seiche" before settling to its normal level.

In turn, the lake waters dramatically affect the local weather. As winter approaches, lake-effect snows commonly occur. The effect is most common in the fall, before the lakes cool and freeze. Cold winds from the north or west pass over the basin, picking up moisture from the relatively warm lakes. The water vapor is then condensed, forming clouds that, in turn, dump heavy snows in coastal zones, especially along eastern Lake Michigan and southern Lake Superior, Lake Huron, Lake Erie, and Lake Ontario. From November 9 to November 12, 1996, 1.2 meters of snow fell along Lake Erie's south shore, paralyzing local communities.

With the exception of Lake Erie, the lake bottoms were scoured by glaciers to depths below sea level. The lakes thus have variable temperatures from the surface to the bottom. As the water temperature changes from season to season, water density is altered. During winter, as ice forms over the lakes, the water beneath the ice remains warmer. As the ice cover breaks up in spring, the deeper warmer water rises, or "turns over," to the surface. It heats up through the summer months, causing stratification of warmer water (called the epilimnion) above colder, denser water. The contrasting water layers are separated by a thermocline, demarcating a rapid temperature transition between the warmer epilimnion and cooler subsurface water.

An issue of concern to scientists regarding the Great Lakes is the impact of the intensification of the greenhouse effect on the lakes' water levels. The increase of greenhouse gases in the atmosphere, especially carbon dioxide, appears to be causing warming at unprecedented rates. Although climatologists differ in their opinions as to the impact of a warmer atmosphere over the Great Lakes, there is general agreement that both evaporation and precipitation will increase and stream runoff will decrease over many years. Based on general circulation models, it appears that lake levels will be from 0.5 meter to 2 meters lower than present levels if the climate continues to warm.

## WETLANDS

Wetlands along the shorelines of the Great Lakes are significant ecological zones located at the meeting of land and lake. Although many wetlands around the basin have been lost or degraded, the remaining habitat has multidimensional functions as part of both upland and aquatic ecosystems. The wetlands are exposed to both short-term (storm surges and seiches) and long-term changes in water levels that constantly alter the biogeography of these habitats. Because of the state of constant water-level change, or "pulse stability," the distribution and types of wetland plants shift dramatically. Thus, a constant renewal of the flora is occurring. Furthermore, because of the flushing action of the rise and fall of lake levels, peat accumulation in Great Lakes wetlands does not commonly occur, as it does in marine settings.

The coastal wetlands serve significant ecological, economic, and social functions. They provide spawning habitat and nursery and resting areas for many species, including fishes, amphibians, reptiles, ducks, geese and other water birds, and mammals. Largely because of the sport fishing industry, these habitats contribute significant revenue to the surrounding states and to the province of Ontario. Furthermore, pollution control and coastal erosion protection are additional benefits provided by these habitats.

## STUDY OF THE GREAT LAKES

The creation and development of the Great Lakes have, in terms of geologic time, occurred relatively recently. Also, modifications such as erosion and deposition of coastal features are continual, active processes. To unravel the events leading to changes in the Great Lakes, scientists use techniques that include varve analysis and radiocarbon dating. A common theme of both techniques is that they express time in numbers of years within a reasonable range of accuracy rather than in a relative or comparative way. Varves consist of alternating light and dark sediment layers deposited in a lake. A light-colored mud is deposited during spring runoff; a dark-colored mud is deposited atop the lighter-colored layer during the following winter, as ice forms and there is less agitation of the lake water. One light and one dark band together represent one year of deposition. Numerous layers can be counted, like tree rings, and the number of years that were required for a sequence to be deposited can be determined. To obtain numerous undisturbed varve layers, researchers use a piston-coring device, which consists of a hollow pipe attached to a cable that is released vertically into the lake. As it falls freely to the bottom, it plunges into the soft sediment. A piston allows the sediment to remain in the pipe as it is raised by an attached cable. The mud can then be extracted from the tube, cut open, and analyzed.

Radiocarbon dating of carbon-rich material such as peat, lime, coral, and even bone material is useful for absolute dating back to about 50,000 years ago. Carbon's abundance in nature, coupled with the youthfulness of the Great Lakes, makes this tool very useful because many glacial, coastal, and sand dune landforms frequently contain some form of carbon suitable for absolute dating.

To map and detect recent changes in the landscape such as coastal erosion or the rate of dune migration, old maps, navigation charts, and aerial photographs are used. Charts and maps of the coastal zone have been available for more than a century, and aerial photographs of the region have been taken since the 1930's. By observing the position of a shoreline on historical sets of detailed aerial photographs over a ten-year period, for example, changes in the shoreline can be detected, and the erosion rate per year can be determined.

Geographic positioning systems can accurately locate the latitude and longitude of a point on a shoreline, store the information, and compare the shoreline position with the position at some future time. Satellite pictures help to detect wetland types and determine acreage; this information can then be compared to a later environmental condition, such as a period of higher lake level, to see if species habitat or acreage have changed.

Because of a geological process known as isostatic rebound, fossil shorelines become uplifted and exposed. Elevations retrieved from older topographic maps reveal how much uplift has occurred. If the age of a relict shoreline can be determined with radiocarbon analysis, the rate of glacial rebound in millimeters per century can then be assessed.

*C. Nicholas Raphael*

**FURTHER READING**
Annin, Peter. *The Great Lakes Water Wars*. Washington, D.C.: Island Press, 2006. Using the Aral Sea of central Asia as the example, discusses the dangerous long-term effects of short-sighted plans to divert water from the Great Lakes.

Bolsenga, S. J., and C. E. Herdendorf. *Lake Erie and Lake St. Clair Handbook*. Detroit, Mich.: Wayne State University Press, 1993. A thorough description of the natural history of Lakes St. Clair and Erie. Includes well-illustrated sections on geology, climate, hydrology, and wetland flora and fauna. Provides numerous maps, tables, and diagrams and a glossary.

Dennis, Jerry. *The Living Great Lakes*. New York: Thomas Dunne Books, 2003. Written for the layperson. Covers the geology, natural history, biology, and industry of the Great Lakes. Chapters are specific to regions or lakes. Discusses the structure of the lakes, from their formation to the current human impact of resource mining, construction of dams and canals, and introduction of invasive species.

Douglas, R. J. W. *Geology and Economic Minerals of Canada*. Report Number I. Ottawa: Geological Survey of Canada, 1970. Details the regional geology of Canada. Chapters on the Canadian Shield and the geology of the Great Lakes are included. Black-and-white aerial photographs illustrate significant glacial landforms. Intended as a reference work for professionals.

Grady, Wayne. *The Great Lakes: The Natural History of a Changing Region*. Vancouver: Greystone Books, 2007. Delivers an extended analytical look at the environment and ecology of the Great Lakes region as in a constant state of environmental flux. Discusses the human impact of invasive species on the ecology of the lakes. Includes further readings, a list of common and scientific names, illustration credits, and indexing.

Holman, J. Alan. *In Quest of Great Lakes Ice Age Vertebrates.* East Lansing: Michigan State University Press, 2001. Examines the fossil record of the Great Lakes and pieces together the biology and geology of the region during the Pleistocene epoch. Limits the use of technical jargon, making the book accessible to the general public.

Le Sueur, Meridel. *North Star Country.* 2d ed. Minneapolis: University of Minnesota Press, 1998. Provides a historical account of the environment, geology, and social aspects of the Great Lakes region. Covers ecology and preservation programs. Appropriate for the high school reader.

Spring, Barbara. *The Dynamic Great Lakes.* Baltimore: Independence Books, 2001. Discusses the history, development, and the many environmental and ecological problems facing the Great Lakes. Encourages appreciation of the Great Lakes ecosystem and the development of solutions to the problems being discussed.

U.S. Environmental Protection Agency. *The Great Lakes: An Environmental Atlas and Resource Book.* Chicago: Great Lakes Program Office, 1995. A free soft-cover publication designed for anyone interested in the lakes. Briefly addresses the physical and cultural environments of the basin. Includes colored maps, tables, and charts.

Ver Berkmoes, Ryan, Thomas Huhti, and Mark Lightbody. *Great Lakes.* Melbourne: Lonely Planet Publishing, 2000. A guide to the human environment of the American states bordering the Great Lakes, focusing on national parks, countryside, and urban attractions.

**See also:** Amazon River Basin; Ganges River; Lake Baikal; Mississippi River; Nile River; Yellow River

# GREENHOUSE EFFECT

*Gases in Earth's atmosphere trap infrared radiation emitted by the Earth, warming the surface in a process called the greenhouse effect. The greenhouse effect occurs naturally but is increasing in significance because of human activities, leading to probable detrimental effects involving climate change.*

## PRINCIPAL TERMS

- **anthropogenic:** influenced by humans; the anthropogenic greenhouse effect, also called the enhanced greenhouse effect, refers to the phenomenon as amplified by human activities
- **biochar:** another name for charcoal, used specifically for usages of charcoal that involve positive effects such as soil improvement or carbon sequestration
- **carbon sink:** a reservoir (either natural or human made) for storing or sequestering carbon
- **fossil fuels:** carbon-rich, energy-storing compounds that, when burned, release greenhouse gases into the atmosphere
- **global warming potential:** a measure of the ability of a greenhouse gas to contribute to global warming, based on its heat-trapping efficiency and its lifetime in the atmosphere
- **greenhouse gas:** a gas in Earth's atmosphere that can absorb and emit infrared radiation
- **near-infrared radiation:** the shortest-wavelength segment of infrared radiation, which has a longer wavelength than visible light; some solar radiation is in the near-infrared frequency range
- **negative emission:** a process that removes carbon dioxide from the atmosphere; technologies include biochar and bioenergy with carbon capture and storage
- **ozone:** a triatomic oxygen molecule and greenhouse gas; absorbs harmful ultraviolet rays from the sun before they reach Earth's surface
- **thermal infrared radiation:** the longest-wavelength segment of infrared radiation, which has a longer wavelength than visible light; the earth emits radiation in the thermal infrared frequency range

## OVERVIEW

The sun's radiation that reaches Earth's atmosphere is made up of ultraviolet, visible light, and near-infrared frequencies, all relatively short wavelengths. Nearly one-half of this radiation reaches Earth's surface; some of it is reflected but much is absorbed and then reemitted by Earth at longer wavelengths, in the thermal infrared range.

In a perfect system, a thermal equilibrium would be reached, but Earth's surface temperature is higher than expected because of a phenomenon known as the greenhouse effect. Certain gases in Earth's atmosphere absorb the energy emitted by the earth and then release that energy in all directions; some of the energy returns toward Earth's surface, and the cycle continues.

The nomenclature of the greenhouse effect is misleading. Whereas real greenhouses absorb sunlight to trap heat, the physical walls of the greenhouse serve to prevent convection, keeping the warm air inside the structure. The environmental greenhouse effect relies on a different process: the absorption and emission of thermal energy. The earth is not the only planetary body to experience the greenhouse effect; Mars, Venus, and Saturn's largest moon, Titan, also show evidence of the same effect.

In 1824, French mathematician and physicist Jean Baptiste Joseph Fourier realized that based on Earth's size and its proximity to the sun, Earth should be colder if its only heat were coming from the sun. Fourier published several articles exploring other potential sources of Earth's extra heat, and one theory involved Earth's atmosphere acting as an insulator. Generally this marks the first time any person described what came to be called the greenhouse effect. In 1863, British physicist John Tyndall revealed experimental evidence proving the existence of the greenhouse effect, particularly regarding the strong infrared-radiation absorption capabilities of water vapor but also the less but still significant absorptions of other atmospheric gases, such as carbon dioxide, ozone, and nitrous oxide.

By the turn of the twentieth century, Swedish physicist and chemist Svante Arrhenius was building on Fourier's and Tyndall's work, focusing specifically on the absorption values for carbon dioxide. Arrhenius predicted that the growing emissions of fossil fuels would ultimately bring about climate change, but he saw this phenomenon in a positive light: He thought it meant the end of ice ages.

If not for the greenhouse effect, Earth's surface temperature would be below the freezing point of water, and life as known would not exist. Whereas earlier researchers such as Arrhenius expected that human influence would push the greenhouse effect beyond natural expectations, but with positive results, many researchers now believe that humans are affecting the system too much, fearing that climate change is heading in a direction detrimental to human life.

## OVERVIEW OF GREENHOUSE GASES

The greenhouse effect depends on the presence of certain gases in the atmosphere that allow the sun's short-wavelength radiation to pass by while absorbing and rereleasing the earth's long-wavelength infrared radiation. These gases are called greenhouse gases.

Water vapor is the greenhouse gas that contributes most, up to 70 percent, to the overall greenhouse effect. On a large scale, water vapor's atmospheric abundance is not significantly affected by humans; it is mainly affected by the climate. While Earth's atmosphere might be nearly 20 percent water vapor in hot and humid regions, in cool and dry regions it might comprise well below 1 percent water vapor. Because its abundance is directly related to heat, water vapor actually contributes positively to the greenhouse effect. As the effect causes the temperature to rise in a particular region, more water vapor exists in the air, which leads to an increase in the greenhouse effect.

Carbon dioxide accounts for about one-fourth of the greenhouse effect, but it is the level of carbon dioxide that is most influenced by human activities. This means that carbon dioxide's increasing atmospheric abundance plays a major role in the future of climate change.

Another greenhouse gas, methane, would actually have a much stronger impact on the greenhouse effect than carbon dioxide if it were present in the same abundance; however, much less methane exists in the atmosphere. Methane is directly responsible for only about 10 percent of the greenhouse effect. It also has an indirect effect, though, because it contributes to the formation of ozone, another greenhouse gas.

Ozone is not plentiful anywhere in the atmosphere, but the largest percentage of it can be found in an area dubbed the ozone layer, which is located 20 to 30 kilometers (12 to 19 miles) above Earth, in the lower stratosphere. The ozone layer absorbs almost all of the sun's ultraviolet radiation, some of which would be harmful were it to reach life-forms, including humans, on Earth's surface. It is important to note that while some media reports group ozone depletion with the issue of greenhouse gases and climate change, the phenomena are quite distinct.

The other major greenhouse gases are nitrous oxide, hydrofluorocarbons, perfluorocarbons, and sulfur hexafluoride. Most of these gases can stay in the atmosphere for a long time; carbon dioxide, for example, generally stays in the atmosphere in the range of three to nine decades. Water vapor, however, lasts only about nine days.

Clouds also are important contributors to the greenhouse effect, although they are not technically a gas. They can absorb and reflect infrared radiation, but they also reflect some solar radiation before it even reaches Earth's surface.

Some of the most abundant atmospheric gases do not significantly contribute to the greenhouse effect. Nitrogen and oxygen, for example, are the two most abundant gases in Earth's atmosphere (78 and 21 percent abundance, respectively), but their diatomic molecular structure is not affected by infrared radiation.

## GREENHOUSE GAS SOURCES AND SINKS

Most greenhouse gases have both natural and human-made sources, especially because the Industrial Revolution (1750–1850) brought about the rise of modern machinery and industrial technology. Similarly, greenhouse gases can be removed from the atmosphere by natural or human-made storage reservoirs known as sinks (or by other processes).

As discussed, the atmospheric abundance of water vapor, unlike other greenhouse gases, is not particularly influenced by humans (except on a small, local scale in agriculture). Water vapor is a product of the water cycle, and it is more abundant in warmer climates.

Carbon dioxide, however, is a product of a variety of human-made sources, particularly the burning of fossil fuels, which are carbon-rich compounds that store energy. Around one-third of fossil fuels being burned are liquid, such as petroleum (and the gasoline derived from it), and about another one-third are solid fuels, such as coal. Since the Industrial Revolution, the burning of both petroleum and coal has been on the rise, the former for transportation and the latter for electricity. The remaining one-third consists of gaseous fuels (such as natural gas), cement production, and several other sources. Deforestation also is a major source of carbon dioxide emissions; as trees are removed on a large scale and not replaced, less carbon dioxide is pulled from the atmosphere for photosynthesis.

Modern agriculture is another major source of greenhouse gases, particularly methane, which is produced by manure management techniques and by enteric fermentation (perhaps better known as cattle flatulence). Additionally, fertilizers are a major source of nitrous oxide.

The natural removal of greenhouse gases from the atmosphere occurs by physical and chemical changes and exchanges. Physical removal can involve either a state change (water vapor condensing or evaporating, for example) or an exchange with a different part of Earth (such as gases being physically mixed into the ocean). Similarly, chemical removal can involve reactions within the atmosphere itself or reactions with something else (removal by plants for photosynthesis, for example).

Increasingly, as humans become more aware of the potentially detrimental effects of civilization—industry in particular—on the greenhouse effect, researchers are finding technological ways to remove greenhouse gases from the atmosphere or to prevent them from reaching the atmosphere in the first place. One such technology, called carbon capture and storage, predictably collects carbon from fossil-fuel power plants and stores it safely, generally underground within large geologic formations. Another technique involves the use of charcoal to sequester carbon in the soil for hundreds or even thousands of years. When charcoal is used for this purpose, it is referred to as biochar.

## THE GREENHOUSE EFFECT AND CLIMATE CHANGE

Humankind's impact on the greenhouse effect is referred to by researchers as the enhanced or anthropogenic greenhouse effect. As discussed, the main human contribution is rising carbon dioxide levels, particularly from the burning of fossil fuels and from deforestation.

Arrhenius and other researchers in his time predicted that the enhanced greenhouse effect would protect humans from future ice ages, but the numbers give modern researchers some cause for alarm

as they try to determine to what extent human actions may be permanently changing Earth's climate for the worse.

In a general sense, the phrase *climate change* refers to significant and lasting weather changes through long periods of time from natural or human-made reasons. *Global warming* technically pertains specifically to increases in Earth's surface temperature. In modern vernacular, though, many nonscientists use *climate change* and *global warming* nearly synonymously.

While some disagreement exists regarding the interpretation of data, most major scientific bodies and industrialized nations recognize that global temperatures have been steadily increasing beyond normal fluctuations since the late nineteenth century and especially since the late twentieth century, and most scientists point at the anthropogenic greenhouse effect as the main reason for this rise in temperatures. As greenhouse gas sources have drastically increased since the Industrial Revolution, the sinks have not been able to keep up. Research by the National Oceanic and Atmospheric Administration shows that carbon dioxide levels have risen 36 percent since the start of the Industrial Revolution and that methane levels have risen 148 percent.

When analyzing the effects of different greenhouse gases on climate change, it is useful to understand the global warming potential (GWP) measure, which takes into account both the efficiency of a greenhouse gas (how much heat it can trap) and its atmospheric lifetime. Carbon dioxide's GWP is standardized to a value of 1, and all other greenhouse gases are calculated relative to carbon, relative to a certain time frame.

If climate change continues in the current pattern, it is expected that sea levels will rise worldwide, resulting in significant changes in precipitation patterns and, ultimately, to desert expansions (desertification) and other habitat changes. Glacial melting and heat waves are also expected, and these changes will likely cause the extinction of various species. These are only a few consequences that demonstrate the growing need for continued research of greenhouse gases and climate change.

*Rachel Leah Blumenthal*

### FURTHER READING

Archer, David. *Global Warming: Understanding the Forecast.* Malden, Mass.: Blackwell, 2007. This text is an accessible yet comprehensive introduction to climate change, covering topics such as the greenhouse effect, blackbody radiation, and the carbon cycle. It will be a comfortable but informative read for those who are new to the topic.

Black, Brian, and Gary J. Weisel. *Global Warming*. Santa Barbara, Calif.: Greenwood Press, 2010. Part of the Historical Guides to Controversial Issues in America series, this guide uses historical context and the concepts of geology, Earth science, and climate science to provide an overview of global warming.

Dow, Kirstin, and Thomas E. Downing. *The Atlas of Climate Change: Mapping the World's Greatest Challenge*. Berkeley: University of California Press, 2006. Introduces climate change with an overview of terms and concepts before moving into causes, effects, and possible solutions. Particular focus on the greenhouse effect and climate systems.

Gore, Albert, and Melcher Media. *An Inconvenient Truth: The Planetary Emergency of Global Warming and What We Can Do About It*. New York: Rodale, 2006. Now an acclaimed documentary, this book by former U.S. vice president Al Gore uses a mixture of scientific research and storytelling to illuminate the dangers of global warming.

Rackley, Stephen A. *Carbon Capture and Storage*. Burlington, Mass.: Butterworth-Heinemann/Elsevier, 2010. An overview of technology available for carbon capture and storage, which is meant to reduce greenhouse gas emissions from industrial processes.

Stille, Darlene R. *The Greenhouse Effect: Warming the Planet*. Minneapolis: Compass Point Books, 2006. A basic overview of the greenhouse effect, covering topics such as the difference between weather and climate, the runaway greenhouse effect, and effects of global warming.

**See also:** *AR4 Synthesis Report*; Atmosphere's Evolution; Atmosphere's Global Circulation; Atmosphere's Structure and Thermodynamics; Atmospheric and Oceanic Oscillations; Atmospheric Properties; Barometric Pressure; Bleaching of Coral Reefs; Carbon-Oxygen Cycle; Climate; Climate Change Theories; Clouds; Cosmic Rays and Background Radiation; Drought; Earth-Sun Relations; Geochemical Cycles; Global Energy Transfer; Global Warming; Hydrologic Cycle; Ice Ages and Glaciations; Icebergs and Antarctica; Long-Term Weather Patterns; Ocean-Atmosphere Interactions; Ozone Depletion and Ozone Holes; Recent Climate Change Research; Remote Sensing of the Atmosphere; Satellite Meteorology; Sea Level; Water and Ice

# GROUNDWATER MOVEMENT

*The flow of water through the subsurface, known as groundwater movement, obeys well-established principles that allow hydrologists to predict flow directions and rates.*

## PRINCIPAL TERMS

- **elevation head:** the elevation of a given water particle above a certain point, usually mean sea level
- **equipotential line:** a contour line connecting points of equal hydraulic head
- **groundwater:** water found in the zone of saturation
- **hydraulic head:** the sum of the elevation head and the pressure head at any given point in the subsurface
- **hydrostatic pressure:** the pressure at any given point in a body of water at rest from the weight of the overlying water column
- **permeability:** the ability of rock, soil, or sediment to transmit a fluid (commonly water)
- **porosity:** the ratio, usually expressed as a percentage, of the total volume of void (empty) space in a given geologic material to the total volume of that material
- **pressure head:** the height of a column of water that can be supported by the hydrostatic pressure at any given point in the subsurface
- **vadose zone:** the region of soil between the surface and the water table in which void spaces contain both air and water
- **velocity head:** the height to which the kinetic energy of fluid motion is capable of lifting that fluid
- **water table:** the upper surface of the zone of saturation
- **zone of saturation:** a subsurface zone in which all void spaces are filled with water

## POROSITY

The movement of water through the earth's subsurface is only one part of a larger circulation system known as the hydrologic cycle. This cycle involves the continuous transfer of water between natural reservoirs within the physical environment, such as the oceans, polar ice caps, groundwater, surface water, and the atmosphere. The main processes in the hydrologic cycle are precipitation, evaporation, transpiration by plants, surface-water runoff, and subsurface groundwater flow. When precipitation falls to the land surface as rain, snow, or other form, some of this water runs off into streams, some evaporates back into the atmosphere, and the remainder soaks into the ground. The water that infiltrates the land surface either is taken up and transpired by plants, or percolates deeper to eventually become groundwater. For water to percolate into the subsurface, interconnected void spaces must be available between particles in the underlying geologic materials. The ratio of void space to the total rock or soil volume is known as porosity. This proportion is usually expressed as a percentage. The higher this ratio, the more void space there is to hold water.

There are various types of porosity. Unconsolidated materials such as soil and sediment have pore spaces between adjacent grains, referred to as intergranular porosity. The ratio of pore space to total volume depends on several factors, including particle shape, sorting, and packing. Loosely packed sediments composed of well-sorted, spherical grains are the most porous. Porosity decreases as the angularity of the grains increases because the particles pack more closely together and the protuberances on the particles due to their irregular shape fill in more of the void space. Similarly, as the degree of sorting decreases, the pore spaces between larger grains become filled with smaller grains, and porosity decreases. Values of porosity for unconsolidated materials range from 10 percent for unsorted mixtures of sand, silt, and gravel to about 60 percent for some clay deposits. Typical porosity values for uniform sands are between 30 percent and 40 percent.

Rocks have two main types of porosity: pore spaces between adjacent mineral grains and voids that are a result of fractures. Rocks formed from sedimentary deposits (such as shale and sandstone) may have significant intergranular porosity, but it is usually less than the porosity of the sediments from which they were derived. This dichotomy is a result of the compaction and cementation that takes place during the process of transforming sediments into rock. Therefore, although sandstone porosities may be as high as 40 percent, they are commonly closer to 20 percent because of the presence of natural mineral cements that partially fill available pore spaces. Igneous and metamorphic rocks are composed of tightly interlocked mineral grains and therefore have little intergranular porosity. Virtually all void space in such rocks is a result of fractures (joints and faults). For example, granite (a dense, igneous rock) usually has a porosity of less than 1 percent, but porosity may reach 10 percent if the rock is fractured.

There are additional types of porosity that occur only in certain kinds of rocks. Limestone, a rock that is soluble in water, especially if the water is acidic, can develop solution conduits, or channels, along fractures and bedding planes. Given enough time, solution weathering may lead to the development of a cave, which has 100 percent porosity. The overall porosity of solution-weathered limestone sometimes reaches 50 percent. Rocks created by volcanic eruptions may contain void space in the forms of vesicles (cavities left by gases escaping from lava), vertical shrinkage cracks developed during cooling (known as columnar joints), and tunnels created by flowing lava (called lava tubes). In extreme cases, the porosity of volcanic rocks may exceed 80 percent.

Underground water includes all water that exists below the land surface, but the subject of groundwater movement is mainly concerned with the water that occurs in the zone of saturation, where all empty spaces are completely filled with water. Between the zone of saturation and the land surface, void spaces contain mostly air, unless a heavy rainfall or a period of snowmelt has just occurred. Water in this upper zone is held under tension by attractive forces between soil particles and water molecules (surface tension forces and molecular polarity effects).

The water table forms the uppermost surface of the zone of saturation and is characterized by having a water pressure equal to atmospheric pressure. It ranges in depth from being very close to the land surface in humid regions to hundreds of meters below the surface in desert environments. In general, the water table mimics the surface topography but with more subdued slopes. When the water table intersects with the surface topography, a surface water feature such as a lake, swamp, river, or spring results. Below the water table, in the zone of saturation, geologic materials are completely saturated, and water pressure (called hydrostatic pressure) increases with depth in the same manner that pressure increases with depth in an open body of water. Water

contained within the zone of saturation is generally called groundwater. It is this zone that supplies water to a well when it occurs in a particular type of geologic formation known as an aquifer.

## GROUNDWATER ENERGY

To understand groundwater flow, it is necessary to examine the forms of energy contained in groundwater. The total energy in any water mass consists of three components: elevation head, pressure head, and velocity head. The elevation head represents the potential energy of the water due to its elevation above mean sea level. The pressure head represents the potential energy of the water due to the hydrostatic pressure of the surrounding fluids. The velocity head corresponds to the kinetic energy of the water resulting from its physical movement. Because groundwater moves relatively slowly, velocity head can usually be neglected, leaving the total energy essentially equal to the sum of the elevation head and the pressure head. This quantity is known as the hydraulic head. Thus the hydraulic head of a given water particle varies directly with its elevation (usually expressed in meters above mean sea level) and its hydrostatic pressure.

The water table can be thought of as a surface with variable hydraulic head. Because water pressure at the water table is always known, since it is by definition equal to atmospheric pressure, the change in hydraulic head across the water table is dependent only upon the variation in elevation. Below the water table, hydraulic head depends on both the elevation and the water pressure, which increases with depth. Therefore, the variation in hydraulic head with depth below the water table reflects the relationship between decreasing elevation head and increasing pressure head. If the increase in hydrostatic pressure exactly offsets the decrease in elevation head, then hydraulic head will not change with depth.

Groundwater moves in response to differences in hydraulic head between two locations. The direction of movement is always from areas of higher hydraulic head toward areas of lower hydraulic head. The change in hydraulic head over a specified distance is known as the hydraulic gradient. Both horizontal and vertical hydraulic gradients can exist.

Horizontal hydraulic gradients are usually defined by the change in water table elevation between any two locations. As water moves through the subsurface, it flows along the steepest hydraulic gradient. Therefore, it is possible to determine the compass direction of groundwater movement from a knowledge of how the water table elevation varies over distance. Because the water table often mimics the land surface, general groundwater flow directions can sometimes be predicted on the basis of surface topography.

Vertical hydraulic gradients describe changes in hydraulic head with depth. As groundwater flows in any given horizontal direction, it may also be rising or sinking, depending on the vertical hydraulic gradient. In locations where hydraulic head decreases with depth below the water table, groundwater flow has a downward component, resulting in recharge areas. Where hydraulic head increases with depth, groundwater flow has an upward component, creating a discharge area. Recharge areas commonly occur in the higher elevations of a particular landscape, and discharge areas usually occur in the valleys near lakes, streams, and swamps. This year-round flow of groundwater from higher to lower elevations permits streams to flow in the dry summer months when there is little surface runoff. In certain situations, the water pressure conditions can cause groundwater to move "uphill" with respect to the surface topography. Therefore, the land surface is not always a good indicator of groundwater flow directions.

Groundwater movement can be divided into local and regional flow. In areas of rugged topography, most groundwater flow is local, meaning that it moves from the hilltops to the nearest stream or lake. In more gentle terrains, however, or in areas where the zone of groundwater movement is very thick, some flow escapes the local system into a deeper, regional system. Thus, water may enter the subsurface at a local zone of recharge and move long distances before surfacing at a regional discharge area. Identifying the boundaries of local and regional flow systems requires detailed information about the horizontal and vertical distributions of hydraulic head over a large area.

### RATE OF FLOW AND PERMEABILITY

The rate at which groundwater moves through the subsurface can also be determined on the basis of scientific principles. Groundwater flow velocities depend on two factors: the hydraulic gradient and the permeability of the geologic materials involved. Considering that differences in hydraulic head are the driving force behind groundwater movement, it is a logical consequence that the hydraulic gradient is related to the rate of flow. All other things being equal, the steeper the gradient, the faster groundwater will move. Hydraulic gradients may change throughout

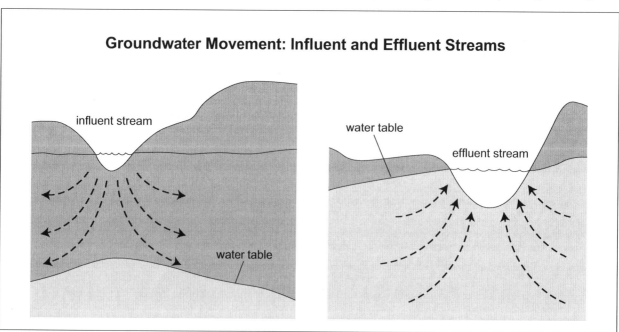

**Groundwater Movement: Influent and Effluent Streams**

influent stream

water table

water table

effluent stream

the year, reflecting the influences of recharge rates. In the spring, when recharge is high, the water table will rise fastest beneath recharge areas, producing a steeper hydraulic gradient and higher flow velocities. As the water table drops throughout the summer, the hydraulic gradient also declines, leading to slower groundwater flow.

Permeability is defined as the ability of porous formations to transmit fluids, and is a property of the geologic material in question. This property depends on both the size of void spaces and the degree to which they are interconnected. Thus, some high-porosity materials such as clay (up to 60 percent porosity) and pumice (a vesicular volcanic rock with up to 87 percent porosity) have very low permeability because the void spaces within them are largely isolated from one another. Materials that have high permeability include sand, gravel, sandstone, and solution-weathered limestone. Rocks with low porosities, such as shale, quartzite, granite, and other dense, crystalline rocks, typically also have low permeabilities, unless they are significantly fractured.

Groundwater is forced to move along tortuous paths through geologic materials as it follows the connecting spaces between voids. Therefore, even in highly permeable materials, groundwater flows much more slowly than does surface water in a river. Whereas the velocity of stream flow may be measured in meters per second, groundwater velocities commonly range from 1 meter per day to less than 1 meter per year (averaging about 17 meters per year in rocks). The highest velocities occur in rocks that are heavily fractured. In the extreme case, groundwater can actually move as a subsurface stream through cavernous limestone or volcanic rock. This situation is not normal, however, and underground streams are much less common than the average person might suspect.

## STUDY OF GROUNDWATER MOVEMENT

The first step in studying groundwater movement in a particular area is to determine the hydraulic gradients that exist. Horizontal gradients are defined by measuring groundwater elevations in wells that intersect the water table. To define vertical gradients that may be present, piezometers are used. Piezometers are essentially vertical pipes that are open at the bottom to allow water to enter. These devices enable the hydraulic head at a particular depth below the water table to be determined. The elevation head is the elevation at the bottom of the piezometer, and the pressure head is the height to which water rises in the piezometer above the intake point.

Contour maps of the variation in hydraulic head are prepared from data collected from a number of wells and piezometers. These maps can show either horizontal variations (such as a map of the water table) or vertical changes in hydraulic head (as shown in a cross-sectional view). The contour lines used on these maps to connect points of equal hydraulic head are called equipotential lines. Groundwater flow directions can be determined once the distribution of known hydraulic head values has been contoured. Flow lines, depicting the idealized paths taken by water particles, are drawn to intersect equipotential lines at right angles, indicating that groundwater moves along the steepest hydraulic gradient. The resulting gridlike pattern of equipotential and flow lines (called a flow net) is a two-dimensional representation of the groundwater flow system.

Flow nets are much better indicators of groundwater flow directions than the land surface topography. Also, from a study of flow nets, it may be possible to determine accurately recharge and discharge areas. In the map view, flow lines will diverge from areas of recharge and converge in discharge areas. In the cross-sectional view, flow lines will have downward components where recharge is occurring and upward components in discharge areas. If hydraulic head does not change with depth, flow lines will be horizontal, indicating that neither recharge nor discharge conditions exist. Flow nets do not give estimations of flow velocities unless permeability values are known for the materials involved. If an estimation of permeability can be obtained, groundwater flow velocity may be calculated by multiplying the permeability value and the hydraulic gradient and dividing this product by the porosity of the geologic formation. This calculation yields an average linear velocity of groundwater flow through the open area provided by void space.

Groundwater flow directions and velocities can also be determined by introducing a "tracer" into the groundwater and monitoring its migration through observation wells. This technique is especially helpful in areas of fractured rock, in which flow patterns are difficult to predict. Tracers are easily detectable substances that will dissolve readily and move with the groundwater without reacting with the geologic materials. Ideally, tracers are safe to use, inexpensive, and easy to detect in low concentrations. Examples of tracers used in groundwater studies include salts (sodium or potassium chloride), fluorescent dyes, and the radioactive isotopes of certain elements (helium, hydrogen, and iodine). The choice of tracer usually depends on the subsurface materials that it must pass through.

In areas where the hydrogeologic conditions are well defined through wells, piezometers, borings, and permeability tests, the groundwater flow system can be studied with the use of computer models. Computers are used to calculate hydraulic head values for the area modeled given the rates of recharge (infiltration) and discharge (by evapotranspiration or discharge to a stream, lake, or well). These computed head values can then be contoured to create a flow net. Although computers are powerful tools, the accuracy of their predictions cannot surpass the accuracy of the information provided for the model.

## SIGNIFICANCE

The understanding of groundwater movement is important to the utilization and conservation of aquifers for water supply. From a quantity point of view, flow rates determine which geologic materials will serve as suitable sources of groundwater. The rates of flow must equal or exceed the desired pumping rate for a well to be successful. In this regard, permeability is the limiting factor, because the drawdown (depletion of water) from a well due to pumping will always create a sufficient hydraulic gradient to favor flow toward the well. If the permeability is too low to support the pumping rate, the geologic formation will become dewatered. For domestic wells requiring only 7 to 19 liters per minute, even moderately impermeable materials may supply sufficient water to be considered as an aquifer. High-yield wells (exceeding 190 liters per minute), however, can be sustained only in very permeable materials, such as sand and gravel, sandstone, and solution-weathered limestone.

Understanding groundwater flow is also important to water quality. As groundwaters move through the subsurface, they dissolve minerals from the geologic materials with which they have contact. Therefore, groundwater that has traveled great distances in a

regional flow system will tend to be the most mineralized and may be the least desirable for a drinking water source. With an increasing number of possible contamination sources, such as landfills, corrupted underground tanks, and accidental spills, the protection of groundwater supplies is crucial. A knowledge of groundwater flow directions and rates aids in the prediction of contaminant migration. It is particularly important to identify areas that recharge regional flow systems because these zones have the greatest potential impact if they become polluted. Municipalities concerned with the preservation of their well fields should undertake wellhead protection studies. The purpose of such studies is to delineate the well field recharge areas that need to be protected from any type of land use that could lead to groundwater contamination. Contamination of a drinking water source can have serious effects. An incident in Walkerton, Ontario, Canada, in 2000, in which surface runoff contaminated municipal water wells with the *E. coli* bacteria, resulted in the deaths of seven people and serious illness in another 2,300 individuals.

Contaminant migration is difficult to predict in fractured rock formations, especially in solution-weathered limestone. Because groundwater movement follows a sometimes random network of discontinuous openings in these settings, water levels may not be related from one well to the next. Therefore, maps of the water table for such areas are often impossible to construct. The flow of water through fractured rock is one of the concerns surrounding the choice of a repository for high-level nuclear wastes, which need to be isolated from the environment for at least ten thousand years.

*David L. Ozsvath*

## FURTHER READING

Ahmed, S., R. Jayakumar, and Abdin Saleh. *Groundwater Dynamics in Hard Rock Aquifers: Sustainable Management and Optimal Monitoring Network Design.* Dordrecht: Springer, 2008. Presents the findings of pilot research on the structure and functioning of an aquifer in a granitic formation in order to characterize the system and its properties with regard to geophysical, geological, and remote sensing applications.

Appelo, C. A. J., and Dieke Postma. *Geochemistry, Groundwater and Pollution.* 2d ed. Boca Raton, Fla.: CRC Press, 2005. Provides a complete and authoritative account of modern hydrogeochemistry. Uses numerous examples and modeling templates to elucidate the physical and chemical principles involved, including the use of PHREEQC code for calculating and simulating field and laboratory data.

Batu, Vedat. *Applied Flow and Solute Transport Modeling in Aquifers.* Boca Raton, Fla.: CRC Press, 2006. A highly technical text discussing the dynamics of aquifer water flow, flow conditions, solute transport, and sorption principles. Covers aquifer modeling fundamentals, both numerical and analytical approaches, in detail. Contains a list of references and indexing. Best suited for aquifer researchers and advanced graduate students with a background in fluid dynamics or groundwater analysis.

Fetter, C. W. *Applied Hydrogeology.* 4th ed. Upper Saddle River, N.J.: Prentice Hall, 2000. Emphasizes the practical aspects of understanding groundwater occurrence and movement. Contains a detailed discussion of the influence geologic conditions have on groundwater occurrence, with special emphasis on groundwater regions in the United States. Addresses the regional groundwater movement within aquifers and contains helpful illustrations of the principles discussed. Includes a glossary of important terms. Suitable for college-level readers.

_____. *Contaminant Hydrogeology.* 2d ed. Upper Saddle River, N.J.: Prentice Hall, 1999. Emphasizes the conditions that lead to the contamination of groundwater, the problems this contamination can cause, and remediation techniques used in attempts to correct these problems. Includes a detailed discussion of the influence that geologic conditions have on groundwater occurrence and contamination. Also addresses the regional groundwater movement within aquifers and contains helpful illustrations of the principles discussed. Includes a glossary of important terms.

Guymon, Garry L. *Unsaturated Zone Hydrology.* Upper Saddle River, N.J.: Prentice-Hall, 1994. Uses a quantitative modeling approach, and combines research and knowledge from a number of disciplines to provide a comprehensive exploration of the movement of water in the unsaturated zone with a focus on the movement of water through the unsaturated zone into the saturated zone and aquifer structures.

Hamblin, Kenneth W., and Eric H. Christiansen. *Earth's Dynamic Systems.* 10th ed. Upper Saddle River, N.J.: Prentice Hall, 2003. Has a good discussion of groundwater movement in Chapter 13, although not limited to the topic of groundwater. The color figures are especially helpful to one unfamiliar with subsurface geology. Suitable for both high school and college-level readers.

Healy, Richard W. *Estimating Groundwater Recharge.* New York: Cambridge University Press, 2010. Discusses many topics in groundwater management and hydrology. Contains practical information, suited for working hydrologists. Discusses management, modeling, and data collection methods. The main focus is groundwater recharging.

Ingebritsen, S. E., Ward E. Sanford, and C. E. Neuzil. *Groundwater in Geologic Processes.* 2d ed. New York: Cambridge University Press, 2006. Widely used for study in the field of groundwater. Suitable for college-level study and professional reference use.

Mahajan, Gautam. *Evaluation and Development of Ground Water.* New Delhi: S. B. Nangia, 2008. While specifically focused on the Indian subcontinent, the underlying principles are nonetheless relevant to North America.

Mays, Larry W. *Ground and Surface Water Hydrology.* Hoboken, N.J.: John Wiley & Sons, 2011. Stresses the fundamental principles of hydrologic processes of both surface and groundwater hydrology, making significant use of online resources.

Montgomery, C. W. *Physical Geology.* 3d ed. Dubuque, Iowa: Wm. C. Brown, 1993. Introductory geology textbook with a good discussion of groundwater movement. Includes full-color figures. Emphasizes the environmental aspects of water supply and pollution. Suitable for high school and college-level readers.

Moore, John E. *Field Hydrogeology.* 2d ed. Boca Raton, Fla.: CRC Press, 2012. Provides practical information on field investigation, aquifer testing, and groundwater quality testing. Discusses planning and report writing in detail. Provides case studies from around the world. Best suited for students learning the planning and investigation of field hydrology.

Nonner, Johannes C. *Introduction to Hydrogeology.* London: Taylor & Francis Group, 2006. Covers all major fields in hydrogeology, as well as the occurrence, behavior, and properties of groundwater.

Well illustrated. Makes use of numerous examples of groundwater cases from around the world. University level.

Randolph, John. *Environmental Land Use Planning and Management.* Washington, D.C.: Island Press, 2004. Describes basic principles and strategies of land-use planning and management. More specific chapters discuss various land features, types, and environmental issues, such as soils, wetlands, forests, groundwater, biodiversity, and runoff pollution. Provides case studies and specific examples.

Saether, Ola M., and Patrice de Caritat, eds. *Geochemical Processes, Weathering, and Groundwater Recharge in Catchments.* Rotterdam: A. A. Balkema, 1997. Looks at the geochemical cycles and processes associated with catchment basins. Discusses the artificial recharge of groundwater in such systems. Slightly technical.

Thompson, Stephen A. *Hydrology for Water Management.* Rotterdam: A. A. Balkema, 1999. Discusses hydrology, groundwater flow, and stream flow. Focuses on the management of water supplies in efforts to keep them free of pollutants and available to the largest number of people possible. Illustrations, maps, index, and bibliography.

Todd, D. K. *Groundwater Hydrology.* 3d ed. New York: John Wiley & Sons, 2004. Emphasizes the practical aspects of groundwater occurrence and movement. Introduces the reader to groundwater utilization and its relationship to the hydrologic cycle. Discusses aquifer types and the occurrence of groundwater within the United States. Suitable for college-level readers.

**See also:** Alluvial Systems; Aquifers; Artificial Recharge; Dams and Flood Control; Floodplains; Floods; Freshwater and Groundwater Contamination Around the World; Groundwater Pollution and Remediation; Hydrologic Cycle; Precipitation; River Flow; Salinity and Desalination; Saltwater Intrusion; Sediment Transport and Deposition; Surface Water; Waterfalls; Water Quality; Watersheds; Water Table; Water Wells

# GROUNDWATER POLLUTION AND REMEDIATION

*Groundwater pollution is any artificially induced change in the natural quality of the water. Groundwater pollution remediation is concerned with the preservation of the resource's beneficial use, which is frequently potable water supply. Groundwater contaminants are usually removed through the use of "pump-and-treat" technologies. These technologies involve initial groundwater extraction, followed by treatment at the surface prior to reinjection or consumption.*

## PRINCIPAL TERMS

- **air stripping:** the process of passing contaminated water through an aeration chamber, causing the organic contaminants to volatilize into the gaseous waste stream
- **aquifer:** a saturated underground rock or sediment formation from which water is typically withdrawn for use
- **carbon adsorption:** the process of pumping contaminated water directly through carbon filters, to capture contaminants by adsorptive binding to the surface of the carbon particles
- **contaminant:** any natural or unnatural component that is introduced into the environment in concentrations greater than those normally present
- **flocculation:** a slow process by which suspended particles are gathered together to form larger particles that can then be removed by physical means
- **ion exchange:** the reversible switching of ions between the water being treated and an ion exchange resin by which undesirable ions in the water are exchanged with acceptable ions on the resin
- **leachate:** contaminated fluid produced by the passage of water through decaying garbage in a landfill
- **nonpoint source:** a large, diffuse source of contamination
- **osmosis:** the passage of a solvent, usually water, from a solution of relatively low solute concentration into a solution of higher solute concentration across a semipermeable membrane
- **point source:** a single, defined source of contamination
- **precipitation:** the chemical conversion of a dissolved soluble material into an insoluble one that subsequently settles out of the solution

## GROUNDWATER CONTAMINATION

Groundwater contamination may occur as a result of either natural or human causes. When human activities cause degradation of natural water quality, however, the term "pollution" is used. There are numerous statutes related to groundwater quality in the United States. One of the most important is the Safe Drinking Water Act of 1974. As a result of this legislation the U.S. Environmental Protection Agency (EPA) issued primary drinking water regulations in 1975, which have since been modified as new information regarding environmental exposure became available. The environmental standards in the regulations list maximum permissible concentrations, based on health criteria, of a number of chemicals, both organic and inorganic. The regulations apply to all public water systems; however, individual groundwater supplies would ideally be of the same quality. Although the concentration limits for chemical contaminants are very low, theoretically synthetic organic chemicals should not be present at all. Other chemical components, such as chloride, copper, iron, zinc, and manganese, have recommended maximum levels based on aesthetic or taste criteria.

Much groundwater pollution is the result of microbiological contamination. The permissible limits for an untreated water supply are 100 colonies of total coliforms (bacteria) per 100 milliliters and 20 colonies of fecal coliforms per 100 milliliters. Disease-causing organisms such as *E. coli* bacteria typically come from animal wastes. Contaminated water may cause typhoid fever, cholera, bacillary dysentery, paratyphoid fever, and even polio and infectious hepatitis. Between 1971 and 1977, there were 192 outbreaks of water-borne diseases in the United States and vastly more in developing countries. More stringent regulatory controls on the use and conservation of groundwater resources in North America since that time have drastically reduced the number of water-borne diseases, although they have not eliminated the problems that can arise due to human error or negligence.

## POINT SOURCES

Groundwater pollution sources are often identified by the designations "point" and "nonpoint." Point sources of groundwater contamination are those that can be identified as a single isolated location from which the polluting material is emanating. Nonpoint sources are those from which a polluting material emanates over a broadly delineated area. There is some overlap in the practical application of these definitions in that a single source for a specified pollutant may be considered a point source even though it extends over a large area—for example, when one field of 100 hectares in area drains excess nitrate into an adjacent waterway in runoff water. Potential point sources thus may include such conditions as confined animal operations, land application of wastewater and sewage, a solid waste landfill, a leaking underground fuel tank, and septic tanks. Industrial point sources can include injection wells, hazardous and radioactive waste sites, mining activity sites, oil and gas production fields, chemical spills, and leaking underground storage tanks. In addition, saltwater intrusion can cause groundwater contamination, as can urban storm sewage waters. Storm runoff from streets and parking areas contains petroleum products, synthetic chemicals, and various metals as residues from automobile traffic, which may contaminate the groundwater system if the storm drains have leaks or if the runoff is not collected. Similarly, sewer lines may also cause groundwater contamination. In many cold climate areas, technical-grade salts used for de-icing roads are common pollutants.

Confined animal operations include poultry operations, milking barns, and feedlots for beef production. With hundreds or even thousands of animals confined in an area, the soil is typically not able to assimilate the animal wastes. Storm runoff can easily pollute both ground and surface waters. Animal wastes inject nitrogen compounds, phosphates, chloride, metals, organic chemicals, and bacteria into the environment and potentially into the groundwater. Nitrate-nitrogen is perhaps the most significant contaminant that may reach the groundwater. Similar contamination is possible from the improper application of fertilizers, from sewage and wastewater, and from poorly functioning septic tanks. Nitrate salts in groundwater effectively promote the growth of algae and the overgrowth of aquatic plants, typically to the detriment of the biological oxygen demand of a water system. Usually bacteria, viruses, and phosphates are removed by interaction with the soil. The many millions of septic tanks and cesspool systems that are

in use throughout North America result in some 11 billion liters of partially treated sewage entering the groundwater system annually. This is a tremendous source of potential contaminants if the system does not operate properly. Septic system failures result from inadequate length of flow through the soil, degradation over time of the physical components of the septic system, or, more commonly, from simply overloading the system.

## HAZARDOUS WASTES

Hazardous injection wells and radioactive and hazardous waste sites may contaminate groundwater with a variety of heavy metal salts and inorganic chemicals, such as arsenic, lead, chromium, and uranium, and with toxic organic compounds. The Love Canal incident has received wide publicity but is only one example of groundwater contamination by a hazardous waste site. Between 1947 and 1953, the abandoned Love Canal in New York was used as a hazardous waste dump. Thousands of drums of chemicals were disposed of in the water and in the banks of the canal. After the site was sold in 1953, the canal was filled with dirt, and schools and houses were built nearby. In 1970, chemicals such as chloroform, benzene, toluene, and many other dangerous and carcinogenic compounds were detected in the air in basements of the houses. Chemicals from leaking and disintegrating drums had contaminated the groundwater and seeped through basement walls. The result was an observed incidence of cancers well above the national average, elevated incidence of several other medical conditions, plummeting property values, and several extensive lawsuits.

Hazardous industrial wastes have sometimes been disposed of by means of deep injection wells that inject the fluids into salty water zones lying below freshwater aquifers. Although the reported cases of groundwater contamination from such sources is low, the potential for contamination exists if the injection well leaks back to the freshwater level. Accidental spills of inorganic or organic chemicals on the surface may soak through the soil into the groundwater. Chemical spills may occur at the site of production, at the site of use, or during transportation.

## LEACHATE AND PETROLEUM PRODUCTS

There are more than 100,000 industrial and municipal landfills in North America disposing of several hundred million tons of waste per year. A city of 1 million people typically generates enough refuse in a year to cover approximately 145 football fields 15 meters deep with garbage. In a landfill or other open repository, the waste materials slowly break down over time as they interact with each other and their environment, releasing their molecular components to water that seeps through and from the material, leaching thousands of different decomposition products from the mass. Leachate produced by the decomposition of the large volumes of buried municipal waste poses a great potential threat to groundwater quality. Although the composition of leachate is variable, it is enriched in dissolved solids (especially chloride and metals) and will consume a large amount of oxygen (measured as chemical oxygen demand), primarily because of its high concentration of undecomposed organic material. Methods for groundwater protection at landfills include surface-water control, covering the waste with soil (clays) that does not transmit significant amounts of precipitation into the landfill, and liners (natural clay or synthetic) that keep the leachate from leaving the burial site. Many large sites include collection points from which leachate can be removed for processing, as well as vent systems for trapping and collecting methane gas produced during the decomposition process.

Petroleum-product storage tanks, the primary feature of the many thousands of service stations in North America, constitute the major type of underground storage tanks. Faulty tanks—resulting from defective construction, improper emplacement, or physical breakdown and rusting—leak these organic chemical materials directly into the soil, where they are free to migrate into the groundwater. Leaky oil wells and distribution pipelines may also contaminate the groundwater with petroleum and petroleum by-products. Millions of liters of gasoline directly enter the environment annually, a bit at a time as the drops that fall from the ends of filler nozzles at each use. Larger spills of gasoline at gas stations and by tanker trucks ruptured in traffic accidents also have contaminated groundwater. Because oils and gasoline are immiscible with (will not mix with) and are less dense than water, most of the pollution from these sources floats on top of the groundwater. Small amounts of the petroleum products, however, can dissolve in the water and contaminate the entire aquifer.

## ACIDIC MINE WATERS AND SALTWATER

Acidic mine waters, rich in metals such as lead, zinc, and cadmium, are common in many ore and coal mining operations. Pyrite (commonly known as fool's gold) is composed of iron and sulfide. When pyrite interacts with air and water under appropriate conditions, sulfuric acid is produced, which dissolves more metals from the rock. Protecting groundwater from contamination produced by mining operations is difficult, especially for deep mines.

Saltwater contamination of groundwater may result from oil and gas wells, as well as from freshwater wells. Most deep oil and gas wells encounter brines (salty water), which are returned to their sources with injection wells. If either the petroleum wells or the injection wells have leaks, saltwater may enter more shallow freshwater aquifers. In many areas of the country, especially in coastal areas, salty water underlies freshwater. If freshwater wells (for example, irrigation wells) pump too much water, the level of the freshwater can be lowered to the point that the underlying saltwater encroaches on the freshwater aquifer recharge zone, contaminating it.

## NONPOINT SOURCES

Nonpoint (diffuse) sources of contamination are equally important. As much as 50 percent of the total water pollution problem has been attributed to nonpoint sources. Although these two sources of pollution are used for discussion purposes, there is often a gradation between point and nonpoint sources. For example, if only a few septic tanks in an area are causing pollution, each of the offending septic tanks would be considered a point source of pollution. The overall area in which those septic tanks are located may be contaminating the groundwater over a broad area, however, so that there are no distinct points from which the pollution emanates, and the contamination would be considered nonpoint source pollution. The contamination of the groundwater in Long Island, New York, is an example of nonpoint source pollution of groundwater from septic tank discharges. The type of soil and the concentration of a large number of septic tanks in a small area have resulted in broad, general contamination of the groundwater by nitrate, ammonia, and detergents. Nitrate-nitrogen and ammonia-nitrogen concentrations in this area have exceeded 20 milligrams per liter, which is well above the acceptable concentration of nitrogen in drinking water. As a result, the water must be treated before use. Atmospheric deposition (both wet and dry) of chemicals downwind of smokestacks in urban areas can cause nonpoint pollution of groundwater over a large area.

Agricultural practices cause a significant amount of nonpoint source pollution. Contamination by pesticides and by fertilizers that incorporate nitrates is most significant because they are spread over large areas and may migrate through the soil into the groundwater. Irrigation waters can also cause contamination in arid areas. The water in wet soil dissolves salts, allowing the salts to move to the surface. As the water evaporates, the dissolved salts and other chemicals are concentrated in the remaining water at the surface, sometimes rendering the soil unsuitable for certain crops. Prolonged periods of wetness due to extensive irrigation or precipitation can then carry the enhanced salt load deep into the soil, where it can then soak into the groundwater system. This water, enriched in dissolved chemicals, may then be pumped up from groundwater sources and used for irrigation again. This cycle may be repeated several times, depending on the flow rate of the groundwater in replenishing the wells, ponds, or streams from which it has been pumped, eventually producing groundwater with very high concentrations of chemicals.

## TREATMENT TECHNOLOGIES

Selection of a groundwater treatment system for groundwater pollution remediation is normally a three-step process consisting of an initial groundwater investigation to determine the type and concentration of contaminant and extent of pollution, the establishment of cleanup goals, and the selection of treatment technology. The last step almost invariably comprises groundwater extraction, surface treatment, and disposal or reinjection, an overall process termed pump-and-treat remediation.

Treatment technologies may be categorized as biological, chemical, or physical. In biological treatment methods, the contaminants are metabolized by aerobic or anaerobic microorganisms. Chemical treatment methods are based on the use of a chemical reactant to immobilize or break down a contaminant and typically include adsorption and precipitation stages. Physical treatment technologies utilize a physical property, such as a contaminant's molecular weight or solubility, as the basis for separating the contaminant from the polluted groundwater. These technologies include air stripping, reverse osmosis, and electrodialysis.

## AIR STRIPPING

The basic concept behind air stripping is mass transfer, whereby the contaminant in water is transferred to a solution in air. Contaminated water is brought into contact with air, either by passing air through the water or by injecting the water into the air as a fine spray. Some of the volatile contaminants become vapors and partition between the air and the water. That is to say, a portion of the volatile components transfer to the air and are removed from the water as the vapors are carried off in the airstream. The transfer of a contaminant is related to its vapor pressure in air relative to its solubility in water. This ratio is formally expressed as Henry's law: $H = C_{ig}/C_{il}$, where $C_{ig}$ is the equilibrium concentration in gas phase (grams per cubic meter), $C_{il}$ is the equilibrium concentration in liquid phase (grams per liter), and $H$ is the Henry's law constant. The Henry's law constant can be used to predict a contaminant's strippability; the higher the constant, the greater the strippability.

As mentioned, water is brought into contact with air either by putting air through water or by putting water through air. Two systems that use the air-through-water methodology are diffused air aeration and mechanical surface aeration. Diffused air aeration is a procedure in which compressed air is injected into a tank of water through a porous base plate or through perforated pipes. In mechanical surface aeration, an impeller creates turbulent mixing of air and water.

In each of the air-through-water systems, the mass transfer effect occurs at the bubble surface. In water-through-air aeration systems, the mass transfer is facilitated by the creation of thin water films or small water droplets. Water-through-air system configurations include crossflow towers and tray aerators, in which the air flows crosscurrent or countercurrent, respectively, to water flowing downward over trays or slats; spray basins, in which water is sprayed from a network of nozzles within a basin into the air in fine droplets to be subsequently collected as they fall back into the basin; and packed towers. Packed towers utilize a countercurrent flow scheme. Air is blown up the tower while water trickles down over an inert (chemically inactive) packing, commonly polypropylene moldings.

## CARBON ADSORPTION

In carbon adsorption, contaminants are removed from water by their attraction and binding to the surface of the activated carbon adsorbent. Intermolecular attractive forces between the carbon atoms and the molecules of various contaminants act rather like magnets, making the molecules adhere loosely to the carbon particle surfaces, or adsorb. Adsorption performance is estimated through the use of a liquid adsorption isotherm test. When contaminated water is mixed with activated carbon, the contaminant concentration decreases to an equilibrium concentration, at which point the number of molecules leaving the surface of the adsorbent is equal to the number of molecules being adsorbed. The relationship between adsorption capacity and equilibrium concentration, known as an adsorption isotherm, is described by the Freundlich equation: $X/M = KC_e^{l/n}$, where $X/M$ is the adsorption capacity, or milligram volatile organic carbon (VOC) adsorbed per gram of activated carbon; $C_e$ is the contaminant concentration at equilibrium VOC milligram per liter; and $K$ and $l/n$ are empirical constants. From the isotherm, the adsorption capacity of the carbon for a contaminant can be estimated using the $X/M$ value that corresponds to the incoming water contaminant concentration.

Activated carbon is produced by high-temperature pyrolysis (the chemical change of a substance by heat) of an organic material to produce a carbon char, followed by partial oxidation at high temperature in an oxygen-poor atmosphere. During partial oxidation, the oxidation occurs along planes within the carbon, creating macropores and micropores, and thereby greatly increasing the surface area of the carbon: The resulting surface area can be up to 1,400 square meters per gram.

During activated carbon treatment, the contaminated water is placed into contact with the carbon for between fifteen and sixty minutes at a surface loading rate of 2 to 17 gallons per minute per square foot. In contact systems using granular carbon, the contaminated water flows either through a fixed bed of carbon or a moving bed, in which the carbon moves down the column under gravity countercurrent to the flow of water. The main advantage of the moving bed is reduced suspended solids removal, although this advantage is rarely a factor in groundwater treatment. The flow of water in fixed-bed systems may be either up or down through the bed, as it does not depend on gravity to induce water flow. System configurations may be a single column or multiple columns arranged in parallel or in series. In powdered carbon systems, the carbon is mixed with the contaminated water, typically in the clarifier of an activated sludge system, and allowed to settle before disposal. Upon exhaustion of granular carbon's adsorptive capacity, the carbon is removed and regenerated in a furnace, where the high temperature destroys the adsorbed organic matter.

## REVERSE OSMOSIS AND ION-EXCHANGE TREATMENT

The term "osmosis" describes the phenomenon in which certain types of membrane will permit the passage of a solvent but restrict the movement of solutes such that water molecules will pass through a semipermeable membrane from a weak solution to a strong solution, eventually equalizing the solute concentrations on both sides of the membrane. In reverse osmosis, pressure (typically 200 to 400 pounds per square inch) is applied to force the water from a contaminated groundwater source through a membrane. Contaminant movement is retarded by the membrane, and purified water is obtained as it passes through to the other side. Cellulose acetate or polyamide membranes in tubular, spiral-wound, or hollow-fin fiber configurations are used.

The electrodialysis process uses electric potential rather than pressure to remove ions from a solution. Two ion-selective membranes, one that is permeable to cations (positively charged ions) and one that is permeable to anions (negatively charged ions), partition the contaminated water from the brine solution and the electrodes. When an electric current is passed across the cell, the cations migrate through the cation-permeable membrane toward the cathode. The anions, however, are prevented from migrating by the membrane. At the anion-permeable membrane, the converse occurs, and, consequently, both cations and anions are removed from the contaminated water. In operational electrodialysis systems, several hundred alternate anion and cation permeable membranes, spaced at 1 millimeter intervals, are placed between a single set of electrodes. The passage of water between the membranes usually takes 10 to 20 seconds, during which time 25 to 40 percent of the ions are removed. Depending upon treatment goals, the water may pass between 1 and 6 membrane stacks.

Precipitation is the process whereby soluble inorganic contaminants are converted, by the addition of reagents, to insoluble precipitates, which are then removed by flocculation and sedimentation. With respect to dissolved metals, precipitation is achieved by increasing the concentration of the anion of a slightly soluble metal-anion salt, usually carbonate, hydroxide, or sulfide.

## ON-SITE TREATMENT

In addition to contaminant pump-out, reverse osmosis, and ion-exchange treatment of inorganic contaminants, and air stripping-based treatment of organic contaminants, two on-site treatments have been under evaluation. On-site nitrate removal involves the use of autotrophic denitrifying bacteria, which are able to catalyze enzymatically the reduction of nitrate to nitrogen gas. Degradation of hydrocarbons by naturally occurring groundwater microorganisms is expected to treat gasoline-related contaminants. Fully implemented pump-and-treat systems, meanwhile, are estimated to yield a total of 70,700 acre-feet per year of remediated groundwater.

## STUDYING GROUNDWATER POLLUTION

Although it is possible to determine if a well is polluted, it is difficult to determine the extent of aquifer pollution. A large number of wells must be drilled and designed to collect water from various levels within the aquifer to evaluate the full extent of aquifer contamination. This approach is obviously expensive and is used only in areas strongly thought to be polluted. Monitoring wells are located near waste disposal sites in order to monitor the quality of the water in the vicinity. The locations and design of the monitoring wells are based on detailed studies of the geology (including soil) and hydrology of the site. Water samples are collected from these wells at regular intervals in order to monitor any degradation of the water. The parameters selected for monitoring are those associated with wastes that are soluble in water. For example, chromium, lead, and copper might be used for monitoring a hazardous waste site that accepts waste from metal-processing plants. Early detection of pollution allows the source of contamination and a relatively small portion of the aquifer to be cleaned up before the entire aquifer is irreparably damaged.

Computer modeling of the transport and deposition of contaminants has become a major method in studying groundwater pollution. These types of studies require not only computer and mathematical expertise but also considerable knowledge of hydrology, geology, and chemistry. The direction and rate of groundwater flow are important. The type of rock and the presence of fractures and faults, which affect the movement of groundwater, are important factors also. Equally important is the solubility of chemicals and their reaction with the soil and aquifer material. These models are helpful in designing monitoring networks for sites. In addition, these models can be useful in designing a remediation program for polluted portions of aquifers.

## IMPORTANCE OF GROUNDWATER QUALITY PROTECTION

High-density populations of organisms, including human beings, usually encounter waste disposal problems. These problems are especially critical for industrialized societies that produce large quantities of municipal and hazardous wastes. Disposal of these large volumes of wastes has often resulted in major contamination of groundwater. Approximately one-half the people in the United States utilize groundwater for drinking water. Approximately 60 percent of these people receive their water from a community system, and the remaining 40 percent have private wells. Groundwater is also important as a source of irrigation water, especially in the western part of the United States. The quality of the groundwater is important for its agricultural use. If the water becomes too salty, plants cannot grow. If the water contains high concentrations of trace metals (for example, selenium), then plants may concentrate the metal and pose a health problem.

Pollution of groundwater is more critical than that of surface water for two reasons. First, it is more difficult to gain access to groundwater; therefore, it is more difficult to determine groundwater pollution and to clean up the contamination. Groundwater pollution cannot be seen and is detected only when a well or a spring becomes noticeably polluted. Second, groundwater movement through aquifers is usually very slow, so remediation of the groundwater will also be slow. Because it is difficult and expensive to clean up an aquifer once it has been polluted, considerable effort should be spent in careful design of installations and monitoring systems, and in land-use planning.

Land-use management is crucial to protecting groundwater quality. For example, if soils in an area are too thin or of the wrong type to allow natural treatment of polluted surface waters from feedlots, these operations should be banned in this area. In some cases, agricultural contamination of aquifers has occurred because too much pesticide or fertilizer was used for the soil and vegetation conditions in an area. Mapping of faults and fractures, which often are zones of increased groundwater movement, can be useful in land-use planning. Limestone areas are often very susceptible to groundwater contamination because the rocks are usually fractured and these fractures may be enlarged as the groundwater dissolves the rock. Caves, sinkholes, and disappearing streams also are often associated with limestone aquifers. These features do not allow natural filtration of recharge water; therefore, the groundwater may be quickly polluted by surface waters.

## CONTEXT

Groundwater supplies approximately half of the United States' drinking water, providing for the domestic needs of about 117 million people and accounting for 35 percent of municipal drinking water supply and 95 percent of rural drinking water supply. Data on the extent and nature of groundwater contamination are limited. Information compiled by the EPA indicates that 20 percent of all drinking water systems and 30 percent of the systems in municipal areas using groundwater as their source show at least trace levels of volatile organic carbons. The true extent of groundwater contamination is, however, probably much greater than that indicated by the EPA data. Potential point sources of pollutants alone include 29,000 hazardous waste sites, 93,000 landfills, and 22 million septic systems. Given possible widespread groundwater contamination, increasing public awareness of the importance of this resource and the health implications of its

contamination, and the increasing prominence of environmental issues on the political agenda, it is clear that the comparatively new technology of groundwater pollution remediation will become an increasingly important component of environmental management.

*Kenneth F. Steele and Richard J. Boon*

## FURTHER READING

Appelo, C. A. J., and Dieke Postma. *Geochemistry, Groundwater and Pollution.* 2d ed. Boca Raton, Fla.: CRC Press, 2005. Thoroughly rewritten and extended edition. Uses a quantitative approach in discussing interactions of groundwater with various mineral and biological contaminants. Well illustrated with figures and tables. Includes numerous examples of chemical and physical principles.

Batu, Vedat. *Applied Flow and Solute Transport Modeling in Aquifers.* Boca Raton, Fla.: CRC Press, 2006. Highly technical. Discusses the dynamics of aquifer water flow, flow conditions, solute transport, and sorption principles. Covers aquifer modeling fundamentals, both numerical and analytical approaches, in detail. Contains a list of references and indexing. Best suited for aquifer researchers and advanced graduate students with a background in fluid dynamics or groundwater analysis.

Brimblecombe, Peter, et al., eds. *Acid Rain: Deposition to Recovery.* Dordrecht: Springer, 2010. Compiles articles from *Water Air, & Soil Pollution: Focus* (2007), discussing acid rain from various perspectives. Discusses agriculture, human impact, ecological impact, wet versus dry deposition, soil chemistry, and surface water quality.

Canter, L. W., R. C. Knox, and D. M. Fairchild. *Ground Water Quality Protection.* Chelsea, Mich.: Lewis, 1987. An excellent summary of all aspects of groundwater pollution. Some chapters are at the college level, but most of those dealing with pollution are suitable for all readers.

Dillon, P. J., ed. *Management of Aquifer Recharge for Sustainability.* Boca Raton, Fla.: Taylor & Francis, 2002. Compiles papers discussing groundwater management and aquifer recharging. Covers topics such as bank filtration, soil aquifer treatment, and rainwater harvesting.

Fetter, C. W. *Contaminant Hydrogeology.* 2d ed. Upper Saddle River, N.J.: Prentice Hall, 1999. Emphasizes the conditions that lead to the contamination of groundwater, the problems this contamination can cause, and remediation techniques used to correct these problems. Includes a detailed discussion of the influence geologic conditions have on groundwater occurrence and contamination. Addresses regional groundwater movement within aquifers and contains helpful illustrations of the principles discussed. Includes a glossary of important terms.

Huang, P. M., and Iskandar Karam, eds. *Soils and Groundwater Pollution and Remediation: Asia, Africa, and Oceania.* Boca Baton, Fla.: Lewis, 2000. A collection of essays that covers a variety of hydrology topics, including groundwater pollution, soil remediation, groundwater purification, and soil pollution. Technical at times, but on the whole provides a good overview of hydrology and groundwater movement.

Moore, John E. *Field Hydrogeology.* 2d ed. Boca Raton, Fla.: CRC Press, 2012. Provides practical information on field investigation, aquifer testing, and groundwater quality testing. Discusses planning and report writing in detail. Also provides case studies from around the world. Best suited for students learning the planning and investigation of field hydrology.

Nyer, Evan K. *Groundwater and Soil Remediation: Practical Methods and Strategies.* Chelsea, Mich.: Ann Arbor Press, 1998. Aims to provide a broad understanding of contaminated groundwater treatment technologies and their application. Focuses on treatments, particularly air stripping and carbon adsorption, for organic contaminants. Suitable for college-level readers.

Randolph, John. *Environmental Land Use Planning and Management.* Washington, D.C.: Island Press, 2004. Describes basic principles and strategies of land-use planning and management. More specific chapters discuss various land features, types, and environmental issues, such as soils, wetlands, forests, groundwater, biodiversity, and runoff pollution. Provides case studies and specific examples.

Schmoll, Oliver. *Protecting Groundwater for Health: Managing the Quality of Drinking Water Sources.* London: IWA Publishing, 2006. Aimed primarily at the prevention of groundwater contamination by analyzing hazards and assessing risks. Describes specific contaminants and their modes of transport in the environment. Suitable for use by all from basic staff to top-level managers involved in water source management projects.

Simon, Franz-Georg, T. Meggyes, and Chris McDonald. *Advanced Groundwater Remediation: Active and Passive Technologies.* London: Thomas Telford Publishing, 2002. Reviews the methodologies of groundwater remediation that were developed through the last half of the twentieth century. Discusses general issues of remediation using passive and active processes and presents a selection of up-to-date research results focused on heavy metal contamination.

Thompson, Stephen A. *Hydrology for Water Management.* Rotterdam: A. A. Balkema, 1999. Thoroughly examines hydrology, groundwater flow, and stream flow. Focuses on the management of water supplies in efforts to keep them free of pollutants and available to the largest number of people possible. Illustrations, maps, index, and bibliography.

Van Beynen, Philip E. *Karst Management.* New York: Springer, 2011. Discusses the management of karst environments. Topics include public education, policy and regulation, and cave and groundwater management. Written for university students, professional geologists, and land management committees.

**See also:** Aquifers; Artificial Recharge; Dams and Flood Control; Floods; Freshwater and Groundwater Contamination Around the World; Groundwater Movement; Hydrologic Cycle; Precipitation; Salinity and Desalination; Saltwater Intrusion; Surface Water; Waterfalls; Water Quality; Watersheds; Water Table; Water Wells

# GULF OF CALIFORNIA

*The Gulf of California, located between the Baja California peninsula and mainland Mexico, is home to an abundant array of land and marine animals. However, overfishing and industrial runoff have led to environmental problems that threaten to destroy the ecological balance of the region.*

## PRINCIPAL TERMS

- *chubasco*: a type of severe storm that occurs in the Gulf of California and along the west coast of Mexico from time to time
- **gill net**: a large, meshed fishing net that allows the head of a fish to pass through, entangling its gill covers and preventing it from passing through or otherwise escaping; because of their size, gill nets often catch many marine animals in addition to the target fish
- **tectonics**: the study of the processes that form the structural features of Earth's crust

## GEOGRAPHY AND GEOLOGY

The Gulf of California, also known as the Sea of Cortez, lies north of the Tropic of Cancer between the east coast of the Baja California peninsula and the west coast of Mexico. The Mexican states of Sonora and Sinaloa on the country's mainland are located on the gulf's eastern shore. To the north, a narrow strip of the Mexican states of Baja California North and Sonora separate the gulf from the southwestern U.S. states of California and Arizona. The mouth of the Colorado River is located on the northern reaches of the gulf. The Mexican states of Baja California North and Baja California South lie to the west of the gulf. The Pacific Ocean borders its southern reaches.

Geologists estimate that the gulf originated more than 4 million years ago, the result of a clash of tectonic plates that caused the Baja California peninsula to drift westward from Mexico's mainland. The area now containing the gulf subsided, allowing ocean water to flow in from the south.

The Pacific Ocean interacts continuously with the waters of the gulf. As the ocean tide rises, seawater floods into it. As it recedes, the water level in the gulf also drops. The tide surge is most noticeable in the north, where the incoming tide floods the surrounding lowlands for miles inland and then departs, leaving miles of sandy beaches and mud flats rich with sea life. The most extreme wave action occurs twice per lunar month, with the advent of the new and full moons.

Over one hundred islands dot the gulf. Most are barren, and few have a supply of potable water. As a result, wildlife is scarce and is limited to only a few of the larger islands. Only three have any human habitation: San José and Isla del Carmen have colonies of salt mine workers, and San Marcos contains a contingent of employees of a gypsum mining operation.

The weather in the gulf remains fairly constant. The occasional storms that occur are mild compared to those found out in the Pacific Ocean. More tempestuous storms, called *chubascos*, occur from time to time, threatening the small boats on the gulf's waters. The watercraft are also threatened by local currents. The islands of the Salsepuedes group, two-thirds of the way north from the ocean entrance, present a particularly dangerous area to sailors in regard to local currents. Temperature can also be a danger. In the mid-summer, especially in July and August, temperatures on the gulf often reach an uncomfortable 38 degrees Celsius.

The Colorado River, for millennia, fed its waters into the gulf. At the beginning of the twentieth century, humans interrupted this natural process by erecting a series of dams and canals along the river's course. Although some freshwater does flow into the gulf from underground sources, today the Colorado River delivers only a relative trickle of surface water to the Gulf of California. The river's output is heavily loaded with pesticides, fertilizers, and human waste as well. The dam system has also reduced the amount of sediment that flows into the gulf, which has resulted in a decline in the amount of nutrients fed into the gulf's waters.

## MARINE LIFE

The Gulf of California's outstanding characteristic lies in the quality and quantity of its sea life. The diversity of its marine occupants varies from great whales to plankton, tiny invertebrates that serve as the basic food source for the gulf's feeders. Its open access to the Pacific promotes a constant interchange of a wide variety of species between the two bodies of water.

Huge finback whales, the second largest of the whale species, travel well up into the gulf. The gray whale, once thought to be on the verge of extinction, migrates from its home in Alaska to Scammon Lagoon on the Pacific side of the peninsula. It is at Scammon that the female gray whales give birth to their offspring. The gray whale migration is now a common sight along the shores of Washington, Oregon, and California during their annual pilgrimage south. The gulf is also home to many other mammals. Dolphins, porpoises, orcas, and sea lions have taken up residence. There is an ample food supply within the gulf for all species. Huge turtles are found in the north, although their popularity as a dish for human consumption threatens their long-term survival.

There are thirty different types of sharks among the estimated 650 different species of fish. Lower on the food chain, in addition to the plankton, a variety of clams, mussels, starfish, flatworms, and sea anemones make their home in the mud flats and shallow pools surrounding the gulf. Overhead, huge flights of birds hunt continuously for marine prey. The gulf is a living biological museum.

## HISTORY

Cultural anthropologists believe that humans first settled in the area surrounding the sea as much as ten thousand years ago. Its abundant seafood and small land animals, coupled with the area's temperate climate, would have been attractive to settlers. Sites uncovered by archaeologists reveal the gulf's early human inhabitants to have had a simple social organization. The clement weather obviated the need for either protective clothing or permanent shelter. They used only the most basic types of tools and occupied themselves with gathering food rather than growing it.

In 1535, a Spanish expedition organized by Hernán Cortés, the conqueror of Mexico, landed at La Paz, now a port on Baja California's southeastern shore. A plan by the Spaniards to establish a permanent colony there ended in failure when conflicts with the local natives broke out. As had been the case with other Spanish expeditions, an attempt by the explorers to exploit or enslave the natives caused the fight.

A more sympathetic group of Spaniards arrived at the gulf about 150 years later. Jesuit missionaries landed at Loreto, farther up Baja's east coast,

to establish Spain's first permanent colony in the area. The Jesuits sought to build a string of missions throughout the largely barren peninsula. They erected some thirty-five churches at one time or another in Baja California, but most failed to last. In most cases, water shortages proved to be insurmountable. Even today, because of its aridity, most of the peninsula remains sparsely populated.

## ECONOMY

The Gulf of California today depends on fishing and tourism, as well as some cattle raising, farming, and mining, for its economic existence. The principal cities, especially those on the Baja peninsula, reflect this. Only in recent years has the Mexican government taken an active role in the gulf's development.

Numerous ports, both on the Baja peninsula and the Mexican mainland, surround the gulf. Many of the peninsula's economic necessities must be imported from the Mexican mainland or the United States. Much of the needed imports arrive by sea. The gulf's many ports provide evidence of the areas's economy. There are six key harbors on the gulf's western reaches and five on the Mexican mainland side.

San Felipe, in Baja California North, is the gulf's northernmost facility. The small village has become a popular destination for fishermen of all types, as well as tourists from the United States and elsewhere. The tremendous tidal surge in the area allows boat owners to beach their craft during one of the monthly high tides. When the water recedes, the boats are left high and dry for both minor and major repairs and maintenance. The craft can then be launched easily at the next high tide. Santa Rosalía, in Baja California South, is the terminal for the gulf's mining operations. A gypsum mine exists on nearby San Marcos Island. A French concession operated a copper mine nearby during the nineteenth century, until it was taken over by the Mexican government. The government closed it down in 1954 but is considering revitalizing the operation as part of its overall plan for the development of the area.

Mulege, in Baja California South, looks very much like an oasis in the desert. It is surrounded by a virtual jungle of date palms and semi-tropical fruits planted originally by the Jesuit order in the early eighteenth century. The port contains several resort hotels catering to both Mexican and

American tourists. Loreto, in Baja California South, is considered the oldest settlement in Baja California. The port also seeks to attract tourists to its resorts. As is the case with Mulege, Loreto has an extensive growth of date palms surrounding it. The town has been destroyed several times in the past by earthquakes, severe *chubascos*, and fire. Loreto is the home of the Museum of the Missions of California.

La Paz is the state capital for Baja South. The port has excellent beaches, a fleet of fishing boats for hire, guides for hunting in the nearby hills, and government-sponsored duty-free stores for visitors. Regular ferry service crossing the gulf itself exists between La Paz and Mazatlán, Sinaloa, on Mexico's mainland. The first-class resort town of Cabo San Lucas is situated on the gulf's entrance to the Pacific. It contains the most luxurious hotels found in Baja California. Campgrounds are also available for tourists traveling with their own accommodations. Game fishing is among the best available throughout the area. Ferry service is available between the port and Puerto Vallarta, Jalisco, on the Mexican mainland.

Puerto Peñasco and Puerto Kino, Sonora, have a modest tourist trade, including many visitors from Arizona. Entrepreneurs have built some small resort hotels near the ports and have laid out campgrounds for travelers with trailers. Sport fishing is the prime attraction at both locations. Guaymas, Sonora, is home to commercial fishermen and acts as the staging area for commercial goods to be shipped inland as well. It has a natural harbor that is spacious and sheltered. Guaymas also exports local products such as cotton and grains from the hinterland. Good hotels and trailer parks are available at the sea front.

A shipping line operates some small freighters between Topolobampo in Sinaloa and La Paz. Shipping is restricted by the shifting sand bars at the entrance to the bay in which the port is situated, but fishing is excellent and a wide variety of game fish can be found in the surrounding waters. Mazatlán, Sinaloa, is the key port for all of northwestern Mexico. The Mexican navy maintains a base at the port, and a number of small warships are in evidence. The largest amount of sea traffic between Mexico and the Baja peninsula passes between Mazatlán and La Paz. An international airport services the area as well. Both hunting and sport fishing are available to the visitor. Good hotels, restaurants, and night clubs can be found in abundance.

## ENVIRONMENTAL PROBLEMS

The marine resources of the Gulf of California include some of the most outstanding varieties of fish and other sea creatures on the face of the globe. The gulf is home to more than nine hundred species of fish and marine animals. The question remains as to how long this unusual concentration of marine riches will continue. Some species of fish have already disappeared from the gulf's waters. Others have reached such low numbers that they are seriously threatened with the same fate. Already lost are cabrillas, black seabass, white seabass, gulf groupers, yellowtails, manta rays, roosterfish, dog snappers, sierras, and vaquetas. The problem lies with overfishing; fish are being taken illegally, often through the bribing of local law enforcement officials.

Commercial shrimp boats from Mexico's mainland have raided the protected areas on the gulf designed to ensure a continuation of the species. Authorities in Baja California are calling for the banning of commercial fishing boats from the mainland, and commercial fishing permits restricted to Baja residents. Many feel that effective control of overfishing is better left in the hands of local authorities.

Drug smuggling operations have also had a negative impact. In one instance, a cyanide-based dye used to mark drug drop-off spots killed both fish and mammals in the area. Government patrols have not been able to stop the smuggling. Gill net fishing, illegal in the gulf, has resulted in an overkill as well. This indiscriminate method of harvesting wholesale lots of fish also scoops up many fingerlings in addition to mature species. One Baja hotel owner stated that the nets have turned the seas around his property into a graveyard for marine life.

Corruption of government authorities is a major problem. The fisheries on the mainland have millions of dollars at stake. Bribes have reached the highest level of government. The solution, in the eyes of most protectors of the environment, is to transfer the surveillance and law enforcement authority to the local level, where control will be in the hands of those who have the most to lose, the locals who make their living from fishing the gulf. During the 1990's, the central government began the process of creating civilian surveillance committees.

In 1996, Mexico's president Ernesto Zedillo declared the bay at Loreto a national park, banning mainland shrimp trawlers from exploiting the area.

As a result, marine life locally is rebounding, and the local fishermen are able to earn a living. The shrimp inside the park are multiplying, and the average size of the shrimp in the catch is growing larger. Loreto could become an example for environmental protection throughout the gulf.

The conflict over environmental protection is not easily solved. Mexican government officials and the Japanese firm Mitsubishi have clashed with environmentalists over a proposed saltworks to be built at the Laguna San Ignacio on the peninsula's Pacific side. A group of dedicated scientists is convinced that the implementation of the plan will have a seriously adverse effect not only on the gray whale migrants who come there to calve but also on a number of other rare marine species making their homes in the area. The government's experts deny that the program will have a negative environmental effect.

*Carl Henry Marcoux*

**FURTHER READING**

Brusca, Richard C., ed. *The Gulf of California: Biodiversity and Conservation.* Tucson, Ariz.: University of Arizona Press, 2010. Contains chapters written by two dozen experts on the Gulf of California environment from both Mexico and the United States, discussing the Gulf from its origins and environmental characteristics to conservation measures.

Case, Ted J., Martin L. Cody, and Exequiel Ezcurra. *A New Island Biogeography of the Sea of Cortés.* New York: Oxford University Press, 2002. Begins with a description of the physical aspects of the Sea of Cortez. Focuses on Sea of Cortez biology. Discusses human impact on the gulf ecosystem and conservation issues. Multiple appendices list the plant life in different regions of the Gulf of California.

Houston, Roy S. *Natural History Guide to the Northwestern Gulf of California and the Adjacent Desert.* Bloomington, Ind.: Xlibris Corporation, 2006. Broadly illustrated with true-to-life images rather than close-ups. Introduces the reader to the physical nature of the region. Offers an overview and guide to the local animal and plant species.

Hupp, Betty, and Marilyn Malone. *The Edge of the Sea of Cortez.* Operculum, 2008. Filled with images from the Gulf of California. Suitable as a coffee-table book, tide pool guide, or introduction to marine ecology. Provides information on the biodiversity of the Gulf of California.

Johnson, Markes E., and Jorge Ledesma-Vazquez. *Atlas of Coastal Ecosystems of the Western Gulf of California: Tracking Limestone Deposits on the Margin of a Young Sea.* Tucson, Ariz.: University of Arizona Press, 2009. Focuses on the role of calcium carbonate in both living and ancient ecosystems of the Gulf of California, linking the health of coastal ecosystems with the evolutionary history of the region. Suitable for geologists, paleontologists, and biological scientists.

Knudson, Tom. "An Overview of the Destruction of the Sea of Cortez." *Sacramento Bee* (December, 1995): 10-13. Investigates the environmental problems developing in the gulf and how these problems have led to the formation of the Sea Watch Foundation, committed to restoring the ecological balance of the Gulf of California.

Miller, Tom, and Elmar Baxter. *The Baja Book.* 16th ed. Santa Ana, Calif.: Baja Trail Publications, 1992. A detailed description and mapping of the ports and fishing grounds in the gulf.

Steinbeck, John. *The Log from the Sea of Cortez.* New York: Penguin Books, 1995. Reports on Steinbeck and biologist Edward F. Ricketts's 6,500-kilometer, six-week scientific expedition to study the gulf's invertebrate marine population.

**See also:** Aral Sea; Arctic Ocean; Atlantic Ocean; Black Sea; Caspian Sea; Gulf of Mexico; Hudson Bay; Indian Ocean; Mediterranean Sea; North Sea; Ocean Pollution and Oil Spills; Pacific Ocean; Persian Gulf; Red Sea

# GULF OF MEXICO

*The Gulf of Mexico is often called the American Mediterranean as it is almost completely enclosed by landmasses. The waters of the gulf yield millions of pounds of food fish and millions of dollars worth of oil and gas each year. It is also the source of such weather systems as destructive hurricanes.*

## PRINCIPAL TERMS

- **dead zone:** a local region within a body of water that does not support any living systems and so is devoid of plant and animal life
- **estuary:** a highly productive, partially enclosed tidal body of saltwater, with freshwater input from a river or from coastal runoff
- **ocean current:** horizontal movement of seawater induced by the wind and affected by rotation of the planet, nearby landmasses, and the temperature and salinity of the water
- **salinity:** a measure of the quantity of dissolved salts in water
- **tidal range:** the difference in height between high tide and low tide at a given point
- **tide:** the periodic, predictable rising and falling of the sea surface as a result of Earth's rotation and the gravitational attractions of the moon and sun

## OVERVIEW

Geographers and oceanographers define the Gulf of Mexico as a Mediterranean-type sea—that is, one that is partially landlocked. The gulf is bounded on the east, north, and west by the United States, from Key West at the tip of Florida to Brownsville, Texas. Mexico forms part of the extreme western reaches of the gulf to Cancun at the tip of the Yucatán Peninsula. The long, east-west-oriented island of Cuba forms the southern boundary of this marginal sea. The Gulf of Mexico opens into the Caribbean Sea between the Yucatán Peninsula and Cuba, and into the Atlantic Ocean between Cuba and Florida.

Water circulation in the Gulf of Mexico is generated by the great volume of water that enters the gulf from the Caribbean Sea by way of the Yucatán Strait. The water circulates in the gulf in a more or less clockwise direction, then exits to the east through the Florida Straits between Cuba and Key West. Loaded with warm, salty water, the current forms the beginning of a great oceanic gyre that includes the Gulf Stream and the North Atlantic Current.

Freshwater flows into the Gulf of Mexico via a number of rivers and streams, the greatest being the Mississippi River. Draining nearly one-half the land area of the United States, this river drops abundant sediments and nutrients into the gulf waters. Other rivers around the rim of the gulf basin add their burden of sediment and nutrients, making the Gulf of Mexico a rich environment supporting a host of marine animals including fish, shrimp, and some whales. The productive gulf waters supply approximately 20 percent of the commercial fish catch of the United States' fisheries, as well as those of Mexico and other Central American nations.

Part of the productivity of the gulf's waters comes from the broad continental shelf that forms the rim of the gulf basin. The sunlit waters over the shelf—which is up to 320 kilometers wide in some areas—support a host of marine organisms that form an elaborate ecosystem, ranging from microscopic plants to large fish and giant whales. At greater depths, oil seeps support fish and shellfish whose food supply is based not on sunlight energy but on energy from oil and gas oozing from the sea floor. The oil and gas deposits in the gulf floor are also tapped as a commercial source of petroleum and petroleum products.

The basin of the Gulf of Mexico is believed by geologists to have formed about 100 million years ago, as tectonic movement brought the continents of North and South America toward their present locations. Over geologic time, the waters in the basin periodically evaporated, leaving vast deposits of salt that formed immense domes. During periods when the gulf basin was flooded by a warm, shallow sea, tiny marine plants and animals called plankton flourished in unimaginable numbers. As the plankton died, their remains settled to the sea floor and were overlain by sediments. Eventually, the organic remains were changed into petroleum and natural gas.

Over long periods of time, large sections of Earth's crust shift in vast, slow-moving tectonic movements. However, researchers have interpreted geologic evidence that indicates the occurrence of a sudden catastrophic event that not only affected the shape of the young Gulf of Mexico but also adversely affected all life on Earth. About 65 million years ago, a giant meteorite impacted near Chicxulub at what is now the northern edge of the Yucatán Peninsula. The impact shattered the continental edge and created devastating geologic and atmospheric disasters that spanned the entire globe. In the process, at least 75 percent of all species of plants and animals then inhabiting the world were extinguished.

## OCEANOGRAPHIC OBSERVATIONS

The waters of the gulf fill a basin covering more than 1.5 million square kilometers. The basin measures about 1,600 kilometers between Florida and Mexico and 1,280 kilometers from New Orleans, Louisiana, to the Bay of Campeche. The floor of the basin features a varied topography, including rises and depressions. The deepest depression is the Sigsbee Deep, almost in the center of the basin, which has been plumbed to a depth of 5,203 meters below the sea surface. The average depth of the gulf basin is 1,430 meters. Numerous knolls and flat-topped underwater mountains called guyots break up the monotony of the abyssal plain.

The gulf's productivity results from a combination of the nutrients swept in by the many rivers and the temperature and salinity regime. Sea-surface temperatures in the summer range between 18 degrees Celsius in the northern gulf to 24 degrees Celsius off the Yucatán Peninsula. In some years, summer highs of 32 degrees Celsius have been recorded. Bottom-water temperatures of about 6 degrees Celsius occur in the southern part of the Yucatán Channel. Salinity of Gulf of Mexico waters, measured in parts per thousand, is lowest near the shore, especially off the mouth of the Mississippi River, which sometimes measures less than twenty-four parts per thousand. Elsewhere in the gulf, salinities as high as thirty-seven parts per thousand have been recorded.

The tidal range varies from place to place in the gulf but is, on average, small. The gulf features diurnal tides, one high and one low tide each tidal day. Depending on where in the gulf they are measured, the tidal range varies between 0.8 meter and 1.2 meters.

Climatically, the region over and around the Gulf of Mexico varies from tropical to subtropical. The heat from the sun and warm water that sweeps in from the Caribbean Sea produce unstable atmospheric conditions over the gulf that help produce more precipitation in the United States than the Atlantic and Pacific Oceans combined. Warm, moist air flowing northward from the gulf, with no topographic boundaries to interfere, often collides with cold, dry continental air flowing southward from Canada. The meteorological disturbances that result are often monumental. These conflicting air masses may produce deluges and floods, dangerous lightning storms, and, occasionally, destructive tornadoes. During the Atlantic hurricane season (from June 1 to November 30), tropical cyclones may form in the gulf or enter from the eastern equatorial Atlantic Ocean, causing massive destruction, flooding, and loss of life. With the effects of global warming becoming more apparent, the frequency and intensity of tornado-spawning thunderstorms and hurricanes also seems to be increasing, though whether this is a directly related effect or a natural cycle is still debated.

The climatic regime of the Gulf of Mexico tends to feature monsoonal weather. In the summer, winds blowing onshore from the gulf bring an abundance of warm, moist air and copious rainfall. In the winter, cool, dry air wafts southward on offshore winds.

## ORGANIC AND MINERAL RESOURCES

Numerous estuaries ring the Gulf of Mexico, forming a sort of buffer zone between the open waters of the gulf and the coastal uplands. The organisms that live in these estuaries exist in ever-changing environments. They are affected by river runoff and the daily tides, which cause the estuarine waters to fluctuate in salinity, temperature, and nutrient levels. Yet they are among the most productive marine areas on the planet. Approximately 60 to 80 percent of marine organisms valued by humans for food or recreation spend at least part of their life cycle in estuaries. Oysters, scallops, shrimp, flounders, redfish, and mullet—nourished by a food chain that begins with the dead and decaying marsh plants—grow fat and abundant. Shrimp, for example, spawn in the gulf, get swept by tides and currents into estuaries to fatten, then move back into the gulf to be caught by shrimp trawlers. The catch, about 242 million pounds annually, is worth about $500 million. The BP oil blowout at the offshore *Deepwater Horizon* drilling rig in the Gulf in 2010, however, effectively destroyed the shrimp fisheries and shellfish industries in the area for an indeterminate length of time, and created a "dead zone" that now occupies a significant portion of the gulf. In that zone, no shrimp, fish, or plant life are supported.

In addition to the living resources of finfish and shellfish, the Gulf of Mexico contains valuable mineral resources. These include sulfur, sand, gravel, shells used for construction, petroleum, and natural gas. In 1937, the first successful offshore gulf oil well was drilled in water 4 meters deep. Since then, more than twenty thousand wells have been drilled in the near-coastal waters of the gulf, some as deep as 1,800 meters. Oil reserves in the gulf are believed to be worth an estimated $75 billion.

Oil has leaked naturally into the waters of the Gulf of Mexico for centuries, producing some curious ecosystems. Researchers from the University of Texas and Pennsylvania State University descended more than 500 meters below the surface in a submersible to explore such an ecosystem. At that depth, where no sunlight penetrates, the researchers observed tube worms, clams, and mussels that were thriving on a diet of bacteria that produced organic energy from petroleum. Instead of producing energy-rich food by photosynthesis with sunlight, as surface plants do, the bacteria produced energy-rich food through the process of chemosynthesis from the petroleum. The presence of fish near the oil seeps suggest that a petroleum-based food chain may have developed in those areas.

## SIGNIFICANCE

The Gulf of Mexico is a productive, biologically rich, highly unique, and economically valuable ecosystem. More than two-thirds of the North American watershed drains into the gulf, mostly through the Mississippi River. Its coastal wetlands provide feeding areas, nursery sites, and breeding grounds for a host of marine and coastal animals. Its waters provide food and recreation for millions of people from Mexico, Central America, and the United States, as well as visitors from other countries. The submerged lands of the gulf contain mineral resources, particularly petroleum, that are vital to the commerce of nations adjacent to the gulf but also to the world at large.

*Albert C. Jensen*

## FURTHER READING

Bartolini, Claudio, Richard T. Buffler, and Abelardo Cantu-Chapa. *The Western Gulf of Mexico Basin: Tectonics, Sedimentary Basins and Petroleum Systems.* Tulsa, Okla.: American Association of Petroleum Geologists, 2001. Focuses on petroleum geology and resources. Offers an in-depth description of the western Gulf of Mexico in that context.

Buster, Noreen A., and Charles W. Holmes. *Gulf of Mexico Origin, Waters and Biota.* College Station, Tex.: Texas A & M University Press, 2011. Discusses the origins of the Gulf of Mexico and the processes that are at work there and its coral reef structures. Uses a great deal of data that had previously been sequestered in industry and government files. Part of a larger series.

Davis, Richard A. *Sea-Level Change in the Gulf of Mexico.* Nansha: Everbest Printing Company, 2011. Examines the effects of rising and falling sea levels in the Gulf of Mexico from a variety of causal factors, with special emphasis on the past twenty thousand years and the melting of the great glaciers of the last ice age.

Felder, Darryl L., and David K. Camp. *Gulf of Mexico Origin, Waters, and Biota.* Amherst, Tex.: Texas A&M University Press, 2009. Chapters discuss the biodiversity of the Gulf of Mexico, beginning with the earliest life-forms and continuing through to chapters on more evolved biota. Over fifteen thousand species are evaluated in this biodiversity assessment.

Gardner, Jim. *Multibeam Mapping in the Outer Continental Shelf Region of the Northwestern Gulf of Mexico.* Washington, D.C.: U.S. Department. of the Interior, Minerals Management Service, Gulf of Mexico OCS Region, 2003. Analyzes bathymetry data and provides surveys of multiple banks along the continental shelf off the Gulf Coast. Available on CD-ROM. A good technical account of research conducted on the continental shelf.

Gore, Robert H. *The Gulf of Mexico: A Treasury of Resources in the American Mediterranean.* Sarasota, Fla.: Pineapple Press, 1992. Offers an overview of the history, geology, geography, oceanography, biology, ecology, and economics of this important body of water. Written in a readable style for high school students and lower-division college students.

Graham, Bob, et al. *Deep Water: The Gulf Oil Disaster and the Future of Offshore Drilling.* National

Commission on the BP Deepwater Horizon Oil Spill and Offshore Drilling, 2011. A government report addressing the disaster in the Gulf of Mexico, the events leading up to it, and repercussions to follow. Focuses on efforts to prevent such disasters in the future. Provides great detail from key players, scientists, and local residents.

Keim, Barry D., and Robert A. Muller. *Hurricanes of the Gulf of Mexico.* Baton Rouge, La.: Louisiana State University Press, 2009. A definitive guide to Gulf of Mexico hurricanes from the 1800's to 2008 with assessment and predictions of future hurricane activity. Suitable for all readers.

MacDonald, Ian R., and Charles Fisher. "Life Without Light." *National Geographic* 190 (October, 1996). Details the findings of two research oceanographers who explored the floor of the Gulf of Mexico in a submersible and discovered a productive ecosystem of mussels and tube worms subsisting on bacteria that thrive on gas and oil seeping from cracks in the seabed. Describes the process of chemosynthesis in a lightless environment. Excellent illustrations. Suitable for high school readers.

Malakoff, David. "Death by Suffocation in the Gulf of Mexico." *Science* 5374 (July 10, 1998): 190-192. Describes large areas in the Gulf of Mexico featuring low dissolved oxygen levels. Explains how these "dead zones" drive away or kill shrimp and fish. Suggests that the phenomenon is caused by fertilizers carried more than 1,600 kilometers from farms in the midwestern United States and

that these fertilizers may also help to produce "red tides." Provides an excellent review of the problem. Suitable for high school readers.

Warrick, Joby. "Death in the Gulf of Mexico." *National Wildlife* 37 (June/July, 1999): 48-52. A readable review of the dead water in the gulf, with some details on the specifics of the adverse water quality. Examines farm fertilizers and other pollutants carried down by the Mississippi River as the cause of the low-oxygen water. Suitable for high school readers.

Weber, Michael, Richard T. Townsend, and Rose Bierce. *Environmental Quality in the Gulf of Mexico: A Citizen's Guide.* 2d ed. Washington, D.C.: Center for Marine Coordination. 1992. Discusses the physical features of the Gulf of Mexico, including geology, climate, and oceanography. Also describes the ecosystems and principal economic activities in the gulf, especially tourism, fisheries, offshore oil and gas production, and hard-mineral mining. Includes suggestions for managing these diverse resources. Suitable for high school students.

**See also:** Aral Sea; Arctic Ocean; Atlantic Ocean; Beaches and Coastal Processes; Black Sea; Caspian Sea; Gulf of California; Gulf of Mexico Oil Spill; Hudson Bay; Hurricane Katrina; Indian Ocean; Mediterranean Sea; North Sea; Ocean Tides; Pacific Ocean; Persian Gulf; Red Sea; Surface Ocean Currents

# GULF OF MEXICO OIL SPILL

*In the United States, most offshore drilling takes place in the Gulf of Mexico, where, on April 20, 2010, an explosion and subsequent oil spill occurred on an offshore rig operated by British Petroleum. The spill introduced about 4.9 million barrels (about 206 million gallons) of oil into the gulf's waters in several months. More than 3 million liters (about 800,000 gal) of chemical dispersants was used to combat the spill.*

## PRINCIPAL TERMS

- **berm:** an artificial ridge or wall, usually made of sand; normally used to prevent flooding but employed after the gulf spill as a barrier to keep oil from polluting beaches
- **blowout preventer:** a large series of safety valves installed at the wellhead of an oil rig to seal the well in case of a problem and to prevent the uncontrolled release of oil
- **boom:** a floating device, usually made of either plastic or cloth, which is linked with other booms to form a flexible barrier intended to prevent oil spills from moving inland
- **dispersant:** a chemical, usually consisting of a surfactant and a solvent, which is used to break up an oil slick into smaller droplets that will then disperse into the water and be weathered more quickly
- **hydrocarbon:** any of a large group of organic compounds containing various combinations of carbon and hydrogen atoms; a common substance in petroleum products, including crude oil and natural gas
- **loop current:** an ocean current that transports warm water from the Caribbean Sea into the Gulf of Mexico
- **surfactant:** a substance that reduces the surface tension of a liquid in which it is dissolved
- **tarball:** a dense, sticky blob of weathered oil that can travel great distances and is difficult to break down
- **weathering:** the disintegration and degradation of a substance, such as oil, through the action of wind, water, and waves
- **wellbore:** a hole drilled into the earth to look for or extract a natural resource, such as oil, water, or gas
- **wellhead:** a general term for the equipment installed at the surface of a wellbore; designed to provide a pressure seal for the oil or other substance being extracted

## OFFSHORE DRILLING

The majority of the world's liquid petroleum (oil) and natural gas resources take the form of deposits known as traps that are buried under incredibly thick layers of dirt and rock. These traps have been formed through millions of years, as organic materials like the bodies of plants and animals are "cooked" by heat and pressure into hydrocarbons: organic compounds containing carbon and hydrogen atoms.

As the world's demand for energy grows and as oil companies begin to exhaust the supply of land-based oil reserves that are easily recoverable, more companies are turning to offshore drilling to increase supply. Offshore drilling is the search for and extraction of petroleum or natural gas reserves from beneath the ocean floor, usually at locations on the continental shelf: the area of seabed that surrounds a large landmass.

In the United States, offshore drilling occurs anywhere from about 60 meters (about 200 feet) to about 320 kilometers (about 200 miles) off the country's coastline. The wellbores, or drilled holes, which have been created in the process have been of varying depth. Some have extended as far as 5.5 km (3.4 mi) below the ocean's surface. Deeper wellbores are becoming increasingly more common as the search for viable traps expands.

To extract oil from undersea deposits, a trap must first be located. This is often done through a technique known as seismic surveying, in which shock waves emitted by compressed air guns and explosives are sent down into the seabed. When the reflected waves bounce back, they are analyzed to measure the specific geophysical properties of the rocks below and to determine the likelihood of oil deposits in that region. Next, a mobile drilling platform is established over a likely site; exploratory drilling then occurs. If core samples reveal what is called a show, or clear evidence of a petroleum deposit, a production well or wells are drilled.

The wells are affixed to an enormous platform, usually attached directly to the ocean floor by a

253

foundation made of metal, concrete, and cables. The entire rig is designed to last for the time it takes to deplete the deposits, which may be ten or twenty years or longer; the structure also must be sturdy enough to weather storms and large enough to accommodate supplies, equipment, and housing for the platform workers.

## DRILLING HAZARDS

The oil industry in the United States maintains that offshore drilling's benefits (which include stabilizing the price of oil and reducing the United States' dependence on foreign petroleum imports) outweigh any serious risks to the environment, particularly in the wake of technological improvements such as advanced safety valves and sensitive temperature and pressure sensors. However, opponents of offshore drilling outline many potential hazards.

The most severe of these hazards is the risk of an oil spill. Even a smoothly functioning rig may have detrimental effects on the environment and may also pose health risks to oil workers because of the danger of fires, electrocutions, and other accidents. Air pollutants such as carbon monoxide, nitrogen oxide, methane, and volatile organic compounds are discharged during routine activities, such as loading and shuttling oil from the rig to shuttle tanks. Heavy metals and other contaminated sediments released in drilling fluids also are likely to be picked up from the surfaces of the rig by ocean currents and dispersed to surrounding areas.

The physical infrastructure associated with offshore drilling platforms, including pipelines and wellbores, also causes permanent changes to the ecosystems that exist at and below the ocean floor (in the benthic and abyssal zones). Finally, when rigging equipment, pipelines, and ships are taken from one place to another, invasive species such as mussels and barnacles can be spread to new environments, where they may disrupt the balance of the native ecosystem.

## THE DEEPWATER HORIZON OIL SPILL

A 2009 report by the Energy Information Administration, part of the U.S. Department of Energy, estimated that about 50 percent of oil consumption in the United States was met by domestic production. About one-third of that production comes from offshore drilling, the majority of which is attributable to oil rigs in the Gulf of Mexico—namely those off the coast of Louisiana, Mississippi, Alabama, and Florida.

At about 10 p.m. Eastern standard time on April 20, 2010, an explosion rocked an offshore oil rig in the Gulf of Mexico that was operated by British Petroleum (BP); the rig was known as the Deepwater Horizon. Deepwater Horizon's wellbore ran about 5.5 km (about 3.4 mi) below sea level.

Because the weight of the rocks above an oil trap creates pressure, and because this pressure increases with depth, deeper wells present greater risks of blowouts, or explosions. The Deepwater Horizon explosion caused gas, oil, and concrete to burst violently up the wellbore onto the deck of the platform, starting a fire. Eleven platform workers were killed and seventeen were injured in the incident. Drilling of the rig's primary wellbore had begun in February of that year, and the final placement of the cement designed to seal it had been completed just before the accident. The fire raged for two days, after which the entire rig sank below the surface of the water.

According to the National Commission on the BP Deepwater Horizon Oil Spill and Offshore Drilling, an independent U.S. presidential commission that studied the disaster, an intricate array of factors led to the explosion and subsequent spill. Technically, the reason the explosion occurred was straightforward: The cement that had been pumped down to the bottom of the wellbore failed to seal the wellbore from the highly pressurized hydrocarbons around it.

Once the wellbore was declared to be completely prepared, the drilling mud inside it was replaced by seawater, reducing the pressure inside the well. When the pressure of the hydrocarbons became greater than the pressure in the well, oil and gas began to flow into the ring-shaped space around the well casing, a phenomenon known as a kick. The stream of volatile hydrocarbons then traveled up to the wellhead, where it burst through the blowout preventer and then ignited. The commission cited pervasive failures of management, communication, and safety procedures as responsible for these technical problems.

## STRATEGIES FOR SEALING THE LEAK

On April 24, underwater cameras detected a leak in the wellbore that was releasing oil into the ocean; estimates of the rate at which oil was spilling from the rig varied dramatically in the next weeks and months, but official figures eventually settled at a rate of approximately 5,000 barrels (about 210,000 gal) of oil per day.

The oil continued to gush into the gulf for several months, even as a large number of different techniques were employed to try to stem the flow. For example, attempts were made to properly activate the blowout preventer: a series of safety valves designed to prevent an explosion of oil, but which were never fully deployed during the disaster. Two relief wells were dug that were used to inject mud and cement into the leaking well; golf balls, rubber, and other objects were also used as attempted plugs. A large box made of steel, known as a containment dome, was lowered over the leak. A tube was inserted into the broken pipe to divert the oil to a ship on the surface. Several different caps were placed on top of the blowout preventer and used to stop the leakage of oil and simultaneously siphon oil and gas to tankers on the surface.

Eventually, a combination of mud and cement, introduced slowly into the well, provided a complete seal. Officials declared the well permanently sealed on September 21, 2010. Ultimately, an estimated total of 4.9 million barrels (about 206 million gal) of oil was released into the gulf. To reduce the amount of oil that washed to shore, floating devices called booms and artificial sand walls called berms were deployed as physical barriers.

## CHEMICAL DISPERSANTS AND THE EFFECTS OF THE WEATHER

Cleanup efforts were occurring at the same time as attempts to seal the well. Chemical dispersants are most commonly used to deal with oil spills. They consist of substances that break up large amounts of surface oil into smaller droplets that disperse; the droplets can then be more easily weathered, or biodegraded. (Oil slicks that are not dispersed present little surface area to which natural underwater microbes like bacteria can attach.) Dispersed oil also is less likely to wash up to coastal wetlands or to form tarballs, which are dense, sticky blobs of weathered oil that can travel great distances and can persist for some time after a spill.

Dispersants typically consist of a solvent, or a component that dissolves other substances, and a surfactant, a surface active agent that reduces the surface tension of a liquid in which it is dissolved. Surfactants consist of long molecules with a chain-like structure. One end of the chain is hydrophilic, meaning that it has a tendency to be drawn to and dissolve in water. The other end is oleophilic, meaning that it has a tendency to be drawn to and dissolve in oil. Normally, oil and water do not mix. As the surfactant molecules attach themselves to water on one end and to oil on the other, they form a kind of bridge between the two substances. This lowers the tension of the surface between them and allows the oil to break up into droplets.

To deal with the Deepwater Horizon spill, BP used more than 3 million liters (about 800,000 gallons) of chemical dispersants. Mainly, the dispersants consisted of a particular dispersant known as Corexit 9500A, one of whose main ingredients was a surfactant called dioctyl sodium sulfosuccinate (DSS). The Corexit and other dispersants were not simply applied at the surface of the ocean; they were directly injected into the flow of oil and gas that was leaking from the wellhead in the Gulf of Mexico, at depths of about 1,200 to 1,500 m (4000 to 5000 ft). This substantial use of dispersants in the deep ocean was unprecedented, and a later analysis revealed that a detectable level of DSS remained in the deep waters of the gulf four to five months later. Scientists did not know if the DSS lingered for long enough, or at high enough levels, to be toxic to deep-water organisms.

One factor that worked in favor of the cleanup efforts was the warm weather in the Gulf of Mexico. Higher water temperatures allow certain volatile compounds in the oil to evaporate faster; higher temperatures also increase the rate at which bacteria can break down the compounds into harmless carbon dioxide and water.

The region's tropical weather systems also caused concern among scientists, however. On the one hand, the strong winds and turbulent seas would help to accelerate the biodegradation and weathering of oil. On the other hand, a hurricane could distribute the oil farther than it would otherwise travel, bringing it farther inland toward coastlines that would otherwise not have seen much oil. If, for example, winds drove oil into the loop current, the oil could have been carried into the Gulf Stream around Florida and partway up the East Coast of the United States. Ultimately, no major hurricanes made contact with the Deepwater oil slick.

## ENVIRONMENTAL IMPACT

Oil spills can have serious effects on marine wildlife and habitats. The spills kill fish and other ocean inhabitants, especially if the spills occur during breeding season when eggs are vulnerable. The gulf spill had a negative impact on the shrimp, crab, and oyster industries in Louisiana for this reason.

Oil spills are notorious for killing sea birds and shore birds. Oil that coats feathers can prevent birds from flying, can destroy their natural temperature regulation mechanisms and waterproofing, and can poison birds as they attempt to clean themselves. Oil that directly reaches marine mammals, such as whales, otters, and seals, has similar harmful effects. Oil also can contaminate the food chain as it washes up to beaches and coastal wetlands like marshes and mangrove forests, making its way into grasses and other plants. This can damage or destroy essential feeding, breeding, and nesting grounds for many species. Also, the cleanup efforts themselves can do harm to delicate wetlands.

After the gulf spill, long stretches of the Louisiana coastline, particularly in the south, did experience heavy surges of oil, which blanketed the marshes there. An estimated eight thousand birds were retrieved and identified as having died as a result of the spill. Scientists believe that up to eight times that number were actually killed.

Multiple analyses performed about one year after the spill found that a surprising amount of the oil that had leaked from the wellhead did not travel with ocean currents outside the gulf, but had instead remained in deep water and been broken down by microbes. Because of favorable weather conditions the worst predictions about how much oil might wash up to the coastline were not fulfilled. In particular, oil never flowed around the tip of Florida up the East Coast of the United States. However, the long-term environmental effects of the Deepwater Horizon spill itself will take years, if not decades, to fully assess, especially when it comes to the impact the disaster had on marine populations such as bluefin tuna, dolphins, starfish, and coral.

*M. Lee*

## FURTHER READING

Fingas, Mervin F, ed. *Oil Spill Science and Technology.* Burlington, Mass.: Elsevier, 2011. A substantial reference book, best suited for undergraduates looking for information about specific topics, such as modeling the movement of oil in water, chemical dispersants, or environmental restoration. Data-rich and up to date.

Hunter, Nick. *Offshore Oil Drilling.* Chicago: Heinemann Library, 2012. Full-color photographs, illustrations, and text boxes fill this brief, student-friendly overview of offshore drilling techniques and associated controversies.

Juhasz, Antonia. *Black Tide: The Devastating Impact of the Gulf Oil Spill.* Hoboken, N.J.: John Wiley, 2011. Based on hundreds of interviews with scientists, government officials, oil workers, fishers, crabbers, and Gulf Coast residents, this is a well-researched, big-picture look at the disaster and its human impact.

Lehner, Peter, and Bob Deans. *In Deep Water: The Anatomy of a Disaster, the Fate of the Gulf, and How to End Our Oil Addiction.* New York: OR Books, 2010. An informative analysis of the causes, course, and implications of the gulf oil spill; accessible to all readers. Includes a timeline, resource guide, and some black-and-white illustrations and images.

McGinnis, Tim. "Seafloor Drilling." In *Drilling in Extreme Environments: Penetration and Sampling on Earth and Other Planets*, edited by Yoseph Bar-Cohen and Kris Zacny. Hoboken, N.J.: Wiley, 2009. A technical but accessible chapter covering the practical technological challenges and solutions related to deep-sea drilling, covering topics such as sampling techniques and robotic drilling. Includes black-and-white photographs, charts, and a list of further reading.

Tainter, Joseph A., and Tadeusz W. Patzek. *Drilling Down: The Gulf Oil Debacle and Our Energy Dilemma.* New York: Copernicus Books, 2011. This look at the economics and politics of the United States' dependence on oil, written by researchers in sustainability and petroleum engineering, presents a technical perspective on the fundamental causes of the gulf oil spill. Student-friendly and filled with color photographs.

**See also:** Gulf of Mexico; Hurricane Katrina; Ocean Pollution and Oil Spills; Remote Sensing of the Oceans; Seawater Composition; Surface Ocean Currents; Surface Water

# GULF STREAM

*The Gulf Stream is a geostrophic surface current that constitutes the northwestern part of the North Atlantic Gyre. It moves huge quantities of water at remarkably fast velocities across vast distances with many geological, physical, and biological repercussions. By itself, however, it is not responsible for the mild climate of Western Europe.*

## PRINCIPAL TERMS

- **Coriolis effect:** an apparent force acting on a rotating coordinate system; on Earth this causes things moving in the Northern Hemisphere to be deflected toward the right and things moving in the Southern Hemisphere to be deflected toward the left
- **geostrophic current:** a current resulting from the balance between a pressure gradient force and the Coriolis effect; the current moves horizontally and is perpendicular in direction to both the pressure gradient force and the Coriolis effect
- **gyre:** the major rotating current system at the surface of an ocean, generally produced by a combination of wind-generated currents and geostrophic currents
- **pressure gradient:** a difference in pressure that causes fluids (both liquids and gases) to move from regions of high pressure to regions of low pressure
- **thermohaline circulation:** a mode of oceanic circulation that is driven by the sinking of denser water and its replacement at the surface with less dense water
- **wind-driven circulation:** the surface currents on the ocean that result from winds and geostrophic currents

## WATER CIRCULATION

It is convenient to consider any systematic movement of water at sea as being part of either a wind-driven circulation system or a thermohaline circulation system. In the former, the linkage with atmospheric movement is direct, the currents are usually at or near the sea surface, and the velocities of the flows are often in the range of several centimeters per second (or several knots). In thermohaline circulation, the driving force is gravity, which causes denser water to sink and flow to the deepest parts of the sea, and the currents are usually much slower. In the North Atlantic Ocean, thermohaline circulation occurs on a vast scale as saline surface water gives up its heat, sinks, and eventually flows across the bottom of both the North and South Atlantic basins.

Wind-driven circulation develops a gigantic clockwise circular motion called the North Atlantic Gyre. The most intense flow of this gyre is along its northwestern boundary and is called the Gulf Stream. Because the flow is circular, it does not actually have a beginning or end. As it is usually geographically defined, however, the Gulf Stream begins in the straits of Florida, where water leaving the Gulf of Mexico and the Caribbean joins with water continuing to go around the gyre. There the Gulf Stream moves about 30 million cubic meters per second past any point. The volume of water entrained in this flow continues to increase, and by the time it reaches Cape Hatteras, there is about 85 million cubic meters moving by any point every second. When the Gulf Stream reaches longitude 65 degrees west, off the Grand Banks, it moves 150 million cubic meters per second.

To put this in perspective, consider that 1 cubic meter of water has a mass of 1,000 kilograms. A 150-pound person has a mass of 68 kilograms. Therefore, 1 cubic meter of water has the mass of about fifteen people. The flow off Cape Hatteras would be equivalent to 1.2 billion people, roughly the population of China, streaming by every second. By the time it gets to the Grand Banks, the flow would be larger by 65 million cubic meters, or 975 million more people. The Gulf Stream truly dwarfs most of nature's other wonders. The total hydrological cycle, which includes all the rain, snow, sleet, and hail that falls on Earth (oceans plus continents), moves an average of only 10 million cubic meters of water per second.

## CORIOLIS EFFECT

The secret to maintaining such huge flows of water lies in their circular nature. A "hill" of warm, low-density water sits over the core of the North Atlantic Gyre. Water on or within this hill is driven by gravity or a pressure gradient to move downslope or away from the center of this hill. This moving water is deflected by the Coriolis effect, which is a consequence of the planet's rotation. In the Northern Hemisphere, the

Coriolis effect causes things moving with a horizontal velocity to move to their right. Therefore, water continuously trying to move out from the center of this hill is deflected to move around the gyre instead. Eventually a balance is achieved between the Coriolis effect and the pressure gradient forces. As a result, the water does not move away from the center of the hill but instead circles it in a clockwise fashion. This kind of current is called a "geostrophic" current.

Earth's spherical shape means that the effects of the Coriolis effect vary with latitude. One result is that the flow within the North Atlantic Gyre varies with location. The hill is not symmetrical but has its steepest slopes on its northwest edge. Because currents must balance slopes, the gyre is most intense there. The center of the gyre is in the western Atlantic not far from Bermuda. The hill also slopes away to the east, but very gradually, so that the southern flows of the gyre are slow and spread out over a very large region.

Driving the gyre and maintaining the Bermuda High are the winds over the Atlantic Ocean. This subtropical atmospheric feature is a region of descending, dry air near the center of the North Atlantic Ocean. After descending, this air pushes out across the ocean. It, too, is deflected by the Coriolis effect, developing into somewhat circular, clockwise winds. Near the center of this system, winds are weak, and precipitation is uncommon. Early sailors, faced with long, dry periods of calm, sometimes made their thirsty horses walk the plank. This is the origin of the term "horse latitudes," sometimes used to describe this region.

With little cloud cover and a subtropical latitude, this region receives intense solar radiation that warms the surface water and causes intense evaporation. This causes the water of this area, the Sargasso Sea, to become extraordinarily saline and very warm. Just as oil floats on vinegar in a salad dressing, this warm, and correspondingly less dense, saline water floats on top of the cooler, denser water below.

It is this warm water that forms the hill driving the Gulf Stream. The warm water tries to spread out and flow over cooler surface waters far from the center of the gyre. The Coriolis effect makes it move around the hill, not down it, and the boundary between this warm, rotating mass of water and the cooler water it is trying to flow over is where the currents are most obvious. This is the Gulf Stream, a distinct boundary between the productive coastal waters, which are green

and teeming with life, and the dark blue, nearly lifeless Sargasso Sea waters.

## GULF STREAM AS A BOUNDARY CURRENT

For decades, schoolchildren have been taught that the mild climate of the British Isles and Western Europe gets its heat from the Gulf Stream. The Gulf Stream is often presented as a river of warm water moving north and then east, eventually to deliver its heat to the European continent. The significance of the Gulf Stream as a source of heat changes when it is recognized for what it is, just a boundary current. It separates the huge hill of warm, saline Sargasso Sea water from the colder surrounding waters.

The Gulf Stream is a very active boundary region with motions driven by the wind and geostrophic currents that surround the huge quantity of surface water, heated by the sun in subtropical high-pressure zones and made especially salty by accompanying evaporation. The Gulf Stream gets all the press, but it is actually this huge quantity of water that conveys heat to Europe.

In the North Atlantic, beyond the northern extent of the circulating gyre, cold winds remove heat from the surface waters. In the process, these winds are warmed and bring pleasant temperatures to Europe. However, by removing heat, these winds cool the saline surface waters until they become dense enough to sink to the bottom of the sea. This is called thermohaline circulation. The sinking waters are replaced by a gradual northward flow of surface water. It is likely that most of this surface water spent time in the North Atlantic Gyre, but its transport northward was independent of that circular motion. If the gyre were to stop tomorrow, the thermohaline circulation would continue, and Europe would stay just as warm as it is today. In fact, some researchers have suggested that the strength of the Gulf Stream actually reduces the warming effects of this thermohaline circulation. If they are correct, were the Gulf Stream to stop, Europe might grow warmer.

Although its role as a heat-delivery system may have been overstated, the Gulf Stream is still an incredibly powerful element of oceanic circulation and, as such, greatly influences the biology and chemistry of the surface ocean. This significance is easily seen where meanders develop on the Gulf Stream, typically beyond Cape Hatteras. Just as meanders can grow and develop on a slow-moving river, they also develop on the Gulf Stream. Whereas a stream meander may form an oxbow lake if the course of the stream closes in upon itself, a meander in the Gulf Stream produces a circulating eddy separated from the rest of the stream when it closes upon itself. These eddies will have a core of warmer, less fertile water and rotate in a clockwise manner if they close off on the southeastern side of a meander. They will have a core of cooler, more fertile water and rotate in a counterclockwise manner if they close off on the northwestern side of a meander. These rings persist for months to a year or more and may establish their own ecosystems during their lifetimes.

The Gulf Stream disperses eggs, seeds, and juvenile and adult organisms. As a chemical agent, it stirs up the surface waters, keeping its warm waters well mixed. As a physical agent, it moves enormous quantities of water. It is clearly a remarkable current and a very important part of the global ecosystem.

## STUDY OF THE GULF STREAM

The Gulf Stream is studied by directly measuring the strength of its currents at different depths and locations and by examining the effects of its currents through monitoring the position of floats released within it and designed to stay at particular depths.

Floating current meters can be moored to anchors at the bottom of the sea. Their depth is controlled by the length of the tether keeping them attached to the anchor. They can record data electronically, storing it in computer memory. When a vessel is at the surface, ready to retrieve the meter and its data, it transmits a special coded sound pulse. This instructs the meter to release itself from the tether and rise to the surface, where it transmits a radio signal allowing the vessel to home in on it for recovery. The data are incorporated in complex computer models that tie together the results obtained from hundreds of current meters deployed during overlapping time periods. Snapshots of the current system can then be obtained, and sequences of these snapshots reveal the behavior of the currents over time.

Floating objects can be released at sea and tracked by satellite. To ensure that these objects are being moved by ocean currents rather than surface winds, they usually have a large parachute or sail deployed in the water beneath them. Because the density of seawater increases with depth, floats are often designed to have neutral buoyancy at a particular depth (neither sinking below, nor floating above that depth). A layer of the ocean (the SOFAR channel) acts as a wave guide for sound waves. Floats in this layer can transmit sounds over tremendous distances, permitting them to be tracked efficiently by a small number of surface ships with sonar receivers suspended into this layer.

Because geostrophic currents are driven by the slopes of the ocean's surface, any technique that can measure those slopes can provide valuable insight into the driving forces behind the Gulf Stream. These slopes are very gradual—total dynamic relief over all the world's oceans is about 2 meters—and, consequently, direct measurement is difficult. Satellite techniques, coupled with computer models to filter out waves, tides, and dozens of other confounding effects, are approaching the point where they will be able to measure this topography. Yet this dynamic topography is generally determined indirectly by measuring temperature and salinity as a function of depth and position. These data are used to determine the density of the seawater as a function of depth and position. By assuming that at some depth the horizontal pressure gradients have disappeared, it is possible to reconstruct the differences in the height of the water column needed to accommodate these variations in density. Then the velocities and directions of the resulting currents can be calculated. When the theoretical models are compared with currents measured by moored meters or revealed by the paths of floating objects, the results are in agreement. This gives strong support to the theoretical concepts underlying the study of ocean currents.

Many of the approaches used to study of the Gulf Stream are exercises in applied mathematics. That computed and measured results agree so well is a triumph of geophysical fluid dynamics.

## SIGNIFICANCE

The North Atlantic Gyre, which has the Gulf Stream as its northwest boundary current, dominates surface flow in the Atlantic Ocean. Voyages of discovery, exploration, conquest, and exploitation all were affected to some extent by this system and the winds that accompany it.

The Gulf Stream, studied since the eighteenth century, has provided the basis for much of what scientists know of ocean currents. Elaborate mathematical constructs, including the entire concept of geostrophic currents, have been developed to describe, analyze, and comprehend this powerful system.

As scientists have learned more about the Gulf Stream, they have discovered many new areas of study, including branches in the stream, counter-currents at the surface and at depth, and fluctuations in flow and velocity with time and place. Some researchers devote their entire careers to studying just the rings of the Gulf Stream, which contain entire ecosystems. Others investigate the location and strength of the Gulf Stream in the distant past, during and even before the ice ages. There is evidence that at times the current has taken different paths across the continental shelf, perhaps scouring out valleys in the ocean floor in the process.

Scientists' understanding of the dynamics of the planet relies on comprehending the transfer of energy from the equator to the poles. The Gulf Stream is an important component of this transfer. As people have become more aware of the fragility of the environment and become more concerned about issues of global climate change, the role of the Gulf Stream and thermohaline circulation in influencing temperatures in Europe and elsewhere on the planet has taken on a new importance.

*Otto H. Muller*

**FURTHER READING**

Allen, Philip A. *Earth Surface Processes*. Oxford: Blackwell Science, 1997. Serves as a clear introduction to oceanography and the earth sciences. Details the processes, properties, and composition of the ocean. Color illustrations, maps, index, and bibliography.

Broecker, Wally. *The Great Ocean Conveyor*. Princeton, N.J.: Princeton University Press, 2010. Discusses ocean currents, focusing specifically on the great conveyor belt. Explains the conception of this theory and the resulting impact on oceanography. Written in an easy-to-follow manner for readers with some background in science, yet still relevant to graduate students and scientists.

Colling, Angela. *Ocean Circulation*. 2d ed. Oxford: Butterworth-Heinemann, 2001. A textbook for the Open University Oceanography course. Well illustrated. Discusses the effects of ocean circulation as reflected in various phenomena.

MacLeish, William H. *The Gulf Stream*. Boston: Houghton Mifflin, 1989. Compelling reading.

Offers a personal approach to the Gulf Stream and many of its effects. Organized as a narrative rather than a scientific tome, making it easy to read and suitable for high school readers. No equations and few technical discussions, but represents an obviously thorough research effort.

Segar, Douglas. *Introduction to Ocean Sciences*. 2d ed. New York: W. W. Norton & Co., 2007. Comprehensive coverage of all aspects of the oceans, their chemical makeup, and circulation. Readable and well illustrated. Suitable for high school students and above.

Teramoto, Toshihiko. *Deep Ocean Circulation: Physical and Chemical Aspects*. New York: Elsevier, 1993. Provides a detailed look at ocean circulation and currents. Provides much information on the chemical processes that occur in the deep ocean. Illustrations, maps, and bibliographical references. College-level text.

Ulanski, Stan L. *The Gulf Stream: Tiny Plankton, Giant Bluefin, and the Amazing Story of the Powerful River in the Atlantic*. Chapel Hill: University of North Carolina Press, 2008. Explores the science and history of the Gulf Stream with regard to the ways that it affects and is affected by the surrounding environment. An interesting compilation of current information on the Gulf Stream. Written for the general public.

Vallis, Geoffrey K. *Atmospheric and Oceanic Fluid Dynamics: Fundamentals and Large-scale Circulation*. New York: Cambridge University Press, 2006. Begins with an overview of the physics of fluid dynamics to provide foundational material on stratification, vorticity, and oceanic and atmospheric models. Discusses turbulence, baroclinic instabilities, wave-mean flow interactions, and large-scale atmospheric and oceanic circulation. Best suited for graduate students studying meteorology or oceanography.

Voituriez, Bruno. *The Gulf Stream*. New York: UNESCO, 2006. Examines complex scientific information to describe the causes and dynamics of the Gulf Stream, through the history of its discovery to current exploration efforts.

**See also:** Atmospheric Properties; Barometric Pressure; Carbonate Compensation Depths; Cyclones and Anticyclones; Deep Ocean Currents; Global Energy Transfer; Hydrothermal Vents; Ocean-Atmosphere

Interactions; Ocean Pollution and Oil Spills; Oceans' Origin; Oceans' Structure; Ocean Tides; Ocean Waves; Sea Level; Seamounts; Seawater Composition; Surface Ocean Currents; Tsunamis; Turbidity Currents and Submarine Fans; World Ocean Circulation Experiment

# H

# HUDSON BAY

*Although considered an extension of the Atlantic Ocean, Hudson Bay, a shallow gulf covering 827,000 square kilometers in the northern part of east-central Canada, is often treated as a separate entity because of its isolation. The area remains a vast, essentially unspoiled wilderness and preserve for wildlife. Uplift of lands surrounding Hudson Bay has provided earth scientists with clues about processes associated with the retreat of ice sheets, especially crustal rebound.*

## PRINCIPAL TERMS

- **Canadian Shield:** the geologic core of North America, extending over north-central Canada, that experienced glaciation during the Pleistocene epoch; characterized by an undulating surface of moderate relief and containing the oldest dated rock formations on the planet
- **craton:** an area of the land surface that has been stable for millions of years
- **entisol:** a weakly developed soil layer that does not exhibit distinct horizons or stratification layers
- **histosol:** soil composed primarily of organic material
- **ice sheet:** a broad, flat glacial mass with relatively gentle relief; ice sheets once covered extensive portions of North America
- **inceptisol:** relatively recent soil deposits that exhibit the first signs of horizon differentiation
- **isostatic rebound:** a process based on the opposing influences of buoyancy and gravity within Earth's crust by which the surface adjusts itself vertically until these forces are balanced and isostatic equilibrium has been reached
- **Pleistocene epoch:** the most recent ice age period, during which Earth experienced cycles of continental glaciation
- **spodosol:** an acidic soil characterized by subsurface accumulations of humus complexed with aluminum and iron

## LOCATION AND DISCOVERY

Ranking twelfth in area among Earth's seas and oceans, the horseshoe-shaped Hudson Bay is approximately 1,370 kilometers long and 1,050 kilometers wide. The southern end of Hudson Bay extends into a smaller bay of almost identical shape—a "bay within a bay"—called James Bay. The Hudson Bay is bounded by the Canadian provinces of Quebec to the east and south, Ontario to the south, Manitoba to the southwest, and Nunavut to the northwest (Nunavut was formed in 1999 by partition of the Northwest Territory). The bay is linked to the Atlantic Ocean by Hudson Strait and to the Arctic Ocean by the Foxe Channel and Roes Welcome Sound. Hudson Bay and its islands are administered by the District of Keewatin, a region within Canada's Nunavut Territory. Its largest islands are Southampton (41,214 square kilometers) and Mansel (3,181 square kilometers), both located in the north. Other large islands include Coats (5,499 square kilometers), located on the southeast side of Fisher Strait, and Akimiski (3,002 square kilometers) within James Bay. Island groups include the elongated Belcher Islands and smaller Nastapoka Islands positioned in the southeastern portion of the bay and the Ottawa Islands to the west of Ungava Peninsula.

Early exploration of the bay was driven by the hope of locating a northwesterly shortcut to Asia. Systematic investigations of such a route began with John Cabot's visit to Canada's eastern shore in 1497. Hudson Bay was discovered by and named for English navigator Henry Hudson. During his first expedition in 1607, Hudson and the crew of his ship, the *Hopewell*, landed on the shores of Greenland and the Svalbard Islands, later turning north in an unsuccessful attempt to find a route to East Asia by way of the Arctic Ocean. He returned the following year to renew the search, passing the Novaya Zemlya Islands in the Barents Sea. Hudson began his last expedition in 1610, during which he reached Hudson Bay and spent three months investigating its eastern shores and islands. Pack ice prevented a departure from the bay, and, after a winter of extreme deprivation, the crew of his ship mutinied. Placed in a boat and set adrift by the mutineers, Hudson and eight others were never seen again.

Further exploration of the bay was carried out by Sir Thomas Button, who reached the western shore of the bay near present-day Churchill, Manitoba, in 1612. The following summer, Button departed from the bay, passing Southampton Island on his way to Hudson Strait. Others who explored the bay included William Baffin in 1615 and Luke Fox and Thomas James in 1631. In 1662, Pierre Esprit Radisson and Médard Chouart de Groseillers became the first to reach Hudson Bay using an overland route. Since its first exploration, the bay has served as a graveyard for many mariners. Numerous sunken vessels, along with the only stone military fortification in the Arctic, are part of the bay's historical legacy.

## GEOLOGY

Hudson Bay is located within a depression of the vast Canadian Shield and is underlain with ancient Precambrian rocks that are more than 500 million years old. Covering 4.8 million square kilometers in Canada and the northern United States, the shield encompasses Labrador, Baffin Island, and portions of Quebec, Ontario, Manitoba, Saskatchewan, the Northwest Territory, Nunavut, Wisconsin, Minnesota, New York, and Michigan. The oldest region within the North American crustal plate, the shield is considered a craton, an area of the land surface that has been stable for many millions of years and perhaps even from the original formation of the continental masses. Radiometric dating of some Canadian Shield formations has yielded an age of 4.5 billion years, which is nearly as old as the planet itself is believed to be. Hudson Bay is bordered by Paleozoic limestone, sandstone, and dolomitic rocks. Rocks of sedimentary origin within the Canadian Shield contain the fossils of some of the earliest forms of life on Earth.

The bay is believed to have been formed by glacial activity during the Pleistocene epoch, which began about 2.4 million years ago. Centered over what is now Hudson Bay, the Laurentide ice sheet stripped away soil and deposited glacial drift as it moved across the landscape. The enormous weight of accumulated ice on the continent depressed that portion of the crust, causing it to sink into the underlying plastic layer of the mantle. As the ice sheet began to melt, numerous remnants were left in the form of till deposits and long, sinuous ridges of sand and gravel called eskers. The retreat of ice sheets of the Wisconsin glacial period began about 13,000 years ago, although large portions of glacial ice remained until about 7,000 years ago. As the ice melted, relatively warm seawater invaded Hudson Bay through Hudson Strait, further shrinking the ice mass. Eventually, the ice sheet was split into the Labrador and Keewatin ice centers, which disappeared completely between 6,500 and 5,000 years ago. As the ice melted, the depressed crust began to slowly undergo isostatic rebound, slowly returning to its natural level as determined by the buoyancy of the continental mass on the mantle layer. Isostatic adjustment within the Hudson Bay region is not complete, and the entire area surrounding the bay continues to rise at a rate of about 0.6 meter per century. Uplifting of the landscape is especially obvious along portions of the coast where lines representing former beaches run parallel to the shore. The rapid rate of rebound suggests that Hudson Bay will become much shallower and may disappear when isostatic equilibrium has eventually been reached.

The underwater physiography of Hudson Bay demonstrates broad contours that are generally concentric with its periphery. The bay averages 128 meters in depth, with a deepest known depth of 183 meters. Its floor is predominantly smooth but is incised with cuts and banks in some places. There are a few submarine troughs and ridges with trends that are controlled by the geologic structure. The basin forming Hudson Bay has a north-south elongation, which is a reflection of its bedrock structure.

Marine currents are the most significant agents of sediment transport within the bay. Because Hudson Bay is covered by snow and ice during several months of the year, the normal movement of sedimentary discharge from streams is inhibited during those periods. In some cases, sediments, coarse gravel, and boulder-sized rocks are carried by moving ice away from shore in a process known as ice rafting. Although concentrated within 30 to 36 kilometers of the shoreline, materials carried by ice rafting are strewn over the entire floor of the bay. In most places within the bay, but particularly in the western half, postglacial sedimentation has become a less important factor than marine erosion in shaping underwater physiography.

## COASTAL LANDSCAPE AND WATER

Typical shoreline areas of the northern portion of the bay are low in elevation, rocky, and indented with numerous inlets and small islands. In other areas, the bay is surrounded by a broad, flat plain. A few higher areas (with elevation exceeding 305 meters) are located to the north and northeast on Southampton Island and Quebec's Ungava Peninsula. The size of Hudson Bay's drainage basin is more than 3,800,000 square kilometers. Shoreline areas on the west and southwest coasts form an extensive region of drowned swampland. Deltas and estuaries are common, and, in some locations, tidal flats extend up to 9 kilometers inland. Tundra surrounding much of the bay is, in essence, a cold desert in which moisture is scarce. Plant cover in the tundra includes grasses, lichens, and a scattering of low-growing shrubs. Muskeg or bogs with small black spruce trees are found in the south, especially around James Bay. Plants must complete their annual cycles during the brief summer in waterlogged environments, which are the result of poor drainage tied to the underlying permafrost. Soils on land areas surrounding the bay include infertile entisols, histosols, inceptisols, and spodosols. A discontinuous region of permafrost extends southward to 54 degrees north latitude on the western side of Hudson Bay and 61 degrees north latitude on the eastern side.

Arctic in nature, waters found within Hudson Bay demonstrate fairly uniform temperatures that average near freezing. Decreasing from the periphery of the bay to the center, temperatures are slightly warmer over underwater shoals and cooler over deeper areas, reflecting seasonal warming near the surface layer. The general pattern of water circulation is counterclockwise. River outflow maintains an influx of freshwater that inhibits saline Atlantic waters from entering Hudson Strait, resulting in the low salinity of Hudson Bay waters, especially in spring and summer months. The difference between high and low tide ranges from just less than 1 meter to greater than 4 meters in the western portion of the bay. Pack ice blankets the water from October to June each year. Strong prevailing winds along most of the shoreline help to separate land from pack ice. The bay has historically been navigable for a short period of time extending from early July to October, though in recent years this period has steadily lengthened, presumably due to the effects of global warming, and a fully closed ice pack has not formed.

Hudson Strait connects Hudson Bay with the Atlantic Ocean to the east. Unlike in the bay, water within the strait is predominantly deep, with recorded depths of up to 500 meters. Undersea cliffs and canyons are found near the strait's shoreline. Water from the West Greenland Current mixes with water passing out of Hudson Bay within the strait, creating conditions that support plankton and a diversity of invertebrate and fish species.

## CLIMATE

Hudson Bay's climate is influenced by cold, dry, and stable continental polar and Arctic air masses. During much of the year, the bay is normally covered with a bleak and inhospitable blanket of snow and ice. A diagonal line running from Chester Inlet in the northwest to the Belcher Islands in the southeast divides Hudson Bay into two climatic regions. To the northeast, the bay has a Köppen classification of ET (tundra), while the southwest is classified Dfc (continental taiga climate). Average January temperatures for most of the land area surrounding Hudson Bay range from −30 to −20 degrees Celsius. In contrast, average July temperatures range from 10 to 20 degrees Celsius. The coldest land area adjacent to the bay is located on its northwest side, extending from Manitoba's border with Nunavut to Southampton Island. Precipitation totals throughout the region surrounding Hudson Bay are highest in the summer. The influence of the bay's continental climate can be illustrated in a comparison with Scotland, which is located at about the same latitude. In contrast to conditions around Hudson Bay, Scotland's maritime climate, moderated by the Atlantic Ocean, supports pastures for cattle and sheep.

## WILDLIFE AND HUMAN OCCUPATION

During the summer months, the fauna of Hudson Bay is dominated by birds and insects, especially mosquitoes and flies. The rocky coastline and islands, along with tidal flats and inland marshes, provide nesting sites for one of the world's largest concentrations of migrating waterfowl and shorebirds. Along with almost one-half of the eastern Arctic's population of lesser snow geese are Canada geese, Brant geese, old-squaws, loons, black guillemots, and common eiders. Also found within the area are gulls, Hudsonian godwits, whimbrels, snowy owls, horned larks, ptarmigan, and the world's largest concentration of peregrine falcons. Caribou, arctic hare, and lemmings are also found on its shores. One of the highest densities of polar bear denning in Canada is found adjacent to Churchill, where a large number of bears gather during the autumn to await the return of the ice and better feeding. More than forty freshwater, Arctic, and subarctic marine fish species are found within the bay. Examples include capelin, Arctic cod, ogac, and Arctic char, as well as salmon, cod, halibut, and plaice. Also inhabiting the bay are seals, whales, dolphins, walruses, and beluga whales.

The earliest occupants of the area were the nomadic Inuit people, who lived in igloos and skin tents. In 1670, the Hudson's Bay Company received a charter from the English crown for exclusive trading rights within the watershed of Hudson Bay. Soon after, trading posts were constructed at the mouths of the Moose and Albany Rivers. Between 1682 and 1713, the French attempted to force the British out of the bay. However, by the terms of the 1713 Treaty of Utrecht, the French handed over all posts in the area to the British. In 1929, the Hudson Bay Railway was completed to the town of Churchill, facilitating shipments of grain produced in Canada's prairie provinces by boat to world markets. During World War II, the United States operated an air base near Churchill on the western side of the bay. The shoreline of Hudson Bay remains sparsely settled, with a few small trading villages located at the mouths of rivers entering the bay in Quebec, Manitoba, and Ontario. Pack ice filling the bay makes villages on its shores inaccessible for much of the year. The shoreline has a population density of fewer than three people per square mile.

*Thomas A. Wikle*

## FURTHER READING

Dickson, R. R., Jens Meincke, and Peter Rhines, eds. *Arctic-Subarctic Ocean Fluxes: Defining the Role of the Northern Seas in Climate.* Dordrecht: Springer, 2008. Describes the Hudson Bay's role as a major oceanic body and one of the largest of Arctic seas. Examines the bay's notable impact on the climate of northern Canada, as well as its global climate influences. Collects much of the evidence needed for the development of climate models that accurately incorporate the influence of the northern seas.

French, Hugh, and Olav Slaymaker. *Changing Cold Environments: A Canadian Perspective.* Oxford: Wiley-Blackwell, 2012. Directed at upper-level undergraduate students. Provides a comprehensive overview of the changing nature of Canada's "cold environments," such as the Hudson Bay region, and discusses the implications of ongoing global climate change for cold environments globally.

Hood, Peter J., ed. *Earth Science Symposium on Hudson Bay.* National Advisory Committee on Research Geological Survey of Canada Paper 68-53 (1968). Although somewhat dated, this compendium of scientific papers on Hudson Bay provides a good overview of the geomorphology of Hudson Bay and the Hudson Bay Lowlands.

Levinton, Jeffery S., and John R. Waldman. *The Hudson River Estuary.* New York: Cambridge University Press, 2006. Discusses the geology, physics, and chemistry of the Hudson Bay, including sedimentary processes of the estuary, as well as microbial and nutrient dynamics and primary production of the estuary. Covers ecology and conservation efforts. Well organized and well indexed.

Middleton, Gerard V., ed. *Encyclopedia of Sediments and Sedimentary Rocks.* Dordrecht: Springer, 2003. Cites a vast number of scientists, who are listed in the author index. Covers biogenic sedimentary structures, Milankovitch cycles, deltas and estuaries, and vermiculite. Provides an index of subjects as well. Designed to cover a broad scope and a high degree of detail useful to students, faculty, and professionals in geology.

Mills, Eric L. *The Fluid Envelope of Our Planet: How the Study of Ocean Currents Became a Science.* Toronto: Toronto University Press Inc., 2009. A detailed account of the history of oceanography, in which the study of current flow in Hudson Bay has played a significant role.

Pienitz, Reinhard, Marianne S. V. Douglas, and John P. Smol. *Long-Term Environmental Change in Arctic and Antarctic Lakes.* Dordrecht: Springer, 2004. Written for students and advanced researchers studying earth, atmospheric, and environmental science. Focuses on paleolimnology as a source of environmental records in research.

Riley, John L. *Flora of the Hudson Bay Lowland and Its Postglacial Origins.* Ottawa: National Research Council (Canada) Research Press, 2003. Discusses recent research on the Hudson Bay lowlands, documenting the region's biodiversity. Covers geological history and ecology of the area. Examines the climate of the wetland. Provides many maps and color illustrations to supplement the text.

Ruddiman, W. F., and H. E. Wright. *North America and Adjacent Oceans During the Last Deglaciation.* Boulder, Colo.: Geological Society of America, 1987. Examines physical processes and causes associated with ice-sheet erosion in North America during Pleistocene glaciations, as well as the long-term history of North American ice sheets. Includes detailed charts, diagrams, and maps.

**See also:** Aral Sea; Arctic Ocean; Atlantic Ocean; Black Sea; Caspian Sea; Gulf of California; Gulf of Mexico; Gulf Stream; Hurricanes; Indian Ocean; Mediterranean Sea; Mississippi River; Monsoons; North Sea; Ocean Tides; Pacific Ocean; Persian Gulf; Red Sea

# HURRICANE KATRINA

*A hurricane is a tropical storm with sustained winds exceeding 64 knots (74 miles per hour) and with heavy rains and the ability to spawn tornadoes. The center of Hurricane Katrina passed southeast of New Orleans, Louisiana, on August 29, 2005, causing significant damage and loss of life. An understanding of the anthropogenic and natural factors contributing to the extent of the damage resulting from Katrina and storms like it is essential to planning and preparation for such storms.*

## PRINCIPAL TERMS

- **atmospheric pressure:** force exerted on a surface by the weight of air above that surface; measured in force per unit area
- **barrier islands:** coastal landforms that run parallel to the mainland; formed by the action of waves and currents; protect the mainland from storms and other weather events
- **Coriolis effect:** natural effect causing objects (including fluids) to appear to skew to the right of their destination in the Northern Hemisphere and to the left of their destination in the Southern Hemisphere
- **eye:** the center of a hurricane; area of lowest surface pressure with calm, clear weather
- **eye wall:** just outside the eye of the hurricane; area with peak winds and rain
- **hurricane:** severe tropical storm with winds greater than 64 knots (74 mph); it originates in the equatorial regions of the Atlantic Ocean or the Caribbean Sea or the eastern regions of the Pacific Ocean; travels north, northwest, or northeast from point of origin; usually involves heavy rains; hurricans are known as cyclones in the Southern Hemisphere and west of the international date line
- **levee:** an embankment built to prevent the overflow of a body of water
- **low-pressure area:** region where the atmospheric pressure is lower than that in surrounding areas
- **Saffir/Simpson hurricane scale:** hurricane category scale that classifies storms according to their winds and effects and their storm surge; categories range from 1 (least severe) to 5 (most severe)
- **storm surge:** an offshore rise in water associated with a low-pressure weather system
- **topography:** study of the earth's features and surface shape

## PROPERTIES OF HURRICANES

Hurricanes form in tropical regions where the ocean temperature is at a minimum 27 degrees Celsius (80 degrees Fahrenheit) to a depth of 46 meters (150 feet) or more. Hurricanes begin as storms that result from warm air rising off the surface of the ocean. This movement leaves an area of low pressure between it and the surface of the earth. More air rushes in to replace the rising air, and that air then heats and rises. This activity creates an area of thunderstorms that begins to spin with the earth's rotation.

If conditions are right and no strong winds exist in the upper atmosphere to disperse the storm, the counterclockwise motion of the storm (in the Northern Hemisphere) will intensify and an eye will form in the center of the hurricane. The hurricane will continue to gain strength as the air above the warm ocean waters continues to rise and cooler air rushes in to fill the space left behind.

A tropical storm becomes a hurricane when its sustained winds reach a speed of 64 knots (74 mph) or more. A hurricane can be up to 965 km (600 mi) wide and can travel at speeds of 8.5 to 17 knots (10 to 20 mph) over the open ocean. Because they get their power from the heat and energy they gather during their contact with warm ocean waters, hurricanes lessen in intensity as they move into cooler waters or over land. Hurricanes can pick up speed and intensity if their paths take them back across warm water.

A hurricane's high winds and rain will increase in intensity as the eye comes near to a given location, with the heaviest rain and wind experienced when the eye wall passes. Shortly after, the calm of the eye arrives with light winds and sunny weather. This calm is followed by the severe winds and rains of the side of the eye wall opposite the one first experienced. Finally, the winds of the outer regions of the hurricane pass.

The damage done by a hurricane is not limited to that caused by its wind and rain. A storm surge, also associated with a hurricane, is a rise in the water levels in front of the hurricane, caused by the low-pressure system associated with the hurricane. The extent of a storm surge varies with the intensity of the hurricane, but a category 1 hurricane includes a storm surge of 1 to 1.5 m (4 to 5 ft), while a category 5 hurricane, such as Katrina, results in a storm surge of more than 5.5 m (18 ft). For cities such as New Orleans, which are at or near sea level, the extent of the storm surge can have serious consequences.

Hurricanes are classified by categories based on a system known as the Saffir-Simpson hurricane scale. This scale takes several attributes of a tropical storm into account and rates the storm's intensity. At the bottom of the scale is category 1, which is a storm with sustained winds of 64 to 82.5 knots (74 to 95 mph) and relatively minimal associated damage. Category 5 hurricanes are expected to cause catastrophic damage from sustained winds of more than 135.5 knots (156 mph) and an extensive storm surge.

## TOPOGRAPHY OF NEW ORLEANS

The city of New Orleans is located in the Mississippi River Delta. To the north of the city is Lake Pontchartrain, the second largest inland body of saltwater in the United States. To the east of the city are Lake Borgne and the Gulf of Mexico. To the south are the Mississippi River and the Gulf of Mexico.

Large areas of New Orleans exist below sea level. Because of this, a series of levees and pumps are in constant action to keep the city dry. A levee is a type of dike or dam that runs along a river, canal, or other water body and acts as a wall to keep water from flowing into a protected area. In New Orleans, many of the levees are large earth embankments; levees, however, can be made from other materials too. Pumping stations ensure the protected areas remain dry. In New Orleans, the Lower Ninth ward (city district) is especially dependent upon the levees and pumps because of its location: south of Lake Pontchartrain, north of the Mississippi River, and at the nexus of several canals.

New Orleans and the rest of the Gulf Coast are not protected by barrier islands. The wetland areas that historically offered some protection have been disappearing at the rate of 50 acres each day. This has been caused by the conversion of more than 80 percent of wetlands to agriculture and urban areas and is associated with controls placed on the Mississippi River. Some of the barrier island losses, however, come from sinking, as the relative sea-level rises with climate change. Because of this loss and the erosion of

the land along the coast, the existing levees and the city of New Orleans itself are sinking farther below sea level.

## HURRICANE PAM SIMULATION

A major port and the largest city and metropolitan region in Louisiana, New Orleans is home to more than 1 million people. The city's topography and resulting vulnerability to hurricanes have long been of concern to those whose role it is to protect the city and its residents and businesses.

Facing more violent hurricanes because of the warming of the earth's oceans, more than 270 officials from all levels of government around the United States met in July, 2004, for a hurricane simulation developed by the Federal Emergency Management Agency and other government and private agencies and groups. The event, which lasted eight days and used computer models developed at Louisiana State University, was designed to test what would happen if a category 3 or stronger hurricane hit New Orleans.

Simulation participants estimated that a storm like the theoretical Hurricane Pam would require the evacuation of more than 1 million residents, would lead to the destruction of more than one-half million buildings, and would lead to the creation of a 23 million cubic meter (30 million cubic yard) debris field along with 181 cubic meters (237 cubic yards) of household hazardous waste. One thousand shelters would be required (only 784 existed at the time), and the shelters would likely remain open for one hundred days. Up to eight hundred searchers and rescue personnel would be needed to locate and assist stranded residents. There also would be a need to relocate patients from vulnerable hospitals and to resupply those hospitals after the hurricane passed.

## HURRICANE KATRINA

In 2005, in preparation for Katrina, the parishes (counties) to the south of New Orleans evacuated on Saturday, August 27. The Plaquemines Parish president declared a mandatory phase I evacuation.

One plan, which included the addresses of persons who would need special help, was put into place. Officials worked to ensure compliance as Katrina neared and the forecast for New Orleans became more severe. No mandatory evacuation was ordered. New Orleans also did not have a command center or alternative means of communication.

Hurricane Katrina reached a category 5 designation while in the Gulf of Mexico. By the time the hurricane came ashore, it was a category 3. The levees were designed to protect the city against the storm surge from a category 3 hurricane, but Katrina was a slow-moving storm that had attained category 5 strength. Significant amounts of water in the form of rain and storm surge reached the area. The surge and flooding from the severe rain increased the burden placed on the levees.

As the front of Katrina passed over New Orleans, power was lost and high winds caused damage to buildings. Initial reports indicated that New Orleans had received little damage. As the southern edge of Katrina passed over New Orleans, reports of damage changed. The counterclockwise winds swept over Lake Pontchartrain. The levee system was strained until some portions gave way. Water rushed over their tops, gushing into "the bowl" that made up the city. The resulting flooding caused widespread damage and trapped residents in their homes; many, especially the poor and elderly, were unable to leave the city for several days after the storm because the water, once trapped inside the bowl, had no way to exit. With pumps clogged and inoperable and with the levees breached, the flooding continued until the U.S. Army Corps of Engineers was able to fortify the levees and return the pumps to operation.

Early reports continued to portray New Orleans as having avoided the worst damages, as news media remained unaware of the levee breaches. Once the breaches became known, it also would become clear that many residents of New Orleans had not evacuated. This was particularly problematic in the Lower Ninth ward, where the flooding was severe. Residents in this ward often lacked private transportation. Many also did not have the money to leave or to pay for a hotel or other place to stay if they were to leave. In the absence of a mandatory evacuation and with no way out of town, they stayed in their homes, counting on the levees to hold. When the levees were breached and water flooded the ward, many residents were trapped in their attics or on their rooftops. They waited to be rescued from the floodwaters, which became contaminated with hazardous materials.

Located slightly above sea level, the Louisiana Superdome provided shelter beginning at 8 a.m. Sunday, August 28. At 10 a.m., New Orleans mayor Ray Nagin issued orders for a mandatory evacuation of the city. He also announced that the Superdome would be a shelter "of last resort." The Superdome did not flood but did sustain damage; two large sections of its roof came loose during the hurricane. Officials, who had estimated that ten thousand residents would take shelter at the Superdome, were faced with more than twenty-five thousand residents in need to shelter. This led to a shortage of supplies and facilities to accommodate them.

Wherever they were, residents found themselves without adequate food and water. The lessons of the Hurricane Pam simulation had not led to actual plans and provisions, which would have helped residents after Katrina.

It would take days for state and federal officials to work out the logistics of assisting those in need. Those who had sheltered at the Superdome were bused to shelter in Houston, Texas. The stay would be lengthy—and would even become known as a diaspora—because for many persons no home awaited their return. Close to two thousand people died from the hurricane and its aftermath, and more than two hundred bodies were never claimed. Areas of the Lower Ninth ward remained damaged for years following Katrina, and much of it never recovered.

## GULF COAST IMPACT

Hurricane Katrina caused widespread death and destruction in New Orleans. It also caused an estimated $96 to $125 billion in damages to the Gulf Coast. Some experts estimate the total economic loss to be about $250 billion. This total includes Katrina's longer-term disruption of oil production in the region, the decimated the sugar crop in Louisiana, and the economic effects of halted operations at the twelve casinos in the Mississippi region. In all, 1,836 people died; several hundred remain unaccounted for.

*Gina Hagler*

## FURTHER READING

Birch, Eugenie L., and Susan M. Wachter. *Rebuilding Urban Places After Disaster: Lessons from Hurricane Katrina*. Philadelphia: University of Pennsylvania Press, 2006. Examines the rebuilding of cities and their surrounding areas after a disaster. Focuses on making cities less vulnerable to disaster, reestablishing economic viability, responding to the ongoing needs of the displaced, and re-creating a sense of place.

Brinkley, Douglas. *The Great Deluge: Hurricane Katrina, New Orleans, and the Mississippi Gulf Coast.* New York: Morrow, 2006. Recounts Hurricane Katrina and its widespread devastation. Covers repercussions of the tragedy and its aftermath and the ongoing crisis in the region.

Cap, Ferdinand. *Tsunamis and Hurricanes: A Mathematical Approach.* New York: SpringerWien, 2006. Discusses the mathematics used in the calculation of the strength of tsunamis and hurricanes. Includes formulas, tables, and results of calculations.

Elsner, James B., and Thomas H. Jagger. *Hurricanes and Climate Change.* New York: Springer, 2009. Papers presented at the Summit on Hurricanes and Climate Change in 2007. Gathers the research and opinions of a variety of experts exploring the question of what effect climate change has had on the power of tropical storms.

Freudenburg, William R., et al. *Catastrophe in the Making: The Engineering of Katrina and the Disasters of Tomorrow.* Washington, D.C.: Island Press/ Shearwater Books, 2009. Explores the roles of humans and technology in the Katrina disaster. Also explores the implications for future large-scale hurricanes.

Mogil, H. M. *Extreme Weather: Understanding the Science of Hurricanes, Tornadoes, Floods, Heat Waves, Snow Storms, Global Warming, and Other Atmospheric Disturbances.* New York: Black Dog & Leventhal, 2007. Examines the science behind extreme weather such as tornadoes, heat waves, hurricanes, and floods. Describes the conditions that must exist and other variables that contribute to extreme weather events.

"Scour Power: Big Storms Shift Coastal Erosion into Overdrive." *Science News* 178, no. 5 (2010): 14. Describes the impact on beachfront homes and businesses when a category 2 hurricane hits. Specifically relates to Hurricane Ike and the Bolivar Peninsula in Texas, but the lessons and discussion are applicable to Katrina and other hurricanes.

**See also:** Atmospheric Properties; Barometric Pressure; Beaches and Coastal Processes; Climate; Climate Modeling; Clouds; Cyclones and Anticyclones; Dams and Flood Control; Deltas; Earth-Sun Relations; Floods; Global Energy Transfer; Gulf of Mexico; Hurricanes; Long-Term Weather Patterns; Monsoons; Remote Sensing of the Oceans; Saltwater Intrusion; Satellite Meteorology; Sea Level; Severe and Anomalous Weather in Recent Decades; Severe Storms; Tornadoes; Tropical Weather; Tsunamis; Weather Forecasting; Wind

# HURRICANES

*Hurricanes are cyclonic storms that form over tropical oceans. A single storm can cover hundreds of thousands of square kilometers and have interior wind speeds of 65 to 230 knots (74 to 200 miles) per hour near its eye. Destruction is caused by wind damage, as well as by storm surge and subsequent flooding.*

## PRINCIPAL TERMS

- **condensation:** the process by which water, or any other substance, changes from a vapor state to a liquid state, releasing heat into the surrounding air; this process is the opposite of evaporation, which requires the input of heat
- **Coriolis force:** an apparent force caused by the rotation of the planet, in which objects moving above Earth's surface (such as the wind) deflect to the right in the Northern Hemisphere and to the left in the Southern Hemisphere
- **knot:** a unit of nautical distance equivalent to 1.86 kilometers or 1.15 miles
- **tropical cyclone:** an area of low pressure that forms over tropical oceans, characterized by extreme amounts of rain, a central area of calm air, and winds that attain speeds of up to 300 kilometers per hour rotating counterclockwise in the Northern Hemisphere and clockwise in the Southern Hemisphere
- **tropical depression:** cyclonic thunderstorms with wind speeds from 36 to 64 kilometers per hour
- **tropical storm:** a thunderstorm with cyclonic winds circulating at speeds of 64 to 118 kilometers per hour
- **vortex:** a mass of air, water, or other fluid that spins about a central axis, capable of reaching high velocities

### ANATOMY OF A HURRICANE

Hurricanes are huge, swirling storm systems that can cover thousands—sometimes hundreds of thousands—of square kilometers, averaging about 600 kilometers in diameter. Often called the "greatest storms on Earth," hurricanes have sustained winds of at least 65 knots (74 miles) per hour with maximum wind speeds of 230 knots (200 miles) per hour. In the Western Hemisphere, these storms are called hurricanes. They are also referred to as typhoons in the western North Pacific, cyclones in the Indian Ocean, and *baguios* in the Philippines. The swirling motion of these storms is counterclockwise, or cyclonic, in the Northern Hemisphere and clockwise, or anticyclonic, in the Southern Hemisphere.

Hurricanes are as individual and unique as fingerprints. Their behavior is difficult to predict, even when satellites are used to track them. Scientists who have studied decades of hurricane data have found some patterns. Hurricanes evolve primarily in specific areas of the west Atlantic, east Pacific, south Pacific, western north Pacific, and north and south Indian Oceans. They rarely move closer to the equator than 4 to 5 degrees latitude north or south, and no hurricane has ever been known to have crossed the equator. They are more common during the warmer months of the year depending on their ocean of origin. In the Northern Hemisphere, hurricanes are most common from May to September; in the Southern Hemisphere, the hurricane season typically ranges from December to May. More recently it has been found that the hurricane season has been increasing in length, in accord with increased warming of ocean waters.

The reasons for these patterns are that hurricanes need warm surface waters, high humidity, and winds from the same direction at a constant speed in order to form. Cyclonic depressions can only develop in areas where the ocean surface temperatures are more than 24 degrees Celsius; the eye structure, which must be present for the storm to be classified as a hurricane, requires surface temperatures of 26 to 27 degrees Celsius to form. This means that hurricanes will rarely develop above 20-degree latitudes because the ocean temperatures are never warm enough to provide the heat energy needed for their formation. In the Northern Hemisphere, the convergence of air that is ideal for hurricane development occurs above tropical waters when eastward-moving waves develop under the fairly constant force of the trade winds. The region around the equator is traditionally called "the doldrums" because there is no consistent direction to wind flow. Hurricanes, needing wind to form, can be found as close as 4 to 5 degrees away from the equator. At these latitudes, the Coriolis effect, an apparent deflecting force associated with Earth's rotation, gives the moving air masses the spin necessary to form hurricanes.

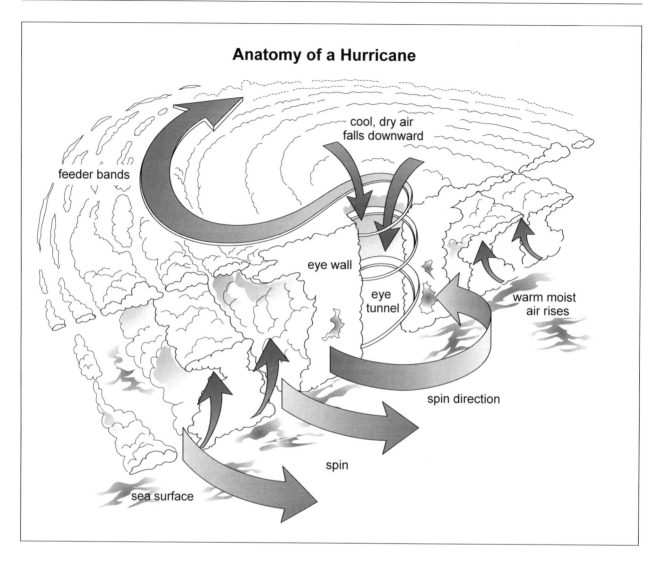

## Anatomy of a Hurricane

In hurricane formation, heat is extracted from the ocean, and warm, moist air begins to rise, forming clouds and causing instability in the upper atmosphere. As the air rises, it spirals inward toward the center of the system. This spiraling movement causes the seas to become turbulent, and large amounts of sea spray are captured and suspended in the rising air. This spray increases the rate of evaporation fueling the storm with water vapor.

As the vortex of wind, water vapor, and clouds spin at a faster and faster rate, the eye of the hurricane forms. The eye, which is the center of the hurricane, is a relatively calm area that has only light winds and fair weather. The most violent activity in the hurricane takes place in the area immediately adjacent to

the eye, called the eye wall. It is in the eye wall that the spiraling air rises and cools, while moisture condenses into droplets that form rainbands and clouds. The process of condensation releases latent heat, which causes the air to rise and generate more condensation. The air thus rises rapidly, creating an area of extremely low pressure close to the storm's center. The severity of a hurricane is often indicated by how low the pressure readings are in the central area of the hurricane.

### EYE FORMATION

As the air moves higher, up to 15,000 meters, the air is propelled outward in a cyclonic or anticyclonic flow, depending on latitude. However, some of the air is forced inward into the eye. The compression of air in the eye causes the temperature to rise. This warmer air can hold more moisture, and the water droplets in the central clouds will evaporate. As a result, the eye of the hurricane becomes nearly cloud-free.

The temperature is much warmer in the eye, especially in the middle and upper levels, than outside of it. Therefore a large pressure differential develops across the eye wall, which establishes the violence of the storm. Waves of 15 to 20 meters are common in the open ocean because of winds around the eye. Winds in a hurricane are not symmetrical around the eye; when one faces the direction that the hurricane is moving, the strongest winds are usually to the right of the eye. The radius of hurricane-level winds (velocities of 119 kilometers per hour or more) can vary from 15 kilometers away from the eye in small hurricanes to 150 kilometers in large hurricanes. The strength of the wind decreases in relation to its distance from the eye.

Depending on the size of the eye, which can range from 5 to 65 kilometers in diameter, the calm period of blue skies and mild winds can last from a few minutes to hours when the storm hits. The calm is deceiving because it is not the end of the storm but only a temporary lapse until the winds from the opposite direction are encountered.

## HURRICANE DAMAGE

Hurricanes, and similar storms that are less intense, are classified by their central pressure and their sustained wind speed. Tropical depressions have wind speeds below 20 to 34 knots per hour, whereas tropical storms have wind speeds of 35 to 64 knots. To be classified as a hurricane, storms must have sustained winds of 65 knots (74 miles) per hour or higher.

The Saffir-Simpson scale further categorizes the intensity of hurricanes into five levels. Category 1 hurricanes are considered weak, with sustained winds of 63 to 83 knots (74 to 95 miles) per hour. They cause minimal damage to buildings but are powerful enough to damage unanchored mobile homes, shrubbery, and trees. Normally they cause coastal road flooding and minor damage to piers. Category 2 hurricanes have winds of 83 to 95 knots (96 to 110 miles) per hour and can easily damage roofs, doors, and windows on buildings. They also cause substantial damage to trees, shrubs, mobile homes, and piers. Flooding of

roads and low-lying areas normally occurs with this level of storm. Category 3 hurricanes are considered to be strong, with winds of 96 to 113 knots (111 to 130 miles) per hour. Such storms are sufficient to destroy mobile homes and can cause structural damage to residences and utility buildings. Flooding from this level of hurricane can destroy small structures near the coast, while larger structures normally sustain damage from floating debris. Flooding from this level of hurricane can extend to 15 kilometers or more inland. Category 4 hurricanes are categorized as very strong, boasting winds of 114 to 135 knots (131 to 155 miles) per hour. These storms can cause major damage to lower floors of buildings and can cause major beach erosion. Residences often sustain roof structure failure and subsequent rain damage. Land lower than 3 meters above sea level can be flooded by storm surge, which may lead to the mass evacuation of residential areas up to 10 kilometers inland. Category 5 hurricanes are severe, classified by having sustained winds greater than 135 knots (155 miles) per hour. These can completely destroy roofs on residential and industrial buildings, and may destroy the buildings themselves. Structures less than 5 meters above sea level can sustain major flood damage to lower floors, and mass evacuations of residential areas 10 to 20 kilometers inland from the shoreline can be required.

Because wind speeds of about 43 knots (50 miles) per hour can break tree branches and cause some damage to structures, hurricane winds are considered to be among the most destructive of natural disasters. The wind force applied to an object increases with the square of the wind speed. A building 30 meters long and 3 meters high that has 100-mile-per-hour hurricane speed winds blowing against it would have 40,000 pounds of force exerted against its walls. This is because a 100-mile-per-hour wind exerts a force of approximately 430 pounds per square meter. If the wind speed was 138 knots (160 miles) per hour, the force against the house would be over 1 million pounds per square meter. Rain blown by the wind also contributes to an increase of the pressure on buildings.

Hurricane Andrew, which hit the United States in 1992, caused wind gusts of at least 280 kilometers per hour in south Florida and caused an estimated $25 billion in damage, making it the most expensive hurricane to hit the United States. In 2005 Hurricane

Katrina caused massive flooding in the city of New Orleans, which is continuing to recover from the effects of that disaster nearly a decade later.

## STORM PATHS

Hurricanes can travel in very sporadic and unpredictable paths. Some will travel in a generally curved path, while others will change course quite rapidly. Their movements can reverse direction, zigzag, veer from the coast back to the ocean, intensify over water, stall, or return to the same area. The paths are affected by pressure systems in the surrounding atmosphere, the prevailing winds, and Earth's rotation. They can also be influenced by the presence of high- and low-pressure systems on the land they encounter. High-pressure areas can act as barriers to hurricanes; if a high is well developed, its outward-spiraling flow will guide the hurricane around its edges. Low-pressure systems tend to attract the hurricane system toward it.

After forming over warm, tropical water, hurricanes generally travel indirectly toward the poles until they lose their energy over cooler waters. An average hurricane can travel from 450 to 650 kilometers per day and more than 1,600 kilometers before it is downgraded to a tropical storm. Rarely, a hurricane can maintain the required wind speed to latitudes as high as 40 degrees.

## STORM SURGE

The greatest cause of death and destruction due to a hurricane comes from the rise in sea level as water is pushed ahead of the high winds, a condition known as storm surge. As the hurricane crosses the continental shelf and moves to the coast, the water level may increase as much as 5 to 7 meters. This is caused by the drop in atmospheric pressure at sea level inside the hurricane, which allows the hurricane's force to pick up the sea while the winds in front of the hurricane will pile the water up against the coastline. This results in a wall of water that can be up to 7 meters tall and 75 to 150 kilometers wide. This wall of water can sweep across the coastline where the hurricane makes landfall. The combination of shallow shore water and a strong hurricane causes the highest surges of water.

If the storm surge arrives at the time of high tide, the water heights of the surge can increase nearly an additional 1 meter. The height of the storm surge

also depends upon the angle at which the storm encounters the mainland. Hurricanes that make landfall at right angles to the coast will cause a higher storm surge than hurricanes that enter the coast at an oblique angle. Often, the slope or shape of the shore and ocean bottom can create a bottleneck effect and cause an even higher storm surge.

Water weighs approximately 2,250 pounds per cubic meter and thus has considerable destructive power. Storm surge is responsible for 90 percent of deaths in a hurricane. The pounding of the waves caused by the hurricane can easily demolish buildings. Storm surges cause severe erosion of beaches and the destruction of coastal highways. Often, buildings that have survived hurricane winds have had their foundations eroded by the sea surge or have been demolished by the force of the pounding waves. Storm tides and waves in harbors can destroy ships. The saltwater that inundates land can kill existing vegetation, and the residual salt left in the soil makes it difficult to grow new vegetation.

The most destructive hurricane in the twentieth century occurred in November 1970 in what is now known as Bangladesh. Of the 500,000 people who were killed, most of them were victims of storm surge. In 1985, the same area was hit by another hurricane, causing the deaths of 100,000 people. In 1989, Hurricane Hugo hit the mainland United States near Charleston, South Carolina. It blew in with sustained winds of more than 113 knots (130 miles) per hour and a 7-meter storm surge. The storm caused economic losses of up to $6 billion and caused damage in North Carolina, Puerto Rico, and the Virgin Islands.

## RAINFALL AND FLOODING

The amount of rainfall received depends on the diameter of the rainband within the hurricane and on the hurricane's speed. One typhoon in the Philippines caused 185 centimeters of rain to fall in a twenty-four-hour period, a world record. Heavy rainfall over a small area that has insufficient drainage can cause flash floods or river floods. Flash floods last from thirty minutes to four hours. The excess water overflows streambeds, damaging bridges, underpasses, and low-lying areas. The strong currents associated with flash floods can move cars off the road, wash out bridges, and erode roadbeds.

Floods in an existing river system develop more slowly. It can take two or three days after a hurricane hits before large rivers overflow their banks. River floods cover extensive areas, last one week or more, and destroy both property and crops. The flood waters eventually retreat, leaving buildings and residences full of mud and ruined furnishings. Rain driven by the wind in hurricanes can cause damage to buildings around windows, through cracks, and under shingles.

## TORNADOES AND OTHER DANGERS

Hurricanes may also spawn tornadoes. The tornadoes associated with hurricanes are usually about one-half the size and power of tornadoes that form in the Midwest, and are of a shorter duration. The area these tornadoes affect is small, usually only 200 to 300 meters wide and about 2 kilometers long. Despite their smaller size, they can be very destructive. Tornadoes normally occur to the right of the direction of the hurricane's movement. Approximately 94 percent of tornadoes occur within 10 to 120 degrees from the hurricane eye and beyond the area of hurricane-force winds.

Tornadoes associated with hurricanes are most often observed in Florida, Cuba, and the Bahamas, as well as along the coasts of the Gulf of Mexico and the South Atlantic Ocean. In 1961, 26 tornadoes were associated with Hurricane Carla; Hurricane Beulah released 115 tornadoes in 1967. Hurricane Camille, one of the most lethal storms to hit the United States, created more than 100 tornadoes in 1969.

Not all dangers due to hurricanes have their origins in the atmosphere. During hurricanes, snakes are driven from their natural habitats by the high inflow of saltwater. They are strong swimmers and can be found along roads, in buildings, and in high, dry places. Many people who are bitten by snakes during hurricanes have problems receiving medical attention for their bites because of breakdowns in communications and transportation.

*Toby Stewart and Dion Stewart*

## FURTHER READING

Ahrens, C. Donald. *Meteorology Today: An Introduction to Weather, Climate and the Environment.* 8th ed. Belmont, Calif.: Thomson Brooks-Cole, 2007. One of the most widely used and authoritative introductory textbooks for the study of meteorology and climatology. Explains complex concepts in a clear, precise manner and supports them with numerous images and diagrams. Discusses hurricanes and the mechanisms that generate them extensively.

Bryant, Edward A. *Natural Hazards.* 2d ed. Cambridge, England: Cambridge University Press, 2005. Provides a sound scientific treatment for the educated layperson. Readers should have a basic understanding of mathematical principles. Presents many case studies. Contains photographs, tables, figures, and a glossary of terms.

Emanuel, Kerry. *Divine Winds: The History and Science of Hurricanes.* New York: Oxford University Press, 2005. Covers a wide variety of topics on a number of hurricane events and related research. Presents events in chronological order with related topics scattered between. Written in a popular style, but still contains portions with "hard science." Includes multiple useful appendices, a list of sources, further reading, credits, and indexing.

Fitzpatrick, Patrick J. *Contemporary World Issues: Hurricanes.* 2d ed. Santa Barbara, Calif.: ABC-CLIO, 2006. A reference work that provides background material on the issues, people, organizations, statistics, and publications related to hurricanes.

Mooney, Chris. *Storm World: Hurricanes, Politics and the Battle Over Global Warming.* Orlando, Fla.: Harcourt, 2007. Provides a clear discussion of the nature of hurricanes and the perceived changes in their nature that are still hotly debated in regard to their relationship with global warming. Very readable, presenting very technical information in a nontechnical manner.

Pielke, R. A., Jr., and R. A. Pielke, Sr. *Hurricanes: Their Nature and Impacts on Society.* New York: John Wiley & Sons, 1997. A very informative book written by a meteorological team. Focuses on the United States, integrating science and societal policies in response to these storms.

Prothero, Donald R. *Catastrophes!: Earthquakes, Tsunamis, Tornadoes, and Other Earth-Shattering Disasters.* Baltimore: Johns Hopkins University Press, 2011. Provides a detailed and clear explanation of the many natural and anthropogenic disasters facing our planet. Each chapter is devoted to a different catastrophe, including earthquakes, volcanoes, hurricanes, ice ages, and current climate changes.

Robinson, Andrew. *Earth Shock: Hurricanes, Volcanoes, Earthquakes, Tornadoes, and Other Forces of Nature.* London: Thames and Hudson, 1993. Provides a good mix of science, individual event summaries, and note-worthy facts and figures. Written for high school students and adults with no background in science.

Rosenfeld, Jeffery. *Eye of the Storm: Inside the World's Deadliest Hurricanes, Tornadoes and Blizzards.* Cambridge, Mass.: Basic Books, 2003. Described as a "must read for students and meteorologists." Provides a detailed inside look at the title phenomena. Offers readers a foundation in the similarities and differences between these atmospheric phenomena.

Schneider, Bonnie. *Extreme Weather: A Guide to Surviving Flash Floods, Tornadoes, Hurricanes, Heat Waves, Snowstorms, Tsunamis and Other Natural Disasters.* Basingstoke: Palgrave Macmillan, 2012. Presents vivid explanations of how, when, and why major natural disasters occur. Discusses floods, hurricanes, thunderstorms, mudslides, wildfires, tsunamis, and earthquakes. Presents a guide of how to prepare for and what to do during an extreme weather event, along with background information on weather patterns on natural disasters.

Sheets, Bob, and Jack Williams. *Hurricane Watch: Forecasting the Deadliest Storms on Earth.* New York: Vintage Books, 2001. Provides a history of the study of hurricanes. Discusses modeling of hurricanes. Includes many excellent appendices of the strongest, deadliest, most damaging hurricanes, as well as appendices of hurricane names, probabilities, and costs. Written for the layperson with some more technical portions, yet still easily accessible.

Tufty, Barbara. *1001 Questions Answered About Hurricanes, Tornadoes, and Other Natural Air Disasters.* New York: Dover, 1987. Provides much more than just excellent answers to commonly asked questions. Has a logical flow, in which questions raised by reading one answer are addressed in the next question. Includes excellent illustrations.

**See also:** Atmosphere's Global Circulation; Atmospheric and Oceanic Oscillations; Atmospheric Properties; Barometric Pressure; Climate; Clouds; Cyclones and Anticyclones; Deep Ocean Currents; Drought; Hurricane Katrina; Lightning and Thunder; Monsoons; Ocean-Atmosphere Interactions; Oceans' Structure; Precipitation; Remote Sensing of the Oceans; Satellite Meteorology; Seasons; Severe Storms; Surface Ocean Currents; Tornadoes; Tropical Weather; Van Allen Radiation Belts; Volcanoes: Climatic Effects; Weather Forecasting; Weather Forecasting: Numerical Weather Prediction; Weather Modification; Wind

# HYDROLOGIC CYCLE

*Water circulates on Earth through a system called the hydrologic cycle. This water cycle functions through vegetation, in the atmosphere, below the ground, and on land, lakes, rivers, and oceans. The sun and the force of gravity provide energy to drive the cycle from ground and surface water to atmospheric moisture that returns to the land and oceans as precipitation.*

## PRINCIPAL TERMS

- **base flow:** that part of a stream's discharge derived from groundwater and interflow seeping into the stream, representing the normal amount of water in that system
- **capillary force:** a phenomenon in which water moves through tiny pores in rock, soil, and other materials, driven by intermolecular attraction between the water and the porous materials
- **evaporation:** the process by which substances, especially water, change from a liquid into a vapor; when a substance changes directly from solid to gas without an intermediate liquid stage, the process is called sublimation
- **infiltration:** the movement of water into and through the soil
- **interception:** the process by which precipitation is captured on the surfaces of vegetation before it reaches the land surface
- **overland flow:** the flow of water over the land surface caused by direct precipitation
- **precipitation:** atmospheric water in the form of rain, hail, mist, sleet, or snow that falls to the earth's surface
- **runoff:** the total amount of water flowing into a stream, including overland flow, return flow, interflow, and base flow
- **soil moisture:** the water contained in the unsaturated zone above the water table
- **transpiration:** the process by which plants give off water vapor through their leaves

## EVAPORATION, CONDENSATION, AND PRECIPITATION

The unending circulation of water on Earth is called the hydrologic cycle. This system is driven by the heat energy received from the sun. Gravity pulls water that falls on the surface back to the oceans to be recycled once again. The total amount of water on Earth is an estimated 1.36 billion cubic kilometers. Most of this vast amount of water, some 97.2 percent, is found in the oceans. The Greenland and Antarctic ice caps and glaciers contain 2.15 percent of Earth's water. The remaining 0.65 percent is divided among rivers (0.0001 percent), freshwater and saline lakes (0.017 percent), groundwater (0.61 percent), soil moisture (0.005 percent), the atmosphere (0.001 percent), and the biosphere and groundwater below 4,000 meters (0.0169 percent). While the percentage of water appears small for each of these water reservoirs, the total volume of water contained in each is immense.

A description of the hydrologic cycle must begin with the oceans, as most of Earth's water is located there. Each year, about 320,000 cubic kilometers of water evaporates from the world's oceans. Evaporation is the process whereby a liquid changes to a gas. Adding energy in the form of heat to the water causes the water molecules to become increasingly active and to move more rapidly, weakening the chemical and physical forces that bind them together. As the temperature of the water increases, water molecules tend to move from the ocean's surface into the overlying air. Factors that influence the rate of evaporation from free water surfaces are solar radiation, temperature, humidity, and wind velocity. It is estimated that an additional 60,000 cubic kilometers of water enters the atmosphere from rivers, streams, and lakes or through transpiration by plants every year. Thus, a total of about 380,000 cubic kilometers of water is evapotranspired on Earth every year.

The amount of water vapor that can be present in the air depends largely on the temperature. At higher temperatures, more vapor can be present. As the vapor-laden air is lifted and cooled at higher altitudes, the vapor condenses to form droplets of water. Condensation is aided by small dust and salt particles or nuclei in the atmosphere. As droplets collide and coalesce, clouds begin to form and precipitation can begin. Wind may transport moisture-laden air long distances, and most precipitation events are the result of three causal factors: frontal precipitation, or the lifting of an air mass over a moving weather front; convectional precipitation related to the uneven heating of Earth's surface, causing warm air currents to rise and cool; and orographic precipitation, resulting from a moving air mass being forced to move upward over a mountain range, cooling the air as it rises. Each year, about 284,000 cubic kilometers of precipitation falls on the world's oceans. This water has completed its cycle and is ready to begin a new cycle. Approximately 96,000 cubic kilometers of precipitation falls upon the land surface each year. This precipitation follows a number of different pathways in the hydrologic cycle. It is estimated that some 60,000 cubic kilometers evaporates from the surface of lakes or streams or transpires directly back into the atmosphere. The remainder—about 36,000 cubic kilometers—is intercepted by human structures or vegetation, is infiltrated into the soil or bedrock, or becomes surface runoff.

### INTERCEPTION, RUNOFF, AND INFILTRATION

Although the amount of water intercepted by and evaporated from human structures—the surfaces of buildings and other artificial surfaces—may approach 100 percent, much urban water is collected in storm sewers or drains that lead to a surface drainage system or that spread water over the land surface to infiltrate the subsoil. Interception loss from vegetation is dependent upon interception capacity (the ability of the vegetation to collect and retain falling precipitation), wind speed (the higher the wind speed, the greater the rate of evaporation), and rainfall duration (the interception loss will decrease with the duration of rainfall, as the vegetative canopy will become saturated with water after a period of time). Broad leaf forests may intercept 15 to 25 percent of annual precipitation, and a bluegrass lawn may intercept 15 to 20 percent of precipitation during a growing season.

When the duration and intensity of the rainfall are greater than the soil's ability to absorb it, the excess water begins to run off, a process termed overland flow. Overland flow will begin only if the precipitation rate exceeds the infiltration capacity of the soil. Infiltration is the process whereby water sinks between soil particles directly into the soil surface or into fractures of rocks. It is dependent upon the characteristics of the soil or rock type and upon the nature of the vegetative cover. Sandy soils have infiltration rates of 3.6 to 3.8 centimeters per hour, and clay rock soils average 2.0 to 2.3 centimeters per hour. Nonporous rock would have an infiltration rate

of zero, and all precipitation would become runoff. The presence of vegetation impedes surface runoff and increases the potential for infiltration to occur.

Water infiltrating into the soil or bedrock encounters two forces: capillary force and gravitational force. A capillary force is the tendency of water in the subsurface to adhere to the surface of soil or sediment particles. This tendency may actually draw the water upward against the downward pull of gravity. Capillary forces are responsible for the soil moisture found a few inches below the land surface.

### COMPLETION OF THE CYCLE

Growing plants are continuously extracting soil moisture and passing it into the atmosphere through a process called transpiration. Soil moisture is drawn into the plant rootlet because of osmotic pressure. The water moves through the plant to the leaves, where it is passed into the atmosphere through specialized leaf openings called "stomata." The plant uses less than 1 percent of the soil moisture in its metabolism, with the remainder being used for the transport of nutrients and metabolites between the roots and leaves, and for temperature regulation. Thus, transpiration is responsible for most water vapor loss from the land in the hydrologic cycle. An oak tree, for example, may transpire 151,200 liters per year.

The water that continues to move downward under the force of gravity through the pores, cracks, and fissures of rocks or sediments will eventually enter a zone of water saturation. This source of underground water is called an aquifer—a rock or soil layer that is sufficiently porous and permeable to hold and transport water. The top of an aquifer, or saturated zone, is the water table. This water is slowly moving toward a point where it is normally discharged into a lake, spring, or stream. Groundwater that feeds and maintains the flow of a stream is called base flow. Base flow is typical of so-called spring-fed streams, and enables such streams to continue to flow during droughts and through cold winter months. Groundwater may flow directly into the oceans along coastlines.

When the infiltration capacity of the soil surface is exceeded, overland flow can begin as broad, thin sheets of water no more than a few millimeters thick called sheet flow. After flowing a short distance, the sheets break up into threads of current that flow in tiny channels called rills. The rills can coalesce into

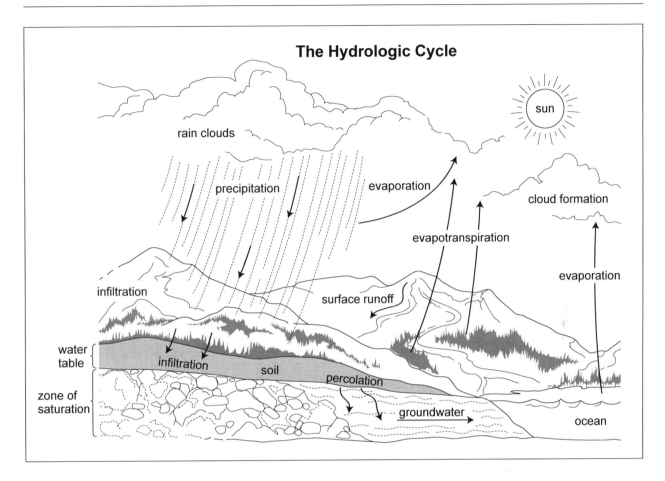

## The Hydrologic Cycle

gullies and, finally, into streams and rivers. While evaporation losses do occur from the stream surface, much of the stream's water is returned to the oceans in this way, completing the hydrologic cycle.

Scientists are interested in how long it takes water to move through the hydrologic cycle. The term "residence time" refers to how long a molecule of water would remain in the various components of the hydrologic cycle. The average length of time that a water molecule would stay in the atmosphere is about one week. Two weeks is the average residence time for a water molecule in a river and ten years in a lake. It would take four thousand years for all the water molecules in the oceans to be recycled just once. Groundwater may require anywhere from a few weeks to thousands of years to move through the cycle. This time period may appear extremely long to humans, yet it suggests that over the course of geologic time, every water molecule on Earth has been recycled millions of times.

## STUDY OF THE HYDROLOGIC CYCLE

Scientists have developed a vast array of mathematical equations and instruments to collect data to quantify the complexities of the hydrologic cycle. The geographic, secular, and seasonal variations in temperature, precipitation, evapotranspiration, solar radiation, vegetative cover, and soil and bedrock type, among other factors, must be evaluated to understand the local, regional, or global hydrologic cycle.

Precipitation, an extremely variable phenomenon, must be accurately measured to determine its input into the hydrologic cycle. The United States has more than thirteen thousand precipitation stations equipped with rain gauges placed strategically to compensate for wind and splash losses. Techniques have been developed to determine the average depth of precipitation falling on a given area or drainage basin. The effective uniform depth method utilizes a rain-gauge network of uniform density to determine the arithmetic mean for rainfall in the area. The isohyetal and polygonal methods are used to determine the arithmetic mean for an area or basin with a nonuniform distribution of rain gauges. The amount of water in a snowpack is estimated by snow surveys. The depth and water content of the snowpack are measured and the extent of the snow cover mapped using satellite photography.

The amount of precipitation lost by interception can be measured and evaluated. Interception is determined by the type of vegetation, the amount of evaporation that occurs during the storm, and the length of the storm. Most often, interception is determined by measuring the amount above the vegetative canopy and the amount received at the surface. The difference is deemed to be the amount lost to interception.

The volume of water flowing by a given point at a given time in an open stream channel, measured in cubic meters per second (CMS), is called discharge. Discharge is determined by measuring the velocity of water in the stream channel with a current meter. The Price meter and the pygmy-Price meter meet the specifications of the U.S. Geological Survey. The cross-sectional area of the stream channel is determined at a specific point and multiplied by the stream velocity to determine discharge. Automated stream-gauging stations are located on most streams to supply data for various hydrologic investigations.

The U.S. National Weather Service maintains about five hundred stations using Class A land pans to measure free-water evaporation. These pans are 122 centimeters in diameter and 25.4 centimeters deep, and they are made of unpainted galvanized metal. Water depths of 17 to 20 centimeters are maintained. The wind velocity is also determined. Errors may result from splashing by raindrops or from birds. Because the metal pan will also heat and cool more rapidly than will a natural reservoir, a pan coefficient must be employed to compensate for this phenomenon. A lake evaporation nomograph is employed to determine daily lake evaporation. The mean daily temperature, wind velocity in kilometers per day, solar radiation, and mean daily dew point are the variables used to determine daily lake evaporation.

The amount of evapotranspiration can be measured using a lysimeter, which is a large container holding soil and living plants. The lysimeter is set outside, and the initial soil moisture is determined. All precipitation or irrigation is measured accurately.

Changes in the soil moisture storage determine the amount of evapotranspiration.

Those techniques are utilized to determine the water budget for different geographic areas. Collectively, these data enable scientists to estimate the total water budget of the hydrologic cycle.

*Samuel F. Huffman*

## FURTHER READING

Dodds, Walter K., and Matt R. Wiles. *Freshwater Ecology: Concepts and Environmental Applications of Limnology.* 2d ed. Burlington, Mass.: Academic Press, 2010. Covers the physical and chemical properties of water, the hydrologic cycle, nutrient cycling in water, as well as biological aspects. Written by two of the leading scientists in freshwater ecology. An excellent resource for college students. Each chapter provides a short summary of main topics.

Fetter, C. W. *Contaminant Hydrogeology.* Grove, Ill.: Waveland Press, 2008. Emphasizes the conditions that lead to the contamination of groundwater, the problems this contamination can cause, and the remediation techniques used to correct these problems. Includes a detailed discussion of the influence that geologic conditions have on groundwater occurrence and contamination. Also addresses regional groundwater movement within aquifers. Contains helpful illustrations of the principles discussed. Includes a glossary of important terms.

Guymon, Gary L. *Unsaturated Zone Hydrology.* Upper Saddle River, N.J.: Prentice-Hall, 1994. Delivers a comprehensive exploration of the movement of water through and within the unsaturated zone, using a quantitative approach to combine research and knowledge from several different disciplines. Suitable for students of engineering and environmental science, as well as for practicing engineers and scientists.

Hamblin, Kenneth W., and Eric H. Christiansen. *Earth's Dynamic Systems.* 10th ed. Upper Saddle River, N.J.: Prentice Hall, 2003. Offers an integrated view of Earth's interior not common in books of this type. Includes excellent illustrations, diagrams, and charts, as well as a glossary and laboratory guide. Suitable for high school readers.

Hunt, Constance Elizabeth. *Thirsty Planet: Strategies for Sustainable Water Management.* New York: St. Martin's Press, 2004. Discusses the hydrologic cycle, components of the cycle, human-induced changes to the cycle, and the repercussions of the growing water crisis. Also covers floods, climate change, and freshwater ecosystem restoration. Provides a broad overview of hydrology topics of study.

Ingebritsen, S. E., Ward E. Sanford, and C. E. Neuzil. *Groundwater in Geologic Processes.* 2d ed. New York: Cambridge University Press, 2006. Extensively revised, well received, and widely used. Focuses on the relationship of groundwater and geologic processes within the field of hydrology. Intended for specialist study in hydrology and hydrogeology.

Lerman, A. "Geochemical Cycles." In *The Oxford Companion to the Earth.* Edited by Paul Hancock and Brian J. Skinner. New York: Oxford University Press, 2000. Provides an excellent overview of geochemical cycles, focusing on the water cycle, sodium cycle, and carbon cycle. Accessible to lay readers.

Lutgens, Frederick K., Edward J. Tarbuck, and Dennis Tasa. *Essentials of Geology.* 11th ed. Upper Saddle River, N.J.: Prentice Hall, 2011. Provides an excellent discussion of the hydrologic cycle and associated processes. Designed for the college student.

Mays, Larry W. *Ground and Surface Water Hydrology.* Hoboken, N.J.: John Wiley & Sons, 2011. Stresses the fundamental principles of hydrologic processes relevant to groundwater and surface water, making extensive use of online resources and discussing newer topics in hydrology such as remote sensing and GIS. Covers all principal climate types. Written at the college level.

Moore, John E. *Field Hydrogeology.* 2d ed. Boca Raton, Fla.: CRC Press, 2012. Provides practical information on field investigation, aquifer testing, and groundwater quality testing. Discusses planning and report writing in detail. Provides case studies from around the world. Best suited for students learning the planning and investigation of field hydrology.

Tarbuck, Edward J., and Frederick K. Lutgens. *Earth: An Introduction to Physical Geology.* 10th ed. Upper Saddle River, N.J.: Prentice Hall, 2010. Offers an excellent discussion of the hydrologic cycle and Earth's water balance. Illustrations, index, and bibliographical references.

Thompson, Stephen A. *Hydrology for Water Management.* Rotterdam: A. A. Balkema, 1999. Provides a thorough account of hydrology, groundwater flow, and stream flow, focusing on the management of water supplies and efforts to keep them free of pollutants and available to the largest number of people possible. Illustrations, maps, index, and bibliography.

**See also:** Aquifers; Artificial Recharge; Barometric Pressure; Clouds; Dams and Flood Control; Floods; Groundwater Movement; Groundwater Pollution and Remediation; Icebergs and Antarctica; Precipitation; Rainforests and the Atmosphere; Salinity and Desalination; Saltwater Intrusion; Surface Water; Waterfalls; Water Quality; Watersheds; Water Table; Water Well

# HYDROTHERMAL VENTS

*Hydrothermal vents are openings on the sea floor where hot water is released. Hydrothermal water forms when cold seawater percolates down into the seabed rock and toward the underlying mantle, where it is heated by hot rocks. Both rocks and water are changed by this interaction. Deposits rich in metals form around these vents, and many ore deposits originally developed in similar environments. Numerous kinds of organisms live around hydrothermal vents, supported by bacteria that derive their energy from the reduction of sulfide to sulfate.*

## PRINCIPAL TERMS

- **basalt:** a typical volcanic rock of the ocean floor, with a relatively low silica content
- **convergent plate margin:** an area where two tectonic plates impinge upon each other with pressure directed toward the area of contact; typically in such regions, the lithosphere is being returned to the mantle at a subduction zone, forming volcanic "island arcs" and associated hydrothermal activity
- **divergent margin:** an area where two tectonic plates are moving away from each other; where the crust and lithosphere form by seafloor spreading
- **endosymbiont:** an animal-hosting autotrophic bacterium, with both host and bacterium enjoying the benefits of symbiosis
- **hydrostatic pressure:** the pressure resulting from an overlying, continuous column of water, approximately 1 bar for every 10 meters of water
- **magma:** molten rock generated by melting of in the earth's mantle
- **midocean ridge:** a region of the sea floor where new oceanic crust is created by seafloor spreading
- **millimole:** a universal relative quantity equal to one thousandth of a mole
- **mole:** the quantity, in grams, of any pure material numerically equivalent to the atomic or molecular weight of that material
- **ophiolite:** a section of oceanic crust and upper mantle that has been thrust out of the ocean floor and up onto the continental crust

## CHARACTERISTICS

The late 1970's discovery of hydrothermal vents and the unique communities that live around them astounded scientists. Studies of these features have provided important and exciting research opportunities for biologists, geologists, and marine chemists. Almost all of the deep sea floor is a cold and quiet place, but this is not true near hydrothermal vents.

In contrast to the homogeneous and stable environment of the deep sea, hydrothermal vents are constantly changing, and scientists have observed many changes in hydrothermal vents and their associated communities in the relatively short period of time that they have been studied. Hydrothermal vents are easily the most spectacular features, both geologically and biologically, on the floor of the abyss. There are almost certainly many more hydrothermal vents remaining to be discovered along midocean ridges, where hydrothermal vents are concentrated.

Hydrothermal vents are hot springs on the ocean floor, the "exhaust pipes" of seafloor hydrothermal

*A black smoker at a midocean-ridge hydrothermal vent.* (National Oceanic and Atmospheric Administration)

systems. These vents are the most accessible parts of seafloor hydrothermal systems, driven by heat sources that lie several hundred meters or even kilometers beneath the ocean floor. Hydrothermal systems form where near-freezing seawater from the bottom of the ocean penetrates deep into the crust along fissures until it comes into contact with hot rock structures. The water may be heated to as high as 400 degrees Celsius before it rises back up through the crust and jets back into the deep sea at the vents. Development of hydrothermal systems requires two things: hot rock structures and a way for seawater to percolate along cracks into the hot zone and return to the surface. How vigorous the resulting hydrothermal system is depends on both these factors. The hottest rock structures are found at midocean ridges and other submarine volcanic centers such as the Loihi seamount off the coast of Hawaii, where frequent eruption of basalt lavas maintains conditions that are most favorable for developing and maintaining hydrothermal systems.

Yet it is not enough for hot rock to lie beneath the surface; there must also be a way for seawater to be brought into contact with the hot rock and kept there until the water is sufficiently heated. This requires just the right configuration of fractured hot rock: If the cracks are too wide or the hot rock is not deep enough below the sea floor, the water will not be heated sufficiently before it returns to the surface. Similarly, if the cracks are too narrow, not enough water will flow through the system to develop a robust hydrothermal system, or the channels may be easily blocked by minor seismic or tectonic movements or by mineral deposits. The sensitivity of deep-sea hydrothermal systems to the sub-seafloor circulation geometry may be why hydrothermal events experience rapid changes and are generally short-lived features.

Seafloor hydrothermal systems are similar to hot springs or geysers found on the continents in that both form as hot rock heats and chemically modifies cold water. Seafloor hydrothermal systems differ, however, in their much higher temperatures. The boiling temperature of water increases as pressure increases, and pressures on the sea floor are several hundred times greater than at the surface. This relationship between pressure and the temperature at which a liquid boils is well known to most cooks, who are familiar with the fact that water boils at a lower temperature at high altitude, where the pressure is

lower, than it does at sea level. Hydrothermal systems on land, such as those of Yellowstone National Park, have a maximum water temperature of not much more than about 100 degrees Celsius. At greater temperatures, the water is superheated and "flashes," or turns instantly to steam. This is what makes a geyser such as Old Faithful erupt, as the water in contact with the hot underlying rock turns to steam and violently forces out the colder, overlying water. The higher pressure of the sea floor prohibits flashing, however, and allows water to be heated to much higher temperatures.

Although most hydrothermal systems are found along the midocean ridges, where magma lies not more than a few kilometers beneath the ocean floor, hydrothermal water is not heated by magma. Water heated by magma would be much closer to magmatic temperatures of 1,200 degrees Celsius. Instead, the water is heated as it passes through hot but solid rock—the solid rock is heated by underlying magma. The upper limit of about 400 degrees Celsius for seafloor hydrothermal systems may reflect the maximum rock temperature at which fractures can form and remain open, or it may reflect separation of the fluid into immiscible fluids. It may be that at temperatures much higher than 400 degrees Celsius, basaltic rock begins to slowly flow, closing any fractures that have formed. Another possibility is that, at pressures corresponding to typical depths of midocean ridges (about 300 bars), seawater separates into two fluids at about 400 degrees Celsius. One phase is enriched in salt relative to seawater, while the other fluid contains a lower concentration of salt than seawater. The more concentrated phase will be the denser phase, and this may be why it is not found among waters issuing from hydrothermal vents. About one-third of the total heat lost from the planet's interior through the sea floor is lost as a result of seafloor hydrothermal systems.

## METAL DEPOSITS

The seawater moving through a seafloor hydrothermal system is chemically changed as it is heated, and it alters the rocks through which it passes as well. All of the magnesium and sulfate in the seawater is absorbed by the hot rock, and large amounts of metals such as manganese, cobalt, copper, zinc, and iron are lost from the hot rock to the circulating water. In fact, magnesium is so completely removed from seawater that the depletion of this element in seawater can only be explained by seafloor hydrothermal activity. Oxygen-rich seawater is transformed into oxygen-poor hydrothermal water, and a high concentration of metals can be dissolved in such water. Seawater, for example, contains negligible magnesium and iron, whereas hydrothermal waters contain up to 1 millimole of magnesium and 6 millimoles of iron per liter.

When hot, chemically transformed water returns to the ocean at hydrothermal vents, the result is often spectacular. Near the vent, the hot, oxygen-poor, hydrogen-sulfide-rich hydrothermal water mixes with cold, oxygen-rich, hydrogen-sulfide-poor seawater. This may cause the metals in the hydrothermal fluid to precipitate. Some of it may precipitate in fractures just beneath the sea floor. These "stockwork" deposits are most likely to be preserved and many ore deposits represent ancient stockwork deposits. Some metals precipitate around the vent itself, forming metal-rich chimneys. These chimneys are typically composed of an outer layer of anhydrite and inner deposits of copper-iron sulfides. These chimneys can have fantastic shapes, resembling spires, columns, cones, and beehives. Scientists visiting these chimneys give them fanciful names: "nail," "fir tree," and "moose" are names given to vent features from one hydrothermal field. Chimneys have been found that are up to 30 meters in diameter and up to 45 meters tall. Their growth rates can be on the order of 1 meter per year, and vent chimneys are thus among the fastest-changing of all geologic phenomena. Fast-growing vent chimneys rapidly become unstable and collapse, then rise again. Cycles of growth and collapse continue as long as the hydrothermal fluids continue to issue from the vent.

Metals that do not precipitate as stockwork or vent chimneys rise with the hydrothermal water issuing from the vent. The hydrothermal water exits the vents in a jet stream due to the pressure exerted from below. Because this water is so hot, it is less dense than seawater and the jet forms the base of a hydrothermal plume. The plume becomes larger as it rises because cold seawater mixes turbulently with the hydrothermal water, and these waters quickly become well mixed. The hottest hydrothermal fluids (those heated to 300 degrees Celsius or more) are sufficiently metal-rich and acidic that they precipitate large quantities of microscopic grains of iron sulfides, zinc sulfides, and copper sulfides upon mixing with seawater. Much of this precipitates as the plume

rises, producing a sulfide "cloud." The rapid precipitation of sulfides darkens the hydrothermal plume, giving it the appearance of smoke. Those vents releasing the hottest fluids have the greatest density of sulfide precipitation and are often referred to as "black smokers." In contrast, vents releasing cooler water (100 to 300 degrees Celsius) do not contain sufficient sulfide or metals in solution to cause this effect. Instead, mixing of these hydrothermal waters with seawater causes white particles of silica, anhydrite, and barite to form, forming a white cloud in the mixing plume and giving rise to the name "white smoker." White smokers may reflect mixing just below the surface, with the result that most of the metals associated with white smoker vents may be deposited just below the sea floor. Because the original hydrothermal vent waters are cooler and thus denser, plumes from white smokers do not rise as far as those from black smokers before obtaining neutral buoyancy and spreading laterally.

Mixing progressively dilutes the hydrothermal water until the mixed water cools to the point that it has the same density as ambient seawater and no longer rises. A tremendous amount of seawater must be mixed with the hydrothermal fluids before the plume attains neutral buoyancy. This may ultimately involve 10,000 or even 100,000 times as much seawater as hydrothermal water. During dilution, the mixture rises tens to hundreds of meters to a level of neutral buoyancy, eventually spreading laterally as a distinct hydrographic and chemical layer, recognizable hundreds or even thousands of kilometers away from the hydrothermal vent. Continued settling of hydrothermal iron and magnesium from these layers is the source of most of the metals in manganese nodules of the abyssal sea floor, far away from the mid-ocean ridges.

Vast quantities of metals are deposited as sulfides around hydrothermal vents, especially as stockwork and in sediments around chimneys. These are particularly rich in iron, copper, zinc, and lead, and the exact composition of these deposits reflects several controlling factors, including temperature of the hydrothermal fluid, water depth, composition of the source rock, and flow regime in the portion of the hydrothermal system that lies beneath the sea floor. Many of the world's great ore deposits seem to be fossil seafloor hydrothermal systems, with the possible difference being that these may have mostly formed at convergent plate margins instead of divergent plate margins, where most modern hydrothermal systems are known. This may be part of the reason why ancient massive sulfide deposits are typically much larger than modern ones.

### STUDY OF HYDROTHERMAL VENTS

To a marine biologist, hydrothermal vents are the oases of the deep sea floor. In contrast to most of the ocean floor, which supports few animals, hydrothermal vents teem with life. In fact, the analogy to a desert oasis falls short because there is much more life around a hydrothermal vent. Life in most of the deep sea is scarce because food is scarce. There is no sunlight, so plants, the basis of most food chains, cannot grow. In contrast, food is abundant around hydrothermal vents; often, life is so crowded that the animals obscure the sea floor.

Vent fields can be divided into three biotic zones reflecting distance from the hydrothermal vents: the vent opening, the near-field, and the periphery. The bulk of the biomass is at the vent openings, where the density of life is so great that it appears to be limited by space, not food. Around hydrothermal vents on the East Pacific Rise and Galápagos Rift, vent openings are dominated by endosymbionts such as tube worms, clams, and mussels. These endosymbionts can grow to great size; some tube worms are 1 meter long with 3-meter tubes, while clams up to 30 centimeters long are common. Other animals live with these, including limpets, polychaete worms, bresiliid shrimp, crabs, and fish. Vent chimneys at several Mid-Atlantic Ridge vents are almost entirely populated by bresiliid shrimp.

Autotrophic bacteria are the primary producers around hydrothermal vents. These use sulfur to convert carbon dioxide, water, and nitrate into essential organic substances in a fashion similar to the way in which plants use sunlight during photosynthesis. This process is called "chemosynthesis." Autotrophic bacteria live within the subsurface plumbing system of the vents, on the sea floor, and suspended in and around the hydrothermal plume itself, sometimes in such abundance that they color the water a milky blue or carpet the sea floor in white or bright yellow mats. Some of these bacteria can tolerate inordinately high temperatures, up to 110 degrees Celsius or more. All the other life around a hydrothermal vent ultimately feeds off the autotrophic bacteria. Some of the autotrophic bacteria are symbiotic, living within larger vent animals in a mutually beneficial relationship. The bacteria provide food, and the animals provide essential inorganic nutrients.

The flow of water around hydrothermal vents controls the distribution of life. Because cold water mixes with the hydrothermal plume, cold seawater flows in from all directions to converge on the rising plume. This means that the sulfur on which the autotrophic bacteria depend is not distributed around the vent. For a similar reason, the animals that feed or depend on the bacteria cannot survive away from the vent opening. The near-field is mostly populated by suspension feeders, animals that capture bacteria and other organisms that drift away from the vent opening. It is presumed that these animals live as close as possible to the vent but are forced to maintain a certain distance because of toxic effects resulting from very high concentrations of heavy metals. Animals living on the vent periphery include scavengers and other types that sustain themselves from bacteria that settle out of the hydrothermal plume.

Most of the animals that live around hydrothermal vents live nowhere else except in other sulfur-rich, reducing environments, such as in the rotting carcasses of whales or in "cold seeps," where cold, chemically altered seawater percolates up through the sea floor. Many vent animals may be ancient, originating in Mesozoic or earlier times. These animals may have been insulated from the effects of surface catastrophes, such as the meteor collision at Chicxulub believed to have killed off at least 75 percent of all life on the planet, including the dinosaurs, about 65 million years ago.

There are a number of fascinating features about the life around hydrothermal vents, but one of the most intriguing is the suggestion that this environment is similar to what existed when life was just developing about 4 billion years ago. It is entirely possible that the first autotrophic life-forms were chemosynthetic, not photosynthetic.

*Robert J. Stern*

### FURTHER READING

Allen, Philip A. *Earth Surface Processes.* Oxford: Blackwell Science, 1997. Serves as a clear introduction to oceanography and the earth sciences. Details the processes, properties, and composition of the ocean. Includes color illustrations, maps, an index, and a bibliography.

Anderson, R. N. *Marine Geology: A Planet Earth Perspective.* New York: John Wiley & Sons, 1986. Intended for college undergraduates. Includes an excellent chapter on metal deposits and marine geology. Accessible to a general, scientifically interested audience.

Humphris, Susan E., et al., eds. *Seafloor Hydrothermal Systems.* Washington, D.C.: American Geophysical Union, 1995. Offers a technically comprehensive overview of physical, chemical, biological, and geochemical interactions. Includes a wealth of pictures, figures, and references. Technical but useful.

Karl, David M., ed. *The Microbiology of Deep-Sea Hydrothermal Vents.* Boca Raton, Fla.: CRC Press, 1995. Looks into the microbiology of extreme and unusual environments such as hydrothermal vents on the deep ocean floor. Gives special attention to what causes the vents, their benefits, and the environments they create. Suitable for the nonscientist.

Kennish, Michael J. *Practical Handbook of Marine Science.* 3d ed. Boca Raton, Fla.: CRC Press LLC, 2001. Covers marine geology topics such as seafloor topography, sediments, plate tectonics, midocean ridges, and hydrothermal vents. Best suited for advanced undergraduates and graduate students. Provides references for each chapter.

McCarthy, Kevin T., Thomas Pichler, and Roy E. Price. "Geochemistry of Champagne Hot Springs Shallow Hydrothermal Vent Field and Associated Sediments, Dominica, Lesser Antilles." *Chemical Geology* 224 (2005): 55-68. Discusses the source of hydrothermal matter and its effects on local sediment. Provides a geochemical analysis of vented fluids. Offers a technical study of hydrothermal vents. Provides examples of current research methods and covers mechanics and processes related to hydrothermal activity.

Monroe, James S., Reed Wicander, and Richard Hazlett. *Physical Geology: Exploring the Earth.* 6th ed. Belmont, Calif.: Thomson Higher Education, 2007. Provides an in-depth discussion of plate tectonics, continental drift, and seafloor spreading, through which hydrothermal vent systems have their origins, set in a question-and-answer format.

Parson, L. M., C. L. Walker, and D. R. Dixon, eds. *Hydrothermal Vents and Processes.* London: Geological Society, 1995. Offers a thorough look into deep-sea and hydrothermal vent ecology. Discusses the sea floor and plate tectonics. Suitable for the high school reader or someone without a background in oceanography. Color illustrations and maps.

Pinet, Paul R. *Invitation to Oceanography.* 5th ed. Sudbury, Mass.: Jones and Bartlett Publishers, 2009. Presents an overview of the world's oceans and the geophysical processes that have both brought about their formation and continue to function today to shape the overall ocean system of the planet.

Rona, P. A., and S. D. Scott. "A Special Issue on Sea-Floor Hydrothermal Mineralizations: New Perspectives." *Economic Geology* 88, no. 8 (1993): 1933-2249. Deals with how mineral deposits form around hydrothermal vents with many examples of individual vent fields from around the world. Includes a complete listing and summary of all known seafloor hydrothermal systems. Intended for the scientist, but abundant pictures, figures, and references make it useful to all who are interested in the topic.

Talley, Lynne D., George L. Pickard, William J. Emery, and James H. Swift. *Descriptive Physical Oceanography: An Introduction.* 6th ed. London: Elsevier, 2011. Discusses the principal aspects of oceanography, including the discovery and functioning of hydrothermal vent systems. Written at the introductory college level.

Van Dover, Cindy Lee. *The Ecology of Deep-Sea Hydrothermal Vents.* Princeton, N.J.: Princeton University Press, 2000. Discusses the geology, chemistry, and biology of hydrothermal vents. Examines the microbial ecosystems and their symbiotic relationships. Emphasizes the ecology of hydrothermal communities. Accessible to graduate students but could easily be used in an advanced undergraduate course.

Vassallo, L. F., et al. "Paleogene Magmatism and Associated Skarn-Hydrothermal Mineralization in the Central Part of Mexico." *Boletin de la Sociedad Geológica Mexicana.* Discusses Au-Ag deposits at hydrothermal vents, fluorite deposits, and Zn-Cu-Pb-Ag skarn deposits. Highly specific; best suited for graduate students and academics in the field of geochemistry.

**See also:** Atmosphere's Global Circulation; Carbonate Compensation Depths; Climate; Deep Ocean Currents; Gulf Stream; Ocean-Atmosphere Interactions; Ocean Pollution and Oil Spills; Oceans' Origin; Oceans' Structure; Ocean Tides; Ocean Waves; Remote Sensing of the Oceans; Sea Level; Seamounts; Seawater Composition; Surface Ocean Currents; Tsunamis; Turbidity Currents and Submarine Fans; Wind; World Ocean Circulation Experiment

# I

# ICE AGES AND GLACIATIONS

*Glaciers are layers of ice that form on Earth's lithosphere where the temperature is sufficiently low to support year-round ice and snow. Extended periods when temperatures drop sufficiently low to support large-scale increases in glaciation are called glacial epochs or ice ages. Earth is now in an ice age that began 2.4 million years ago and has involved twenty or more fluctuations between glacial and interglacial periods. Estimates indicate that Earth is undergoing cycles of glaciation that occur every eleven thousand years.*

## PRINCIPAL TERMS

- **cryosphere:** portion of the earth's surface in which the year-round temperature remains constant enough to support permanent ice and snow
- **eccentricity:** variation in the shape of the earth's orbit around the sun, ranging from circular to elliptical
- **glacial:** short-term period of glaciation, generally lasting for less than 1 million years and alternating with interglacial periods
- **glacial epoch:** an extended period of global temperature reduction and glaciation that generally lasts for millions of years and includes internal glacial and interglacial periods
- **glacier:** buildup of frozen ice on some portion of Earth's lithosphere
- **interglacials:** periods of reduced glacial coverage that alternate with glacials within a global glacial epoch
- **isostacy:** equilibrium between the lithosphere of the earth and the liquid layer of rock in the inner layers of the strata
- **obliquity:** long-term variations in the tilt of the earth relative to the sun; varies through a cycle of 42,000 years
- **Pleistocene-Quaternary glaciation:** current ice age beginning approximately 2.4 million years ago
- **precession:** variation in the angle of Earth's rotational tilt relative to an astronomical point of reference; shifts through a cycle of 26,000 years

### GLACIERS OF MODERN EARTH

The part of the earth in which the temperature is permanently below the freezing point of water (0 degrees Celsius, or 32 degrees Fahrenheit) is known as the cryosphere. Glaciers are large bodies of ice that form within the cryosphere and are considered permanent by the standard of the human life cycle.

Glaciers are formed primarily from layers of snow that have become compacted by the weight of overlying snow and by the pull of gravity to recrystallize into ice. Glaciers today exist primarily in the far Northern and Southern Hemispheres past the snow line, which is the latitudinal mark beyond which the ambient temperature remains below the freezing point of water.

The portion of the earth covered in permanent ice changes in time according to cycles that affect Earth's global temperature. Multimillion-year periods of prolonged reduced temperature, during which glaciation spreads, are sometimes called glacial epochs or ice ages.

The presence or absence of glaciers has a dominant effect on Earth's geography and climate. Ocean depth is one factor that is highly dependent on the amount of water that is frozen in glacial zones. Estimates show that the world's glaciers in Antarctica and Greenland hold enough frozen water to raise the earth's ocean levels by more than 60 meters (200 feet), thereby drastically altering the amount of land available for habitation.

Glaciers also can influence sea levels through their affect on isostacy, which is the gravitational equilibrium between the earth's crust and the mantle below. The crust of the earth floats along the mantle because the crust is less dense and the mantle is more fluid (caused by heat from the earth's core). As a portion of the crust becomes heavier, that portion sinks into the mantle, causing a depression in the earth's surface and causing water levels to rise in the surrounding area.

During the last ice age, which occurred between 30,000 and 100,000 years ago, portions of North America and Eurasia were covered in thick glacial ice, creating a deep depression in the mantle. As most of this ice melted, the mantle and crust began to return to equilibrium, a phenomenon known as isostatic rebound. As the crust rose, sea levels receded in many parts of the Northern Hemisphere.

Global warming is a trend caused by human activity, a reduction in vegetation worldwide, and an increase in greenhouse gases. This warming phenomenon is noted by the level of glacial retreat, or the melting of glacial ice, which has been measured since the nineteenth century. One focus of the study of glaciers and glaciation is to develop a more complete understanding of climate cycles on Earth and of the future of Earth's environmental evolution.

## SHORT-TERM GLACIAL CYCLES

The Cenozoic period, which is the current geologic age, began approximately 70 million years ago. Evidence from geologic sources indicates that climate change and glaciation have occurred on a relatively regular cycle in this period of Earth's history.

Earth is in an ice age called the Pleistocene-Quaternary glaciation, which began more than 2.5 million years ago. During an ice age, short-term fluctuations in climatic variables lead to periods of relatively low temperatures with increased glacial build up called glacials. These glacial periods alternate with interglacial periods, in which temperatures increase and glacial ice retreats as the climate becomes warmer. Geologic evidence indicates that in the Pleistocene-Quaternary glaciation, the earth has experienced twenty or more alternating glacial and interglacial periods. The earth is now in an interglacial period.

Geologists believe that glacial and interglacial periods are partially related to changes in Earth's solar orbit. In the current geologic period, called the Holocene, the earth orbits the sun in a circular pattern. However, the shape of the earth's solar orbit gradually shifts through thousands of years, alternating between circular and elliptical or oval-shaped patterns. Changes in the orbital patterns (called Earth's eccentricity) can have a major impact on the amount of solar radiation received on the earth at various times of the year.

In addition, the earth is tilted on its axis with respect to the sun at an angle of 23.45 degrees; this tilt is called its axis of rotation. This axis also "wobbles" over a 42,000 year cycle between 21 to 24.5 degrees; this wobble is called the obliquity of Earth's rotation. Finally, the earth's axis of rotation, relative to fixed points in the cosmos, shifts slowly in a 26,000 year cycle, a shift that is known as the earth's precession. While the earth's North Pole is pointed toward the star known as Polaris, or the North Star, the earth's axis will shift gradually toward the star known as Vega.

Serbian geophysicist and engineer Milutin Milanković used calculations of Earth's precession, obliquity, and eccentricity to calculate the cyclic pattern of the planet's orbital relationship with the sun; he then attempted to compare these data with information on climate change. From these data, Milanković developed what came to be called Milankovitch cycles, which are repeating patterns of heating and cooling related to Earth's relative distance from the sun and its angular orientation.

The Milankovitch cycles predict that Earth will experience major changes in climate on three separate cycles—100,000 years, 41,000 years, and 21,000 thousand years—corresponding closely to the patterns in Earth's orbit. Geologic and climatologic evidence indicates that Milankovitch cycles may be largely responsible for the alternating glacials and interglacials during the current ice age.

## LONG-TERM GLACIAL CYCLES

While Milankovitch cycles are helpful in explaining short-term variations in climate, geologic evidence indicates that the earth has experienced periods of long-term glaciation, or glacial epochs that cannot be explained by orbital variation. Some glacial periods last for hundreds of millions of years and involve glaciation far more extensive than that of the current Quaternary ice age.

One of the first glacial epochs to be studied occurred between 2.2 and 2.4 billion years ago, in the Paleoproterozoic era, and is thought to have lasted 200 million years or longer. Though this ancient ice age, called the Makganyene (or Huronian) glaciation, is poorly understood, its global changes in tectonic movement and volcanism have been identified as potential causal factors. Geologists believe that the Makganyene glaciation might have marked the first time that most or all of the earth was covered in glacial ice, a hypothesis that is now called the Snowball Earth theory. Global glaciation results when contributing factors such as tectonic movement and atmospheric composition converge to create a positive feedback loop that allows glaciers to spread until they cover vast portions of the planet's surface.

The period from 850 to 630 million years ago, constituting a major portion of the Neoproterozoic period (1 billion to 540 million years ago), is sometimes called the Cryogenian period because of the two major glacial epochs that occurred during this span, which included the most extensive glaciations in the known history of the earth. The Marinoan (635 million years ago) and Sturtian (710 million years ago) glaciations left widespread geologic evidence around the world, indicating that glacial ice was present at all latitudes and covered the earth in thick layers.

During the Paleozoic era (540–340 million years ago) there existed two glacial epochs, both corresponding to major changes in the biosphere. The Andean-Saharan glaciation occurred between 460 and 430 million years ago, spanning the Ordovician and the Silurian periods, and corresponds with a mass extinction of life on Earth, after which terrestrial plants began to spread across the surface. The Karoo ice age (360 and 260 million years ago) spanned part of the Mississippian and Pennsylvanian periods and is partially explained by the spread of terrestrial plants, which removed greenhouse gases from the atmosphere and caused global cooling.

Geologists believe that glacial epochs are largely the result of tectonic shifting on the earth's surface. Recent evidence suggests that glacial epochs occur when larger continents break into smaller continents, a process known as rifting. Increases in tectonic rifting might affect climate by altering levels of silicate weathering, which is the process by which atmospheric carbon dioxide is removed from the atmosphere through interaction with minerals dissolved from the earth's crust into the oceans.

Silicate weathering increases with temperature. Thus, as continents move over the equator, where they are exposed to higher levels of solar radiation, silicate weathering increases, thereby leading to a reduction in atmospheric carbon dioxide and the beginning of a cooling cycle. Rifting and tectonic

convergence are part of the theoretical supercontinent cycle, which suggests that the continents converge on a cycle of 300 to 600 million years.

## EVIDENCE FOR GLACIATION

Ice-core samples help geologists learn about the climate of the distant past. Ice in glaciers is built of successive layers of snow that have become compacted to form ice. By taking vertical samples of glacial ice, geologists can examine the layers in the ice and can discern information about climate from the differences between and among layers.

Water vapor contains different molecular varieties of water, based on the types of isotopes contained within water molecules. Isotopes are variations of an element, such as hydrogen or oxygen, which have the same number of protons but differ in the number of neutrons. While these isotopes are of the same element, variations in the number of neutrons cause each isotope to behave differently in certain chemical reactions.

In general, the standard variety of water is $H^{16}OH$, which contains an isotope of oxygen known as oxygen-16 ($^{16}O$), considered the most common oxygen isotope. However, water also occurs in the formula $H^{18}OH$, which contains the oxygen-18 ($^{18}O$) isotope. In addition, some water molecules contain deuterium, which is an isotope of hydrogen, and may therefore occur in the chemical formula HOD. When water vapor rises before condensing because of heat, it tends to lose deuterium and oxygen-18 more readily than it will lose the oxygen-16 isotope. This means that colder temperatures favor the accumulation of ice with greater proportions of deuterium and oxygen-18. Therefore, by measuring the occurrence of various isotopes, geologists can measure changing temperatures in ancient ice samples. Ice cores taken from Antarctica preserve temperature-induced isotope variations from 400,000 years or earlier.

In addition to isotope chemistry, ice cores preserve a variety of other environmental elements representing the time when the ice was first deposited. Levels of dust and debris captured in glacial ice indicate that dry air and heavy winds dominated during glacial periods. In addition, geologists have found trapped gas inclusions within ice cores containing carbon dioxide and methane from the Quaternary ice age. Measurements indicate that these gases were more abundant during warmer periods and declined during glacial periods, providing evidence that greenhouse gas levels are a major determinant in glaciation.

In some cases, geologists have found acids related to volcanic activity in glacial ice. Increased volcanism occasionally preceded periods of glacial activity because gases released by volcanoes can block solar radiation, leading to more pronounced differences between seasons.

*Micah L. Issitt*

## FURTHER READING

Dawson, Alastair G. *Ice Age Earth: Late Quaternary Geology and Climate.* New York: Routledge, 1996. College-level text covering climate change during the Quaternary period. Also covers scientific methods used to examine climate change utilizing data taken from soil samples.

Grotzinger, John, and Thomas H. Jordan. *Understanding Earth.* New York: W. H. Freeman, 2009. Covers aspects of Earth's history and geochemical development. Includes discussions of Earth's major glacial periods and information about the study of climate change through geologic evidence.

Hambrey, Michael J., and Jurg Alean. *Glaciers.* 2d ed. New York: Cambridge University Press, 2004. Examines current and past glaciers. Contains a discussion of the processes that lead to the formation of glaciers and their effect on Earth's climate and biodiversity.

Monroe, James S., Reed Wicander, and Richard Hazlett. *Physical Geology.* 6th ed. Belmont Calif.: Thompson Higher Education, 2007. Detailed discussion of geologic principles and research methods. Examines research methods used to investigate glaciation and climate change.

Rapp, Donald. *Ice Ages and Interglacials: Measurements, Interpretations, and Models.* New York: Springer, 2009. Detailed text covering the formation, progression, and environmental impact of glacial periods. Provides coverage of the Snowball Earth period and the interglacial periods leading to Earth's present conditions.

Stanley, Steven M. *Earth System History.* New York: W. H. Freeman, 2004. Overview of information regarding Earth's history and geochemical cycles. Includes descriptions of Earth's glacial and interglacial periods.

**See also:** *AR4 Synthesis Report;* Atmospheric Properties; Climate; Climate Change Theories; Earth-Sun Relations; Geochemical Cycles; Global Energy Transfer; Global Warming; Greenhouse Effect; Hydrologic Cycle; Icebergs and Antarctica; Long-Term Weather Patterns; Precipitation; Recent Climate Change Research; Sea Level; Water and Ice

# ICEBERGS AND ANTARCTICA

*About 98 percent of the surface of Antarctica—the coldest, driest, and southernmost continent on Earth—is covered in a thick, dense layer of ice. Some of this ice moves to the Southern Ocean surrounding the continent and breaks into icebergs. Antarctica is of profound interest to scientists because of its impact on the world's oceans and climate and because of its role as a rich source of data about Earth's history.*

## PRINCIPAL TERMS

- **ablation:** the removal of material from a glacier, ice shelf, or other mass of ice through evaporation, melting, or splitting
- **albedo:** a measure of the proportion of incoming light or radiation that is reflected from a surface, such as snow, ice, or water; also known as reflectivity
- **calving:** the breaking away of a smaller piece of ice from a larger one
- **glacier:** a river of freshwater ice that is massive enough to be put into motion by gravity; usually contains ice, air, rock, and some water
- **iceberg:** a large mass of freshwater ice that has broken from an ice shelf or a glacier; floats in a body of water
- **ice core:** a cylinder-shaped piece of ice that is collected by drilling into a glacier; can be used to analyze the history of Antarctica's climate
- **ice shelf:** a large, flat sheet of freshwater ice formed from a glacier or an ice sheet; floats in a body of water
- **ice stream:** a rapidly moving current of freshwater ice flowing from an ice sheet and moving more quickly than the ice that surrounds it; carries ice from the ice sheet
- **nipping:** process in which ice pushes forcibly against the edge of a ship
- **pack ice:** large, mobile masses of frozen, floating seawater that are not attached to a landform; also known as sea ice

### ANTARCTICA, ICE SHEETS, AND ICE CORES

Antarctica is the southernmost of Earth's seven continents. Surrounded by the Southern Ocean, Antarctica is situated largely within the Antarctic Circle, the parallel of latitude that runs 66.5622 degrees south of the equator.

Antarctica is the only continent with no permanent human residents. In 1959, an international treaty was signed to establish Antarctica as a scientific preserve, to be used for peaceful purposes only. The treaty does not, however, recognize or dispute claims on any part of Antarctica as a national territory; it also does not allow the making of new territorial claims. About thirty countries operate scientific research stations on Antarctica, among them the United States, Australia, India, France, Germany, and Japan.

Antarctica spans about 14 million square kilometers (about 5.4 million square miles) of land area, slightly less than 1.5 times the area of the United States. Antarctica is the continent with the lowest surface temperatures, the greatest amount of wind, and the least precipitation. Much of Antarctica is considered a desert because the continent receives on average only about 5 centimeters, or 2 inches, of rain per year.

Almost the entire surface of Antarctica, about 98 percent, is covered in a thick and heavy layer of ice, which measures up to about 4.8 kilometers (3 miles) in depth in some places. This ice, which occupies about 29 million cubic kilometers of total volume (about 7 million cubic miles), represents approximately 90 percent of the world's total ice and is known as the Antarctic polar ice sheet. (A second, smaller, ice sheet covers most of the island of Greenland in the Arctic Circle surrounding the North Pole.) The Antarctic ice sheet contains enough frozen water that if it were to melt completely, it would cause a global rise in sea level of about 60 meters (200 feet).

Specific weather conditions are required for an ice sheet to form. Winter snowfall should not melt completely when summer temperatures rise. If snow persists throughout the year for thousands of years, each season's snowfall forms a new layer on the previous season's layer; these layers eventually freeze into ice. As new snow falls, its weight also compresses the existing ice, making the ice sheet even more dense and thick.

Because the ice sheet comprises thousands of layers of snow, it presents a detailed historical record of the environment and climate at the time of each snow deposit. By drilling into the sheet to collect samples known as ice cores, scientists can access that record. Ice cores have been used to determine, among other things, the temperature at which each snowfall occurred. A measure called the oxygen isotope ratio is used as a stand-in for air temperature. Other measurements that can be taken include dust levels, greenhouse gas levels (such as carbon dioxide and methane), hydrogen peroxide, and sulfate.

By analyzing these data, scientists have been able to map a picture of climate change throughout Earth's history, pinpointing when historic ice ages and interglacial, or warmer, periods occurred. Ice core data also have confirmed that the concentration of greenhouse gases in Earth's atmosphere has increased significantly since the late eighteenth century, the beginning decades of the Industrial Revolution.

## ICE SHELVES, ICE CALVING, AND ICEBERGS

Because of the force gravity exerts on its own weight, the massive Antarctic ice sheet is in a state of extremely slow but constant movement. Glaciers and ice streams, which are both flowing masses of ice, ooze from the inland portions of the ice sheet into the Southern Ocean.

In the Southern Ocean ice breaks off and forms a multitude of floating, frozen masses that hug the coastline. These masses are known as ice shelves. Ice shelves are dynamic, not static, in nature. They can gain more ice from the ice sheet, but they also can lose ice to melting or calving.

Ice calving is a form of ablation, a term used to describe any of several processes through which some icy material is removed from a larger whole. In calving, a large block of ice suddenly splits from the margin of another mass of ice, usually a glacier or an ice shelf, and breaks away. Calving occurs when fractures in the ice grow and spread toward the margin, causing the ice to become thinner and more brittle. A variety of factors contribute to the likelihood of a calving event, including the extent to which an ice shelf has melted below the water line, the speed at which glaciers and ice streams flow, and the temperature of the ice itself.

Calving occurs mostly where ice meets or stands in water. The resulting calved blocks become known as icebergs, which have an important place in the Antarctic ecosystem. Many seabirds live on their surfaces, and krill, fish, and phytoplankton form communities below the icebergs. (Ships traveling in the

Southern Ocean and the North Atlantic are at risk of being nipped, or struck, by floating icebergs.)

One study has found that when icebergs melt far from the coastline, they introduce trapped nutrients, including iron and krill, to the water. This process speeds the growth of algae and other photosynthesizing phytoplankton, which take up carbon dioxide from the atmosphere. The effects of global warming have been causing ice shelves to disintegrate and to calve icebergs more frequently, and scientists predict that this trend will continue. If icebergs do contribute to the removal of carbon dioxide, they may add an important additional feedback mechanism to the complex, dynamic global climate system.

## SEA ICE AND THE CLIMATE

The seawater surrounding Antarctica freezes at a slightly lower temperature than the freshwater that makes up its ice sheet, glaciers, ice shelves, and icebergs. Typically, seawater freezes at -1.9 degrees Celsius (28.58 degrees Fahrenheit); normally, water freezes at 0 degrees C (32 degrees F).

The ice crystals that form when seawater freezes do not contain salt, so they are lighter and less dense than the unfrozen water. This ice floats on the surface, acting as a barrier that prevents heat and gases from traveling between the atmosphere and the ocean. Even when sea ice breaks into large pieces, known as pack ice, the ice tends to be driven together into a nearly continuous mass that provides excellent cover. This prevents the atmosphere above the Southern Ocean from cooling drastically, making sea ice a critical control factor in keeping global climate stable.

Sea ice is important for another reason. It has a bright, light surface that reflects back into space about 80 percent of the sunlight that strikes it. This process is known as the albedo effect. (A surface's albedo is a measure of how reflective it is.) The surface of the ocean itself is much darker, and it has an albedo that is closer to about 10 percent. As a result, when sea ice melts, a dramatically greater amount of sunlight and radiation is absorbed by the ocean, which heats up, creating a cycle that leads to even more melting.

Decades of scientific data have shown that sea ice in the Arctic north has become thinner and has shrunk more each summer since about 1980, as temperatures there slowly warm. Antarctic sea ice is reacting somewhat differently. Around the Antarctic Peninsula, a long, thin, stretch of land that extends outside the Antarctic Circle, higher temperatures have led to a reduction in the extent of sea ice cover. This is also the region where two major ice shelves, known as the Larsen B and the Ross Ice Shelves, disintegrated (calved into icebergs). In the rest of the continent, sea ice has actually increased slightly since the start of the twenty-first century, partly because the circumpolar current buffers warm water from the tropics from reaching it.

*M. Lee*

## FURTHER READING

Bennett, Matthew M., and Neil F. Glasser. *Glacial Geology: Ice Sheets and Landforms.* Hoboken, N.J.: John Wiley and Sons, 2009. A concise, beginner-friendly overview of the geologic concepts required to understand the icy landforms of Antarctica. Written for undergraduates but accessible to advanced high school students, each chapter includes text boxes, color photographs, illustrations, and suggestions for further reading.

Joughin, Ian, and Richard B. Alley. "Stability of the West Antarctic Ice Sheet in a Warming World." *Nature Geoscience* 4 (2011): 506-513. Analyzes satellite observations and historical data to estimate the magnitude of the contribution of west Antarctic ice sheet melting to past sea level rises, and to model its potential future impact.

McGonigal, David. *Antarctica: Secrets of the Southern Continent.* London: Francis Lincoln, 2009. An illustrated volume that covers Antarctic geology, geography, wildlife, historical expeditions, and scientific endeavors. Stunning photographs and clear, absorbing writing make this an excellent introduction to the continent for high school students and older.

Ravindra, Rasik, et al. "Antarctica." In *Encyclopedia of Snow, Ice, and Glaciers,* edited by Vijay P. Singh, Pratap Singh, and Umesh K. Haritashya. New York: Springer, 2011. A brief but comprehensive overview of Antarctica's ice sheet, glaciers, icebergs, sea ice, and ice cores, complete with color photographs and figures.

Thomas, David N., and Gerhard Dieckmann, eds. *Sea Ice.* Ames, Iowa: Blackwell, 2010. Examines the role of sea ice in the global ecosystem, particularly in relation to the impact climate change is having on sea ice in the Antarctic and other regions. Fully illustrated with color photographs. A technical collection best suited for college students with some grounding in climate science, oceanography, or geology.

Walker, Sally M. *Frozen Secrets: Antarctica Revealed.* Minneapolis: Carolrhoda Books, 2010. This highly readable book, written for high school students, focuses on the work of Earth scientists. Explains the techniques used to analyze ice cores, bedrock, water, and even samples of air, and covers what is known about the continent's past and future. Includes a glossary.

**See also:** Climate; Climate Change Theories; Earth-Sun Relations; Geochemical Cycles; Global Energy Transfer; Global Warming; Greenhouse Effect; Hydrologic Cycle; Ice Ages and Glaciations; IPCC; Precipitation; Recent Climate Change Research; Remote Sensing of the Oceans; Sea Level; Water and Ice

# IMPACTS, ADAPTATION, AND VULNERABILITY

*Earth's climate is undergoing change brought about by industrialization. As researchers explore ways to slow, if not halt, the rate of this climate change, others are focusing on the effects this climate change will have on the planet. A growing field in environmental science is the study of where these conditions will be most severe and of the areas that are most vulnerable to these adversities. This discipline also is concerned with analyzing the ways in which civilization can adapt to the climate changes.*

## PRINCIPAL TERMS

- **El Niño:** meteorological condition in which the waters of the tropical, eastern Pacific Ocean are warmed by the atmosphere
- **La Niña:** meteorological condition in which the waters of the eastern, tropical Pacific Ocean are cooled by a lack of radiation from the atmosphere
- **nor'easter:** massive, rotating Atlantic coastal storm that produces large amounts of precipitation and strong winds
- **supercell thunderstorm:** severe type of storm characterized by strong updrafts that produce torrential rain, hail, and tornadoes

## BASIC PRINCIPLES

One of the most pressing issues facing humanity in recent years is that, after more than two hundred years of industrialization, the pollution emitted into the atmosphere and the water is contributing to an increase in Earth's temperature. Ultimately, this pollution will cause significant changes in the planet's climate. Most scientists agree that such significant climate change has already begun.

While it is important to pursue ways to slow the rate of climate shift, it also is important to study the effects, the areas of particular vulnerability, and the avenues available to adapt to such changes. This area of scientific study developed rather recently, during the latter part of the twentieth century. As evidence of climate change began to surface, interest in this field accelerated. The study of impact, adaptation, and vulnerability remains largely fragmented because case studies that have been conducted have been specific to the conditions of target regions. Scientists have not established a uniform set of standards, terminologies, and principles, which means that the studies conducted are not easily melded with similar research conducted elsewhere.

Increasingly, governmental and intergovernmental organizations have developed programs that specifically target climate change. The United Nations Framework Convention on Climate Change (UNFCCC), for example, was established in 1992 to slow the rate of global climate change caused by greenhouse gas emissions. The UNFCCC also has launched a study assessing the vulnerability of poorer, developing nations to climate change.

The applications for this field are varied. Some studies focus on a particular area of change, such as rising sea levels or severe weather. Other studies are concerned with the effects of weather patterns on the world's poor. Some research has attempted to estimate the financial costs of different areas of climate change for political economic systems.

## THE IMPACT OF CLIMATE CHANGE

A wide range of phenomena are likely to occur because of climate change. The most immediate (and often the most high-profile) of these is severe weather.

Among the types of meteorological events that could increase in both volume and severity are blizzards, hurricanes and typhoons, supercell thunderstorms (strong storms characterized by strong updrafts that produce torrential rain, hail, and tornadoes), and nor'easters (massive, rotating Atlantic coastal storms that produce large amounts of precipitation and strong winds). Climatologists are becoming increasingly concerned with data that suggest that the Pacific Ocean phenomenon known as El Niño is changing and that this change could trigger more severe weather.

El Niño is a warm surface water trend that occurs in the central and eastern Pacific Ocean. This warm, moist air is carried into the atmosphere and contributes to the formation of storms and precipitation-generating fronts. Scientists have discovered that patterns for El Niño (which is cyclical and occurs every four to twelve years) are changing because of global warming and are thus producing longer-lasting El Niño patterns that will likely intensify the number and strength of storm systems.

Climate change also may be extending the duration of La Niña, another Pacific phenomenon. La Niña is characterized by a cooling in the Pacific's surface temperature (enhanced by a lack of solar radiation). Under La Niña, cooler, drier air is captured in the atmosphere and reduces the number of precipitation events. As a result, La Niña contributes to droughts. With evidence suggesting that La Niña is, like El Niño, influenced by climate change in such a way that it is prolonged, the future impact could be devastating for the agricultural sectors and for impoverished nations where food and water supplies are already limited.

In addition to meteorological phenomena, climate change is affecting the world's oceans. The emission of greenhouse gases has caused an increase in the earth's temperature, which in turn has caused the planet's glaciers and ice sheets to melt at a faster rate. This melting process (which can be accelerated with even the smallest shift in temperature) will cause the earth's sea levels to rise—according to some, by more than 1 meter (3 feet)—and cause beach erosion and damage to arable land.

## THE ECONOMIC IMPLICATIONS OF CLIMATE CHANGE

Researchers are further examining, in economic terms, the potentially destructive effects of severe weather and rising sea levels. By linking the issue of climate change to business and economics, a wider range of persons from all backgrounds can understand and appreciate the issue.

In 2008, for example, the University of Maryland's Center for Integrative Environmental Research released a study of the areas in which climate change is affecting U.S. states. For example, Georgia could spend an extra $17 million annually for its transportation infrastructure (such as the trucking industry and roadway maintenance) because of the increase in severe storms along that state's highway system. Meanwhile, the shipping industry of Ohio could lose as much as $556 million annually because of the ongoing drop in water levels in Lake Erie. Because of shallower waters, most freight ships used on Lake Erie would need to lighten their capacities by about 270 tons, leading to a loss of about $30,000 per ship.

Research on the economic effects of climate change has increasingly focused on one of the most vulnerable geographic regions in the industrialized world: the city. Cities are heavily concentrated with infrastructure, buildings, and people. Therefore, the effects of climate change (such as severe weather, smog, and heat and cold waves) affect a larger number of people and property than they do in rural areas. Heat and cold waves create spikes in energy use and costs, severe weather slows public transportation, and poor air conditions cause increases in health problems for the elderly and for other persons with respiratory ailments. The economic effects on vulnerable urban centers include higher insurance rates, greater health care costs, and increased maintenance expenses, plus losses in employee productivity.

## ADAPTATION

Most scientists agree that the concept of climate change is not theoretical. Rather, it is a trend that is occurring now, as sea levels are increasing and as more severe weather events are causing greater damage to property.

It is possible that with a concerted effort to curb greenhouse gas emissions, the climate change that is occurring could be slowed and even reversed. Then again, it would take much time to halt the process of climate change. In the interim, scientists and policymakers agree that it is important to examine ways in which civilization can adapt to these changes.

In some cases, municipalities are creating emergency management protocols that protect the elderly and other vulnerable residents in the event of a heat wave. Other areas are building and fortifying seawalls. Many governments are assessing and, where necessary, retrofitting infrastructure (such as water and sewer systems and roads and bridges) to account for flooding.

Adapting to climate change has become a focus of all levels of government. Global nongovernmental organizations and international governmental organizations (including the United Nations) have, since the late twentieth century, been proactive not only in tracking climate change but also in seeking the installation of adaptive measures. For example, the Intergovernmental Panel on Climate Change, a collaborative effort of the UN Environment Program and the World Meteorological Organization, launched a comprehensive study on managing the risks of extreme weather. The study features nearly six hundred pages of information, such as past periods of climate change, models predicting future climate shifts, and areas that are particularly vulnerable to climate shifts. Based on this information (which targets all audiences, including policymakers and others who have a limited scientific background), the report provides insight into the best approaches to adjust to the phenomena.

The approach to pursuing measures that adapt an area to the effects of climate change is similar to the insurance practice known as risk management. Under this concept, issues such as rising sea levels, severe storms, and drought are analyzed carefully. Models for the potential effects of such issues are then generated, providing an illustration of the varying levels of severity in which climate-change-related phenomena may occur.

Using these data, researchers next examine the many factors that may make a particular sector more vulnerable to these phenomena. Researchers calculate the target region's economic, social, geographic, and ecologic factors that foster sensitivity to extreme events. Furthermore, this pursuit also involves the examination of how prepared the target area is to adapt to these events. Based on this information, and on a thorough examination of the phenomena resulting from climate change trends, risk management practitioners then develop a set of recommendations for clients to help them and their constituents adapt to climate change. The decision makers for whom these studies are prepared may, in turn, by accepting and following these recommendations, save lives and property as climate change continues.

*Michael P. Auerbach*

## Further Reading

Hallegatte, Stephane, Fanny Henriet, and Jan Corfee-Morlot. "The Economics of Climate Change Impacts and Policy Benefits at City Scale: A Conceptual Framework." *Climatic Change* 104, no. 1 (2011): 51-87. Discusses the vulnerabilities of urban centers to climate change. Generates a conceptual framework for cities to assess and adapt to climate change effects.

Kotir, Julius H. "Climate Change and Variability in Sub-Saharan Africa: A Review of Current and Future Trends and Impacts on Agriculture and Food Security." *Environment, Development,* *and Sustainability* 13, no. 3 (2011): 587-605. Describes sub-Saharan Africa as one of the regions most vulnerable to climate change because of its population's heavy reliance on agriculture. This change, according to Kotir, is already occurring, causing decreases in crop production and reductions in the amount of available food for the poor.

McCarthy, James, et al., eds. *Climate Change 2001: Impacts, Adaptation, and Vulnerability.* New York: Cambridge University Press, 2001. A comprehensive set of findings on the causes and effects of climate change. A large volume of studies compiled by the Intergovernmental Panel of Climate Change in 2001.

National Assessment Synthesis Team, ed. *Climate Change Impacts on the United States: Foundation Report.* New York: Cambridge University Press, 2001. Features the combined efforts of policymakers, scholars, researchers, and nonprofit organizations. The National Assessment Synthesis Team discusses the climate changes that are already occurring in the United States and the ways Americans can adapt to those changes.

Pelling, Mark. *Adaptation to Climate Change: From Resilience to Transformation.* New York: Routledge, 2010. Calls upon readers to take proactive steps to safeguard against the effects of climate change. Claims that steps taken to adapt to climate change are overly defensive and suggests that humanity make substantive efforts to enact policies and social changes to meet these growing challenges.

Pham, Tien Duc, David Gerard Simmons, and Ray Spurr. "Climate Change-Induced Economic Impacts on Tourism Destinations: The Case of Australia." *Journal of Sustainable Tourism* 18, no. 3 (2010): 449-473. Discusses the regional effects of climate-change-related conditions (such as storms, temperature waves, and the reduction of vegetation and water levels) on Australia's billion-dollar tourism industry.

**See also:** *AR4 Synthesis Report;* Atmospheric Properties; Beaches and Coastal Processes; Bleaching of Coral Reefs; Carbon-Oxygen Cycle; Climate; Climate Change Theories; Climate Modeling; Drought; Earth-Sun Relations; El Niño/Southern Oscillations (ENSO); Floods; Global Energy

Transfer; Global Warming; Greenhouse Effect; Hurricanes; Hydrologic Cycle; IPCC: Long-Term Weather Patterns; Monsoons; Observational Data of the Atmosphere and Oceans; Ocean-Atmosphere Interactions; Ozone Depletion and Ozone Holes; Recent Climate Change Research; Remote Sensing of the Atmosphere; Remote Sensing of the Oceans; Satellite Meteorology; Sea Level; Severe and Anomalous Weather in Recent Decades; Severe Storms; Tornadoes; Tropical Weather

# INDIAN OCEAN

*The Indian Ocean shares the broad ecological and oceanographic features of the other major oceans of the world, but it possesses many unique and interesting characteristics. In general terms, the Indian Ocean's size means that both its shores and its depths are so varied that it is difficult to describe it as a single geographical unit. Its westernmost tides arrive at the shores of a continent that bears little resemblance to the shores of its easternmost limits. Most of the truly unique characteristics of the Indian Ocean, however, lie beneath its surface in the form of extensive mountain ridges and, in one case at least, a very deep rift that represents an unparalleled underwater world of its own.*

## PRINCIPAL TERMS

- **Gondwanaland:** an ancient supercontinent that geologists theorize broke into at least two large segments; one segment became India and pushed northward to collide with the Eurasian landmass, while the other, Africa, moved westward
- **gyres:** circular patterns in the movement of surface currents that create nearly self-contained local subsections within the larger pattern of a typical ocean current
- **Java Trench:** one of the deepest areas of the Indian Ocean, located off the southern coast of Java in Indonesia; it is a form of geological canyon created by the upward thrust of mountain ridges from the ocean floor
- **monsoon:** a seasonal movement of winds into and out of the Indian Ocean region caused by variations of atmospheric pressure over the Indian Ocean and the interior land mass of Asia
- **Somali Current:** a seasonally reversing current that moves between the eastern coasts of Africa and the Arabian Peninsula

## GEOLOGICAL ORIGINS

The geological origins of the Indian Ocean make it unique among the world's major oceans. Compared with the Pacific and Atlantic, the Indian Ocean is considerably smaller (covering an area of about 73 million square kilometers, in contrast to the Atlantic's nearly 84 million and the Pacific's nearly 166 million square kilometers). It is also of more recent geological origin than the Atlantic, Pacific, or Arctic Oceans. Geologists specializing in the evolutionary history of Earth and plate tectonics estimate that about 150 million years ago, a giant southern continent called Gondwanaland began to break apart. The movement of segments both westward (to what became the African continent) and northeastward (to what became India, a central section of

which became known as Gondwana) took at least 100 million years. The collision of the Indian subcontinent with the Eurasian landmass about 50 million years ago brought about the violent upheaval of the Himalaya Mountains that continues in the present day. One of the effects of this phenomenon was to define new shorelines of the "youngest" of the world's major oceans.

The final product of these major geological upheavals was an oceanic body extending over the area between Australia in the east and Africa to the west. Its northernmost point corresponds to the Tropic of Cancer, where the Indian subcontinent joins the Eurasian landmass. From India's western coast to the southeastern tip of Arabia, the waters of the Indian Ocean form the Arabian Sea. On the opposite side of India, the Bay of Bengal and the Andaman Sea extend eastward to the coasts of Southeast Asia (Myanmar, Thailand, the Malay Peninsula, and the Indonesian Archipelago). If one includes the two smaller subsidiary seas, the Persian Gulf and the Red Sea, the Indian Ocean extends even farther north, to 30 degrees north latitude. To the south, the Indian Ocean technically goes as far as Antarctica. Two features of the Indian Ocean's floor define the point at which it separates the Atlantic and Pacific Oceans: the Atlantic Indian Basin and the South Indian Basin.

## ASSOCIATED SEAS AND MAJOR RIVERS

Although the formation of the Indian Ocean created several important seas and gulfs as distinct subsections of its total surface, there are fewer such bodies here than in the other oceans of the world. One should contrast general geographical denominations such as the Bay of Bengal (comprising most of the area east of India and touching the coasts of Southeast Asia) or the Arabian Sea (separating western India from the coasts of the Arabian Peninsula) with the geographical uniqueness of the Red Sea and the Persian Gulf. The latter two are, in fact, clearly separated from the main body of the Indian Ocean by narrow straits—the Mandab Straits and the Straits of Hormuz, respectively. Both the Red Sea and the Persian Gulf have ecologies that are very different from that of the main body of the Indian Ocean. This is not the case for the two other semi-contained gulfs off the Arabian coast, the Gulf of Oman and the Gulf of Aden. The area known as the Great Australian Bight is simply the slightly curved central southern coast of Australia and is therefore even less circumscribed than the Bay of Bengal west of India. The Andaman Sea lies north of the Indonesian Archipelago and is enclosed geographically from the Bay of Bengal by a line of islands (actually an extension of the Indonesian islands) called the Nicobar and Andaman Islands.

Several major rivers pour large amounts of freshwater into the Indian Ocean. Such rivers are probably much older than the Indian Ocean itself, even though their pattern of flow was different in earlier geological ages. The Zambezi in East Africa, the Indus in northwest India, and the Ganges in northeast India were all probably flowing from their respective continental freshwater sources toward what eventually became the Indian Ocean's coastline. Each of these has had, over the long period since the formation of the ocean, a notable effect on the configuration of the coast where it empties into the ocean. Freshwater currents have, for example, cut actual canyons into the continental shelf area adjacent to the coast. In the case of the Ganges, an immense zone of sediment has built up, affecting both marine life and local currents in its delta area in the Bay of Bengal.

## OCEAN DEPTHS AND SUBMARINE GEOLOGICAL FEATURES

The average depth of the Indian Ocean is in the range of 3,636 to 3,940 meters. Several extremely deep but limited areas, notably the Java Trench to the south of Indonesia, are nearly twice as deep. The continental shelf along the coasts of the Indian Ocean is generally narrower than that of the other oceans, averaging 122 kilometers before deeper waters begin. The area west of the Indian coast, off the major city of Mumbai, is an exception. There, the continental shelf extends almost 325 kilometers into the ocean.

The floor of the Indian Ocean is crisscrossed by a number of underwater mountain ranges. Although notable ridges exist, its underwater topography is nowhere near as complex or spectacular as the eastern half of the Pacific, where extensive archipelagos with many small islands and some very large island formations (Japan, the Philippines, New Guinea, and New Zealand, for example) predominate. The most concentrated area of subsurface mountains in the Indian Ocean is centered near 30 degrees longitude, which is about halfway between the west coast of India and the Gulf of Aden, south of Arabia. A number of small but historically important islands mark high points along the mountain ridges between 30 and 60 degrees longitude. Mauritius and the Seychelles are examples of these. The huge island of Madagascar (588,000 square kilometers) is the most prominent surface example of the complex north-to-south submarine mountain systems located east of the African coast. The first of these is the Mauritius Ridge, marked on the surface at its southernmost point by Mascarene Island (due east from Madagascar) and by the Seychelles Islands to the north. The next range, the Carlsberg Ridge, is longer than the Mauritius Ridge, but none of its peaks emerge to form islands.

Finally, a long ridge extends due south off the southwest coast of India. This range is marked at the surface by the Laccadive and Maldive Islands. Again, like Mascarene Island and the Seychelles near Madagascar, the Maldives are dwarfed by the single major island just to the southeast of the tip of India, Sri Lanka (formerly Ceylon). Sri Lanka, however, is the tailing part of the Indian subcontinental landmass rather than the tip of a submarine mountain range.

Beginning with the 90-degree east longitude line and moving toward the eastern shores of the Indian Ocean, the topography of the ocean floor is quite different from that of the western half. First, the very name of one subsurface range, the Ninety East Ridge, suggests a very regular pattern extending from north to south. The Ninety East Ridge, discovered only as recently as the 1960's, has gained the distinction of being the longest and straightest underwater mountain range in the world. Unlike the other ridges of the eastern basin of the Indian Ocean, most notably those south of the Indonesian coast, the Ninety East Ridge appears to be seismically inactive.

Between 90 degrees east longitude and the western shores of Australia one finds the third-deepest point in the Indian Ocean, the Wharton Basin, measuring nearly 6,364 meters deep. Farther south, off the southwestern tip of Australia, is the Diamantina Deep. Neither of these deep points is, however, associated with the rapid fall-off from mountain ridges to deep valleys that is characteristic of ocean trenches. Trenches are characteristic of the eastern rim of the Pacific Ocean, but only one such phenomenon occurs in the Indian Ocean. The Java Trench off the southern coast of Indonesia is more than 6,060 meters deep. Pioneer scientists examining flora and fauna in this area of what were then still undiscovered ocean trenches hypothesized that Indonesia was very close to the dividing line between tectonic plates. They observed that plants and animals to the east of Java appear to be biologically isolated in their evolution from species farther west.

The long, curved pattern of the ranges that constitute the Indonesian Archipelago actually extends far beyond the northern tip of Sumatra. Its peaks can be found in the chain of islands known as the Nicobar and Andaman Islands. The presence of these island chains west of the coasts of Thailand and Myanmar helps define the Andaman Sea area east of the main body of the Bay of Bengal.

## TIDES AND CURRENTS

The immense size of the Indian Ocean means that tidal phenomena are variable, both in type and in the volume of tidal movement registered. The most common tides are semidiurnal (occurring twice daily). These are characteristic of the subequatorial eastern shores of Africa and, farther north and much farther east, in the Bay of Bengal. Australia's southwest coast, which is roughly opposite the subequatorial eastern coast of Africa, has an entirely different tidal pattern. Australia's Indian Ocean shores experience diurnal (once per day) tides that are extremely light by comparison to those of other coasts.

The Indian Ocean is the only ocean in the world with asymmetric reversing surface currents. Asymmetric conditions apply when currents in the northern half of the ocean are moving in a different direction from those of the southern half. The complexity of Indian Ocean currents close to the surface goes well beyond the relatively simple question of north-south asymmetry. One finds, for example, that wind conditions contribute to the creation of gyres, circular or spiral movements that break the broad pattern of the surface current into localized segments.

It is particularly important to note that broad patterns of currents in the Indian Ocean reverse according to the season. Currents, like so many other factors determining the overall ecology of the Indian Ocean, are affected in large part by the major monsoonal wind and weather conditions that are characteristic of this region of the world.

Probably the most famous current in the Indian Ocean is the so-called Somali Current, which moves, in certain seasons of the year, in a fairly rapid clockwise direction from the northeast coast of the Horn of Africa. In the summer season, this current goes as far as the coast of India. At its farthest point moving east in the summer months, it meets the southwesterly monsoon current off the Indian subcontinent. During the winter, the direction of the Somali Current reverses, a situation that created near-ideal seasonal sailing conditions for centuries for ships sailing between the Arabian peninsular zone, particularly the Persian Gulf, and the Horn of Africa.

## CLIMATIC ZONES AND WIND PATTERNS

Because of the great north-to-south distance covered by the Indian Ocean and its associated seas, there are several quite distinct climatic zones according to geographical location. The most famous, and most important for sustaining seasonal agriculture in the entire Indian Ocean area, is the so-called monsoon zone, which runs from 10 degrees north of the equator to about 10 degrees south. The region between 10 and 30 degrees south of the equator is the zone of what has traditionally been called the trade winds. It is the predictability of steadily blowing southeast winds in this wide region (as distinct from the area of the Somali Current) that made maritime communication, and therefore trade, between the opposite shores of the Indian Ocean possible.

Farther south, near the global band of the Tropic of Capricorn (running through the island of Madagascar on the western shores and Australia on the east) lies the subtropical to temperate zone, between 30 and 45 degrees latitude. Some 20 degrees south of the Tropic of Capricorn, climatic conditions begin to show the temperate cooling influence of the extreme southern extent of Indian Ocean waters leading to the last climate zone. From 45 degrees latitude southward to the Antarctic ice cap, the beginnings of sharply cold antarctic waters mark the end of the gradual transition separating some of the world's hottest climates in the Indian Ocean proper from the extreme cold of the southern zone of the globe. Here, three of the world's four oceans almost literally fuse in the Antarctic ice cap.

## MONSOON WIND CONDITIONS

Generally stated, monsoon conditions are semiannual reversing wind patterns. Extensive areas of high pressure "empty" their air in the direction of equally vast low-pressure zones. When this happens, winds moving across water carry the moisture they pick up, which is typically precipitated as rain before they reach their low-pressure destinations. In the case of Indian Ocean monsoons, the widespread heating of the landmass in the Northern Hemisphere during summer creates conditions of low atmospheric pressure over Asia. This low-pressure zone becomes an attractive force for masses of air that are pressed downward by high-pressure conditions over Australia. The resultant winds that move in a northwesterly direction across Southeast Asia and the Indian subcontinent bring with them a much-needed monsoon season of heavy rains that lasts until the particular atmospheric conditions cease to apply. Generally, the monsoon rainy season is predictable, but the arrival of torrents of rain to quench the dry agricultural fields of south Asia and Southeast Asia is not guaranteed. When the typical monsoon wind pattern develops but insufficient moisture is collected to bring rains, areas that depend on monsoon waters can face serious drought for at least one year, as there is no chance of humid air movements over the landmass once the directional wind pattern created by Asia's summer heating ends.

## NATURAL RESOURCES

The Indian Ocean contains many key minerals that are extracted to supplement the local economies of several of the countries along its shores. Manganese is found in several areas off South Africa and Australia. Other minerals include tin and chromite. Mineral wealth in the water is, however, overshadowed by the vast petroleum reserves concentrated in one of its seas: the Persian Gulf. These are estimated to be the largest oil reserves in the world. Other locations where petroleum wealth is important are on the island of Sumatra and in its offshore waters. Similar intermediate potential for petroleum production is found in the Red Sea and off the shores of western India.

The fishing industries that depend on the Indian Ocean are varied. Many depend on the phenomenon of upwellings, movements of water from lower depths that carry phytoplankton—a basic food source for many fish species—close to the surface. The most common commercialized fisheries seek large schools of sardines, mackerel, and anchovies. The principal single species fished in many areas of the Indian Ocean is shrimp.

## SIGNIFICANCE

The Indian Ocean and its associated seas represent one of the most diversified ecological marine environments in the world. This applies both to its coastlines and to the world beneath its surface. The African landmass along the ocean's western shores reveals a variety of natural characteristics that hardly resemble what one finds along its eastern shores. Dry coastal regions characterize both the southern and northern reaches of the African landmass, with some of the driest deserts in the world extending from the Somalian Horn of Africa into the Red Sea and around the southern shores of Arabia into the Persian Gulf. Although these conditions continue all the way along the northwestern coastline to the Gujarati coast of India, the Indian Ocean ecosystem is clearly distinct once one passes the Indian subcontinent. From the Bay of Bengal southward along the eastern shores of the Indian Ocean, the environment becomes increasingly tropical, passing through some of the most extensive tropical rainforests in the world, especially in the Indonesian Archipelago. Indeed, parts of the west coast of Australia, generally thought to be nearly a desert environment, exhibit a tropical ecology that contrasts notably with the dry climates of the African coastline of the Indian Ocean. With this physical ecological diversification comes an enormous variation in plant and animal life along the ocean's shorelines and in the relatively shallow waters of its continental shelf. Studying the diversity of the Indian Ocean's biological ecology and submarine geology is like observing multiple different worlds on one segment of the planet.

*Byron D. Cannon*

## FURTHER READING

American Museum of Natural History. *Ocean.* New York: Dorling Kindersley Limited, 2006. Discusses the geology, tides and waves, circulation and climate patterns, and the physical characteristics of the ocean. Covers marine biology and ocean chemistry, along with a discussion of icebergs and polar ocean circulation. Provides brief overviews of topics, offering an excellent starting point for anyone learning about oceans and marine ecology. Includes numerous images on each page, an extensive index, a glossary, and references.

Charabe, Yassine, and Salim al-Hatrushi. *Indian Ocean Tropical Cyclones and Climate Change*. Dordrecht: Springer, 2010. Explores the effect of climate change on tropical cyclones, especially in the Indian Ocean, the Arabian Sea, and the Bay of Bengal, where populations are dense along the coastlines. Covers everything from forecasting to disaster management.

Clift, Peter D., and R. Alan Plumb. *The Asian Monsoon*. New York: Cambridge University Press, 2008. Discusses the economic, ecological, and sociological issues resulting from monsoons in Asia. Describes factors that may influence monsoons, such as ocean dynamics and the planet's orbit. Includes references, a further reading list, and an index.

Lighthill, James, and Robert Pearce, eds. *Monsoon Dynamics*. New York: Cambridge University Press, 2009. Discusses atmospheric influences on monsoons and oceanic influences. Discusses tropical climatology, such as drought and rainfall patterns, temperature oscillations, and wave dynamics on a global scale; then focuses on the dynamics in India, the Indian Ocean, and eastern Africa. A technical text best suited for researchers, graduate students, and professional meteorologists.

Mukhopadhyay, Ranadhir, Anil K. Ghosh, and Sridhar D. Iyer. *The Indian Ocean Nodule Field: Geology and Resource Potential*. Volume 10 of the *Handbook of Exploration and Environmental Geochemistry*. Edited by M. Hale. Amsterdam: Elsevier, 2012. Discusses volcanism, sedimentation, tectonics, and geology of the Central Indian Ocean Basin. Also covers resource management and ferromanganese formations. Includes many images and diagrams.

Neprochnov, Y. P., et al., eds. *Intraplate Deformation in the Central Indian Ocean Basin*. Bangalore, India: Geological Society of India, 1998. Presents studies of local features of underwater plate tectonics that attempt to uncover signs of change in geological activity in the Indian Ocean Basin. Describes the relatively recent movement of very large segments of tectonics plates and their effects on the formation of the Indian Ocean. Cites arguments suggesting that the geological formation of the ocean's origins is not actually over.

Pearson, Michael Naylor. *The Indian Ocean*. New York: Routledge, 2003. Discusses the physical elements of the Indian Ocean as the backdrop to its history and influence from ancient times to the present day.

Rao, P. V., ed. *The Indian Ocean: An Annotated Bibliography*. Delhi: Kalinga, 1998. Highlights literature dealing with the various islands of the Indian Ocean. Covers ecological scientific data gathering, both in the realm of hydrology and marine life, and includes comparable studies of plant and animal life in various land environments.

Wiggert, Jerry D. *Indian Ocean Biogeochemical Processes and Ecological Variability*. Danvers, Mass.: American Geophysical Union, 2009. Written primarily for specialists. Discusses a variety of processes that function in the Indian Ocean and their ramifications.

**See also:** Aral Sea; Arctic Ocean; Atlantic Ocean; Black Sea; Caspian Sea; Gulf of California; Gulf of Mexico; Hudson Bay; Mediterranean Sea; North Sea; Pacific Ocean; Persian Gulf; Red Sea

# IPCC

*The Intergovernmental Panel on Climate Change (IPCC) is an international group affiliated with the United Nations and charged with reviewing scientific data related to Earth's climate. Every few years, the group releases reports containing its findings and recommendations for possible future action.*

## PRINCIPAL TERMS

- **anthropogenic global warming:** rises in global temperatures caused by human activity
- **climate change:** change in Earth's climate that persists for roughly ten years or longer
- **great conveyor belt:** a system of ocean circulation in which heat is carried to warm certain parts of the earth; also called great ocean conveyor belt
- **greenhouse gas:** gases within the earth's atmosphere that absorb heat and help maintain the planet's climate
- **ice core record:** ice sheet samples used for chemical analysis of temperature, atmospheric and volcanic activity, and precipitation
- **radiative forcing:** a measurement of how much the greenhouse gas in a given area will affect the level of radiation maintained; measured in watts per square meter
- **remote-sensing satellites:** satellites that carry instruments to monitor and collect climate data while orbiting Earth
- **solar irradiance:** energy from the sun's rays that reach Earth; measured in units of power over time (typically watts per second)
- **tree-ring data:** data gathered from tree rings that reflect climate data from the past; thicker rings indicate the presence of light and nutrients that allow more growth
- **variability:** with respect to climate, refers to small-scale changes, such as those in a few years
- **volcanic aerosols:** volcanic ash released into the atmosphere; can affect the amount of solar radiation reaching Earth through components such as sulfur dioxide

## FOUNDING AND PURPOSE

The IPCC was founded in 1988 by the UN Environment Programme and the World Meteorological Organization (WMO) to review and assess data about the earth's climate. The IPCC is tasked with reviewing scientific evidence for climate change, but it does not conduct its own research or monitoring. Instead, the panel reviews the scientific literature in a variety of areas tied to global climate and educates the public about its findings.

The IPCC includes 194 nations and is open to members of the United Nations and WMO. IPCC reports include input from hundreds of scientists in relevant disciplines. The IPCC has completed four assessments, in 1990, 1995, 2001, and 2007. A fifth assessment is slated for publication in 2013 and 2014.

The reports from the IPCC are typically regarded as marking scientific consensus on global climate trends, and they have guided policymaking at the national and international level. In 2007, the IPCC and former U.S. vice president Al Gore were awarded the Nobel Peace Prize for their educational efforts on climate change.

## PAST REPORTS

IPCC reports highlight a number of different areas of climate research. Additionally, the assessment reports are relied upon for guidance in setting global policy.

An early example of such reliance was the Kyoto protocol, which, in using the first two IPCC reports, set benchmarks for reducing greenhouse gas emissions. Beyond simple recommendations, the Kyoto protocol was considered a binding agreement. The protocol was initially signed in 1997, and provisions became effective in 2005. Ultimately, the hope was to reduce greenhouse gas emissions by 5 percent each year from 2008 through 2012, with plans for a new treaty at that time. The success of the treaty, however, remains unclear. The United States, which produced the most greenhouse gas at the time of the treaty's signing, did not sign the protocol. (China has since surpassed the United States in greenhouse gas emissions.)

The scientific weight of the reports mandates the collection of a great deal of data. In studying climate change, the most accurate information comes from the recent past. Since about 1980, satellites have been available to help gather data on factors such as temperature, water levels, and weather patterns.

However, analyzing the climate requires a longer view. To do this, scientists need to collect data from much further back on these same parameters.

One of the best-known charts in the IPCC reports is often termed the *hockey stick graph.* The chart shows global temperatures from the year 1000 through the year 2000. The report is named for the shape of the graph, which is relatively straight from the beginning until the middle of the twentieth century, when it curves upward sharply.

The hockey stick graph was generated as part of a 1999 study by Michael Mann and colleagues at the University of Massachusetts, Amherst, and from the University of Arizona, using data from that paper and one published in the journal *Nature* the previous year. The graph appeared in the summary of the report on the scientific evidence for climate change in the third assessment of the IPCC (published in 2001) as one piece of evidence indicating a trend of rising global temperatures.

Although the authors admitted uncertainty on data before the year 1400, they reconstructed average temperatures using ice cores (in which temperature can be measured based on the ratio of oxygen isotopes preserved in a given year) taken from polar ice sheets and from tree rings and coral, along with existing historical records. The graph contains a visible measure of uncertainty surrounding individual years but maintains its upward-sloping shape at the end.

Whereas temperature is one way of assessing whether the planet is warming, prediction models require many other measurements, both to explain predictions and to gain an understanding of the effect that a changing temperature might have. Oceanic factors are monitored in climate assessments, including currents, sea levels, and surface temperatures. Remote-sensing satellites play a major role in taking these measurements.

Ocean currents play a major role in heat transfers throughout the planet, in a system known as the great ocean conveyor belt, or sometimes just the great conveyor belt. Water at cooler temperatures is denser, so warmer water at the surface may be carried in different directions by a combination of winds and the earth's rotation. For example, the North Atlantic current carries heat from the tropics to the coast of northern Europe. It is for this reason that the annual temperatures of the United Kingdom and Ireland are comparable to those of the northeastern United States, even though these European countries are at higher latitudes. When the water reaches a high enough latitude and cools, the water becomes denser and sinks.

Melting ice caps, the result of heat in the polar regions, could cause that water to cool sooner and thus disrupt oceanic circulation patterns. Changes in those oceanic currents about twelve thousand years ago led to a significantly cooler climate for that region that lasted for a millennium.

A U.S. satellite known as the Advanced Very High Resolution Radiometer, launched by the National Oceanic and Atmospheric Administration, uses infrared imaging to monitor temperatures on the ocean's surface and therefore track the movement of these ocean currents. Other satellites measure sea levels to see if they have risen, possibly from an influx of water from melting polar ice. These satellites are equipped with a device known as an altimeter that measures the satellites' distances from the earth's surface. The device works by communicating with geographic positioning system satellites and with multiple points on the ground in addition to its distance from the surface of the ocean.

IPCC reports also examine possible causes of climate change, which include levels of certain molecules in the atmosphere. The IPCC's listing of greenhouse gases encompasses a number of gases that can absorb heat and potentially contribute to global warming. These gases include carbon dioxide ($CO_2$), methane ($CH_4$), nitrous oxide ($N_2O$), ozone ($O_3$), and chlorofluorocarbons (CFCs) and hydrochlorofluorocarbons (HCFCs). Many of these occur naturally in the earth's atmosphere or are produced as waste products of organic processes such as respiration, but they affect temperature at higher concentrations.

It is important to note that not all atmospheric pollutants are greenhouse gases. Sulfur dioxide ($SO_2$), for example, reflects rather than absorbs heat. At one point, high levels of sulfur dioxide pollution had some scientists concerned about global cooling, although reduced levels of the gas have reduced its impact.

## RECENT PANEL FINDINGS

*The Fourth Assessment of the IPCC* (sometimes abbreviated as *AR4*) comprises a combination of reports from three working groups, in this case covering the scientific basis for global warming, the impact of climate change and what was most likely to be affected, and potential responses to mitigate the effects of climate change. In the report, the IPCC deemed the warming of Earth's climate to be "unequivocal," citing changes in ice and snow coverage, sea levels, and air and water temperatures.

Sea levels rose in the twentieth century, about 1.8 millimeters (about 0.75 inch) per year between 1961 and 2003. Those changes were tied to a loss of polar ice and glaciers and snow on land, a loss that can be attributed to higher temperatures. The IPCC report did note some inconsistencies, possibly because of a lack of evidence and possibly from an inability to get proper data. For example, the sea ice in Antarctica has not been reduced as might be expected, and observations of tropical storms did not indicate any clear increase or decrease in the number of these storms occurring annually.

Three greenhouse gases in particular raised concerns in the report. Carbon dioxide levels had risen 13 parts per million between 1998 and 2005, while nitrous oxide levels had risen by 5 parts per billion in that same period. This means that those gases increased radiative forcing by 13 percent and 11 percent, respectively, during that time. Methane had increased by 11 parts per billion, but that did not make a major change in the level of radiative forcing it caused. Authors of the report attribute the carbon dioxide increase to increases in fossil fuel use, whereas the methane and nitrous oxide increases were said to be from agricultural production.

In part because of the uncertainty surrounding predictions and measurements in climate change, the report also included an explanation of language to express degrees of certainty. In the report, researchers indicated as "very likely" that instances of frosts and cold temperatures had decreased at night, meaning the researchers were more than 90 percent certain that the statement was accurate. The researchers also indicated as "likely" that heat waves and heavy rains had increased, meaning the researchers were between 66 and 90 percent certain that those statements were accurate. Additionally, the report states as "very likely" that the second half of the twentieth century was warmer than any fifty-year period since 1500, with surface temperatures having risen 1 degree Celsius (34 degrees Fahrenheit) between 1900 and 2000.

The report also concluded that based on evidence from the climate record, the levels of those gases in the atmosphere that were produced by human activity were higher than at any point in the past 250 years. Climate change from greenhouse gases caused by anthropogenic activity was rated by the report as "very likely." The idea of anthropogenic warming on individual continents (aside from Antarctica) was considered "likely." Without intervention, the IPCC report concluded, greenhouse gases in the atmosphere will continue to increase and there will be an accompanying rise in global temperatures of 0.2 degree C (32 degrees F) per decade.

The change in sea levels and temperature is expected to have a negative effect on the world's population. Rising sea levels are expected to flood coastal areas and potentially destroy ecosystems, flood settlements, and disrupt food production and water resources.

The IPCC report goes on to discuss the need to reduce greenhouse gases while acknowledging some of the challenges. Those emissions are not equally distributed, so challenges will need to be met with attention to allowing economic development without repeating greenhouse gas emissions that have accompanied such development in the past. The report calls for further international cooperation along the lines of the Kyoto protocol, in which nations are bound to preset targets.

## UNCERTAINTY AND THE IPCC

In many sciences, the key to establishing facts is to reproduce experiments to reduce the likelihood of being wrong through error. In climate science, such reproduction is not possible. Because of this limitation, many factors need to be considered when trying to assess possible climate change and its possible effects.

While controversies exist regarding the IPCC and climate change, many of these controversies are a matter of degree rather than outright opposition. Many factors influence and accompany rising temperatures and affect the earth's climate. Built into all of these measurements is some small degree of uncertainty, particularly in areas in which the data points are estimated rather than directly measured. It is for this reason that rules for expressing degrees of uncertainty are included in the most recent report.

Mann's hockey stick graph, for example, includes large shaded areas around the line showing global temperature, indicating the broad range where the temperature likely was in a given time period. Because of these uncertainties and because of the need to revisit the ideas, new reviews must be conducted and new reports written every few years.

*Joseph I. Brownstein*

**FURTHER READING**

Bernstein, Lenny, et al. *Climate Change 2007: Synthesis Report.* Geneva: IPCC, 2008. Presents a summary of the findings of the three working groups for the Fourth Assessment Report.

Cantón-Garbín, Manuel. "Satellites Oceans Observation in Relation to Global Change." In *Earth Observation of Global Change: The Role of Satellite Remote Sensing in Monitoring the Global Environment,* edited by Emilio Chuvieco. New York: Springer, 2008. Provides an overview of how climate assessments are made and of how climate change is distinguished from temperature variation.

Intergovernmental Panel on Climate Change. Working Group I. *Climate Change 2007: The Physical Science Basis.* New York: Cambridge University Press, 2007.

_____. Working Group II. *Climate Change 2007: Impacts, Adaptation, and Vulnerability.* New York: Cambridge University Press, 2007.

_____. Working Group III. *Climate Change 2007: Mitigation of Climate Change.* New York: Cambridge University Press, 2007. These three reports make up the most recent IPCC review of available information on climate change and subsequent recommendations.

Mann, Michael E., Raymond S. Bradley, and Malcolm K. Hughes. "Global-Scale Temperature Patterns and Climate Forcing over the Past Six Centuries." *Nature* 392 (1998): 779-787.

_____. "Northern Hemisphere Temperatures During the Past Millennium: Inferences, Uncertainties, and Limitations." *Geophysical Research Letters* 26, no. 6 (1999): 759-762. These papers examined global temperatures between the years 1000 and 2000, finding a drastic increase at the end of the twentieth century; source of the noted hockey stick graph.

**See also:** *AR4 Synthesis Report;* Climate; Climate Change Theories; Climate Modeling; Global Energy Transfer; Global Warming; Greenhouse Effect; Hydrologic Cycle; Impacts, Adaptation, and Vulnerability; Long-Term Weather Patterns; Observational Data of the Atmosphere and Oceans; Ocean-Atmosphere Interactions; Ozone Depletion and Ozone Holes; Recent Climate Change Research; Remote Sensing of the Atmosphere; Remote Sensing of the Oceans; Satellite Meteorology; Severe and Anomalous Weather in Recent Decades; World Ocean Circulation Experiment

# L

## LAKE BAIKAL

*Lake Baikal in southeastern Siberia is the deepest lake in the world and the eighth largest. It has some of the cleanest, coldest freshwater in the world and, until recent times, its waters have been among the most pollution free on Earth.*

### PRINCIPAL TERMS

- **endemic:** found in a particular locality and no other
- **phytoplankton:** tiny plants that make up the lowest part of the food chain
- **plankton:** forms of microscopic life that drift or float in water
- **rift:** a crack or split in the crust, typically in a continental mass
- **zooplankton:** tiny animals that feed on phytoplankton

### GEOLOGICAL DEVELOPMENT AND CHARACTERISTICS

Lake Baikal is found in the center of the east Siberian region of Russia. Most of the region, called taiga, is covered by forests of pine and larch trees, though much of the area is also covered with grasslands. The climate is harsh, and there are many areas of permanently frozen subsoil in the region.

Lake Baikal (also spelled "Baykal") is geologically the oldest lake in the world. It is located about 80 kilometers north of Mongolia. Baikal is the world's eighth-largest lake in area, covering between 30,000 and 31,000 square kilometers. It is 640 kilometers long and 50 kilometers in width at its widest part. No lake is deeper than Baikal. It contains about one-fifth of all the freshwater on Earth and about four-fifths of Russia's freshwater supply.

The lake occupies the deepest continental depression in the world, accounting for its great depth. It is surrounded by mountains that rise to 2,000-meter peaks. The mountain ranges around the lake include the Khamar-Daban, the Baykal, and the Barguzin. Two submerged mountain ranges are found beneath Baikal's surface. They cut across the width of the lake and separate it into three distinct basins. The northern basin depression reaches to 983 meters below the surface, the southern to 1,436 meters, but the greatest depths are in the central basin, which goes down to 1,740 meters. The northern edge of the central basin is bordered by the Akademichesky Range, the peaks of which rise nearly 0.6 kilometer from the floor. Some peaks break through the surface to create islands, the largest being Olkhon, which covers 837 square kilometers and has a mountain peak reaching 795 meters.

Lake Baikal was formed about 80 million years ago in the late Mesozoic era, when dinosaurs still lived on Earth. At that time, the region was made up of a broad basin of shallow lakes and marshes with a subtropical climate. Over time, tectonic and seismic activity in the region formed gigantic mountains and deep valleys. About 25 million years ago, water rushed into the deepest basin and began to fill it up. Seismic activity has continued in the region. There have been more than thirty major recorded earthquakes since the eighteenth century. In 1861, a particularly deadly quake killed 1,300 people in a village on the eastern shore. The quake created a new rift fracture in the crust, and billions of gallons of water rushed in to create a new part of the lake, Proval Bay.

Baikal is fed by 336 rivers, including the 1,134-kilometer-long Selenga River, which comes from northern Mongolia. The Selenga carries about one-half of all the water received by Lake Baikal. Other major tributaries are the Turka, the Barguzin, the Upper Angara, the Kichera, and the Goloustnaya. The tributaries drain an area of about 540,800 square kilometers. Only one river runs out of the lake, the 1,600-kilometer-long Angara. It eventually reaches the 3,830-kilometer-long Yenisei, which flows all the way north to the Arctic.

Baikal is among the cleanest lakes in the world. The lake's water is clear to about 40 meters because it contains few minerals and very little salt. Baikal's water level moves up and down about 0.6

to 1 meter per year, with the highest water in August and September and the lowest in March and April. Melting rain and snow from the surrounding mountains account for most of the change in water level. Rainfall averages about 30 centimeters per year. Other sources of the waters of Baikal are the great underground streams that feed into it.

Baikal is usually free of ice only about 110 days per year. The lake is usually frozen over from January to late in May, though in the north the ice usually does not melt until early June. The ice reaches depths of 1 to 2 meters before spring arrives. Water on the surface reaches about 8.8 degrees Celsius by August, though at lower depths the temperature remains a relatively constant 2.75 degrees. Because of its immense size, the lake has a great influence on the climate of Baykalia, as the surrounding area is called. Summers along the lakeshore are usually much cooler than in the surrounding area, while winters tend to be warmer by the lake and colder away from the lake. June temperatures average around 17.6 degrees Celsius in Irkutsk, a city about 58 kilometers from the lake. Irkutsk, on the Angara River, is the region's largest city, with a population of about 587,000 in 2010. Altogether, about 2,500,000 people live in Baykalia.

## FLORA AND FAUNA

Despite its generally cold climate, the lake is rich in life, and many forms of plants and animals thrive beneath the ice even in the coldest months of winter. Baikal is home to more than 1,500 species of animals, including more than three hundred types of birds, and six hundred species of plants. Conditions are unusually superb for aquatic life, and the great age of the lake has allowed more than sufficient time for a great diversity of species to become established. The lake has some 1,300 species of plants and animals that are found only in its waters and nowhere else, including the Baikal seal, a freshwater shrimp, and the golomyanka fish, which gives birth to live young. There is also a fish, the *Comechorus baicalensis*, that lives at more than 900 meters below the surface. Other important and unique species include the Baikal oilfish, many varieties of mollusks, and the fish found in greatest numbers, the Baikal whitefish, a member of the salmon family.

Plant life is abundant; even in winter, sufficient sunlight breaks through the surface ice to keep plants alive at depths of 10 to 45 meters. Tiny plants,

phytoplankton and algae, bloom by the millions at these depths, providing food for rotifers, copepods, and other kinds of freshwater zooplankton, the animal components of plankton. These lowest but important parts of the food chain can usually be seen only through a microscope and are found by the hundreds of millions in lakes and streams.

Baikal is unusual because of the vast amount of life found at its great depths. Most lakes are nearly lifeless below 300 meters, but because of the circulation of Baikal's waters, enough oxygen is carried to depths of 900 meters to support a variety of species. The lake can be divided into three major depth zones. The first are the shallow coastal waters, which extend to depths of from 18 to 21 meters and which are home to hundreds of species of plants and animals familiar to other Siberian lakes. However, several species of caddis flies found in the shallow waters of Baikal are endemic; that is, they are found there but nowhere else. These flies are adapted to bouncing from wave to wave in the lake. They have small wings shaped like paddles, and legs designed for swimming. Their bodies can float on the water. These "flies" have become so well adapted to the water that they no longer have the ability to fly.

The second zone, the intermediate depths, ranges from 22 meters to 200 meters below the surface. Farther down is the abyssal zone, where life-forms become even more unusual. In these two zones, 80 percent of the species are endemic. In the 2.2-degree Celsius temperatures of the deepest region, there is eternal darkness; hence, many of the species have no eyes or only a minimal sense of sight. One kind of crustacean has a segmented body with seven pairs of legs. This amphipod has small indentations on the sides of its head where its eyes used to be found. The eyes disappeared over time, and now the animal finds its way through the use of long antenna-like structures on its body. Many other species in the abyssal zone have made similar adaptations to the absence of light.

Of the fifty species of fish in Baikal, about one-half are found nowhere else. The largest is the lake sturgeon, 3.3 meters in length and weighing up to 230 kilograms. The sturgeon are highly prized by local fishermen as a source of food and also for the caviar, or eggs, that the species produces. Sturgeon were almost fished out of existence, but recent limitations imposed by the Russian government have helped to restore the population.

Other species of fish include the omul, belonging to the salmon family, which reaches more than 0.3 meter in length and can weigh up to 4 kilograms. The omul is a predator, eating other fish. It is usually inactive until water temperatures reach 12 to 15 degrees in the summer. Omuls live in the upper zone, usually not more than 4 meters below the surface, feeding on smaller fish and zooplankton.

The omul is the most important commercial fish in the lake. When it is taken from the icy waters, it sends out a piercing cry as air is expelled from its bladder. The shriek sounds like a bitter complaint; Siberian fishermen say that anyone who whines or complains "cries like an omul."

The golomyanka, endemic to Baikal, has no scales and grows to about 20 centimeters in length. More than one-third of its body weight is oil. The golomyanka feeds at the surface at night, eating zooplankton and algae. The fish has little tolerance for heat, however, and when temperatures reach above 7 degrees Celsius, it will die. When its body washes ashore, its fat melts, leaving behind little but skin and bone. The golomyanka has another unique property: It gives birth to live larvae rather than laying eggs as most fish do, and the female can produce up to two thousand babies during the autumn breeding season. Frequently, the female's belly contains so many larvae that it explodes, killing the mother and most of the offspring.

The only mammal that feeds in the water is the Nerpa seal, one of only two freshwater seals in the world. The Nerpa's closest relatives live more than 3,000 kilometers away in the Arctic. This seal lives among the rocks on the northeastern coast of Baikal. In the summer, it migrates south, where it sleeps on the shore and feeds on omuls and golomyanka fish. The population of Nerpa seals is estimated at about 25,000. Scientists believe that the seals originated in the Arctic and made their way to Lake Baikal by way of various tributary rivers about twelve thousand years ago. The Baykal hair seal is not as numerous as the Nerpa but is also found on the coasts of various islands in the lake.

## STUDY OF LAKE BAIKAL

Once among the cleanest lakes on Earth, Baikal is beginning to be endangered by industrial wastes, despite its great size. There are no major cities right at the lake, but economic development has brought increased pollution to the area. In the 1960's, two paper mills were built on the southern shore, and government officials in the Soviet Union, eager to expand Siberia's economic growth, called for the construction of more mills. Protests from the Soviet Academy of Sciences helped stop the construction of two plants and prompted a redesign of another to make it less polluting. Still, in 1977, scientists studying Baikal's biology became alarmed at the decline in zooplankton. Especially serious was the killing of millions of *Epischura*, a tiny shrimplike species that is a major food source for larger fish. Quick action led to some changes in levels of chemical pollutants expelled into the lake by the paper mills. The crisis illustrated how even slight changes in the quality of the lake's water could eventually lead to disaster.

Industrial development began in 1905, after the completion of a railroad along the rocky southern shore. The new train replaced the existing ferry boat that previously was the only way across the lake. The advent of the great 14,580-kilometer trans-Siberian route in 1916 opened the region to contact with Russian territory from Moscow in the west to Vladivostok on the Pacific coast in the east. Still, the region had only a small population of mostly Mongolian peoples. The most important of these were the Buryats, population 550,000, a mostly Buddhist people who had settled in the area of Lake Baikal one thousand years before. It was not until 1938, however, with the completion of the Baikal-Amur trunk line that thousands of farmers and settlers came into the region. Since then, major industries have developed in the area; in addition to paper mills, mining, ship building, fishing, and timber cutting have emerged as local industries.

All the cities and industries of the area lie along the trans-Siberian railroad. Irkutsk is the center of the region and the major industrial city. Ule-Ude, east of Lake Baikal, and Chita are two other cities of more than 100,000 people. The trans-Siberian railroad has been the lifeline of this eastern part of Siberia since its completion. Ule-Ude has a rail link to the Mongolian capital of Ulaanbaatar. Since 1954, both cities have been linked to Beijing and northern China.

Another problem created by the paper mills results from the cutting of huge numbers of trees to make into pulp. The loss of trees dramatically increased soil erosion, which led to severe problems of runoff and contamination of Baikal's once crystal-clear waters. In 1969, the lake became part of a water-conservation area decreed by the Soviet government. Regulations prohibiting logging in certain critical areas were imposed, and strict regulations ordering the treatment of industrial pollutants were issued to reduce the amount of contaminated water entering the lake from the paper mills. Unfortunately, these regulations were not effectively enforced, and economic development was usually placed before environmental concerns by local officials. Tougher regulations and stricter enforcement of existing laws have been promoted by defenders of the environment, but a lack of funding has limited the enforcement of antipollution laws.

Because of the harshness of the climate, agriculture is possible only in limited areas around Lake Baikal. Oats are grown in the region, however, and cattle, sheep, and horses are raised there. Food industries are limited to meatpacking, fish canning, and the processing of some dairy products, chiefly cheese and milk. Mining has a long history in the region, as there is a relatively wide range of metals and minerals such as gold, tin, graphite, mica, and zinc. There are large coal deposits west of Lake Baikal along the Angara River. The Angara is also the home of the giant Bratsk hydroelectric plant, which supplies power to Irkutsk.

*Leslie V. Tischauser*

## FURTHER READING

Bogdanov, Uri. *Mysteries of the Deep: From the Depths of Lake Baikal to the Ocean Floor.* Moscow: Progress Publishers, 1989. Includes illustrations and pictures of some of the more unusual animals that live in the abyssal zone of the lake. Also provides information on the geological history of the area.

Di Duca, Marc. *Lake Baikal: Siberia's Great Lake.* Chalfont St. Peter, England: Bradt Travel Guides, 2010. Provides full coverage of activities, wildlife, culture, and religion, as well as practical information on traveling in this diverse region where the Russian and Mongol cultures converge.

Hirschmann, Kristine. *Extreme Places: The Deepest Lake.* San Diego: KidHaven Press, 2002. A children's book that presents the fundamental geology and ecology of Baikal Lake. A starting point for basic facts on Baikal Lake. Contains good information and images.

Kashiwaya, Kenji. *Long Continental Records from Lake Baikal.* Tokyo: Springer-Verlag Tokyo, 2003. Based on the study by the Baikal Drilling Project. Provides information on global climatic and environmental changes for as long as 12 million years and includes discussions on glacial and interglacial transitions that directly link to the present and future environment.

Koptyug, Valentin A., ed. *Sustainable Development of the Lake Baikal Region: A Model Territory for the World.* New York: Springer, 1996. A collection of essays and lectures that evaluates the economic and industrial development of Lake Baikal and the surrounding environment. Illustrations and bibliographical references.

Kozhova, Olga Mikhaeilovna, and L. R. Izmestseva, eds. *Lake Baikal: Evolution and Biodiversity.* Leiden: Backhuys, 1998. Examines the evolution and development of the biological systems and the ecosystem of Lake Baikal, with historical context provided. Somewhat technical but suitable for advanced high school students. Illustrations, maps, index, and twenty pages of bibliography.

Matthiessen, Peter. *Baikal: Sacred Sea of Siberia.* San Francisco: Sierra Club Books, 1995. Combines travelogue-style writing with historical information about the natural evolution and subsequent human-caused deterioration of Lake Baikal. Written from an environmentalist perspective. Color photographs and maps.

Minoura, Koji. *Lake Baikal: A Mirror in Time and Space for Understanding Global Change Processes.* Amsterdam: Elsevier, 2000. Compiles articles from the 1998 symposium discussing the geology, biology, chemistry, and ecology of Lake Baikal. Organized into three parts: paleoenvironment and natural history, physiochemistry, and evolution and biodiversity. Expertly melds together multiple disciplines.

Sergeev, Mark. *The Wonders and Problems of Lake Baikal.* Translated by Sergei Sumin. Moscow: Novosti Press Agency, 1989. Contains many photographs of wildlife, fish, and plants of the region. Discusses the hazards and costs of industrial pollution and its impact on the lake. Also has a brief history of the geology and history of Baykalia. A detailed map is included.

Symons, Leslie. *The Soviet Union: A Systematic Geography.* New York: Routledge, 1990. Contains some useful information on Lake Baikal and its effects on East Siberia. Includes many detailed maps describing the region's mineral, agricultural, and industrial wealth. Also contains a brief description and history of the impact and influence of the trans-Siberian railroad.

**See also:** Amazon River Basin; Drainage Basins; Ganges River; Great Lakes; Lakes; Mississippi River; Nile River; Waterfalls; Yellow River

# LAKES

*Lakes are geologically short-lived features, and sediments deposited in lakes (called lacustrine sediments) have constituted only a tiny fraction of the sedimentary rocks on Earth. Nevertheless, lake sediments are important sources of information about past climates. Several important economic resources—including oil shales, diatomaceous earth, salt and other evaporites, some limestones, and some coals—originate in lakes.*

## PRINCIPAL TERMS

- **allogenic sediment:** sediment that originates outside the place where it is finally deposited; sand, silt, and clay carried by a stream into a lake are examples
- **biogenic sediment:** sediment that originates from living organisms
- **clastic sediments:** sediments composed of durable minerals that resist weathering
- **clay:** a mineral group whose particles consist of structures arranged in sandwichlike layers, usually sheets of aluminum hydroxides and silica, along with some potassium, sodium, or calcium ions
- **clay minerals:** any mineral particle less than 2 micrometers in diameter
- **endogenic sediment:** sediment produced within the water column of the body in which it is deposited; for example, calcite precipitated in a lake in summer
- **mineral:** a solid with a constant chemical composition and a well-defined crystal structure
- **mineraloid:** a solid substance with a constant chemical composition but without a well-ordered crystal structure
- **plankton:** plant and animal organisms, most of which are microscopic, that live within the water column
- **seston:** a general term that encompasses all types of suspended lake sediment, including minerals, mineraloids, plankton, and organic detritus

## GEOLOGICAL ORIGIN OF LAKES

Several geologic mechanisms can create the closed basins that are needed to impound water and produce lakes. The most important of these mechanisms include glaciers, landslides, volcanoes, rivers, subsidence, and tectonic processes.

Continental glaciers formed thousands of lakes by the damming of stream valleys with moraine materials. Glaciers also scoured depressions in softer bedrock, and these later filled with water to form lakes. Depressions called kettles formed when buried ice blocks melted. Mountain glaciers also continue to produce numerous small, high alpine lakes by plucking away bedrock. The bowl-shaped depressions that occur as a result of this plucking are called cirques; lakes that occupy cirques are called tarns. Sometimes a mountain glacier moves down a valley and carves a series of depressions along the valley that, from above, look like a row of beads along a string. When these depressions later fill with water, the lakes are called paternoster lakes, the name (Latin for "our father") coming from their similarity to beads on a Christian rosary.

Landslides sometimes form natural dams across stream valleys. Large lakes then pond up behind the dam. Volcanoes may produce lava flows that dam stream valleys and produce lakes. A volcanic explosion crater may fill with water and so produce a lake. After an eruption, the area around the eruption vent may collapse to form a depression called a caldera. Some calderas, such as Crater Lake in Oregon, fill with water. Rivers can produce lakes along their valleys when the loop of a meandering channel finally is enclosed by sediment and leaves behind an oxbow lake, isolated from the main channel. Sediment may accumulate at the mouth of a stream, and the resulting delta may build, bridging across irregularities in the shoreline, to create a brackish coastal lake.

Natural subsidence creates closed basins in areas underlain by soluble limestones or evaporite deposits. As the underlying limestone is dissolved away, the ground above collapses into the cavity, forming a sinkhole that may later fill with water. Finally, large-scale downwarping of tectonic plates has produced some very large lakes. Large basins form when the crust warps or sinks downward in response to deep forces. The subsidence produces very large closed basins that can hold water. A few immense lakes owe their origins to tectonic downwarping.

## SEDIMENTATION

With few exceptions, most lakes exist in relatively small depressions and serve as the catch basins for sediment from the entire watershed or drainage basin around them. The natural process of sedimentation ensures that most lakes fill with sediment before long periods of geologic time have passed. Lakes with areas of only a few square kilometers or less will fill within a few tens of thousands of years. Very large lakes and inland seas may endure for more than 10 million years. Human-made lakes and reservoirs have unusually high sediment-fill rates in comparison with most natural lakes. Human-made lakes may fill with sediment within a few decades to a few centuries.

Lake sediments come from four sources: allogenic clastic materials that are washed in from the surrounding watershed; endogenic chemical precipitates that are produced from dissolved substances in the lake waters; endogenic biogenic organic materials produced by plants and animals living in the lake; and aeolian or airborne substances, such as dust and pollen, transported to the lake in the atmosphere.

Allogenic clastic materials are mostly mineral in nature, produced when rocks and soils in the drainage basin are weathered by mechanical and chemical processes to yield small particles. These particles are moved downslope by gravity, wind, and running water to enter streams, which then transport them to the lake. Clastic materials also enter the lake via waves, which erode the materials from the shoreline, and via landslides that directly enter the lake. In winter, ice formed on the lake can expand and push its way a few centimeters to 1 meter or so onto the shore. There, the ice may pick up large particles, such as gravel and cobbles. When the spring thaw comes, waves can remove that ice, together with its enclosed particles, and float it out onto the lake. The process by which the large particles are transported out on the lake is called ice-rafting. As the ice melts, the large clastic particles drop to the bottom. They are termed dropstones when found in lake sediments. A landslide into a lake or a flood in a stream that feeds into the lake can produce water heavily laden with sediment. The sediment-laden water is denser than clean water and therefore can rush down and across the lake bottom at speeds sufficient to carry even coarse sand far out into the lake. These types of deposits are called turbidite deposits.

Endogenic chemical precipitates in freshwater lakes commonly consist of carbonate minerals (calcite, aragonite, or dolomite) and mineraloids that consist of oxides and hydroxides of iron, manganese,

and aluminum. In some saline and brine lakes, the main sediments may be carbonates, together with sulfates such as gypsum (hydrated calcium sulfate), thenardite (sodium sulfate), or epsomite (hydrated magnesium sulfate), or with chlorides such as halite (sodium chloride) or more complex salts. Of the endogenic precipitates, calcite is the most abundant. Its precipitation represents a balance between the carbon dioxide content of the atmosphere and that of the carbon dioxide dissolved in the lake water.

Diatoms are distinctive microscopic algae that produce a frustule (a kind of shell) made of silica glass that is highly resistant to weathering. When seen under a high-powered microscope, diatom frustules appear to be artwork, looking like beautiful and highly ornate saucer- and pen-shaped works of glass. A tiny spot of lake sediment may contain millions of them.

A lake's sediment may contain from less than 1 percent to more than 90 percent organic materials, depending upon the type of lake. Most organic matter in lake sediments is produced within the lake by plankton and consists of compounds such as carbohydrates, proteins, oils, and waxes that are made up of carbon, hydrogen, nitrogen, and oxygen, with a little phosphorus. Plankton has an approximate bulk composition of 36 percent carbon, 7 percent hydrogen, 50 percent oxygen, 6 percent nitrogen, and 1 percent phosphorus (by weight). Plankton includes microscopic plants (phytoplankton) and microscopic animals (zooplankton) that live in the water column. Lakes that are very high in nutrients (eutrophic lakes) commonly have heavy blooms of algae, which contribute much organic matter to the bottom sediment. Terrestrial (land-derived) organic material such as leaves, bark, and twigs form a minor part of the organic matter found in most lakes. Terrestrial organic material is higher in carbon and lower in hydrogen, nitrogen, and phosphorus than is planktonic organic matter.

Airborne substances usually constitute only a tiny fraction of lake sediment. The most important of such material is pollen and spores. Pollen usually constitutes less than 1 percent of the total sediments, but that tiny amount is a very useful component for learning about the past climates that have existed on Earth. Pollen is among the most durable of all natural materials. It survives attack by air, water, and even strong acids and bases. Thus, it remains in the sediment through geologic time. As pollen accumulates in the bottom sediment, the lake serves as a kind of recorder for the vegetation that existed around it at a given time. By taking a long core of the bottom sediment from certain types of lakes and identifying the various pollen grains that it contains, a geologist may look at the pollen changes that have occurred through time and reconstruct the history of the climate and vegetation in an area.

Volcanic ash thrown into the atmosphere during eruptions enters lakes and forms a discrete layer on the lake bottom. When Mount St. Helens erupted in 1980, it deposited several centimeters of ash in lakes more than 160 kilometers east of the volcano. Geologists have used layers of ash in lakes to reconstruct the history of volcanic eruptions in some areas. Although dust storms contribute sediment to lakes, such storms are usually too infrequent in most areas to contribute significant amounts. In addition to wind-blown dust, a constant rain of tiny particles enters the atmosphere from space as micrometeorites, some of which reaches the surface and becomes a component of lake and seafloor sediments.

## WATER CIRCULATION

Lake waters are driven into circulation by temperature-induced density changes and by wind. Most freshwater lakes in temperate climates circulate completely twice each year and so are termed dimictic lakes. Circulation exerts a profound influence on water chemistry of the lake and the amount and type of sediment present within the water column. During summer, lakes become thermally stratified into three zones. The upper layer of warm water (epilimnion) floats above the denser cold water and prevents wind-driven circulation from penetrating much below the epilimnion. The epilimnion is usually in circulation, is rich in oxygen (from algal photosynthesis and diffusion from the atmosphere), and is well lighted. This layer is where summer blooms of green and blue-green algae occur and calcite precipitation begins. The middle layer (thermocline) is a transition zone in which the water cools downward at a rate of greater than 1 degree Celsius per meter. The bottom layer (hypolimnion) is cold, dark, stagnant, and usually poor in oxygen. There, bacteria decompose the bottom sediment and release phosphorus, manganese, iron, silica, and other constituents into the hypolimnion.

Sediment deposited in summer includes a large amount of organic matter, clastic materials washed in during summer rainstorms, and endogenic carbonate minerals produced within the lake. The most common carbonate mineral is calcite (calcium carbonate). The regular deposition of calcite in the summer is an example of cyclic sedimentation, a sedimentary event that occurs at regular time intervals. This event occurs yearly in the summer season and takes place in the upper 2 or 3 meters of water. On satellite photos, it is even possible to see these summer events as whitenings on large lakes, such as Lake Michigan.

As the sediment falls through the water column in summer, it passes through the thermocline, into the hypolimnion, and onto the lake bottom. As it sits on the bottom during the summer months, bacteria, particularly anaerobic bacteria (those that thrive in oxygen-poor environments), begin to decompose the organic matter. As this occurs, the dissolved carbon dioxide increases in the hypolimnion. If enough carbon dioxide is produced, the hypolimnion becomes slightly acidic, and calcite and other carbonates that fell to the bottom begin to dissolve. This is essentially the same process that occurs at the carbonate compensation depths of the oceans. The acidic conditions also release dissolved phosphorus, calcium, iron, and manganese into the hypolimnion, as well as some trace metals. Clastic minerals such as quartz, feldspar, and clay minerals are not affected in such brief seasonal processes, but some silica from biogenic material such as diatom frustules can dissolve and enrich the hypolimnion in silica. As summer progresses, the hypolimnion becomes more and more enriched in dissolved metals and nutrients.

Autumn circulation begins when the water temperature cools and the density of the epilimnion increases until it reaches the same temperature and density as the deep water. Thereafter, there is no stratification to prevent the wind from circulating the entire lake. When this happens, the cold, stagnant hypolimnion, now rich in dissolved substances, is swept into circulation with the rest of the lake water. The dissolved materials from the hypolimnion are mixed into a well-oxygenated water column. Iron and manganese that formerly were present in dissolved form now oxidize to form tiny solid particles of manganese oxides, iron oxides, and hydroxides. The sediment therefore becomes enriched in iron, manganese, or both during the autumn overturn, with the amount of enrichment depending upon the amount of dissolved iron and manganese that accumulated during summer in the hypolimnion. Dissolved silica is also swept from the hypolimnion into the entire water column. In the upper water column, where sunlight and dissolved silica become present in great abundance, diatom blooms occur. The diatoms convert the dissolved silica into solid opaline frustules.

As circulation proceeds, the currents may sweep over the lake bottom and actually resuspend 1 centimeter or more of sediment from the bottom and margins of the lake. The amount of resuspension that occurs each year in freshwater lakes is primarily the result of the shape of the lake basin. A lake that has a large surface area and is very shallow permits wind to keep the lake in constant circulation over long periods of the year.

As winter stratification develops, an ice cover forms over the lake and prevents any wind-induced circulation. Because the circulation is what keeps the lake sediment in suspension, most sediment quickly falls to the bottom, and sedimentation then is minimal through the rest of winter. If light can penetrate the ice and snow, some algae and diatoms can utilize this weak light, present in the layer of water just below the ice, to reproduce. Their settling remains contribute small amounts of organic matter and diatom frustules. At the lake bottom, the densest water (that at 4 degrees Celsius) accumulates. As in summer, some dissolved nutrients and metals can build up in this deep layer, but because the bacteria that are active in releasing these substances from the sediment are refrigerated, they work slowly, and not as much dissolved material builds up in the bottom waters.

When spring circulation begins, the ice at the surface melts, and the lake again goes into wind-driven circulation. Oxidation of iron and manganese occurs (as in autumn), although the amounts of dissolved materials available are likely to be less in spring. Once again, nutrients such as phosphorus and silica are circulated out of the dark bottom waters and become available to produce blooms of phytoplankton. Spring rains often hasten the melting, and runoff from rain and snowmelt in the drainage basin washes clastic materials into the lake. The period of spring thaw is likely to be the time of year when the maximum amount of new allogenic (externally derived) sediment enters the lake.

Spring diatom blooms continue until summer stratification prevents further replenishment of silica to the epilimnion. Thereafter, the diatoms are succeeded by summer blooms of green algae, closely followed by blooms of blue-green algae. Silica is usually the limiting nutrient for diatoms; phosphorus is the limiting nutrient for green and blue-green algae.

## DIAGENESIS

After sediments are buried, changes occur. This process of change after burial is termed diagenesis. Physical changes include compaction and dewatering. Bacteria decompose much organic matter and produce gases such as methane, hydrogen sulfide, and carbon dioxide. The "rotten-egg" odor of black lake sediments, often noticed on boat anchors, is the odor of hydrogen sulfide. After long periods of time, minerals such as quartz or calcite slowly fill the pores remaining after compaction.

One of the first diagenetic minerals to form is pyrite (iron sulfide). Much pyrite occurs in microscopic spherical bodies that look like raspberries; these particles, called framboids (from the French *framboise*, meaning "raspberry"), are probably formed by bacteria in areas with low oxygen within a few weeks. In fact, the black color of some lake muds and oozes results as much from iron sulfides as from organic matter. Other diagenetic changes include the conversion of mineraloid particles containing phosphorus into phosphate minerals such as vivianite and apatite. Manganese oxides may be converted into manganese carbonates (rhodochrosite). Freshwater manganese oxide nodules may form in high-energy environments such as Grand Traverse Bay in Lake Michigan.

## STUDY OF LAKES

Scientists who study lakes (limnologists) must study all the natural sciences, including physics, chemistry, biology, meteorology, and geology, because lakes are complex systems that include biological communities, changing water chemistry, geological processes, and interactions among water, sunlight, and the atmosphere.

Modern lake sediments are collected from the water column in sediment traps (cylinders and funnels into which the suspended sediment settles over periods of days or weeks) or by filtering large quantities of lake water. Living material is often sampled with a plankton net. Older sediments that have accumulated on the bottom are collected with dredges and by piston coring, which involves pushing a sharpened hollow tube (usually about 2.5 centimeters in diameter) downward into the sediment. Cores are valuable because they preserve the sediment in the order in which it was deposited, from oldest at the bottom to the most recent at the top. Once the sample is collected, it is often frozen and taken to the laboratory. There, pollen and organisms may be examined by microscopy, minerals may be determined by X-ray diffraction, and chemical analyses may be made.

Varves are thin laminae that are deposited by cyclic processes. In freshwater lakes, each varve represents one year's deposit; it consists of a couplet with a dark layer of organic matter deposited in winter and a light-colored layer of calcite deposited in summer. Varves are deposited in lakes where annual circulations cannot resuspend bottom sediment and therefore cannot mix it to destroy the annual lamination. Some lakes that are small and very deep may produce varved sediments; Elk Lake in Minnesota is an example. In other lakes, the accumulation of dissolved salts on the bottom eventually produces a dense layer (monimolimnion), which prevents disturbance of the bottom by circulation in the overlying fresher waters. Soap Lake in Washington State is an example. Because each varve couplet represents one year, a geologist may core the sediments from a varved lake and count the couplets to determine the age of the sediment in any part of the core. The pollen, the chemistry, the diatoms, and other constituents may then be carefully examined to deduce what the lake was like during a given time period. The study is much like solving a mystery from a variety of clues. Eventually, the history of climate changes of the area may be learned from the study of lake varves.

*Edward B. Nuhfer*

## FURTHER READING

Dennis, Jerry. *The Living Great Lakes.* New York: Thomas Dunne Books, 2003. Written for the layperson. Covers the geology, natural history, biology, and industry of the Great Lakes. Discusses the structure of the lakes from their formation to the current human impact of resource mining, construction of dams and canals, and introduction of invasive species.

Dodds, Walter K., and Matt R. Wiles. *Freshwater Ecology: Concepts and Environmental Applications of Limnology.* 2d ed. Burlington, Mass.: Academic Press, 2010. Covers the physical and chemical properties of water, the hydrologic cycle, nutrient cycling in water, as well as biological aspects. Written by two of the leading scientists in freshwater ecology. An excellent resource for college students as each chapter has a short summary of main topics.

Grady, Wayne. *The Great Lakes.* Vancouver: Greystone Books, 2007. Discusses the natural history, geology, ecology, and conservation of the Great Lakes. A chapter discusses the human impact of invasive species on the ecology of the lakes. Includes further readings, a list of common and scientific names, illustration credits, and indexing.

Håkanson, Lars, and M. Jansson. *Principles of Lake Sedimentology* Caldwell, N.J.: Blackburn Press, 2002. A reference for professionals in the field of lake sedimentology, though parts of it may be accessible to high school students. Focuses on lake sediments in detail. Provides methods of sampling and discusses the influence of lake type and shape on the sediments formed in the lake, the circulation of lake waters, the chemistry of sediments, and the pollution of lakes.

Imberger, Jeorg, ed. *Physical Processes in Lakes and Oceans.* Washington, D.C.: American Geophysical Union, 1998. Details the origins, processes, and phases of oceans, lakes, and water resources, as well as the ecology and environments surrounding them. Illustrations and maps.

Kaye, Catheryn Berger, and Philippe Cousteau. *Going Blue.* Minneapolis, Minn.: Free Spirit Publishing, 2009. Discusses conservation issues related to oceans, lakes, rivers, and other bodies of water, as well as topics such as pollution, watershed management, coral bleaching, and ocean acidification. Also provides a guideline for action with multiple chapters discussing how students can get involved in water conservation. Lists many resources to help the reader find more information or get started on projects.

Kolumban, Hutter, Yongqi Wang, and Irina P. Chubarenko. *Physics of Lakes: Foundation of the Mathematical and Physical Background.* Dordrecht: Springer, 2011. Assumes a certain level of mathematical knowledge on the part of the reader, but the mathematical concepts presented are not beyond ready comprehension, beginning with the fundamental equations of lake hydrodynamics and progressing to angular momentum and vorticity.

Lerman, Abraham, Dieter M. Imboden, and Joel R. Gat, eds. *Physics and Chemistry of Lakes.* New York: Springer-Verlag, 1995. Offers a nice introduction to limnology by examining the geological, chemical, and physical properties of lakes. Suitable for the nonscientist. Illustrations, index, and bibliography.

Magnuson, John J., Timothy K. Kratz, and Barbara J. Benson. *Long-Term Dynamics of Lakes in the Landscape: Long-Term Ecological Research on North Temperate Lakes.* Oxford: Oxford University Press, 2006. Intended for most American limnologists and many ecologists, summarizing twenty years of observation and research at the North Temperate Lakes long-term research site.

Pringle, Laurence. *Rivers and Lakes.* New York: Time-Life Books, 1985. Examines lacustrine (lake) and fluvial (river) environments. Suitable for general readers. Part of the Time-Life Planet Earth series.

Stumm, Werner, ed. *Chemical Processes at the Particle-Water Interface.* New York: John Wiley & Sons, 1987. A highly technical reference book designed for specialists and graduate students. Focuses on the chemical interactions that occur between sediment particles and the surrounding lake waters, the process of clotting of particles, and the role of particle surfaces in removing trace metals from water. Discusses lake sediments in a manner that may be understood by undergraduates who have had rigorous courses in introductory chemistry and geology.

U.S. Environmental Protection Agency. *The Great Lakes: An Environmental Atlas and Resource Book.* Chicago, Ill.: Great Lakes Program Office, 1995. A free soft-cover publication designed for anyone interested in the lakes. Briefly addresses the physical and cultural environments of the basin. Includes colored maps, tables, and charts.

Wetzel, R. G., ed. *Limnology.* 3d ed. San Diego: Elsevier Science, 2001. A well-written textbook typical of those used by undergraduates and graduates in introductory limnology courses. Covers physical, biological, and chemical aspects of lakes. Assumes a knowledge of high school algebra, chemistry, physics, and biology.

**See also:** Alluvial System; Beaches and Coastal Processes; Dams and Flood Control; Deep-Sea Sedimentation; Deltas; Desert Pavement; Drainage Basins; Floodplains; Floods; Freshwater and Groundwater Contamination Around the World; Hydrologic Cycle; Reefs; River Bed Forms; River Flow; Sand; Sediment Transport and Deposition; Surface Water; Weathering and Erosion

# LIGHTNING AND THUNDER

*Lightning is the discharge of accumulated static electricity from a thundercloud as a gigantic electrical arc between the cloud and the ground or another cloud. It is accompanied by intense heating and explosive expansion of air, producing the sonic boom called thunder. Lightning may cause property damage and death, but it also contributes to the nitrogen absorption in the soil necessary for plant life and may have even helped to start life on Earth.*

## PRINCIPAL TERMS

- **ball lightning:** a rare form of lightning appearing as luminous balls of charged air
- **bead lightning:** lightning that appears as a series of beads tied to a string
- **convection:** vertical air circulation in which warm air rises and cool air sinks in a cyclic manner
- **corona discharge:** a continuous electric discharge from highly charged, pointed objects that produces the luminous greenish or bluish halo known as St. Elmo's fire
- **cumulonimbus:** thunderstorm clouds that develop vertically by strong convection in the atmosphere and, due to high-altitude wind shear, often form a top shape reminiscent of an anvil
- **dart leaders:** surges of electrons that follow the same intermittent ionized channel taken by the initial stepped leader of a lightning stroke
- **graupel:** ice particles between 2 and 5 millimeters in diameter that form by a process of accretion in a cloud
- **ionosphere:** an electrically active region of the upper atmosphere from about 80 to 800 kilometers above the surface of Earth that contains a relatively high concentration of ions (charged atoms or molecules) and free electrons
- **return stroke:** the luminous lightning stroke that propagates upward from the ground surface toward the base of a cloud as electrons surge downward and a positive current flows to the cloud
- **stepped leader:** an initial discharge of electrons that proceeds in a series of steps from the base of a thundercloud toward the ground

## EARLY LIGHTNING STUDIES

The search for understanding of lightning has been long and circuitous, and even today there are conflicting theories. In ancient civilizations, lightning and thunder were viewed as manifestations of the power and wrath of the gods. Lightning bolts were the favored weapon of Zeus, the leader of the Greek gods. His Roman counterpart Jupiter subdued monsters with thunderbolts forged by Vulcan, a lesser god believed to have his forge located below Mount Etna, a volcano on the island of Sicily that is often active. The numerous lightning discharges that occur in the dust and ash clouds above volcanoes may have been the source of this belief. The book of Exodus in the Bible relates that before Moses climbed Mount Sinai to receive the Ten Commandments, "there was thunder and lightning, with a thick cloud over the mountain, and a very loud trumpet blast."

A scientific understanding of lightning began in earnest around the year 1750, in part with the work of Benjamin Franklin. Three years earlier, Franklin had experimented with the electrical charging of pointed objects connected to the ground and found that such points could draw a charge from another charged object when placed near it. In 1749, he wrote a letter to the English Royal Society, speculating on the electrification of clouds and suggesting that lightning is an electric discharge to the ground that favors trees, spires, chimneys, and masts. He also compared the snap of an electric spark with the thunder produced by lightning. In 1750, he wrote the Royal Society suggesting that a pointed metal rod extending several feet above a building and connected by a wire to the ground could protect the building from lightning damage, even drawing a charge silently from the cloud.

Along with his idea of the lightning rod, Franklin proposed an experiment for collecting electricity from clouds by placing a properly insulated metal rod above a high tower or steeple and drawing sparks from its lower end. The Comte De Buffon performed this experiment successfully in France in May 1752, but when the German scientist George W. Richmann attempted to repeat it in Russia in 1753, he was struck and killed by a bolt of lightning. In October 1752, Franklin performed his famous kite experiment with a pointed wire fastened to the top of the kite and a metal key at the lower end of the string, which was held by a silk ribbon for insulation. Standing under a shed to keep the silk ribbon dry, Franklin was able to draw sparks from the key and collected enough electric charge to show that it was the same as ordinary electricity. In experiments with a metal rod above his house, he found that the charge at the bottom of thunderclouds was usually negative but sometimes positive, suggesting the complex nature of cloud charging.

## THEORIES OF THUNDERCLOUD CHARGING

Lightning occurs when separate regions in a cumulonimbus thundercloud contain opposite electrical charges. Several theories have been proposed to account for this separation of charge. The precipitation hypothesis, first proposed by the German physicists Julius Elster and Hans Geitel in 1885, assumed that raindrops and ice pellets in a thunderstorm are pulled down by gravity past smaller water droplets and ice crystals suspended in the cloud. The hypothesis suggested that collisions between these particles cause the larger particles to gain a negative charge (in much the same manner as when charge transfers to shoes from a rug), and the smaller ones become positively charged. Updrafts within the cloud then sweep the smaller positive particles up near the top of the cloud, while the larger negative particles accumulate near the lower part of the cloud.

A second theory, the convection hypothesis, was developed independently by the French scientist Gaston Grenet in 1947 and by the American scientist Bernard Vonnegut in 1953. They assumed that cosmic rays from outer space ionize the air above a cloud, separating positive and negative charges, while corona discharges around sharp objects on the land surface produce positive ions. Warm air then carries these positive ions upward by convection until they near the top of the cloud, where they attract the negative ions formed by cosmic rays. These negative ions enter the cloud and attach to water droplets and ice crystals, which are then carried by downdrafts at the edges of the cloud to the lower regions, charging the cloud negatively at the bottom and positively at the top.

A theory developed in the 1980's combines the effects of convection and precipitation with the microphysics of ice particles and is able to explain Franklin's observations that the bottoms of clouds are sometimes positive. It assumes that charging occurs as millimeter-size ice particles called graupel fall through a region of supercooled droplets (liquid water below 0 degrees

Celsius) and ice crystals. When these droplets collide with the graupel, they freeze on contact, releasing latent heat. This heat keeps the surface of the graupel warmer than the surrounding ice crystals. Several studies at American and British universities have shown that negative charge is transferred to graupel below a critical temperature of about –15 degrees Celsius. When the graupel and crystals come into contact, positive ions are transferred to the ice crystals, which are then carried by updrafts to the top parts of the cloud. The graupel and larger hailstones fall, depositing a strong negative charge near the critical temperature level in the cloud. Below this level, positive charge is acquired by the graupel, causing some weaker positive regions near the bottom of the cloud.

## CLOUD-TO-GROUND LIGHTNING

The normal fair-weather atmosphere is characterized by a positively charged upper atmosphere, called the ionosphere, and a negative charge on the land surface. As a thundercloud develops with strong convection updrafts from warmer surface regions, its strong negative charge repels electrons in the ground below, causing it to become positively charged relative to the cloud. The positive charge becomes most concentrated on objects that protrude above the ground, such as trees, poles, and buildings. When the difference in charge between some point on the ground and the cloud builds up to an electric potential of about 1 million volts per meter, the air begins to ionize sufficiently to overcome air's normal electrical resistance, and current begins to flow. High-speed camera studies have revealed that the resulting lightning stroke is a complex series of events rather than a single straightforward electrical arc event.

In cloud-to-ground lightning, a local electric potential of more than 3 million volts per meter develops within the cloud along a path of some 50 meters. This causes a discharge of electrons to rush toward the base of the cloud and then toward the ground in a series of steps, carrying a negative current of a few hundred amperes downward. Each step in the discharge covers about 50 to 100 meters, with stops of about fifty microseconds between steps as charge sufficient to achieve the next step accumulates. This initial discharge is called the "stepped leader" and produces only a faint glow nearly invisible to the human eye. As the stepped leader approaches to within about 50 meters of the ground, a region of positive charge moves up into the air through any susceptible object (usually elevated) to meet it.

When the stepped leader of electrons meets the upward-moving positive charge, an immense quantity of electrons flows to the grounding point, starting from the bottom of the ionized channel in a "return stroke" equivalent to an upward positive current of some 10,000 amperes. The luminous return stroke, basically an electrical arc several centimeters in diameter, generates heat and light in a bright flash lasting less than two-tenths of a millisecond, too fast for the eye to resolve the motion. This leader-and-stroke process is often repeated in the same ionized channel several times at intervals of about two-hundredths of a second. These subsequent surges of electrons from the cloud, called "dart leaders," proceed downward more rapidly because of the lowered resistance of the path, each followed by a less energetic return stroke. The entire process lasts less than one-half of a second, too short to distinguish individual strokes. Some dart leaders that do not reach the ground appear as lighter forked flashes.

## VARIED FORMS OF LIGHTNING AND THUNDER

Lightning and thunder take on a variety of shapes and forms. Sometimes a dart leader is met by a return stroke along a new conducting path from the ground, causing a forked lightning bolt that strikes the ground in more than one place. In fewer than 10 percent of cases, a positively charged leader emanates from the cloud; even more rarely, a leader starts from tall objects on the ground and meets with a return stroke from the cloud. If the wind moves the ionized channel fast enough between strokes, a broader streak called ribbon lightning can be produced, or a series of bright streaks called bead lightning results when the channel appears to break up. A rare phenomenon called ball lightning consists of an electric discharge in the form of a slowly moving basketball-size luminous sphere that has been variously observed to either explode or simply decay.

Studies have shown that about 80 percent of lightning occurs within clouds where it is hidden from view, and occasionally even strikes between clouds. Lightning inside of clouds can illuminate them with flashes of light called sheet lightning. Lightning from distant thunderstorms that can be seen but is too far away for its thunder to be heard is called heat lightning. Since 1989, three new rare types of lightning

have been videotaped jumping from the tops of clouds into the atmosphere. These consist of bright bluish, cone-shaped bursts called "blue jets" that rise more than 20 kilometers from the cloud center, clusters of dim reddish bursts called "red sprites" rising to heights of more than 50 kilometers, and doughnut-shaped bursts of light called "elves" some 400 kilometers wide and 100 kilometers above the cloud tops.

Lightning usually produces thunder. A lightning discharge heats the air along the conducting path so quickly (a few microseconds) and to such a high temperature (several times that of the surface of the sun) that the air expands explosively. This produces a shock wave of compressed air that decays into an acoustic wave within a couple of meters. Because sound travels almost 1 million times more slowly than light, thunder is heard after the lightning is seen. Sound travels at about 330 meters per second, so the distance to the flash can be estimated as 1 kilometer for every three seconds between the lightning flash and the thunder. Within about 100 meters from lightning, thunder sounds like a clap followed by a loud bang. At greater distances, thunder often rumbles as it emanates from sections of the lightning stroke (about 5 kilometers long) at different distances from the observer. Because of temperature differences and turbulence in the air, thunder is seldom heard beyond about 20 kilometers.

## SIGNIFICANCE

Lightning has both destructive and beneficial effects. Estimates indicate that more than forty thousand thunderstorms and 8 million lightning flashes occur daily around the world. Tropical landmasses experience thunderstorms on more than one hundred days each year, while they rarely occur in polar regions or dry climates. In the United States, the average number of thunderstorm days each year varies from ninety in Florida to forty in the Midwest and fewer than ten along the West Coast. Annually in the United States, lightning causes about one hundred deaths and starts about ten thousand forest fires, with timber losses and property damage of about $80 million or more. Lightning can be most dangerous near elevated places and isolated trees. The best protection is secured inside automobiles and buildings, out of contact with any conducting surfaces.

Lightning is essential for life on Earth in that it provides a source of plant fertilizer and ozone. Its electrical action combines nitrogen and oxygen in the atmosphere to form nitric oxide, which then dissolves in precipitation and is brought to the surface as a nitrate. Hundreds of millions of tons of nitrates are produced by lightning each year and absorbed by the soil, where the nitrates help to nourish plant life and food crops. Studies at the National Center for Atmospheric Research, using a computer model called MOZART, showed that lightning is probably the main source of ozone in the upper atmosphere. The ozone layer protects life on Earth by blocking ultraviolet radiation from the sun.

Lightning also helps to maintain the potential difference of some 300,000 volts between the surface and the ionosphere, which reflects the radiowaves used for long-distance communication. This voltage causes a fair-weather current of about 2,000 amperes to discharge from the surface at a rate of a few microamperes per square kilometer. Lightning balances this flow by returning electrons to the surface. Lightning may have even been the spark that stimulated the development of life through the formation of organic molecules on primitive Earth, as well as being a likely source of the first fire used by humans for protection and survival.

*Joseph L. Spradley*

**FURTHER READING**

Ahrens, C. Donald. *Essentials of Meteorology: An Invitation to the Atmosphere.* Belmont, Calif.: Brooks/Cole Cengage Learning, 2012. Discusses various topics in weather and the atmosphere. Covers tornadoes, hurricanes, thunderstorms, acid deposition and other air pollution topics, humidity, cloud formation, and temperature.

_____. *Meteorology Today: An Introduction to Weather, Climate, and the Environment.* 8th ed. Belmont, Calif.: Thomson Brooks-Cole, 2007. Contains a good description of the current understanding of lightning and thunder with several good colored photographs and diagrams. A college-level text.

Barry, James Dale. *Ball Lightning and Bead Lightning: Extreme Forms of Atmospheric Electricity.* New York: Plenum Press, 1980. Provides a comprehensive description of two unusual forms of lightning supported by numerous photographs and drawings and an extensive bibliography.

Betz, H. D., U. Schumann, and P. Laroche, eds. *Lightning: Principles, Instruments, and Applications.* New York: Springer, 2009. Presents a number of articles on the physics principles influencing lightning occurrences. Covers lightning detection and observation extensively and discusses mapping and analysis of lightning. Includes topics such as air chemistry and dynamics, infrasound, supercells, and lightning protection. Uses many mathematical models throughout.

Dunlop, Storm. *The Weather Identification Handbook.* Guilford, Conn.: Lyons Press, 2003. Provides a simple and useful guide to various atmospheric objects, patterns, dynamics and phenomena. Begins with identification of cloud formations, followed by optical phenomena such as rainbows, and discusses various weather patterns and events. Best used as a reference for the weather enthusiast as an introduction to the study of meteorology.

Few, Arthur A. "Thunder." *Scientific American* 233 (July, 1975): 80-90. Describes extensive research on lightning, with a special emphasis on its relation to thunder, including several illustrations and diagrams.

Friedman, John. *Out of the Blue: A History of Lightning: Science, Superstition and Amazing Stories of Survival.* New York: Delta/Random House, 2009. Geared to general readership. Examines the phenomenon of lightning as a social force throughout history and into the present-day scientific comprehension of the nature of lightning.

Golde, R. H., ed. *Lightning.* London: Academic Press, 1977. Covers lightning in history and research by experts in various fields. Includes many photographs and diagrams.

Rakov, Vladimir A., and Martin A. Uman. *Lightning: Physics and Effects.* New York: Cambridge University Press, 2007. Intended for those with a more advanced understanding of the nature of lightning. Covers all aspects of lightning, including the physics of how lightning functions and methods of protection from lightning.

Schneider, Bonnie. *Extreme Weather: A Guide to Surviving Flash Floods, Tornadoes, Hurricanes, Heat Waves, Snowstorms, Tsunamis and Other Natural Disasters.* Basingstoke: Palgrave Macmillan, 2012. Presents vivid explanations of how, when, and why major natural disasters occur. Discusses floods, hurricanes, thunderstorms, mudslides, wildfires, tsunamis, and earthquakes. Presents a guide of how to prepare for and what to do during an extreme weather event, along with background information on weather patterns and natural disasters.

Uman, Martin A. *The Lightning Discharge.* Mineola, N.Y.: Dover Publications, 2001. Offers an extensive monograph by an expert in the field that discusses and vividly describes most of what is known about lightning.

Williams, Earle R. "The Electrification of Thunderstorms." *Scientific American* 259 (November, 1988): 88-99. A well-illustrated discussion of the causes of charge separation in clouds and the lightning discharge.

**See also:** Atmosphere's Global Circulation; Atmospheric and Oceanic Oscillations; Atmospheric Properties; Barometric Pressure; Climate; Clouds; Cyclones and Anticyclones; Drought; Floods; Gulf of Mexico; Hurricanes; Meteorological Satellites; Monsoons; Ocean-Atmosphere Interactions; Pacific Ocean; Precipitation; Satellite Meteorology; Seasons; Severe Storms; Tornadoes; Van Allen Radiation Belts; Volcanoes: Climatic Effects; Weather Forecasting; Weather Forecasting: Numerical Weather Prediction; Weather Modification; Wind

# LONG-TERM WEATHER PATTERNS

*Much work in meteorology is dedicated to the accurate prediction of short-term weather patterns. However, an important part of meteorology is the pursuit of information on long-term trends and conditions. This field involves the study of consistent weather conditions that develop and continue through months and years. The field also entails an understanding of certain phenomena that contribute to these extended weather periods. Research on long-term weather patterns can help scientists understand past climate changes and predict future shifts, enabling society to better prepare for such weather changes.*

## PRINCIPAL TERMS

- **Arctic oscillation:** long-term weather pattern in which the different air pressures in the Arctic and middle latitude regions cause varying weather conditions
- **El Niño:** meteorological condition in which the waters of the tropical, eastern Pacific Ocean are warmed by the atmosphere
- **La Niña:** meteorological condition in which the waters of the eastern, tropical Pacific Ocean are cooled by a lack of radiation from the atmosphere
- **Madden-Julian oscillation:** intraseasonal tropical wave that travels around the globe, causing monsoons and other high-water storms and also suppresses them
- **trade winds:** winds at the level of the ocean surface that blow from the east to the west in the tropical Pacific

## EL NIÑO

A carefully watched long-term weather pattern is the phenomenon known as El Niño, also known as El Niño Southern Oscillation (ENSO). El Niño is a cyclical event in which warm ocean water (which is heated at the equator) moves eastward from the western Pacific Ocean because of stagnation in the trade winds (which blow from east to west along the equator).

Normally, the trade winds send the solar-heated water westward, where the resulting humidity creates weather patterns that travel around the world. With El Niño, the warm water moves eastward and creates new weather systems and (because of El Niño's distinctive sea-level pressure signature) a shift in the manner in which the global atmosphere circulates.

El Niño has different effects on different regions of the world. For example, evidence suggests that El Niño contributes to the generation of powerful, tornado-producing storms in the prairie regions of the United States and Canada. For other regions, such

as the Atlantic seaboard, El Niño can mean more precipitation in the form of snowstorms, nor'easters (large rotating Atlantic coastal storms capable of high winds and heavy precipitation), and other systems. In South America, ENSO can cause storms that produce severe flooding.

In some regions, El Niño can lead to less precipitation. For example, in the Rockies and southwestern Canada, El Niño often means less snow than usual. Meanwhile, in Australia, El Niño patterns frequently cause droughts. ENSO is even known to reduce the number and severity of hurricanes, which form in the typically warm waters off western Africa.

El Niño can last between two and seven years. While it is a cyclical event, its appearance is typically erratic. For example, the longest-lasting El Niño event recorded during the modern era lasted from 1990 to 1995. However, since 1995, the El Niño cycle has been shorter, lasting one or two years on average.

## LA NIÑA

When ENSO periods come to a close, the trade winds intensify because of the difference in surface-level air pressure. The winds blow the warmer water westward and from the west coasts of North and South America.

As the warm water is carried west, cooler water from beneath the surface of the eastern Pacific is drawn to the surface. As is the case during El Niño, the cooling water again causes a change in atmospheric circulation, creating changes in the jet streams (a global band of strong air currents several miles above the earth's surface). As a result, weather patterns change. This pattern is known as La Niña.

La Niña is considered the latter half of the ENSO cycle. Like El Niño, La Niña's effect on the weather varies, based on the region. For example, the warmer water in the western Pacific leads to wetter weather conditions in Indonesia and Australia. In the United States, colder weather frequently exists in the northwest region, while the southern and mid-Atlantic see warmer and drier conditions.

La Niña also results in changes in severe weather patterns. For example, the generation of major storms that produce hail and tornadoes is reduced by the cooler, drier air manifest during La Niña. However, La Niña disrupts the wind shear (the difference in wind speed and direction in two close areas of the atmosphere) that can hinder the development of tropical storms in the Atlantic. The result of this disruption is an increase in the number and severity of Atlantic hurricanes.

## ARCTIC OSCILLATION

ENSO is not the only atmospheric cycle. Another long-term weather pattern, the Arctic Oscillation (AO) system, involves two areas of atmospheric pressure. These two patterns are located at the polar latitude (above the Arctic Circle) and at the middle latitude (between southern Florida and the Arctic Circle).

There are two phases in which the AO is manifest. These phases have alternated frequently during the last century, with each phase lasting between ten and forty years. During the negative phase, higher-than-normal pressure is in place in the Arctic regions, while lower pressure exists at the middle latitudes (such as most of North America and Europe). These conditions cause cold air to move into the middle latitudes, which leads to colder-than-average winters in these regions. During the positive phase, however, the patterns are reversed: Ocean storms and wetter weather are drawn into the northernmost regions of North America and Europe, while the middle latitudes remain drier and warmer.

A number of factors contribute to the change in phases within the AO cycle, including the pressures associated with rising and lowering sea levels, water temperatures, and even greenhouse gases. AO is closely related to two other regional oscillation patterns: the North Atlantic Oscillation and the northern annual mode, which affect winter weather patterns in the North American and Northern European and Asian regions, respectively.

## MADDEN-JULIAN OSCILLATION

Still another type of long-term weather pattern is the Madden-Julian Oscillation (MJO). The MJO is intraseasonal (it can develop within thirty, sixty, or ninety days of a given season), occurring across the planet's tropical regions (particularly in the Indian and Pacific Oceans). The MJO is global as well, traveling in a wave in the atmosphere.

Depending on the phase of the MJO, this long-term weather pattern is responsible for the generation and suppression of heavy tropical rainfall (including the high precipitation that accompanies monsoons). The MJO's phase of high-volume precipitation first begins in the Indian Ocean and then proceeds east toward the western and eastern Pacific. As it approaches the cooler waters of the eastern Pacific, the MJO tends to become less laden with precipitation, as the heat from warmer water dwindles. The MJO does show some limited life in the tropical Atlantic but becomes noticeable again as it re-approaches the Indian Ocean.

The MJO is one of the more vexing long-term weather patterns to study and predict because of its relatively slow movement and development along the tropical path. The key to its ability to develop and strengthen storm systems is the warm waters over which it passes. Through the use of satellite sensors, scientists are therefore attempting to pinpoint the locations of these pockets of warm water, developing models that may one day enable them to better understand and predict MJO movements.

## MODELS FOR FORECASTING LONG-TERM PATTERNS

Meteorologists and climatologists alike see a great deal of value in the analysis and prediction of long-term weather patterns. A season-by-season weather forecast can help scientists prepare for weather patterns.

For example, tracking La Niña during the winter season can help researchers forecast enhanced or subdued storm systems in a given geographic area. For this reason, scientists are working to better track and predict the more erratic AO and MJO. Furthermore, any evidence of above- or below-average precipitation is helpful for the residents of these regions in preparing for the effects of these long-term weather patterns.

To track and predict long-term weather patterns, scientists utilize a number of approaches. First, they gather data on surface-level conditions (such as water temperatures and wind speeds) and weather patterns collected from satellite sensors, airborne remote sensors, and ground-based systems. Using these data, they generate mathematical equations that represent each of the physical processes at work. For example, scientists have developed a series of mathematical equations designed to assign values to circulation anomalies within the MJO; such equations help researchers better understand how such elements influence the MJO's atmospheric dynamics.

Once the data have been compiled and collated, they are used to form computer models that can predict future conditions created by a given long-term weather pattern. One example comes from Japan, whose Ministry of Science utilized the Earth simulator supercomputer—until recently, the fastest computer in the world—to compile global weather data, including long-term weather patterns, ocean currents, and other factors, into a massive model. Scientists believe that, once this model is brought online, they will be able to predict droughts, severe storms, and other weather and climate conditions as far as thirty years into the future.

Other computer models focus on a particular long-term weather pattern and analyze its individual elements, such as regional cold temperature zones along an ENSO track or the relationship between atmospheric circulation and surface-level conditions. For example, the National Weather Service in the United States operates a climate prediction center, which compiles a wide range of sensor data on such elements as surface and water temperature, wind velocity, precipitation, and air currents (and on anomalies in these conditions based on seasonal averages). Based on this information the center generates seasonal short- and long-term forecasts, tracking weather systems based on the influences of ENSO, AO, and MJO and other patterns.

*Michael P. Auerbach*

## FURTHER READING

Ahrens, C. Donald, and Perry J. Samson. *Extreme Weather and Climate.* Independence, Ky. Brooks Cole, 2010. Provides a nontechnical review of the different types of severe weather, such as hurricanes, tornado-producing storms, and flooding. Discusses the long-term weather patterns that produce such weather.

Allen, Robert J., and Charles S. Zender. "Forcing of the Arctic Oscillation by Eurasian Snow Cover." *Journal of Climate* 24, no. 24 (2011): 6528-6539. Describes the factors that contribute to the Arctic Oscillation, including sea level, greenhouse gases, and warm seawater. Argues that snow cover,

particularly in northern Europe and Asia, plays an important role in this process.

Bai, Xuezhi, et al. "Severe Ice Conditions in the Bohai Sea, China, and Mild Ice Conditions in the Great Lakes During the 2009/10 Winter: Links to El Niño and a Strong Negative Arctic Oscillation." *Journal of Applied Meteorology and Climatology* 50, no. 9 (2011): 1922-1935. The authors conducted a comparative study of the Great Lakes and China's Bohai Sea, focusing on the creation of large amounts of ice caused by negative phases in the Arctic Oscillation.

Chang, Chih-Pei, and Mong-Ming Lu. "Intraseasonal Predictability of Siberian High and East Asian Winter Monsoon and Its Interdecadal Variability." *Journal of Climate* 25, no. 5 (2012): 1773-1778. Discusses scientific efforts to predict severe weather in East Asia based on the positive and negative phases of Arctic oscillation.

D'Aleo, Joseph S., and Pamela G. Grube. *The Oryx Guide to El Niño and La Niña.* Westport, Conn.: Greenwood Press, 2002. Examines the causes and effects of the various stages of ENSO. In addition to analysis of the history, economic impacts, and natural forces creating ENSO, reviews attempts to more effectively predict this phenomenon.

Mogil, H. Michael. *Extreme Weather: Understanding the Science of Hurricanes, Tornadoes, Floods, Heat Waves, Snow Storms, Global Warming, and Other Atmospheric Disturbances.* New York: Black Dog & Leventhal, 2007. A detailed analysis of the science of severe weather and climate change. Provides a review of the long-term weather patterns that produce this weather and suggestions about how society can prepare for future weather disasters.

**See also:** *AR4 Synthesis Report;* Atmospheric Properties; Beaches and Coastal Processes; Climate; Climate Change Theories; Climate Modeling; Drought; Earth-Sun Relations; El Niño/Southern Oscillations (ENSO); Floods; Global Energy Transfer; Global Warming; Greenhouse Effect; Hurricanes; Hydrologic Cycle; Impacts, Adaptation, and Vulnerability; IPCC; Monsoons; Observational Data of the Atmosphere and Oceans; Ocean-Atmosphere Interactions; Recent Climate Change Research; Satellite Meteorology; Sea Level; Severe and Anomalous Weather in Recent Decades; Severe Storms; Tornadoes; Tropical Weather

# M

## MEDITERRANEAN SEA

*The Mediterranean Sea, located between southern Europe and North Africa, is one of the most historically important and largest seas in the world. However, environmental problems caused by tourism and industrial discharge threaten to destroy the ecological balance of the region.*

### PRINCIPAL TERMS

- **continental shelf:** the submerged offshore portion of a continent, ending where water depths increase rapidly from a few hundred to thousands of meters
- **eddy:** a smaller, rotating current or turbulence that forms from the movement of a primary water or wind current
- **Mediterranean climate:** a pattern of weather conditions characterized by a long, hot, dry summer and a short, cool, wet winter
- **salinity:** a measure of dissolved salt content in oceanic water controlled by the difference between evaporation and precipitation and by the water discharged by rivers
- **semidiurnal:** having two high tides and two low tides each lunar day

### LOCATION AND HISTORY

The Mediterranean Sea separates Africa geographically from the continents of Europe and Asia. It is surrounded by southern Europe, western Asia, and North Africa, and extends about 3,700 kilometers from east to west and almost 1,600 kilometers from north to south at its widest point. Because of the irregular shoreline, the width varies considerably but averages almost 600 kilometers. The European coast includes Spain, France, Italy, Slovenia, Croatia, Bosnia and Herzegovina, Montenegro, Albania, Greece, the islands of Malta and Cyprus, and the European part of Turkey. The Asian portion continues with the Asian part of Turkey, Syria, Lebanon, and Israel, while the African part includes Egypt, Libya, Tunisia, Algeria, and Morocco. Its total area is about 2.5 million square kilometers.

The Mediterranean Sea is connected to the Atlantic Ocean by the Strait of Gibraltar in the west,
to the Red Sea by the Suez Canal in the southeast, and to the Black Sea by the Strait of Bosporus in the northeast. The Strait of Gibraltar is approximately 20 kilometers wide at the boundary, while the Bosporus varies in width from as little as 0.8 kilometer to about 3.2 kilometers at the Black Sea boundary between Rumeli Burnu in European Turkey and Anadolu Burnu in Asiatic Turkey. The Suez Canal is the only human-made channel of the three and was first opened in the late 1860's to speed naval transportation between Europe and the countries of the Persian Gulf and the Indian Ocean. It is almost 150 kilometers long, 82 meters wide, and about 10 meters deep.

Historically, the Mediterranean Sea has been one of the most important as well as one of the largest seas on Earth. Its name is derived from the Latin words *medius* (middle) and *terra* (land), indicating its perceived identity to ancient cultures as the "sea in the middle of the world." The Mediterranean area has also been known as "the cradle of Western civilization" because of the advanced civilizations that flourished in the region.

All three straits and canals have long been considered critical strategic points for commercial, political, naval, and military reasons, and great military powers have, historically, been interested in controlling them. Such powers included the Phoenicians, the Greeks, the Romans, the Persians, the Byzantines, the Arabs, the Turks, the British, the French, the Italians, the Russians, and others. The Romans referred to the Mediterranean as *mare nostrum* (our sea); Catherine the Great of Russia had dreams about controlling the Dardanelles in the late 1790's; British admiral Horatio Nelson understood that blocking Gibraltar was the only way to beat Napoleon's expansion; Winston Churchill realized that controlling the Dardanelles was critical during World War II; Benito Mussolini attempted to control the Balkan Peninsula and North Africa to revive the ancient Roman Empire; and the British and French declared war against Egyptian President Gamal Abdel Nasser after he nationalized the Suez Canal.

## GEOGRAPHY

The present coastline of the Mediterranean is irregular and indented with numerous small gulfs and capes. Geological studies indicate that during the Triassic period, some 200 million years ago, a sea, known as Tethys Sea, extended as far east as Turkmenistan and the Aral Sea territory and as far north as what is now the Danube valley in central Europe. The area north of the Tethys Sea was called Laurasia, while the area south of it was known as Gondwanaland. That much larger and wider body of water is believed to have been contracted by tectonic factors such as faulting, folding, continental drift, and volcanism as the continental masses moved from their orientation in the supercontinent of Pangaea toward their present global distribution. Beginning in the Cretaceous era (65 million years ago) and ending during the Miocene period (13 million years ago), the smaller Aral, Caspian, Black, Marmara, and Mediterranean seas were created out of that much larger sea basin. Moreover, the two natural outlets of Gibraltar and the Dardanelles appear actually to have been completely landlocked by land bridges that were eventually worn off by the tides, allowing the waters of the outer ocean to inundate the Mediterranean basin. There is also an indication that the islands situated off the eastern Greek mainland were once attached to what is now mainland Greece and Turkey.

The Italian peninsula splits the Mediterranean into two basins of approximately equal size, the western and the eastern, which are connected by the Strait of Sicily and the Strait of Messina. The Strait of Sicily, which lies between Cap Bon in Tunisia and the Italian island of Sicily, is almost 130 kilometers wide, while the Strait of Messina, which lies between Punta Sottile in Sicily and the Italian mainland, is no more than 3.3 kilometers across.

Each of the two basins is also subdivided into four smaller bodies of water. The western basin is composed of the Alboran Sea to the south of Spain, the Balearic or Iberian Sea between mainland Spain and the Islas Baleares, the Ligurian Sea just north of the French island of Corsica, and the Tyrrhenian Sea, which is surrounded by Corsica, Sardinia, Sicily,

and the western part of mainland Italy. The eastern basin includes the Adriatic Sea, which lies between the western parts of some of the former territories of Yugoslavia and the eastern part of mainland Italy all the way down to Capo Santa Maria di Leuca and above the Greek island of Corfu. The Ionian Sea extends southward from the Adriatic Sea, to the east of Sicily and on the western shore of Greece. The Aegean Sea stretches over the eastern Greek shores as far north as the Dardanelles; the Sea of Marmara connects to the Aegean Sea by the Dardanelles and to the Black Sea the Bosporus Strait.

The Mediterranean has many islands, several of which are of volcanic origin. Most of them lie in the eastern basin in the Adriatic Sea and, in particular, the Aegean Sea. The Aegean has more than two thousand islands that form small basins and narrow passages with an irregular coastline and topography. The largest islands are Sicily, with an area of 25,818 square kilometers; Sardinia, with an area of 24,138 square kilometers; Cyprus, with an area of 9,282 square kilometers; Corsica, with an area of 8,762 square kilometers; and the Greek island of Crete, with an area of 8,424 square kilometers. Other significant islands in the western basin include Alboran Island and the Balearic islands (Menorca, Mallorca, Ibiza, and Formentera) near Spain, and Elba, the Pontine islands, Ischia, Capri, Stromboli, and the Lipari islands near Italy. In the eastern basin are Malta off the coast of Tunisia, as well as the Greek Ionian, Cyclades, and Aegean islands. The floor of the Mediterranean is generally made up of carbonate-rich yellow mud, which covers the blue mud floor.

Among the most important rivers that flow into the Mediterranean are the Spanish Ebro, the French Rhone-Saone-Durance, and the Italian Arno, Tiber, and Volturno. Others include the Po and the Tagliamento in Italy, the Vardar in Greece and Macedonia, and the Nestos in Bulgaria and Greece. North Africa has only the Nile, which is the only large river that flows toward the north. Over the years, several deltas have slowly formed at the mouths of the rivers that flow into the sea. The most important one is the Nile Delta, which is the most fertile land in Egypt. More than five thousand years ago, Memphis, the Pharaonic capital of Egypt located close to the current capital, Cairo, was a port.

The coastline is often covered with small hills, while the high-altitude mountains lie farther inland.

The presence of volcanoes such as Vesuvius and Etna in Italy have been responsible for the loss of thousands of lives over the years but have also contributed to the fertility of the neighboring land.

## WATER FLOW

The total water area of the Mediterranean is about 2.5 million square kilometers, with an average depth of about 1,500 meters. Generally, the eastern basin is deeper, with the greatest depth of about 5,450 meters in the Ionian Sea, approximately 57 meters off the Greek mainland. The western basin's deepest spot lies in the Tyrrhenian Sea at about 3,600 meters. The sill depth in both the straits of Sicily and Gibraltar is about 275 meters. The Adriatic Sea is the shallowest of all seas because it is an overextension of the continental shelf. The water of the western basin is a little cooler and fresher than that of the eastern basin. The average surface temperature in the west is as low as 11 degrees Celsius in February and as high as 23 degrees Celsius in August. The corresponding temperatures in the eastern basin are 16 degrees and 26 degrees Celsius. Dissolved salts increase the density of seawater. Seas that are isolated and have few rivers flowing into them are traditionally saltier. This effect is particularly enhanced by arid climates, where evaporation of surface water is greater than its replenishment by precipitation.

The Mediterranean, which fulfills all these requirements, has a sill at the entrance of Gibraltar. As a result, water from the lower-density, but colder, Atlantic Ocean flows in at the surface. At the same time, the Mediterranean water flows over the sill into the ocean, where it sinks to a much lower depth until it reaches levels of the same density. Although a considerable degree of mixing takes place at the upper level because of the initial flow of salty water to the Atlantic, once the water reaches the high-density region, it spreads horizontally and can be followed far into the Atlantic. In general, however, the deep outflowing current does not equal the intake of surface water. The measurement and comparison of salinities of the Atlantic and the Mediterranean waters in the Strait of Gibraltar provide data that give a good measurement of the outflow of water. Taking into consideration also the evaporation rates, the outflow has been estimated to be in the neighborhood of 35 to 63 million cubic feet per second. Temperature and wind variability have an impact on the actual outflow.

Generally, the salinity of the western Mediterranean basin is less than that of the eastern basin. The water mass flowing from the Black Sea through the Bosporus and the Dardanelles to the north of the Aegean Sea is estimated to be close to 200 cubic kilometers per year. The Dardanelles boundary condition results in the spreading of the outflow to the Western Aegean, as well as to the north, along the Greek mainland coast. Black Sea waters appear to be able to reach the central Aegean; their influence is greatest in summer and fall during the time of maximum discharge through the Dardanelles in September and minimum salinity in July. Moreover, a westerly current enters the area between Asia Minor and the Greek island of Rhodes, which reverses in spring and summer. Along the northern coast of Crete, easterly currents appear to move all year long. The tide that is attributed to the Atlantic Ocean appears to lose strength in the Strait of Gibraltar.

The Mediterranean tides are generally semidiurnal, with wave systems that have nodal lines that extend from Barcelona, Spain, to Bejaia, Algeria, and from Turkish Korfezi to Libyan Tobruk. The tide range in most other places does not surpass the 1 meter mark. Strong hydraulic currents occur at the ends of the two straits of Sicily and Messina because of the formation of high water in exactly opposite phases. These are the same waves that made the Messina strait such a dangerous place to navigate in ancient times—this area was known as Scylla and Charybdis in Homer's *Odyssey*. In contrast, the Adriatic has a progressive tide that has the highest range of 1 meter.

Historically, the study of currents was investigated because of the observed continuous inflow of water through the strait of Gibraltar. Scientists became aware of the fact that the evaporation of water over the Mediterranean was larger than the precipitation and river water outpouring. As a result, the possibility of underground canals that would bring back the water to the Atlantic was theorized. It is estimated that Gibraltar provides more than 97 percent of the water entering the Mediterranean, while precipitation, the Black Sea, and river flow provide the rest. In contrast, of the water lost by the Mediterranean, 93 percent leaves through Gibraltar, more than 6 percent is evaporated, and less than 0.5 percent is lost to the Black Sea. There is no doubt that if the Atlantic water inflow through Gibraltar did not occur, the

Mediterranean Sea would be a large, salty pond supporting little or no life, similar to the Dead Sea.

## CLIMATE AND PLANT LIFE

The Mediterranean climate is transitional between the harsh weather of central Europe and the desert-dominated conditions found in North Africa. Because of the decreased humidity, the overall climate is comfortable for its inhabitants. A small degree of sleet and snow precipitation takes place, especially in the mountains, and is supported by occasional polar or Arctic invasions of cold winds. However, the monthly temperature averages never drop below freezing, which is particularly favorable to crops. The southern parts of North Africa and the eastern basin, which are closer to the equator, experience hotter and drier summers because they lie far from the Atlantic Ocean's influence.

Overall, the eastern basin is roughly 6 degrees of latitude nearer to the equator than the western basin. In particular, the Sahara Desert in North Africa makes the greatest parts of Egypt, Libya, and Algeria practically uninhabitable. Often there are three- to four-month spans where no precipitation occurs in the eastern basin, while the western basin territory experiences numerous storms and substantial amounts of rains.

In general, the term "Mediterranean climate" refers to a long, hot, dry summer and a short, cool, wet winter. Several locations on Earth experience the same conditions, such as Los Angeles in California, Santiago in Chile, Capetown in South Africa, and several cities of southwestern Australia. The mild, dry summers and the beautiful scenery of many areas in the Mediterranean prompt a great number of tourists to visit the beaches of the French and Italian Rivieras, southern Italy, the Adriatic coast, the coasts of Lebanon, and the islands of Greece. The unique climate supports characteristic crops and vegetation. The dry weather favors the growing of the long-rooted olive trees and oaks, as well as the dominance of the dense scrub 2 to 4 meters in height known as *maquis* in France and *macchia* in Italy. All lose very little water by evaporation and have a thick bark and small, thin leaves. Interestingly, the climate helps to grow fruit that is characteristic of several other places on Earth, such as grapes and oranges, as long as there is enough water during the dry season.

## SIGNIFICANCE

The Mediterranean Sea's impact on civilization has also added to problems that the human race is encountering. The ecosystem has been substantially eroded by the destruction of forests, which has been taking place for more than thirty centuries. The Phoenicians, Athenians, Romans, Byzantines, and Turks were significant naval powers and used timber for ship construction. The citizens of Athens, Rome, Jerusalem, Alexandria, and Constantinople routinely consumed large numbers of trees in the building of their cities.

Tourism and industry are responsible for one of the greatest concerns, pollution. The use of iron, steel, concrete, petrochemicals, and other industrial products has affected the air quality of cities such as Rome, Athens, Marseilles, and Barcelona. The presence of millions of tourists has led to the need for more sanitary facilities and for recycling projects. Often the natives encounter the great problem of raw sewage and other refuse that overwhelms the beaches.

Finally, oil spills have also added to environmental pollution. The 1975 United Nations-sponsored Mediterranean Action Plan (MAP) convinced most Mediterranean nations to sign a pact to eliminate toxic spillage and chemical waste from the Mediterranean. In 1982, the MAP Coordinating Unit was established in Athens, followed in 1995 by the adoption of MAP Phase I, Barcelona Resolution, Priority Fields of Activities for Environment and Development in the Mediterranean Basin (1996-2005) and the Protocol Concerning Specially Protected Areas and Biological Diversity in the Mediterranean. Beginning in 2014, the Horizon 2020 initiative aims to depollute the Mediterranean Sea by 2020, targeting industrial emissions and municipal and urban waste waters, which make up 80 percent of pollutants entering the sea.

*Soraya Ghayourmanesh*

## FURTHER READING

Chester, R., and Stefano Guerzoni, eds. *Impact of Desert Dust Across the Mediterranean.* Norwell, Mass.: Kluwer Academic Publishers, 1996. Compiles papers on the effect of the Saharan dust transport to the Mediterranean Sea. Covers the modeling of the Saharan dust transport, the chemistry and mineralogy of the dusts and their effect on precipitation, the contribution of the dust to marine sedimentation, the aerobiology of the dusts, and the Saharan dust's impact on the different climates.

Dumont, Henri, Tamara A. Shiganova, and Ulrich Niermann, eds. *Aquatic Invasions in the Black, Caspian, and Mediterranean Seas.* Boston: Kluwer Academic Publishers, 2004. Provides an overview of the links between the Black Sea, Caspian Sea, and Mediterranean Sea. Discusses the physical connections and the resulting movement of invasive species due to shipping. Examines the invasion of jellies into the Caspian Sea. Provides a comparison to other invasive species, including invertebrates, plankton, and ctenophores.

Faranda, Francesco, Letterio Guglielmo, and Giancarlo Spezie. *Mediterranean Ecosystems: Structures and Processes.* Milan: Springer-Verlag Italia, 2001. A collection of multidisciplinary papers from the First National Congress of Marine Sciences and the works undertaken by the Italian science community. Focuses on Mediterranean ecosystems.

Jeftic, L., et al., eds. *Climatic Change and the Mediterranean: Environmental and Societal Impacts of Climatic Change and Sea-Level Rise in the Mediterranean Region.* New York: E. Arnold, 1992. Discusses the climate change of the Mediterranean, with sections on changes in precipitation, hydrological and water resources, and land degradation in such areas as the Nile Delta, the Thermaicos Gulf, and the Lion Gulf. A chapter deals exclusively with the predictions of relative coastal sea-level change in the Mediterranean based on archaeological data.

Johnson, Robert G. *Secrets of the Ice Ages: The Role of the Mediterranean Sea in Climate Change.* Minnetonka, Minn.: Glenjay Publishing, 2002. Provides scientific studies on the influence of the Mediterranean Sea on past ice ages. Well suited for undergraduates studying palaeoclimatology, geology, or palaeoecology. Easy to follow and contains many illustrations, bibliographies, and indexing.

Kalin Arroyo, Mary T., Paul H. Zedler, and Marilyn D. Fox, eds. *Ecology and Biogeography of Mediterranean Ecosystems in Chile, California, and Australia.* Vol. 108. New York: Springer-Verlag, 1995. Presents and analyzes the economorphological characteristics of Mediterranean-like vegetation in Chile, California, and Australia.

Lionello, P. *The Climate of the Mediterranean Region: From the Past to the Future.* London: Elsevier, 2012. Reviews the climate evolution of the Mediterranean region over the past two thousand years, with some predictions for the twenty-first century. Includes consideration of socioeconomic problems. Intended for geologists, oceanographers, and climatophiles.

Rundel, Phillip W., F. M. Jaksic, and G. Montenegro, eds. *Landscape Disturbance and Biodiversity in Mediterranean-Type Ecosystems.* New York: Springer-Verlag, 2010. Compares worldwide ecosystems that are similar to the Mediterranean one. Such ecosystems include those found in South Africa, California, southwestern Australia, and central Chile.

Saliot, Alaine. *The Mediterranean Sea.* Berlin: Springer-Verlag, 2005. Presents the current knowledge of the chemical processes and pollution in the Mediterranean Sea with a focus on environmental management. A reference resource for both students and scientists.

Woodward, Jamie. *The Physical Geography of the Mediterranean.* Oxford: Oxford University Press, 2009. Explores the distinctive physical geography of the Mediterranean region, where three continents meet, with regard to climate, ecosystems, landscapes, and hazards, tracing their development through time and examining processes and issues existing throughout the region.

Wurtz, Mauritzio. *Mediterranean Pelagic Habitat: Oceanic and Biological Processes, An Overview.* Malaga: International Union for Conservation of Nature, 2010. Produced by international specialists in Mediterranean oceanographic and climatic studies, detailing the processes functioning in the Mediterranean Sea.

Zahran, M. A. *Climate-Vegetation: Afro-Asian Mediterranean and Red Sea Coastal Lands.* New York: Springer, 2010. Examines coastal habitats of the Mediterranean and Red Sea. Discusses the interaction between climate and vegetation. Discusses sustainable development of coastal deserts. Very informative without being too technical. The subject matter is quite specific, making this text most useful to graduates or researchers studying climate-vegetation interactions or Afro-Asian Mediterranean and Red Sea coastal ecology.

**See also:** Aral Sea; Arctic Ocean; Atlantic Ocean; Beaches and Coastal Processes; Black Sea; Caspian Sea; Ganges River; Gulf of California; Gulf of Mexico; Hudson Bay; Indian Ocean; North Sea; Oceans' Origin; Pacific Ocean; Persian Gulf; Red Sea

# MISSISSIPPI RIVER

*The longest river in North America, the Mississippi River's drainage basin covers a major portion of the continent, with important historical and present-day effects.*

## PRINCIPAL TERMS

- **alluvium:** a deposit of soil and mud formed by flowing water
- **hydrology:** the study of the events of circulation and the properties of water and earth
- **sediment:** matter that settles to the bottom of a body of water; may be sand, silt, or pieces of broken-up rock

### GEOLOGICAL HISTORY AND CHARACTERISTICS

The Mississippi River fascinates and amazes both residents and travelers. In North America, its 3,804-kilometer length (of which about 2,592 kilometers is navigable) is surpassed only by those of the Missouri River in the northern United States and the Mackenzie River in western Canada. The Mississippi has often been called the "Nile of North America." The comparison is appropriate, as both rivers have deltas that hold fertile land and cliffs that dot their edges. Together with its two main tributaries, the Ohio and the Missouri, the Mississippi drains an immense area of thirty-one states and two Canadian provinces into the Gulf of Mexico. Altogether, the river drains an area 3.2 million square kilometers. It is the fourth-largest drainage basin in the world, exceeded only by the Amazon, Congo, and Nile River basins. The Amazon basin, by far the world's largest, is more than twice as large as the Mississippi basin, draining an area of approximately 7 million square kilometers. The Congo basin is not much larger than the Mississippi's, draining some 3,457,000 square kilometers. The Nile basin drains about the same area as the Congo, some 3,349,000 square kilometers.

The Mississippi River drains lands with greatly differing geologies and environments, as the river and its tributaries flow over a wide variety of igneous, metamorphic, and sedimentary rocks. The sedimentary rocks are easily erodible, and the large amount of material constantly held in suspension by the Mississippi's waters helps explain its muddy appearance. The river flows through lands that have widely differing climates and vegetation covers. It drains diverse territories such as the complex mountain ranges in the West, the arid high plains, and the humid southern woodlands.

The geology of the Mississippi Valley has formed two distinct drainage basins, the Upper and Lower Mississippi Valleys, with the mouth of the Ohio River as the dividing point. The Lower Mississippi would flood seasonally if it were not regulated by human activity. In the upper valley, the river is a shallower stream enclosed by high, rocky mountains that are geologically unlike the hills that create the borders of the lower valley. The lands over which the Lower Mississippi flows have alluvial surfaces and are only a few thousand years old. This area is marked by inliers, which are outcrops of a formation completely surrounded by another

*The Mississippi River.* (PhotoDisc)

layer. In the lower valley, the Mississippi is also a deeper river than it is in the upper valley.

The Pleistocene buildup of ice in the Northern Hemisphere dropped sea levels throughout the world. When these glaciers began to melt, massive amounts of water and sediment were released and flowed toward the mouth of the Mississippi River, which was once a few hundred meters lower than it is today. The surface deposits that cover the lower valley are barely one thousand years old, and there is enough diversity in the alluvium of the lower valley to allow archaeologists to determine how long humans have been in the area. The prehistory of the very large Lower Mississippi River Valley before the appearance of Europeans is important because the present flora and densely wooded areas are not like those the first human beings in the region saw thousands of years ago.

The variety of backswamps, natural levees, and crests along the Mississippi create a river sequence that looks like a winding chain. Overall, the Mississippi River system can be categorized into six main basins: the Upper Mississippi, the Missouri, the Ohio, the Arkansas-White, the Red-Ouachita, and the Lower Mississippi. Most of the area of the Lower Mississippi basin, about 65,000 square kilometers, is flat, while the rest is made up of rolling hills. The lower basin stretches from the mouth of the Ohio River at Cairo, Illinois, to the Mississippi's mouth below New Orleans. The majority of the Mississippi River floodplain is in the lower valley. Swamplands and timberlands are more common than cleared sections. The Lower Mississippi is a murky stream that is alluvial below Cape Girardeau, Missouri, because this part of the river receives the discharges of a number of smaller tributary streams, the basins of which are almost wholly within the alluvial valley of the river below Cape Girardeau.

## SEDIMENT DEPOSITION

Like several other great rivers of the world, the Mississippi has formed an immense and fertile delta that has been created by the gradual deposition of sediment. Investigators disagree somewhat in their estimates of the river's depositions, but most believe that it amounts to about 260 million tons annually. The sediments largely consist of clay silt and fine sand, with different clays accounting for about 70 percent of the total sedimentary load.

These clays form the subsoils of most of the lower southern states and vary greatly in their erodibility. The banks and beds of the Mississippi and its tributaries have also greatly complicated the work of hydrologists and engineers trying to control the flow of these rivers. Since at least Cretaceous times, the drainage basin of the Mississippi River system has been delivering sediments to the Gulf of Mexico. The deposition of materials was enhanced by the thawing of the glaciers in the Jurassic period. During this stage, sediment-filled streams submerged the Lower Mississippi valleys.

Throughout geologic time, the sites of maximum deposition (depocenters) have shifted along the Gulf Coastal Plain. The bulk of the sediments that make up the Gulf Coast geosyncline, which is a portion of the crust subjected to downward warping forces over time, has been derived in part from ancestral Mississippi River drainages. Over long periods of geologic time, the river has constantly fed sediments to the receiving basin, thus building up thick layers of deltaic sediments that have gradually enlarged the coastal plain shoreline seaward.

Deltas form in lobes. In modern times, the seaward progression of the different Mississippi delta lobes has constructed a deltaic plain that has a total area of 28,678 square kilometers. Of this area, 23,992 square kilometers is subaerial, or periodically under water. In the past 7,000 years, the site of maximum sedimentation of delta lobes has shifted and has occupied many positions. One of the earliest deltas, the Sale Cypremort, created a deltaic lobe along the western borders of the Mississippi plain. This deltaic lobe was exceptionally widespread and thin. After approximately 1,200 years of build-out, the site of maximum deposition changed to another delta lobe, the Cocodrie system. A similar procedure of unseating developed, and the river abandoned the Cocodrie deposition, beginning a new delta lobe, the Teche. This process continued until the modern delta achieved its present form some 600 to 800 years ago. The result is a complex of deltaic arrangements stretching some 745 kilometers along the Louisiana coast and inland nearly 260 kilometers. The Mississippi Delta is geologically the most recent.

The soil deposits of the alluvial valley are of uneven thickness. Basins along the river at the north end of the valley hold only a few meters of modern river deposits, whereas more than 60 meters of deposits exists beneath the flat-lying coastal marshlands near the mouth of the river. These modern deposits rest on a substratum of heavier sand and gravel. The top layers of land along the Lower Mississippi are extraordinarily fertile because of large amounts of organic material and nutrients contained in the sediment arriving from the north.

## MEANDER

Rivers that flow through flat, sedimented areas, especially deltas, rarely have straight channels. In rivers that flow through mountainous areas, the steep slopes and the hardness of the rock force waters to take the path of least resistance downhill, which often is a straight line. When rivers cross plains that have gentle slopes, however, the turbulence in the streams is constantly dislodging fragments of softer material and creating new channels, which are usually extremely winding as they cut back and forth through the landscape over time. The channel can then form a sinusoidal pattern that in places comes almost full circle. This process is called "meandering." The meander is the natural response of a river by which it adjusts to the slope of the land through which it flows. Meanders represent the path of minimal stress for a river, and are dependent on the amount of energy in the river's flow. Meandering extends the stream length and reduces the slope of the river's path. During high flow in a freely meandering stream running over alluvial soil, the exterior concave bank bends erode, and the bed scours and deepens. At the same time, the eroded matter scoured from the bed and banks is dropped on the point bar in the next bend downstream. With the low flows, earth removal takes place in the bends. A meandering river can change the location of its main channel or split into branches, thus creating islands.

As meandering continues and as concave banks continue to recede, old river bends become extremely elongated; a narrow strip of land is often all that remains between the two bends of the river. When a flood causes a river to exceed the capacity of its channel, water flows over the narrow neck at a high velocity, thus eroding a new channel, or cutoff, across the strip of land. Eventually, as the new cutoff channel becomes wider and deeper, the old channel becomes separated from the river by deposition on bars, and it then becomes an oxbow lake. A fully developed stream will maintain much the same length over time,

as the decrease in length caused by cutoffs is equal to the added length produced by the development of other meanders. Each cutoff produces a new meander whose evolution climaxes in another cutoff.

The effects of a cutoff on a river and its surroundings can be dramatic. The cutoff shortens the length of the river bottom, which changes both the river-bottom profile and the water-surface profile. Immediately after the cutoff is made, a drawdown curve, or dip, in the water profile occurs in the area of the cutoff. This causes local erosion of the river-bottom upstream of the cutoff to start to take place. The increased erosion upstream will cause a temporary overloading of the river downstream from the cutoff and will result in deposition of the sediment that the river cannot carry. This effect, though temporary, may last for a long time, since the full length of the graded part of the river upstream involves the removal of tremendous quantities of bed material. During this period, the river downstream of the cutoff is overloaded. As a result of the consequent temporary aggravation, river stages and groundwater tables will rise. Because of the reduction in the river's length and loss in storage ability, flood peaks will increase in height. These downstream effects, although temporary, may well outweigh the long-term advantageous effects of the cutoffs that occur upstream.

A successful employment of a human-made cutoff can be found on the Lower Mississippi River, where a channel improvement project that began in 1932 resulted in a lowering of flood levels by approximately 2 meters without creating extreme instability in the river channel. For a cutoff to be permanently successful, however, the hydraulic gradients must be such that the river can preserve them more or less naturally, without requiring periodic human-made adjustments at numerous intervals. To prevent the river from scouring upstream and depositing downstream, the cutoff needs to be planned in such a way that the pilot cutoff channel diverts only the swift top water of the river, while the sediment-laden bottom water stays in the old river channel. This is best done by constructing the pilot channel in line with the upstream river channel and making the old river meander the branch channel of the new route.

## DELTA WETLANDS

The flatness of a delta, the meandering of the river channel, and the great fertility caused by the continual deposition of new organic material mean that river deltas are natural wetlands. Marshes and swamps are areas of low, flat-lying wet earth open to daily or seasonal flooding, although they are not the same things. Physically, marshes and swamps have common traits; they differ primarily in the kind of vegetation that flourishes on the land. Marshes are covered with grasses and other grasslike plants such as cattails. Swamps commonly include hardwood forests. The Mississippi River Delta supports some of the most important wetlands in the world, and it includes about 40 percent of all coastal wetlands in the contiguous forty-eight states.

The Mississippi delta wetlands support plant and animal populations that can adjust to shifting water levels. Annual floods supply the system with sediments containing dissolved organic matter and nutrients. Many species of fish either feed or spawn in the wetlands. At the mouths of the rivers that drain the wetlands are estuaries that support marine life. The same flooding that brings life-supporting elements to the floodplain forests also flushes organic material of all kinds to the sea. This matter supports the food chains of the estuaries, which are especially complex because of the great gradations of temperature and salinity that exist where fresh Mississippi River water meets the salty water of the Gulf of Mexico.

A natural part of the life cycle of a river delta is flooding, a phenomenon caused by an influx of water that surpasses the capability of the river channel to carry it away. Floods are caused by the accumulation of runoff water from melting snow, rainfall mixed with snow melt, glacial melts in mountain rivers, or rainfall alone. Although modern floods often result in loss of life and heavy damage to property, presettlement floods mainly served to rearrange the topography of riparian lands.

During annual floods, river stages will normally top adjacent natural levees. However, flood inundation is not continuous along all portions of a river; in many instances, during a single flood, various crevasse splays will form along low points in the natural levee. These small crevasses will be maintained for a year or more, until enough material is deposited in them to build them back up to the level of the natural levee. They will then cease to be active.

In a natural delta, distributary channels serve as flumes that direct a portion of the water from the parent river system to a receiving basin. With continued enlargement of the distributary channels, a point is reached at which the channel is no longer able to maintain its gradient advantage, and the process of channel abandonment begins. In the Mississippi Delta, where tidal flooding is not usually a factor, the distributary channels commonly fill with fine-grained material mixed with peats and transported organic debris. One of the major features of the Mississippi River basin is the large amount of silting caused by the formation of crevasses, which break off from the main distributaries and deposit materials in the numerous interdistributary bays. Eventually, deposits form in shallow bays between or adjacent to major distributaries and extend themselves seaward through a system of radial channels, similar in appearance to the veins of a leaf.

In essence, then, it is important to understand that the presettlement Lower Mississippi basin was a system that had functioned well throughout geologic time. The Mississippi River meanders carried runoff from its vast watershed to the Gulf of Mexico, and the sediments brought from the north created vast tracts of land that now form much of the gulf states. The sediments continually enriched this land, which supported an immense diversity of life by depositing new fertilizing organic materials. The river's floods did not destroy land but rather enlarged and enriched it. Once floodwaters overflowed the Mississippi's natural levees, they flowed into distributaries or formed crevasses that directed them into the vast area of wetlands that surrounded the main channel.

*Loralee Davenport*

## FURTHER READING

Changnon, Stanley A., ed. *The Great Flood of 1993: Causes, Impacts, and Responses.* Boulder, Colo.: Westview Press, 1996. Examines the flood damage of 1993 and its social and environmental effects. Also looks at the factors that made the flood so devastating, as well as prevention methods put in place as a result. Illustrations and bibliographical references.

Collier, Michael. *Over the Rivers: An Aerial View of Geology.* New York: Mikaya Press, 2008. Discusses the dynamic landscape of the rivers. Explains the processes that shape the landscape and its influence on humans. Written in the popular style, making it easily accessible to the general public. Filled with bits of information and extraordinary photographs. Presents multiple examples drawn from the Mississippi River.

Darby, Stephen, and David Sear. *River Restoration: Managing the Uncertainty in Restoring Physical Habitat.* Hoboken, N.J.: John Wiley & Sons, 2008. Begins with theoretical and philosophical issues of habitat restoration that provide a strong foundation for decision making. Addresses logistics, planning, mathematical modeling, and construction stages of restoration. Covers post-construction monitoring and long-term evaluations to provide a full picture of the habitat restoration process. Highly useful for anyone involved in the planning and implementation of habitat restoration.

Foster, J. W. *The Mississippi Valley: Its Physical Geography.* Chicago: S. C. Griggs, 1869. Describes the geography of the Mississippi River in the 1860's. Provides an interesting historical perspective.

Mathur, Anuradha, and Dilip da Cunha. *Mississippi Floods: Designing a Shifting Landscape.* Boston: Yale University Press, 2001. Written by a landscape architect and an architectural planner. Combines scientific, engineering, and natural views of the Mississippi River in an effort to increase human understanding of the relationship between the river and the society through which it flows.

National Research Council. *Drawing Louisiana's New Map: Addressing Land Loss in Coastal Louisiana.* Washington, D.C.: National Academies Press, 2006. Produced by the U.S. National Research Council. Presents a critical assessment of the environmental status of the Mississippi River Delta in Louisiana with respect to anthropogenic modification, erosion, and its potential future under rising sea levels.

_____. *Mississippi River Water Quality and the Clean Water Act: Progress, Challenges and Opportunities.* Washington, D.C.: National Academies Press, 2008. A governmental publication that discusses the geology and history of the Mississippi River basin as the backdrop for a broader discussion of its management under the auspices of the Clean Water Act. Presents proposals for water quality testing, monitoring, protection, and cooperation among the multiple states along the river. Discusses and organizes research conducted along the river and comparisons drawn between the river and Chesapeake Bay for presentation to the EPA.

Shaffer, James L., and John T. Tigges. *The Mississippi River: Father of Waters.* Charleston, S.C.: Arcadia Publishing, 2000. Provides a narrative overview of the social and economic place occupied by the Mississippi River throughout the history of North America.

Wohl, Ellen. *A World of Rivers.* Chicago: University of Chicago Press, 2011. Describes the Amazon, Ob, Nile, Danube, Ganges, Mississippi, Murray-Darling, Congo, Chang Jiang, and Mackenzie Rivers. Offers excellent figures loaded with well-organized information. Discusses the natural history, anthropogenic impact, and the future environment of these ten great rivers. Bibliography is organized by chapter.

**See also:** Amazon River Basin; Deltas; Ganges River; Great Lakes; Gulf of Mexico; Hurricane Katrina; Lake Baikal; Lakes; Nile River; Yellow River

# MONSOONS

*Monsoons are seasonal wind systems that reverse directions biannually and are crucial for the economic stability and agricultural productivity of affected geographic areas.*

## PRINCIPAL TERMS

- **austral:** referring to an object or occurrence that is of the Southern Hemisphere
- **boreal:** referring to an object or occurrence that is of the Northern Hemisphere
- **convection:** heat transfer by the circulating movement that occurs in fluid materials as warmer, less dense material rises above cooler, denser material
- **geostrophic:** descriptive of wind that occurs when the Coriolis force is in exact balance with the force of a horizontal pressure gradient and therefore blows in a straight line
- **Intertropical Convergence Zones (ITCZ):** low-pressure areas where southern and northern trade winds meet
- **orography:** study of mountains that incorporates assessment of how they influence and are affected by weather and other variables
- **oscillation:** variation of some physical property or condition between two opposing states, much like the rising and falling of a wave between its maximum and minimum heights
- **troposphere:** the level of the atmosphere closest to the ground, extending from the surface to an altitude of 11 kilometers
- **trough:** a long and relatively narrow area of low barometric pressure
- **vortex:** the central locus of a whirling liquid or gas, about which the fluid mass circulates

## ORIGINS OF MONSOONS

The term "monsoon" originated from the Arabic word *mausim*, meaning season, used by sailors to comment about changing winds above the Arabian Sea. Solar heat produces winds that shift to the north and south according to the sun's position each season. Scientists cite geological evidence that suggests conditions favorable for monsoons began millions of years ago when the Indian subcontinent collided with the Asian plate, eventually creating the Himalaya Mountains and the Tibetan Plateau. Warm and cool ocean water and landmasses affect atmospheric circulation, creating recurring wind systems that sweep over large regions. The majority of monsoons occur in tropical areas adjacent to the Indian Ocean, although seasonal winds also affect Africa, northeast Asia, Australia, and North and South America. Monsoons are described by geographical terminology; the two predominant patterns are called the Asian-Australian and American monsoon systems. Monsoons are also referred to as boreal or austral, according to their location in the Northern or Southern hemisphere, respectively.

For thousands of years, monsoons have been incorporated into the literature, folklore, and religious rituals of numerous cultures. The monsoon motif is a universal symbol for rebirth and fertility as well as devastation and death. Monsoons are predictable to the extent that it is known that the monsoons probably will occur during specific seasons. However, the winds are often erratic, being delayed or appearing prematurely and precipitating extreme or minute amounts of rainfall and sometimes bypassing regions entirely. Usually, between April and October, monsoon winds develop in the southwest, shifting direction to originate from the northeast between October and April. Monsoons are a prolonged series of winds and not restricted to a single storm.

Monsoons manipulate the planet's climate, often proving to be beneficial and occasionally detrimental. The winter monsoon blowing from land to sea is usually associated with dryness, while the summer monsoons storming from the sea onto land produce torrential, sustained rainfall that is vital for agricultural activity. Sufficient precipitation assures growth of ample crops to nourish domestic populations, to export for economic profits, and to provide employment for farm laborers. The absence of monsoon rains can result in famines and impoverishment.

## MONSOON DYNAMICS

For several centuries, scientists have analyzed monsoons to determine the physical forces that generate the winds and regulate their behavior. Researchers have agreed on basic explanations regarding fundamental aspects of monsoons, such as their relationship to atmospheric and oceanic conditions and how monsoons tend to vary instead of conforming to exacting standards. Scholars continue to seek answers to complex questions about monsoons, especially concerning fluctuations displayed in time periods ranging from seasons to decades. Such information might enable meteorologists to predict possible occurrences and outcomes that could impact both local and global populations and economies. Researchers want to understand more about the onset of monsoons and their active and break periods thought to be caused by shifting troughs. The actual beginning of a monsoon is often disputed, with some scientists saying that increased humidity indicates the monsoon's start, while many say precipitation or formation of a vortex signals the beginning.

Monsoons are caused by sea and land breezes that create temperature and air-pressure differences between landmasses and bodies of water. Land absorbs heat to a different extent than water does, and the difference between land and water temperature is one factor in the instigation of monsoons. Areas in low latitudes near the equator undergo circulatory and precipitation changes because of temperature deviations on adjacent continents and seas. The amount of solar radiation emitted each season affects the temperature and air pressure over continents. During the Northern Hemisphere's summer, when it tilts toward the sun, high-pressure systems move from the cooler ocean into the land's low-pressure area. As the landmass cools during winter, its high-pressure air mass moves from the land to the low-pressure air mass over the warmer ocean.

The jet stream moves south during winter and north in the summer, transporting air masses to and from monsoon regions. In winter, the Siberian high-pressure system causes air to circulate clockwise. Winds move from the northeast, down the Himalaya Mountains, and across land cooled during shortened days toward the sea, creating dry conditions and causing monsoon rains in Indonesia and Australia. By summer, the winds reverse direction because the land is heated by increased solar energy. Moving counterclockwise, the winds carry moisture from the sea in the southwest toward land. For example, the Somali jet stream near Africa moves across the equator to the Arabian Sea and alters the direction of ocean currents, which causes cold water to rise from the depths and surface temperatures to drop. Humid winds shift into India, rising when they reach the Himalaya Mountains. As the winds are lifted over the Tibetan Plateau, the air cools sufficiently to become saturated, triggering thunderstorms and convectional rainfall. The variation and strength of the Indian monsoon also seems to be related to the Southern Oscillation, which is characterized by a reversal of air pressure at opposite ends of the South Pacific Ocean at irregular intervals of three to seven years. The mechanics of this relationship are not yet well understood.

Scientists have designated three monsoon circulation patterns. The lateral component indicates a monsoon that circulates across the equator. Transverse circulation moves from North Africa and the Middle East into southern Asia. The Walker circulation, also a transverse pattern, moves across the Pacific Ocean. The size, shape, coastal positions, and elevation of landmasses influence these patterns, and the shifting of the low-pressure Intertropical Convergence Zone (ITCZ) between north and south can determine how monsoon systems move. These factors contribute to variations in monsoon intensity and duration. In areas such as Australia, monsoon winds do not rise over or descend from mountains, and high-pressure masses undergo geostrophic adjustment as they ascend over monsoon troughs. Orography influences the nature of monsoons globally. African and Australian monsoons tend to be weaker because they are not elevated as high as those lifted by Asian winds. Because the Rocky Mountains and the Sierra Madre lift air masses somewhat like the Himalaya, though not as high, some researchers claim that a monsoon circulation pattern occurs in North America, causing increased precipitation in Mexico and the southwestern United States every summer.

**IMPACT ON CIVILIZATION**

The effects of monsoons permeate the civilizations of regions where they occur. Approximately 60 percent of people worldwide are economically dependent on the climate affected by the Asian-Australian monsoon system. Humans, animals, and plants rely on monsoons to provide essential moisture. The absence of monsoon rains can cause droughts and famine, killing millions of people by starvation. Any fluctuation in the monsoon cycle, whether within one season, year, or decade, can be detrimental. At least 2 billion people rely on monsoons to irrigate rice and wheat crops. In India, grain production is essential to feed the growing population that expands annually at a rate greater than agricultural yields increase. At least one-fourth of the Indian economy is based in agriculture, and 60 percent of laborers are engaged in agricultural employment. Monsoon rain is critical to maintain this balance.

Ironically, monsoons also flood areas and drown people, cause landslides that destroy communities, and inundate crops. Millions of acres often remain underwater for months. Several thousand people die annually during monsoons, and thousands more become homeless. Many people are reported missing after a monsoon deluge. Flood waters wash away unstable dams, buildings, and graves in cemeteries. Extreme humidity is stifling for most people. Humanitarian and relief agencies such as the International Red Cross and Red Crescent Society provide emergency food rations and shelter. Shipping along trade routes in monsoon regions is often both helped and hindered by winds.

Folklore, proverbs, and prayers provide insights into personal experiences with monsoons. Natives and visitors to monsoon regions have documented their encounters with monsoons. Some tourists purposefully travel to Asia during monsoons because hotels and businesses are not crowded, prices are lower, and reservations are easier to secure. However, some visitors note the inconvenience of always carrying an umbrella, traveling on monsoon-damaged roads, and encountering storm-related delays. Astronomers can observe celestial phenomena because of clear skies during the winter monsoons. Oceanographers also consider the monsoons useful because nutrient-rich waters rise to the surface, allowing scientists to study how plants, animals, the sea, and the atmosphere exchange carbon dioxide.

## Modeling Monsoons

Because monsoons are essential for affected populations to thrive despite the wind's uncertain behavior, scientists have initiated programs of cooperative research regarding monsoons. They hope to collect sufficient data to develop computer models that can predict when monsoons will occur and how they will impact landmasses. Such forecasting efforts have often frustrated researchers because of the monsoons' capricious nature. Speculation is primarily hindered by intraseasonal oscillations between the active and passive precipitation phases of monsoons. The Center for Ocean-Land-Atmosphere Studies (COLA) examines the relationship of the ocean, the atmosphere, and heat sources. While empirically based forecasts have often proved more reliable than modeled simulations during the twentieth century, scientists seek to perfect their experimental methods.

Researchers recognize that regional topographic differences and varying hydrodynamic situations impede monsoon modeling attempts. Incorporating information about variables such as the temperature of the sea surface and snow cover, scientists try to understand how these factors alter monsoon behavior. They also want to comprehend how the monsoons affect ocean and atmosphere interactions in addition to how the water-air pressure system relationship influences monsoons. Studies have been designed to expand knowledge about the role of convection in monsoons. Researchers are also interested in studying the impact of monsoons on climates inside and outside the monsoons' immediate zone. Scientists disagree whether monsoons affect or are affected by El Niño; studies have been done to evaluate global precipitation data and drought conditions to determine any correlations.

Monsoon variability hinders forecasting and potential benefits to agriculture based on information concerning the onset of rainfall. For example, with advance knowledge, farmers could plant crops that require less water in case a weaker monsoon is forecast. Monsoons, however, do not always begin when expected, and variations can happen within the yearly cycle or fluctuate over several years. Using different global circulation models, researchers seek to comprehend the fundamental physical processes of monsoons, then apply this knowledge to create specific models representing a seasonal cycle based on their observations, statistics, and hypotheses. Interpreting results produced by these models compared with satellite data, researchers become aware of how monsoons vary according to the landmasses they traverse and according to unique oceanic and atmospheric conditions.

Realizing these factors are linked, researchers create models to consider numerous variables concurrently but realize that more sophisticated modeling of factors such as sea-surface temperature, solar radiation, water vapor, cloud cover, soil moisture, and terrain is necessary to improve prediction methods.

Such models must accurately simulate a monsoon's average seasonal rainfall and behavior, taking into account how it varies within one season and one year. The models must also consider anomalies with other documented monsoons in that area to isolate how external conditions such as altered topography may affect the monsoon's internal dynamics.

## FUTURE RESEARCH

Monsoons fascinate the researchers who strive to acquire more complex understanding of the seasonal wind systems. Scientists hope to predict monsoons more precisely, because the winds impact global economies and populations. Meetings of international monsoon experts have been scheduled to share knowledge of monsoon cycles, coordinate research methods, and set future goals. Primarily, scientists want to explain why monsoons vary in behavior and how such variables as sea-surface temperature and location of land-based heat sources interact with and influence the monsoons' fluctuations. Researchers also want to explore the relationship between monsoons and the El Niño/Southern Oscillation phenomenon.

Scientists realize that to achieve accurate predictive techniques, full understanding of sea-surface temperature anomalies that affect monsoon circulation must be obtained through an analysis of oceanic processes that currently remain vague. Future models will assess these temperature patterns in different geographic regions. Models will also further evaluate water surface fluxes, land surface coverings of snow and vegetation, and global warming, all of which affect temperatures and alter monsoon cycles. Until such comprehension is attained through enhanced model simulations designed to analyze numerous dynamic variables simultaneously, predictions will be minimized, slowing endeavors to manage the winds, and monsoons will continue to affect life both advantageously and harmfully in tropical zones.

*Elizabeth D. Schafer*

## FURTHER READING

Ahrens, C. Donald. *Essentials of Meteorology: An Invitation to the Atmosphere.* Belmont, Calif.: Brooks/Cole Cengage Learning, 2012. Discusses various topics in weather and the atmosphere. Covers tornadoes, thunderstorms, acid deposition and other air pollution topics, humidity, cloud formation, and temperature.

Bamzai, A. S., and J. Shukla. "Relations Between Eurasian Snow Cover, Snow Depth, and the Indian Summer Monsoon: An Observational Study." *Journal of Climate* 12 (1999): 3117-3132. Studies the relation between the Indian summer monsoon and the snow cover in the Himalaya Mountains and other nearby regions.

Chang, Chih-Pei. *East Asian Monsoon.* Hackensack, N.J.: World Scientific Publishing, 2004. Reviews recent research on the East Asian monsoon system, considering its impact on one-third of the world's population and the global climate system, and its connections with west Pacific and Australian monsoons and the El Niño/Southern Oscillation.

Chang, Chih-Pei, Yihui Ding, and Ngar-Cheung Lau. *The Global Monsoon System: Research and Forecast.* Hackensack, N.J.: World Scientific Publishing, 2011. A current review of the science of monsoon research and forecasting, building on the concept that monsoon systems in various parts of the world are, in fact, parts of an integrated global monsoon system.

Cheung, Chan Chik. *Synoptic Patterns Associated with Wet and Dry Northerly Cold Surges of the Northeast Monsoon.* Hong Kong: Royal Observatory, 1997. A short, somewhat technical report that discusses monsoon patterns in the Hong Kong region of China. Bibliography.

Clift, P. D., R. Tada, and H. Zheng. *Monsoon Evolution and Tectonics: Climate Linkage in East Asia.* Bath, England: The Geological Society, 2010. Presents a series of specialist reports on studies relating variations of the Asian monsoon system on orbital and tectonic timescales, investigating its relationship to such phenomena as the uplifting of the Tibetan Plateau and the Himalaya range.

Douglas, Michael W., Robert A. Maddox, Kenneth W. Howard, and Sergio Reyes. "The Mexican Monsoon." *Journal of Climate* 6 (August, 1993): 1665-1677. Discusses the phenomenon of summer rainfalls in North America, which resemble Asian monsoons. Illustrated with maps. Sources listed.

Fein, Jay S., and Pamela L. Stephens, eds. *Monsoons.* New York: Wiley, 1987. Includes comprehensive meteorological and geological data, in addition to a discussion of cultural responses to monsoons throughout history. Illustrations, bibliography, and index. Useful for high school students.

Hodges, Kip. "Climate and the Evolution of Mountains." *Scientific American* 295 (2006): 72-79. Presents a study in the Himalayas addressing the effects of climate on geological structures. Discusses plate tectonics, monsoons, mountain building, and erosion. Accessible to the layperson.

Lighthill, James, and Robert Pearce, eds. *Monsoon Dynamics.* New York: Cambridge University Press, 2009. Discusses atmospheric influences on monsoons and oceanic influences. Discusses tropical climatology, such as drought and rainfall patterns, temperature oscillations, and wave dynamics on a global scale. Also examines the dynamics in India, the Indian Ocean, and eastern Africa. A technical text best suited for researchers, graduate students, and professional meteorologists.

McCurry, Steve. *Monsoon.* New York: Thames and Hudson, 1988. A brief personal account of an author's travels during a monsoon that includes historical and cultural insights. Contains many photographs documenting people and places affected by monsoons. Readers of all ages can use this book to supplement more scholarly texts.

Oshima, Harry T. *Strategic Processes in Monsoon Asia's Economic Development.* Baltimore: Johns Hopkins University Press, 1993. Examines how climatic conditions caused by monsoons influence employment and other socioeconomic concerns. References and index. Accessible to high school readers.

Vesilind, Priit J. "Monsoons." *National Geographic* 166 (December, 1984): 712-747. A narrative supplemented with photographs that describes the author's experiences during monsoons and explains the scientific principles of monsoons. Suitable for high school readers.

Wang, Bin. *The Asian Monsoon.* Chichester: Praxis Publishing, 2006. Presents a comprehensive interdisciplinary view of the Asian monsoon with regard to a number of different cause-and-effect relationships, including human effects on the monsoon weather pattern.

Webster, P. J., et al. "Monsoons: Processes, Predictability, and the Prospects for Prediction." *Journal of Geophysical Research* 103 (June 29, 1998): 14,451-14,510. Explores potential methods to model variables to forecast future monsoons. Illustrations and references.

**See also:** Atmosphere's Structure and Thermodynamics; Atmospheric and Oceanic Oscillations; Atmospheric Properties; Barometric Pressure; Climate; Clouds; Cyclones and Anticyclones; Drought; Lightning and Thunder; Ozone Depletion and Ozone Holes; Precipitation; Satellite Meteorology; Seasons; Severe Storms; Tornadoes; Tropical Weather; Van Allen Radiation Belts; Volcanoes: Climatic Effects; Weather Forecasting; Weather Forecasting: Numerical Weather Prediction; Weather Modification; Wind

# N

## NILE RIVER

*The Nile River is an interesting and important hydrologic system. Its complex geologic evolution began with the drying up of the Mediterranean Sea about 6 million years ago. The Nile carries water northward across the Sahara Desert from high-rainfall regions of equatorial Africa. It nurtured one of the great civilizations of ancient times and sustains millions of people in modern Egypt.*

### PRINCIPAL TERMS

- **cataract:** rough water or a waterfall in a river, generally obstructing navigation
- **hydrology:** the science that deals with the properties, distribution, and circulation of water on land
- **mantle plume:** a rising current of extra-hot magma in the mantle
- **monsoon:** a seasonal air current system; in the Northern Hemisphere's winter, dry winds flow from the continents to the ocean, whereas in summer, moist winds flow from the ocean to the continents and cause heavy rains
- **pyroxene:** a rock-forming mineral commonly found in igneous rocks such as basalt or gabbro
- **rift:** a portion of the earth's crust where tension has caused faulting, producing an elongated basin; rifts fill with sediments and, sometimes, volcanic rocks
- **tectonic:** relating to differential motions and deformation of the earth's crust, usually associated with faulting and folding of rock layers

### GEOGRAPHY AND CHARACTERISTICS

The Nile River is the longest river on Earth, extending some 6,650 kilometers from Lake Victoria to the Mediterranean Sea. Only the Amazon River in South America approaches its length. The river follows a generally south-to-north path as both the source of the White Nile in equatorial Africa and its mouth on the southern shore of the Mediterranean Sea lie within 1 degree of longitude. The Nile crosses 35 degrees of latitude, a distance comparable to the width of the continental United States, and flows across a greater variety of regions than any other river.

In spite of its great length and large drainage basin (approximately 3.4 million square kilometers, or about 10 percent of the area of Africa), the Nile carries relatively little water. During the twentieth century, annual flow ranged from a low of 42 cubic kilometers in the drought year of 1984 to a high of 120 cubic kilometers in 1916. This comparatively small flow results from the fact that no water is added to the river for the final one-half of its journey to the sea, while a significant amount of the water is lost to evaporation. Most rivers merge with increasingly large streams as they approach the sea, joining their waters into an ever-swelling stream. The Nile flows instead through the Sahara Desert, the largest and most desolate tract of land on Earth, matched only in size and desolation by the icy wastes of Antarctica.

Passage through the Sahara increases the importance of Nile water while reducing its volume. Its greatest value is reached as it nears the sea, where it has for more than five thousand years nurtured agriculture in Egypt. The rich soil, near-constant sunshine, and abundant water of the Nile Valley in Egypt combine to produce one of the most productive agricultural regions of the world. The ancient civilizations of Egypt were built on this firm economic foundation. Without the Nile, Egypt would be as barren as the rest of the Sahara, but with the Nile, Egyptian farms can produce two or even three crops every year. Egyptian civilization was nurtured by the Nile and protected from invasion by the sea to the north and the desert to the east and west. Egypt—and, consequently, much of world culture—truly is the "gift of the Nile."

### WHITE NILE AND BLUE NILE

Dramatic changes in the amount of water flowing through the Nile occur annually, reflecting the fact that two independent streams, the Blue Nile and the White Nile, join at Khartoum, the capital of North

Sudan. The White Nile issues from Lake Victoria, the second-largest freshwater lake on Earth. It subsequently tumbles down a series of falls and rapids to Lake Kioga and Lake Albert before spilling out into a huge swamp known as the Sudd in South Sudan. The Sudd was the great barrier to European explorers seeking the source of the White Nile. The vast swamps caused explorers traveling in boats up the Nile to lose their way, forcing them to turn back or perish, often having contracted malaria or some other endemic disease carried by insects of the swamps. As a result, the source of the White Nile was discovered in 1862 by John Speke, who traveled west from Zanzibar to Lake Victoria instead of following the upstream course of the Nile River.

In the Sudd, the White Nile—there known as the Bahr el Gebel—is joined by several tributaries. From the west flows the Bahr El Arab, and the Sobat flows in from the east. The vastness of the Sudd results in huge losses due to evaporation, amounting to about 50 percent of the water flowing into the Sudd. The Sudd also acts as a buffer, so that greater water flow into the Sudd causes the swampy area to expand, which in turn results in increased evaporation. Efforts to plan for expected increases in water demand, especially in Egypt, have focused on reducing these losses by building a 350-kilometer diversion, the Jonglei Canal, around the Sudd, the project having been conceived in the 1940's. Civil war in southern Sudan stopped its construction in 1983, but growing environmental concerns rendered it unlikely that the canal would be quickly finished even if the war ended. The present state of separation of South Sudan from North Sudan to form two independent nations has made completion of the Jonglei Canal unlikely. The combination of relatively constant rainfall over the Lake Victoria region, the vastness of Lake Victoria, and evaporative losses in the Sudd result in a relatively constant flow of water down the White Nile to Khartoum. North of Malakal, the Nile flows over a flat stretch in a well-defined channel without swamps; the river drops only 8 meters in elevation over its 800-kilometer length from Malakal to Khartoum.

In contrast to the White Nile, the Blue Nile and its little sister, the Nahr 'Atbara, are seasonally affected streams. For about one-half of the year they flow as feeble trickles, contributing negligible water to the Nile from January to June. In mid- to late summer, the monsoons sweep large amounts of moisture

evaporated from the Indian Ocean toward the Horn of Africa. As the water-laden air flows up over the Ethiopian highlands, it cools, and torrential rains fall. These rains swell the Blue Nile so much that its waters overwhelm the smaller White Nile at Khartoum and cause it to flow back upstream. Between 70 and 80 percent of the Nile's total annual water budget results from the flood phase of the Blue Nile and Nahr 'Atbara Rivers. These raging rivers also erode large amounts of black sand and silt from the basaltic highlands of Ethiopia and carry these sediments north. In his 1899 book *The River War*, Winston Churchill described the Nile in flood: "As the Nile rises its complexion is changed. The clear blue river becomes thick and red, laden with the magic mud that can raise cities from the desert sand and make the wilderness a garden." The historic fertility of the Nile Delta and Valley in Egypt owe as much to the new layer of this rich soil added to the inundated fields as to the deep soaking. This annual flooding and delivery of sediment no longer occurs downstream of the Aswan High Dam, which was constructed in the 1960's to ensure that the water needs of Egypt could be met in spite of drought in the Nile headwaters.

Like the White Nile, the Blue Nile issues from a large lake, Lake Tana. It initially flows to the southwest but progressively turns to the southwest and then northwest as it descends through the Blue Nile Canyon. It leaves the mountains near the border between South Sudan and Ethiopia and continues on a northwestward track across the Sudanese lowlands until it joins the White Nile at Khartoum.

## CATARACT NILE, GREAT BEND, AND EGYPTIAN NILE

For a distance of 1,850 kilometers from Khartoum to Aswan, the Nile is defined by two features: the cataracts and the great bend. The cataracts are sections where the river tumbles over rocky outcroppings, creating serious obstacles to navigation. There are six "classical" cataracts, but in reality there are many more—as much as 565 kilometers of this portion of the Nile, referred to as the Cataract Nile, is affected by cataracts. The cataracts are also significant because they define river segments where granites and other resistant rocks come down to the edge of the Nile. The floodplain in this region is narrow to nonexistent, and opportunities for agricultural development are correspondingly limited. These two factors—the navigational obstacles and the restricted floodplain—are the chief reasons that the land around this part of the Nile is thinly populated and why the historic border between Egypt in the north and Nubia or Sudan in the south was found not far from the First Cataract at Aswan.

The great bend is one of the most unexpected features of the Nile. For most of its course, the Nile flows inexorably to the north. At one location in the heart of the Sahara, however, it turns southwest and flows away from the sea for 300 kilometers before resuming its northward journey. This deflection of the river's course is the result of tectonic uplift of the Nubian Swell that has taken place during the past few hundred thousand years. This uplift is also responsible for the cataracts. If not for this geologically recent uplift, these rocky stretches would have been eroded away long ago by the abrasive action of the sediment-laden Nile.

The northernmost segment is the Egyptian Nile, extending for 1,200 kilometers between Aswan and the Mediterranean Sea. It consists of two parts, the Nile Valley and the Delta. The Nile Valley consists of the broad floodplain that is imprisoned between steep limestone or sandstone hillsides. The boundary between the lush valley floor and the flanking desert is stark and sudden; a visitor can stand with one foot on the black mud, brought 3,000 kilometers from Ethiopia, and the other foot on desert sand or barren limestone. The floodplain widens progressively to the north until it opens up just north of Cairo into the Delta. The mouth of the Nile is where the term "delta" was first applied. The ancient Greeks noted the triangular shape of the land around the Nile's mouth and the similarity that it had to the fourth letter of the Greek alphabet. The Nile splits into two branches at the south end of the Delta: the western, or Rosetta, branch (where the famous Rosetta Stone was discovered) and the eastern, or Damietta, branch. Up to the time that the Aswan High Dam was built, the Delta continued to grow as annual floods laid down their loads of silt. These sediments are now deposited in Lake Nasser, with the result that the sediment-starved Delta is slowly sinking and its shoreline is retreating due to erosion by the waters of the Mediterranean Sea.

## GEOLOGICAL FORMATION OF THE NILE

The evolution of the Nile is an important area of scientific research. The river consists of a series of steeper and flatter segments, and this is thought to indicate that several independent drainage systems previously existed in the region now drained by the Nile. Much is known about when and how the course of the Nile came into being, but much remains to be learned.

A critical event in the formation of the Nile was the evaporation of the Mediterranean Sea about 6 million years ago. Because of the climatic zone in which it lies, more water evaporates from the Mediterranean than is supplied by the rivers that flow into it. This water deficit requires that replenishing seawater flow into the Mediterranean from the Atlantic. When tectonic movement closed the Strait of Gibraltar as the African and European continental masses collided, the influx of Atlantic Ocean water was stopped and the Mediterranean slowly dried up. The Dead Sea of Israel and Jordan, at about 400 meters below sea level, is presently the deepest spot on the continents, but the Mediterranean sea floor, lying as much as 3,000 meters below sea level, was 0.6 kilometer deeper and one thousand times more vast than the Dead Sea region. This may have been the greatest sea ever to evaporate completely, and the event profoundly affected the streams that flowed into it. A north-flowing river existed in what is now Egypt; as sea level dropped, the stream became steeper and steeper as it cut down into relatively soft limestones. The enhanced erosive power allowed its upper tributaries to extend into the headwaters and capture upstream drainages. The increased water from the captured streams further increased the stream's erosive power, further stimulating the expansion of the drainage system upstream. This led to the development of the "Eonile" (the "dawn Nile"), which flowed through a huge canyon that was deeper than the Grand Canyon of Arizona and many times longer. This canyon is buried beneath the silt beds of the Egyptian Nile that have accumulated over the intervening millions of years, but it cannot be traced south of Aswan.

In time, the barrier at Gibraltar ruptured, and a tremendous waterfall brought Atlantic seawater to refill the Mediterranean basin. The "Grand Canyon of Egypt" became a drowned river valley or estuary, similar to the fjords of Norway but very different in origin. Slowly, this estuary filled with sediments brought in by rivers flowing from the south, and a landscape not too different from the present was established by 3 million years ago.

## STUDY OF THE NILE

It is much more difficult to discern the development of the Nile upstream from Aswan, but there are some clues. Lake Victoria did not exist prior to about twelve thousand years ago. Before this time, the streams of the Ugandan highlands flowed west to join the Congo, which drains into the equatorial Atlantic. Recent tectonic activity lifted and tilted the region, forming the lake and directing its overflow to the north. Similarly, recent tectonic activity changed the effects of the Nubian Swell and the Bayuda Uplift, with the result that northward flow across Nubia is sometimes permitted and sometimes blocked. Probably the oldest parts of the Nile drainage are those associated with the Sudd. These follow the axes of sediment-filled rifts that formed more than 65 million years ago and that have continued to sink and to fill with sediments since that time.

The Ethiopian highlands began to form about 30 million years ago as a result of tremendous volcanic activity, as a mantle plume punctured the crust. However, the contribution of the distinctive black sediments from these highlands is not recognized in the Egyptian Nile until about 650,000 years ago. This may be the result of increased rainfall on the Ethiopian highlands accompanying the development and intensification of monsoonal circulation in the recent past. Monsoonal circulation is caused by the change in position of atmospheric low-pressure cells, which lie over the equatorial Indian Ocean during northern winter and over south-central Asia during northern summer. The result is that cold, dry winds blow southward from Asia during winter, but warm, moist winds blow northward from the sea toward Asia in summer. The westward deflection of summer winds resulting from the Coriolis effect brings part of the moisture-laden air currents over Ethiopia, where the air cools as it rises. Cool air can hold less moisture than warm air, so clouds and then rain form as the monsoon rises over the Ethiopian Plateau. This brings the long, drenching rains in Ethiopia that cause the annual Nile flood. The monsoonal circulation has intensified over the last few millions of years as a result of continued uplift of the Tibetan Plateau. The mystery of the Nile flood that puzzled the ancient Egyptians can thus best be understood by knowing about mountain-building events occurring thousands of kilometers away. Water from the Ethiopian highlands may not have reached Egypt because the Nubian Swell acted as a barrier, perhaps deflecting the water to the west. It may be that only with the additional water provided as a result of the intensifying monsoon was the upstream Nile able to erode its way through the Nubian Swell and continue north to the Mediterranean Sea.

The geological history of the Nile is mostly inferred from sedimentary deposits in the Delta and Egyptian Nile. Scientists know that great river systems carried sediments—preserved today as the Nubian Sandstone—north from central Africa as long ago as the Cretaceous period, about 100 million years ago, but the course of these rivers is poorly known. No link can be established between the rivers of the Cretaceous Period and the present-day Nile. The sea invaded Africa from the north toward the end of the Cretaceous period, and a large portion of northeast Africa was covered by a shallow sea during much of the early Tertiary period, about 70 million to 40 million years ago. River deposits from the late Eocene and early Oligocene (about 35 million years ago) are known to the west of the present Nile, but these sediments did not travel far, indicating that they came from a relatively small river. This may have been the precursor of the stream that carved the great canyon following evaporation of the Mediterranean Sea about 6 million years ago.

*Robert J. Stern*

## FURTHER READING

Adar, Korwa G., and Nicasius A. Check. *Cooperative Diplomacy, Regional Stability and National Interests: The Nile River and Riparian States.* Pretoria: Africa Institute of South Africa, 2011. Addresses the colonial and postcolonial agreements and treaties affecting the ten nations through which the Nile River flows, focusing on the 2010 Agreement on the River Nile Basin Cooperative Framework, the objective of which is to establish a durable legal regime in the region.

Al-Atawy, Mohamed-Hatem. *Nilopolitics: A Hydrological Regime.* Cairo: American University in Cairo Press, 1996. Offers a brief but detailed discussion of the politics and conservation methods surrounding the Nile River watershed, ecology, and water supply. Map and bibliography.

Awulachew, Seleshe Bekele. *The Nile River Basin: Water, Agriculture, Governance and Livelihoods.* London: Taylor & Francis, 2012. Describes the history of the Nile River basin, as well as current and future challenges and opportunities faced by the people living in the ten riparian nations of the Nile River.

Churchill, Winston. *The River War.* New York: Longmans, Green, 1899. Provides an account of the Mahdi's rebellion in Sudan and the British expedition against it during the late nineteenth century. Accounts of military operations can be skimmed over, but the trials of navigation up the Nile provide the best account of the river upstream from Aswan. Provides a good understanding of why Egypt stops near Aswan. Suitable for the general reader.

Collins, Robert O. *The Nile.* Harrisonburg, Va.: R. R. Donnelley & Sons, 2002. Focuses on the dynamic interplay between human and environmental forces along the Nile River throughout recorded history. Chronicles the past, present, and future of the great river. A bit dated by the course of events in the region over the intervening years since its publication, but of interest for the sake of historical comparisons.

Darby, Stephen, and David Sear. *River Restoration: Managing the Uncertainty in Restoring Physical Habitat.* Hoboken, N.J.: John Wiley & Sons, 2008. Begins with theoretical and philosophical issues with habitat restoration to provide a strong foundation for decision making. Addresses logistics, planning, mathematical modeling, and construction stages of restoration. Post-construction monitoring and long-term evaluations provide a full picture of the habitat restoration process. A highly useful text for anyone involved in the planning and implementation of habitat restoration.

Guadalupi, Gianni. *The Discovery of the Nile.* New York: Stewart, Tabori, and Chang, 1997. Allows the reader to explore the entire length of the Nile River through foldout color maps and illustrations. Suitable for all levels.

_____. *The Nile: History, Adventure, and Discovery.* Vercelli: White Star, 2008. Covers the history of the Nile, presenting the expeditions of the Egyptians, French, and British.

Hillel, D. *Rivers of Eden.* New York: Oxford University Press, 1994. Offers a wonderful account of the history of water use in the Middle East and the problems that the future holds for the region. Written by an environmental scientist and hydrologist. Based on sound technical considerations, but enlivened with a generous spice of history, religion, and personal experience. Suitable for the general reader.

Howell, P. P., and J. A. Allan, eds. *The Nile: Sharing a Scarce Resource.* Cambridge, England: Cambridge University Press, 1994. Offers an overview of the history of the Nile, with special emphasis on how its flow has changed in the past. Discusses the future utilization of the river and the problems facing planners. A comprehensive, excellent text, suitable for college students and professionals.

Melesse, Assefa M. *Nile River Basin: Hydrology, Climate and Water Use.* Dordrecht: Springer, 2011. Examines scientific studies and issues such as hydrological modeling, water management, and policy matters related to the hydrological processes functioning amid changing land use, as population places increasing pressure on natural resources in the Nile River basin.

Wohl, Ellen. *A World of Rivers.* Chicago: University of Chicago Press, 2011. Covers the Amazon, Ob, Nile, Danube, Ganges, Mississippi, Murray-Darling, Congo, Chang Jiang, and Mackenzie Rivers in individual chapters. Provides valuable figures loaded with straightforward and well-organized information. Discusses natural history, anthropogenic impact, and the future environment of these ten great rivers. Bibliography is organized by chapter.

**See also:** Alluvial Systems; Amazon River Basin; Aquifers; Deltas; Drainage Basins; Floodplains; Floods; Ganges River; Great Lakes; Groundwater Movement; Lake Baikal; Mississippi River; River Bed Forms; River Flow; Sediment Transport and Deposition; Yellow River

# NITROGEN CYCLE

*The nitrogen cycle is the shift between different chemical forms of nitrogen through biologic, physical, and geologic processes on Earth. Nitrogen is an essential element for all living things. It is a building block of biologic molecules such as proteins and nucleic acids. The majority of nitrogen on the planet is in the form of molecular nitrogen in the air. Only certain bacteria can convert nitrogen into biologic molecules that occur mainly inside living cells. Humans are interfering with the nitrogen cycle by making nitrogen fertilizers and by oxidizing atmospheric molecular nitrogen through the extensive burning of fossil fuels.*

## PRINCIPAL TERMS

- **ammonia:** a colorless and highly toxic gas with a strong odor; the odor of ammonia is frequently detected in stables or in sewage; salts of ammonia are used as fertilizer for plants
- **ammonium ion:** produced as a waste of such animals as fish, during decomposition of organic nitrogen wastes by bacteria, and by metabolism of some bacteria; forms after dilution of ammonia in water; acidic and toxic to humans because it interferes with respiration
- **biogeochemical cycle:** cycling of chemical elements such as nitrogen, carbon, and phosphorus
- **enzyme:** biologic catalyst made of proteins
- **eutrophication:** process in which water bodies (rivers, ponds, lakes, and oceans) receive excess nutrients (mainly nitrogen and phosphorus) that stimulate abundant growth of algae and plants
- **food web:** the complex web of feeding relationships in nature
- **nitrate:** the ion of nitric acid; an essential nutrient for plants
- **nitric acid:** nitrogen-containing strong acid, used by medieval alchemists to separate gold from silver; now used in the manufacturing of dyes, plastics, and drugs and in laboratories
- **nitrite:** the ion of nitrous acid; source for some microorganisms; extremely hazardous to humans, especially babies
- **nitrogen:** a key chemical element on Earth; a colorless and odorless component of air

## CYCLING OF NITROGEN ON EARTH

The majority of Earth's chemical elements are circulating through biologic, physical, chemical, and geologic processes. These processes operate in circles and are called biogeochemical cycles.

Nitrogen is one of the key elements in human activities and in biologic, physical, chemical, and geologic processes. Estimates show that more than 20 million tons of nitrogen exists on every square mile of the planet. The atmosphere contains up to 78 percent molecular nitrogen ($N_2$), and this nitrogen is mainly cycling through biologic processes.

Four major nitrogen-transformation (biologic) processes exist in nature: nitrogen fixation, ammonification, nitrification, and denitrification. Mineralization is the only geologic process that is involved in the circulation of nitrogen. The main mineral sources of nitrogen on Earth are Bengal saltpeter ($KNO_3$) in India and other Asian countries and Chile saltpeter ($NaNO_3$) in South America. Natural gas also contains nitrogen.

Vast amounts of nitrogen are circulated by physical and chemical processes. Nitric oxide ($NO$) is formed in the air from $N_2$ and $O_2$ (molecular oxygen) during thunderstorms by lightning. Nitric oxide oxidizes further to nitrogen dioxide ($NO_2$) and later reacts with water to form nitric (or nitrous ($HNO_3$) acids. Acids fall to the ground during rain and form nitrates ($NO_3^-$) and nitrites ($NO_2^-$) in the soil (acid rain).

Living things require nitrogen as a component for proteins, nucleic acids (deoxyribonucleic acid, or DNA, and ribonucleic acid), and other organic compounds. Nitrogen is often a limiting plant nutrient. Plants take up nitrogen from soil mainly as ammonium ions, nitrate, or nitrogenous organic compounds and incorporate nitrogen into organic molecules such as proteins or nucleic acids. The nitrogen then follows food webs from plant eaters (herbivores) to decomposers (mainly microbes). Animals use only organic forms of nitrogen.

## NITROGEN FIXATION

The utilization of molecular nitrogen ($N_2$) by particular bacteria is called nitrogen fixation. Some of these bacteria (*Rhizobium*) live in symbiosis with certain legume plants and others are free-living bacteria such as cyanobacteria or *Azotobacter*. Legume plants include soybeans, clover, alfalfa, beans, and pears. Symbiotic nitrogen-fixing cyanobacteria provide nitrogen to other plant species such as the water-fern *Azolla* and liverworts and cycads.

The nitrogen fixation or reduction of $N_2$ to $NH_3$ (ammonia) is a complicated, multistep process ($N_2$ + $8e^-$ + $8H^+$ + 16ATP → $2NH_3$ + $H_2$ + 16ADP + 16P). Ammonia produced by this process is further converted to proteins, nucleic acids (DNA), and other nitrogen-containing organic molecules ($NH_3$ → nitrogenous organic molecules: proteins, nucleic acids, and so forth).

The nitrogen fixation is catalyzed by the enzyme nitrogenase. Nitrogenase is sensitive to molecular oxygen ($O_2$). Nitrogen-fixing organisms possess a number of morphological and biochemical modifications designed to protect enzymes from oxygen inactivation. For example, the bacterium *Rhizobium* controls the oxygen level in cells by the protein leghemoglobin, which catches oxygen. In the case of cyanobacteria, there are specialized cells (called heterocysts) for nitrogen fixation. Heterocysts show high rates of respiration, which ultimately reduces oxygen levels in these cells.

Nitrogen fixation is an energy-consuming process, which explains why cyanobacteria normally have only 5 to 10 percent of heterocysts among their cells. To maintain nitrogen fixation, other cyanobacterial cells (vegetative cells) work to generate enough energy for heterocysts.

All life on Earth depends on nitrogen fixation because the main reservoir of nitrogen on Earth is in the air as molecular nitrogen ($N_2$). The main path of nitrogen from the air into biologic nitrogen-containing molecules of different organisms is through nitrogen fixation. Nitrogen fixation also is of enormous importance to agriculture because it supports the nitrogen needs of many crops. This process was discovered by Russian microbiologist Sergei Winogradsky.

Apart from natural nitrogen fixation, the industrial Haber-Bosch process converts molecular nitrogen to ammonia. In this process nitrogen fertilizers are made for agriculture. Haber-Bosch is an energy-consuming route, and the process of manufacturing nitrogenous fertilizers consumes up to 50 percent of the energy input in modern agriculture.

## AMMONIFICATION

Ammonification is the process of making ammonia or ammonium ions ($NH_4^+$) by living things. Ammonium ions are produced as a waste of such animals as fish and during decomposition of organic nitrogen wastes by bacteria and by metabolism of some bacteria. Bacteria, for example, can convert nitrate into ammonia in soils or in the human gut.

Globally, only a small amount (15 percent) of nitrogen reaches the atmosphere as ammonia, compared with $N_2$ and $N_2O$. The majority of ammonium ions are quickly consumed in soil and water by microorganisms and plants. At different points in the food web, ammonium ions are returned to the environment.

## NITRIFICATION

Nitrification is caused by the sequential action of two separate groups of soil bacteria: the ammonia-oxidizing bacteria (the nitrosifyers) and the nitrite-oxidizing bacteria (nitrifying bacteria). These bacteria obtain energy by consuming nitrogen compounds and can feed only on inorganic compounds. The end product of nitrification is nitrate, a valuable nitrogen source for plants.

Nitrification is a two-step process. Nitrosifyers, such as the bacterium *Nitrosomonas*, convert ammonium ion into nitrite first ($NH_4^+$ + $O_2$ → $NO_2^-$ + $H_2O$ + $H^+$). Later, nitrifying bacteria, such as the bacterium *Nitrobacter*, oxidize nitrite into nitrate ($NO_2^-$ + $O_2$ → $NO_3^-$).

Nitrosifyers and nitrifying bacteria are common in soil and water. They live especially in areas where ammonia is present in high amounts, such as sites of ammonification and in wastewater and manure. Nitrification does not contribute significantly to agriculture. Although liked by plants, nitrate is not always available for plants in soils. Nitrate is quickly consumed by microorganisms during denitrification. Additionally, one species of Archaea (microorganisms similar to bacteria) undergoes nitrification by oxidizing ammonia in the oceans.

## DENITRIFICATION

The conversion of nitrate into gaseous nitrogen compounds such as $N_2O$, NO, and $N_2$ by different bacteria in soils is called denitrification, or nitrate reduction. Bacteria use nitrate as a substitute for oxygen during respiration and convert it to different nitrogenous compounds according to the following chain of reactions: $NO_3^-$ → $NO_2^-$ → NO → $N_2O$ → $N_2$.

Eventually, nitrogen is released into the atmosphere as $N_2O$ and NO or as $N_2$. Simultaneously, bacteria decompose significant amounts of organic matter within the soil.

Denitrification has a negative effect on agriculture, as it removes nitrogen from soils. In contrast, denitrification can be useful in wastewater treatment.

## HUMAN INTERFERENCE IN NITROGEN CYCLE

Human interference in the nitrogen cycle can be significant. Human activities are generally responsible for adding excessive amounts of inorganic nitrogen into the water or soil. Adding nitrogen into water causes a rapid cultural eutrophication of water bodies.

Most of the inorganic nitrogen in water comes from soil fertilization, industrial and domestic wastes, septic tanks, feed-lot discharges, domestic-animal wastes (including birds and fish), and discharges from car exhausts. Eutrophication occurs regularly in nature but does so slowly, often through hundreds or thousands of years.

During human eutrophication of water bodies, algae grow fast and choke waterways and consume large amounts of dissolved oxygen. Some algae also produce toxins. This rapid and uncontrollable growth of algae—a process that produces algal blooms or red tides—causes decay and, eventually, the destruction of aquatic ecosystems. Fish and other aquatic organisms die by the thousands, suffocated by oxygen depletion or killed by the action of toxins. Algae need inorganic nitrogen ions as a nitrogen source for their growth (for making proteins).

Ammonium and nitrate ions are the main human nitrogen pollutants in water. The removal of inorganic nitrogen from industrial and domestic wastewater is required to protect water quality. Inorganic nitrogen ions can be removed from water by physicochemical and biologic methods.

Human contamination of soil by inorganic nitrogen is another substantial problem that occurs through nitrogen fertilization of soil. In such cases, nitrogen derives from industrial nitrogen fixation or from using leguminous plants (soy, beans, peas, and alfalfa).

In addition, extensive burning of fossil fuels by humans converts atmospheric nitrogen ($N_2$) into nitrogen oxides. Nitrogen oxides react in the atmosphere with the ozone ($O_3$) to make nitric acid. Nitric acid is one of the components of acid rain, which damages soils and forest richness by destroying communities of organisms. Thus, the burning of fossil

fuels by humans contributes to acid rain and to the destruction of the ozone layer in the atmosphere.

*Sergei A. Markov*

## FURTHER READING

Chang, Raymond, and Jason Overby. *General Chemistry: The Essential Concepts.* 6th ed. Columbus, Ohio: McGraw-Hill, 2010. A comprehensive book on general chemistry that includes chapters on the nitrogen cycle and nitrogenous compounds.

Graham, Linda E., and Lee W. Wilcox. *Algae.* Upper Saddle River, N.J.: Prentice Hall, 2000. A popular book on algae, including their role in the nitrogen cycle and nitrogen fixation.

Madigan, Michael T., et al. *Brock Biology of Microorganisms.* 13th ed. Boston: Benjamin Cummings, 2012. Several chapters of this classical textbook on microbiology describe in detail microbial processes such as nitrogen fixation, ammonification, nitrification, and denitrification. Also examines nitrifying and nitrogen-fixing bacteria.

Markov, Sergei A., Michael J. Bazin, and David O. Hall. "Potential of Using Cyanobacteria in Photobioreactors for Hydrogen Production." In *Advances in Biochemical Engineering/Biotechnology,* edited by A. Fiechter. Vol. 52. New York: Springer, 1995. Detailed scientific review of cyanobacteria and their role in nitrogen fixation. Process of nitrogen fixation is explained from biochemical and physiologic points of view.

Nebel, Bernard J., and Richard T. Wright. *Environmental Science: Towards a Sustainable Future.* Englewood Cliffs, N.J.: Prentice Hall, 2008. A popular textbook on ecology and environment that describes the nitrogen cycle and human interference with the cycle. Also examines the control of hazardous nitrogen chemicals.

Reece, Jane B., et al. *Campbell Biology.* 9th ed. Boston: Benjamin Cummings, 2011. A general biology textbook that explains biologic molecules that contain nitrogen and their role in the metabolism of living things. Separate chapters cover the nitrogen cycle, human interference in the cycle, and different biologic processes involving nitrogen, such as the excretion of animal-produced nitrogenous wastes.

White, David. *The Physiology and Biochemistry of Prokaryotes.* New York: Oxford University Press, 2011. Describes bacteria and explains in detail bacterial metabolism of nitrogen compounds such as nitrogen fixation, nitrification, denitrification, and ammonification.

**See also:** Acid Rain and Acid Deposition; Atmospheric Properties; Carbon-Oxygen Cycle; Geochemical Cycles; Hydrologic Cycle; Lightning and Thunder; Ozone Depletion and Ozone Holes; Rainforests and the Atmosphere; Remote Sensing of the Atmosphere; Severe Storms

# NORTH SEA

*The North Sea, located between the United Kingdom and continental Europe, is one of the most economically important bodies of water in the world. The impact of human activity on the North Sea is of major concern of environmentalists.*

## PRINCIPAL TERMS

- **Baltic Sea:** the body of water between Scandinavia and Eastern Europe
- **bank:** an elevated area of land beneath the surface of the ocean
- **fjord:** a steep-sided narrow inlet eroded into the face of seaside cliff, typical of Scandinavia but found throughout the world
- **Norwegian Sea:** the body of water north of the North Sea
- **strait:** a narrow waterway connecting two larger bodies of water
- **trench:** a long, narrow, depressed area in the ocean floor

## PHYSICAL CHARACTERISTICS

The North Sea is an arm of the Atlantic Ocean located between the islands of Britain and the mainland of northwestern Europe. It is bordered by the island of Great Britain to the southwest and west, by the Orkney Islands and the Shetland Islands to the northwest, by Norway to the northeast, by Denmark to the east, by Germany and the Netherlands to the southeast, and by Belgium and France to the south.

To the north, the North Sea opens to the Norwegian Sea. To the south, it is connected to a narrow waterway known as the Strait of Dover, located between southeast England and northwest France. The Strait of Dover connects the North Sea to a wider waterway known as the English Channel, located between southern England and northern France. The English Channel opens to the Atlantic Ocean. To the east, the North Sea is connected to a strait known as the Skagerrak, located between Norway and Denmark. The Skagerrak is connected to a strait between Sweden and Denmark known as the Kattegat, which opens to the Baltic Sea.

The North Sea covers an area of about 570,000 square kilometers. It contains about 50,000 cubic kilometers of water. The North Sea is generally shallow, with an average depth of about 94 meters. By comparison, the Atlantic Ocean has an average depth of about 3,930 meters. The southern part of the North Sea is the most shallow, with the northern part growing deeper as it approaches the much deeper Norwegian Sea.

The floor of the North Sea is rough and irregular. In the southern part, where the water is often less than 40 meters deep, many areas of elevated underwater land known as banks are shifted and reworked by tides and currents. These moving banks often present a hazard to navigation. The Dogger Bank, a large bank located roughly in the center of the North Sea, is only about 15 to 30 meters below sea level.

Several areas of greater-than-average depth, known as trenches, are located in the North Sea. In the otherwise shallow waters of the south, a trench known as Silver Pit reaches a depth of about 97 meters. Not far north of the Dogger Bank, a trench known as Devils Hole reaches a depth of more than 450 meters. The deepest part of the North Sea is the Norwegian Trench, a large trench that runs parallel to the southern coast of Norway. The Norwegian Trench is between 25 and 30 kilometers wide, with depths ranging from 300 to 700 meters.

The coastline of the North Sea varies from rugged highlands in the north to smooth lowlands in the south. The coast of Norway is mountainous and broken, with thousands of rocks and small islands, and frequently indented by steep-sided inlets called fjords. The coasts of Scotland and northern England are high and rocky, but less broken. The coasts of middle England and the Netherlands are low and marshy, and in places have been isolated from the sea by human-made dykes. The coasts of southern England, France, and Belgium are low and sandy.

## GEOLOGICAL HISTORY, HYDROLOGY, AND CLIMATE

The shape and size of the North Sea have varied greatly over time. At the end of the Pliocene Epoch, about 1.6 million years ago, the southern half of the North Sea was part of the mainland of Europe. At that time, the Rhine River, which now empties into the North Sea in the southern part of the Netherlands, ran to a point about 400 kilometers north of London. The Thames River, which now empties into the North Sea east of London, continued eastward until it met the Rhine River.

During the Pleistocene epoch, from about 1.6 million to 10,000 years ago, vast ice sheets advanced and retreated several times and deposited a thick layer of clay on the bottom of the North Sea. At its greatest extent, the ice covered the entire North Sea. About 8,000 years ago, the ice retreated for the last time. A few hundred years later, the rising waters of the North Sea broke through the land bridge connecting England and France, forming the Strait of Dover. The modern coastlines of the North Sea were formed about 3,000 years ago.

The movements of the ice sheets were largely responsible for the rugged floor of the North Sea. The Dogger Bank and smaller banks in the southern part of the North Sea were created when the ice deposited large amounts of earth and stones in a particular area. Some of the trenches in the North Sea are believed to be located in areas where ancient rivers emptied into the North Sea when it was much smaller.

Water enters the North Sea from the Atlantic Ocean by way of the Strait of Dover and the Norwegian Sea. This water is relatively warm and salty. It is heated by the warm North Atlantic Current, which moves north along the western side of the British Isles and enters the Norwegian Sea. Colder, less salty water from the Baltic Sea enters by way of the Skagerrak, creating a counterclockwise current in the North Sea.

Freshwater enters the North Sea from the Thames, the Rhine, and other large rivers. The salt content of the North Sea varies from about thirty-four to about thirty-five parts per thousand. Higher concentrations are found off the coast of Great Britain, and lower concentrations are found off the coast of Norway.

The average temperature of the surface water in the North Sea in January varies from about 2 degrees Celsius east of Denmark to about 8 degrees Celsius between the Shetland Islands and Norway. In July, the average temperature varies from more than 15 degrees Celsius along the coast from the Strait of Dover to Denmark, to about 12 degrees Celsius between the Shetland Islands and the Orkney Islands.

The average air temperature varies from between 0 and 4 degrees Celsius in January to between 13 and 18 degrees in July. Winters are stormy, and gales are frequent. The average difference between low tide and high tide is between about 4 meters and 6 meters along the coast of the British Isles and the southern coast of the European mainland. Along the northern coast of the European mainland, the difference is usually less than 3 meters.

## HUMAN ACTIVITIES

Because of its long coastline and the many rivers that empty into it, the North Sea has long been an area of important human activity. The exchange of people, goods, and ideas made possible by the North Sea had a profound influence on the cultural development of northwestern Europe during the Renaissance.

In modern times, the North Sea is one of the busiest shipping areas in the world. Adding to its economic importance is the fact that it provides the only waterway between the Baltic Sea and the Atlantic Ocean. Because of the North Sea, the Netherlands and the United Kingdom are among the world's leading nations in the volume of cargo carried by sea. The Europoort complex at Rotterdam, in the Netherlands, handles more cargo than any other seaport in the world. Other major ports located on the North Sea include Antwerp, Belgium; Dunkirk, France; London, England; and the three ports of Hamburg, Bremerhaven, and Wilhelmshaven in Germany.

The accessibility of the North Sea also made it an early area of scientific research. The *Challenger* expedition, launched by the British in 1872, began a new era in oceanography. In 1902, the International Council for the Exploration of the Seas was founded in Denmark for the purpose of studying the North and Baltic Seas. It has since compiled the longest record of marine ecological conditions in the world. In more recent years, a large number of marine laboratories, research centers, and scientific vessels have been active in the area.

The North Sea has also long been an area of land reclamation and flood-control projects, particularly in the Netherlands. For centuries the Dutch have reclaimed land from the North Sea by building dikes. In the 1930's, a dike 30 kilometers long was built in the Netherlands, creating a large freshwater lake. Three-fifths of an area of land formerly under the North Sea was then reclaimed as farmland. After abnormally high tides flooded a large part of the Netherlands in 1953, the Dutch built a large flood-control system on rivers emptying into the North Sea. The British built a similar system on a smaller scale on the Thames River in 1984.

Fishing has long been a major activity in the area. The constant mixing of shallow water in the North

Sea provides a rich supply of nutrients to plankton—microscopic organisms that support a wide variety of commercially important fish. The main species caught are cod, haddock, herring, and saithe. Lesser quantities of plaice, sole, and Norway pout are also caught. Sand eel, mackerel, and sprat are caught for the production of fish meal.

The major fishing nations in the region are Norway, Denmark, the United Kingdom, and the Netherlands. In 1983, the nations of Europe created the Common Fisheries Policy. This arrangement establishes the amount of each species that each nation may catch in the open waters of the North Sea. This policy does not apply to fish caught in coastal waters, which are considered to belong to a particular nation.

## NATURAL GAS AND OIL

New economic resources were discovered in the North Sea in the second half of the twentieth century. In 1959, the first known source of natural gas in the North Sea was identified. This source was an extension into the sea of a large natural gas field in the northeastern part of the Netherlands. By the end of the 1970's, a large number of natural gas production sites were developed in the North Sea. These sites are located primarily along a line running east and west for about 150 kilometers between the Netherlands and England. A smaller number of natural gas sources are located in the central and northern regions of the North Sea.

The first nation to obtain oil from the North Sea was Norway, which began operating its first offshore oil well in 1971. The United Kingdom began extracting oil from the North Sea in 1975. The oil wells are primarily located in the northern and central regions of the North Sea. By the 1980's, offshore oil wells were in operation from north of the Shetland Islands to an area about 650 kilometers to the south.

Oil is found in the North Sea in a basin of sedimentary rock thousands of meters thick. Almost all the world's supply of oil is found in similar basins. Such basins are believed to exist in areas where the outer layer of the earth has been stretched and thinned. This causes a basin to be formed beneath the outer layer, which then collects sediments for millions of years. If the sediments are subjected to certain temperatures and pressures, organic matter within them

is transformed into oil. Seismic measurements using controlled explosions have confirmed that the outer layer of land beneath the central portion of the North Sea is about half as thick as the same layer elsewhere. The stretching and thinning of this area took place millions of years ago, when Europe and the island of Great Britain were drifting apart.

## SIGNIFICANCE

The most critical issue facing the nations that border the North Sea is the impact of human activity in the area. Pollution from ships and from land-based operations is a major concern. Although international agreements on limiting pollution of the North Sea have been in place since 1969, enforcement remains difficult. Events such as the Ecofisk oil well blowout, which spilled more than 30 million liters of oil into the North Sea in 1977, point out the importance of controlling contamination.

An important problem is the disposal of offshore equipment that is no longer in use. As long ago as 1958, international agreements stated that all such equipment must be removed from the sea. However, because of the extremely high cost of removing large installations, new agreements were made in 1989 that allowed such equipment to be disposed of at sea. The agreement required that the water where the equipment is disposed be at least 100 meters deep and that at least 35 meters of water remain above the disposed debris. This policy met with worldwide controversy in 1995, when the Shell company decided to dispose of an unused oil well known as the Brent Spar in this way. Under intense pressure from environmentalists, Shell reversed this decision and made plans to remove the Brent Spar entirely.

Less dramatic, but possibly even more important, is the problem of oil pollution from muds used to lubricate drills in offshore oil wells. Studies have indicated that oil from this and other sources is far more widespread than previously thought. The most significant impact was on an organism known as the burrowing brittle-star. The digging of this animal brings oxygen into sediments, encouraging other organisms to grow there. The burrowing brittle-star is also an important food source for many fish. The number of burrowing brittle-stars per square meter fell from more than one hundred in unpolluted waters to zero within 1 or 2 kilometers of oil wells. Such losses may have been a major factor in the decrease of the numbers of North Sea fish such as cod, whose population fell by two-thirds between 1980 and 1995.

Rose Secrest

## FURTHER READING

Flemming, Nicholas Coit, Council for British Archaeology, and English Heritage. *Submarine Prehistoric Archaeology of the North Sea: Research Priorities and Collaboration With Industry.* York: Council for British Archaeology, 2004. Brings together comparative archaeological evidence from Norway, Denmark, Germany, the Netherlands, and the United Kingdom to describe Paleolithic, Mesolithic, and Neolithic settlement in areas that are now beneath the waters of the North Sea.

Gaffney, Vincent, Kenneth Thomson, and Simon Fitch, eds. *Mapping Doggerland: The Mesolithic Landscapes of the Southern North Sea.* Oxford: Archaeopress, 2007. Discusses various techniques of topographical and seismic mapping in reference to the North Sea. Contains a description of methodologies, followed by an evaluation of results from multiple studies and multiple sites within the North Sea. Presents tectonics, the paleolandscape, the paleoecosystem, and land management of the southern North Sea.

Glennie, Kw. *Petroleum Geology of the North Sea: Basic Concepts and Recent Advances.* 4th ed. New York: John Wiley & Sons, 2009. Examines the depositional history and stratigraphy of the hydrocarbon-related rock units of the North Sea. Focuses on the appraisal and development of existing North Sea petroleum resources, rather than exploration for new resources.

Harvie, Christopher. *Fool's Gold: The Story of North Sea Oil.* London: Hamish Hamilton, 1994. A well-researched, detailed account of the discovery and production of oil in the North Sea. Deals largely with the environmental, economic, and political consequences of extracting oil from the region.

Ilyina, Tatjana. *The Fate of Persistent Organic Pollutants in the North Sea: Multiple Years Model Simulations of γ-HCH, α-HCH and PCB 153.* Berlin: Springer-Verlag, 2007. Examines the use of FANTOM, an ocean circulation model that has been tested for use in predicting the distribution and activity of persistent organic pollutants in the North Sea.

Ozsoy, Emin, and Alexander Mikaelyan, eds. *Sensitivity to Change: Black Sea, Baltic Sea, and North Sea.* Boston: Kluwer Academic Publishers, 1997. The proceedings of a workshop sponsored by the North Atlantic Treaty Organization (NATO) held in Bulgaria in 1995. Discusses the impact of human activity and climate change on the North Sea and other major arms of the Atlantic Ocean.

Shennan, Ian, and Julian E. Andrews, eds. *Holocene Land-Ocean Interaction and Environmental Change Around the North Sea.* Special Publication No. 166. London: Geological Society of London, 2000. A compilation of articles covering sedimentation, fossil dating, geochemistry, and sea-level change of the North Sea during the Holocene. Presents research findings from the Land-Ocean Evolution Perspective study.

Smith, Norman J. *North Sea Oil and Gas, British Industry and the Offshore Supplies Office.* Vol. 7. Edited by John Cubitt. Boston: Elsevier, 2011. Discusses the development of the North Sea Petroleum industry, British economics, and case studies.

_____. *The Sea of Lost Opportunity: North Sea Oil and Gas, British Industry and the Offshore Supplies Office.* Oxford: Elsevier, 2011. Describes the reasons that British industry has never reaped the full benefit of North Sea oil and gas discoveries despite the support of the British government.

Warren, E. A., and P. C. Smalley. *North Sea Formation Waters Atlas.* London: Geological Society, 1994. Offers an extremely detailed description, with numerous maps, of the physical structure of the North Sea. Includes information on the location of natural gas and oil fields.

Ziegler, Karen, Peter Turner, and Stephen R. Daines, eds. *Petroleum Geology of the Southern North Sea: Future Potential.* London: Geological Society, 1997. A scientific account of the possibility of producing oil from regions of the North Sea not usually associated with oil fields. Intended for advanced students.

**See also:** Aral Sea; Arctic Ocean; Atlantic Ocean; Black Sea; Caspian Sea; Deep Ocean Currents; Gulf of California; Gulf of Mexico; Hudson Bay; Indian Ocean; Mediterranean Sea; Ocean Tides; Pacific Ocean; Persian Gulf; Red Sea; Saltwater Intrusion; Surface Ocean Currents

# NUCLEAR WINTER

*Fallout consists of radioactive particles that settle out of the atmosphere, carrying hazardous radioactivity far from the site of its release. Nuclear winter is a model of the projected consequences of nuclear war, which may result in drastically reduced surface temperatures over several seasons or years. This model has dramatically altered thinking on the ability of humanity to survive a major nuclear exchange.*

## PRINCIPAL TERMS

- **aerosol:** a suspension of solid or liquid particles in a gas
- **greenhouse effect:** various mechanisms for increasing the absorptive capacity of radiated energy by the atmosphere
- **half-life:** the time it takes for any initial quantity of a radioactive isotope to decay to one-half of that initial amount
- **isotope:** atoms of an element having the same number of protons but a different number of neutrons
- **thermal budget:** the balance of incoming and outgoing radiative energy on Earth

## FALLOUT

A nuclear explosion occurs when the nuclei of a critical mass of susceptible isotopes interact to either break apart into smaller nuclei through nuclear fission or fuse together to form more massive nuclei through the process of nuclear fusion. In less than one-millionth of a second after detonation, a nuclear explosive changes from solid material into a rapidly expanding fireball. Eventually, it cools enough so that the bomb vapor condenses into the solid particles that incorporate the radioactive ashes of the nuclear explosive. Fallout is the name given to the radioactive particles that rain down from the debris cloud of a nuclear explosion. The length of time it takes for fallout to reach the ground depends largely on the size of the particles and how high they are lofted. Thus, fine particles can remain in the stratosphere for years, circling the globe many times. The longer these particles remain aloft, the weaker the radioactivity will be when the fallout reaches the ground. Half-lives of radioactive isotopes from nuclear weapons range from a fraction of a second to billions of years, but all radioactive isotopes obey a simple rule: For each factor-of-seven increase in time, the intensity of radioactivity decreases by approximately a factor of ten.

Short-lived isotopes are intensely radioactive for a brief time; long-lived isotopes are only weakly radioactive, but they are active for a long time. Those with intermediate half-lives are the most dangerous fallout isotopes. Two other important considerations are the type of radiation emitted and the human body's affinity for certain elements. Fallout radiation is usually gamma rays or beta particles. Gamma rays are similar to X rays, but they have higher energies and are more penetrating.

Radioactive fallout can also be produced by a nuclear reactor accident, as happened when the Soviet reactor at Chernobyl exploded in 1986. Because of operator error coupled with a poor design, two chemical explosions lifted the 1,000-ton cover plate from the reactor core, blasted radioactive particles more than 7 kilometers into the air, and started several fires in and around the reactor. Emissions from the fires formed a radioactive cloud 1.5 kilometers high at one point. As the fire burned, radioisotopes continued to be released over a period of several days. More recently, in 2011, a tsunami resulting from a major subsea earthquake off the coast of Japan severely damaged several reactors at the Fukushima nuclear power station. This caused the cooling systems to fail, and station workers were forced to use seawater to cool the reactor cores. This action not only rendered the cores unusable, but also produced several by-product explosions that further increased the damage and caused the release of radioactive contaminants into the atmosphere. The greatest fear in regard to nuclear explosions, whether from bombs or industrial accidents, is that the rapid environmental changes that could result would initiate an artificial ice age that has been termed "nuclear winter."

## EARTH'S THERMAL BUDGET

The theory of "nuclear winter" proposes that massive quantities of smoke and dust injected into the atmosphere by multiple nuclear detonations would drive temperatures on Earth below the freezing point of water for extended periods and would diminish the amount of sunlight reaching the surface, creating a period of prolonged cold and darkness. Atmospheric circulation could distribute the smoke and dust veils from even a moderate-sized nuclear exchange to make the effects global, rather than confined only to the regions of the detonations. Secondary effects from radioactive fallout and toxic substances carried by the smoke and dust veils also would be extremely widespread. The combined effects of nuclear winter could be sufficient to cause the eventual extinction of plant and animal life, and possibly of humanity itself.

The key factor in the nuclear-winter scenario is that the combined effects of smoke and dust injected into the atmosphere by multiple nuclear detonations could severely reduce the amount of sunlight reaching Earth's surface, throwing the "thermal budget" of the ecosystem into imbalance and sending surface temperatures well into the subfreezing range. Over time and across latitudes, the thermal budget of Earth is more or less in balance, with the amount of thermal radiation arriving from the sun being roughly equal to the amount of energy radiated back into space. (Low latitudes characteristically receive more thermal radiation than they emit, and higher latitudes less, creating large-scale atmospheric and oceanic circulation patterns that balance the thermal budget and account for major weather patterns.)

Radiation from the sun reaches Earth mostly in the visible light wavelengths of the electromagnetic spectrum, but thermal radiation emitted from the ground surface back into space is mostly in infrared wavelengths. If Earth did not have an atmosphere, the thermal budget would balance out at a level significantly below the freezing point of water. Fortunately, components of the atmosphere—primarily water vapor and carbon dioxide molecules—are excellent absorbers of infrared radiation. They reradiate at least half of the infrared energy they absorb back toward the ground surface rather than out into space. The process is sufficient to increase the average temperature of a planet by several degrees, and has been termed the "greenhouse effect." Consequently, it helps to keep ambient temperatures on most places on Earth above the freezing point of water, making possible a biological regime based on liquid water. Any planetary-scale effect that would prevent sunlight from reaching the surface would diminish or even eliminate the greenhouse effect of those gases in the atmosphere, allowing temperatures to drop to the subfreezing levels that would otherwise exist.

## DIMINISHED SUNLIGHT

Nuclear winter is thus a kind of "anti-greenhouse" effect. The greenhouse effect is produced by the small proportion of infrared-absorbing carbon dioxide, methane, water vapor, and other gases in the atmosphere, which increases ambient temperatures on Earth by reducing the amount of infrared energy being emitted back into space from the ground surface. In the nuclear-winter scenario, changes in the atmosphere reduce the amount of incoming solar radiation reaching the surface by absorbing greater quantities of solar radiation in the upper atmosphere, with no reduction in surface infrared radiation to maintain a thermal balance. Surface temperatures would drop significantly and, because any appropriate atmospheric agent responsible for such a reduction would have to absorb primarily visible light, continuous darkness would result at the surface.

While dangerous radioactive materials can be injected into the atmosphere by the failure of nuclear power installations, the "nuclear winter" scenario is much more closely associated with the war-time exchange of nuclear weapons. Depending upon the extent of a general nuclear exchange, 100 million to 300 million metric tons of smoke and soot particles might be injected into the atmosphere, enough to diminish sunlight reaching the ground surface by more than 90 percent if it were distributed uniformly throughout the atmosphere. The smoke would come from immense firestorms generated by detonations over urban or large industrial centers, so powerful that they would generate hurricane-like convective winds similar to those experienced by the victims of the Allied incendiary bombing of Dresden, Germany, during World War II.

The convective energy of these firestorms would drive the smoke and soot clouds beyond the dense troposphere and up into the stratosphere. The higher these clouds were injected, the longer they would be likely to persist, because smoke and dust are cleansed from the atmosphere mostly by rainfall, and there is little water vapor in the stratosphere. Research suggests that this material would remain aloft long enough to become distributed throughout the stratosphere, with global rather than merely regional repercussions. Firestorms in modern urban centers, burning not only structures and material life but also huge quantities of plastics, light metals, and a host of synthetic chemicals of still unknown potential, would be far more dangerous over much greater areas than anything previously experienced. Additional smoke accumulation would come from forests and grasslands ignited by nuclear explosions.

## EFFECTS OF NUCLEAR WINTER

The horrific loss of human life from the immediate effects of a nuclear conflict remains in the nuclear-winter scenario, but the secondary, longer-term losses are even more sobering. Deprived of heat and light from the sun, much of Earth would experience a state of subfreezing temperatures and continuous darkness, even at midday, lasting many years. Among the survivors of the nuclear holocaust, deaths from cold and exposure, and extraordinary psychological stress, would continue to increase the fatality rate in the aftermath of nuclear war. Over several successive seasons, the result would be the collapse of the natural food chain, the destruction of most plant life and the animal life that feeds on it, and mass starvation. All remaining order would likely break down as survivors fought for food, water, and shelter in a world that would no longer have the advanced infrastructure and technology that is taken for granted today.

Interruption of sunlight to the surface would result in a complete disruption of the process of photosynthesis in plants and lead inevitably to the mass failure of food crops around the world. Wild plant species would also be affected, and those susceptible to small changes in the environment would be quickly driven to extinction. Large-scale extinction of vertebrate species dependent on plant matter for food would soon follow. At sea, the phytoplankton population that anchors the food chain for many large species would collapse; many of the larger species of aquatic life would face extinction as well.

Moderate-size thermonuclear explosions are capable of perforating the ozone layer of the atmosphere, and a significant number of such explosions could lead to massive ozone depletion. In addition to radioactive fallout from the explosions themselves—a hazard now judged to be more serious than previously believed, as a result of the long-term suspension of matter in the atmosphere—survivors of a nuclear war could be exposed to harmful ultraviolet radiation from the sun that is ordinarily blocked by atmospheric ozone.

The impact of radioactive fallout, once presumed to be localized in the vicinities of the nuclear explosions themselves, would become global and prolonged as these substances circulate in the upper atmosphere. Water and food sources hundreds or thousands of miles distant from actual detonations would be contaminated by the atmospheric circulation of radioactive matter and toxic substances from burning cities. Human populations would also decline, making the species vulnerable to eventual extinction.

Possible consequences of nuclear winter include long-term climatic disruption. For example, deposition of dust and soot over much of Earth's surface could lower the albedo, or reflective value, of the surface, with unknown climatic consequences over time. Coastal areas—in contrast to continental interiors, where convection would be almost nonexistent for years—might suffer storms of unprecedented magnitude, as relatively warm, humid air currents from over the oceans collided with chilled air masses on land.

## MAGNITUDE OF NUCLEAR WAR

Phenomena associated with the nuclear-winter scenario may vary considerably depending upon the magnitude of a nuclear exchange and the nature of selected targets; catastrophic results are possible far short of a full-scale thermonuclear war. In a "medium-range" exchange, for example, involving some 5,000 megatons of detonations, including one thousand ground bursts—the sort of exchange that might be envisioned should one superpower attempt to eliminate the first-strike capability of another—the resulting smoke and dust veils could create a global atmospheric inversion that would essentially stop wind currents near the surface. As the smoke and dust absorbed visible sunlight and blocked it from reaching the surface, the upper atmosphere would be heated as much as 30 to 80 degrees Celsius. As the ground cooled, these hot layers would rise and expand, blanketing the planet in an unbroken cloud of smoke and dust. Even several months after the exchange, the lowest levels of the atmosphere would receive only enough radiation to drive very weak convective currents. This temperature inversion would inhibit the rise of moist air from the surface and thereby inhibit cleansing rainfall, increasing the time that the smoke and dust veils would remain suspended in the atmosphere.

357

One of the most significant features of the nuclear-winter model is the ability of its climatic consequences to involve the tropics through atmospheric circulation. Tropical plant and animal species are far more vulnerable than others to even minor shifts in temperature or rainfall. Prolonged cold and darkness over the tropics could result in the extinction of many species on the planet. At the very least, destruction of the forest regions of the tropics, major absorbing agents for solar energy, would further alter Earth's thermal balance toward long-term subfreezing conditions. Widespread destruction of forests would also lead to further large-scale fires.

The nuclear-winter model now predicts much more severe radiation hazards, spread over wider areas, than those estimated by earlier projections. Most radioactive isotopes in a nuclear explosion condense onto aerosols and dust sucked in by the initial fireball. Inasmuch as the model predicts very large quantities of dust aerosols in the atmosphere for longer periods of time, the dangers of transport or dispersal of radioactive materials would multiply, particularly with respect to intermediate and prolonged fallout. Although not immediately lethal, this fallout could lead to chronic radiation sickness, depressed immune systems, genetic defects, and death for millions from cancer or other radiation-induced causes.

## STUDY OF NUCLEAR WINTER

Formulation of the nuclear-winter scenario, and its widespread discussion in the scientific, military, and political communities, resulted largely from the work of a research team appointed in 1982 by the National Academy of Sciences (NAS). The NAS requested that Richard B. Turco, Owen B. Toon, Thomas P. Ackerman, James B. Pollack, and Carl Sagan (a team subsequently known by their last initials as "TTAPS") investigate the possible effects of dust raised by nuclear detonations. The TTAPS team comprised scientists who already had investigated a variety of atmospheric effects of dust particles, including the consequences of volcanic explosions, the behavior of dust storms on Mars, and theories of mass extinction caused by asteroid or comet impacts in the geologic past.

Computer modeling represents the only feasible means of studying the consequences of a nuclear exchange. The TTAPS researchers used advanced models of atmospheric circulation, particle microphysics, and radiative-convective behavior. Also used were models of nuclear-exchange scenarios of varying size and geographical complexity. Other researchers, including Paul J. Crutzen at the Max Planck Institute for Chemistry, in Germany, and John W. Birks, at the University of Colorado, persuaded the TTAPS team to include in their research the possible effects of smoke from massive fires caused by nuclear explosions.

Prior to the TTAPS research, little was known about potential large-scale smoke-and-dust pollution except as the consequence of volcanic explosions. The TTAPS team was able to study firsthand results of the explosion of El Chichón in 1982, which generated a significant dust veil and possibly some short-term surface temperature reduction in some areas. The much larger explosion of Krakatau in 1883 caused visible atmospheric effects around the world and, possibly, minor surface-temperature changes. Historically, several volcanic explosions far larger than Krakatau are known from the eighteenth and nineteenth centuries, some of them comparable in energy to several-multimegaton nuclear detonations. Yet none generated more than minor multiyear climatic effects.

The TTAPS team concluded that earlier attempts to judge the possible effects of nuclear explosions based on records and observations of volcanic explosions were misleading. No matter how large, volcanic events were episodic and localized. The clouds of soil-and-dust particles, or aerosols, created by these explosions consist mostly of relatively large and bright, light-scattering particles that can create dazzling displays but do not absorb much visible light. By including in their model the smoke and soot from fires generated by nuclear explosions, however, the TTAPS team identified the potential source of aerosols of very dark and very fine particles with average diameters smaller than the typical wavelength of infrared radiation. These smoke and soot particles would have a very high absorptive capacity for visible light but would not correspondingly increase infrared absorption in the atmosphere, thus bringing about surface darkness and temperature reduction.

One of the most unsettling results of the TTAPS research was to show that a full-scale nuclear exchange would not be needed to bring about the most serious nuclear-winter effects. Presuming that such an exchange would entail a total explosive yield of about 25,000 megatons, nuclear winter could be triggered by much smaller conflicts, in the range from 3,000 to 5,000 megatons, depending upon the nature of the targets. Prior to the 1963 international treaty banning atmospheric testing of nuclear weapons, scenarios of nuclear conflict centered on massive atmospheric explosions of thermonuclear weapons in the 20 -to 40-megaton range, principally over cities and refinery areas. Many of these weapons were to have been delivered by strategic bombers.

Subsequent development of ballistic missiles using multiple reentry warhead configurations of 1 to 5 megatons each, directed initially at the enemy's own missile silos, greatly increased the anticipated number of ground-level detonations in a nuclear war scenario, thereby increasing the expected amount of material injected into the atmosphere by the explosions. Ironically, the strategic use of larger numbers of considerably smaller-yield warheads makes a nuclear-winter catastrophe even more likely, even if the nuclear exchange were limited to "surgical" strikes concentrated on missile silos and military installations.

*Ronald W. Davis*

## FURTHER READING

Badash, Lawrence. *A Nuclear Winter's Tale: Science and Politics in the 1980s.* Boston: Massachusetts Institute of Technology, 2009. Discusses the rise and fall of the concept of nuclear winter in the arena of research, public relations, and politics of the 1980's.

Byrne, John, and Steven M. Hoffman, eds. *Governing the Atom: The Politics of Risk.* New Brunswick, N.J.: Transaction, 1996. Researches the environmental and social aspects of the nuclear industry. Pays careful attention to government policies that have been implemented to monitor safety and accident-prevention measures.

Chernousenko, Vladimir M. *Chernobyl: Insight from the Inside.* Berlin: Springer-Verlag, 1991. An interesting collection of personal anecdotes, pictures, reports, evaluations, and recommendations regarding the Chernobyl incident. Begins with an account of the accident and carries through to the medical problems caused by fallout. Faults the government for being more concerned with managing the news than with dealing with physical problems.

Diermendjian, Diran. *"Nuclear Winter": A Brief Review and Comments on Some Recent Literature.* Santa Monica, Calif.: RAND Corporation, 1988. An independent study of the first few years of debate about nuclear winter and policy implications. Covers some of the research literature outside the United States.

Ehrlich, Paul R., et al. *The Cold and the Dark: The World After Nuclear War.* New York: W. W. Norton, 1984. A summary of the TTAPS research for the general public.

_____. "Long-Term Biological Consequences of the Nuclear War." *Science* 222 (1983): 1293-1300. Confirms and elaborates upon the TTAPS team projections of the dangers of nuclear winter to plants and animals (and, hence, human food supplies) and projects likely conditions over several different time intervals. Together with a TTAPS article preceding it in the issue, announced the nuclear-winter scenario, and debate about it, to the scientific community.

Fanchi, John R. *Energy: Technology and Directions for the Future.* Burlington, Mass.: Elsevier Academic Press, 2004. Written for scientists and engineers. Includes several explanations using advanced calculus in the context of concepts and techniques used in all major energy components.

Fanchi, John R., and Christopher J. Fanchi. *Energy in the 21st Century.* 2d ed. Hackensack, N.J.: World Scientific Publishing Co., 2011. Presents a broad critical examination of the various energy opportunities in the present century, with a detailed discussion of nuclear power and the Chernobyl reactor failure. Includes numerous "points to ponder" and review exercises. For general readership.

Fisher, David E. *Fire and Ice: The Greenhouse Effect, Ozone Depletion, and Nuclear Winter.* New York: Harper & Row, 1990. Juxtaposes the twin hazards of global warming through the greenhouse effect and possible global winter resulting from nuclear war. Calls for more environmental and foreign-policy responsibility to avoid these dangers.

Greene, Owen. *Nuclear Winter: The Evidence and the Risks.* Cambridge, England: Polity Press, 1985. An independent confirmation of the TTAPS research that had a significant impact in Europe.

Harwell, Mark. *Nuclear Winter: The Human and Environmental Consequences of Nuclear War.* New York: Springer-Verlag, 1984. Offers a general survey of nuclear-winter conditions, with particularly good treatment of the potential hazards of widespread distribution of toxic materials by fires in bombed cities.

Izrael, Urii Antonievich. *Radioactive Fallout After Nuclear Explosions and Accidents.* Oxford: Elsevier Science, 2002. Focuses on studies of the composition of radioactive fallout, the formation of aerosol particles that transport the radioactive products, the analysis of the external radiation doses resulting from nuclear explosions and/or accidents, and restoration and rehabilitation of contaminated land areas.

*Nuclear Winter and National Security: Implications for Future Policy.* Maxwell Air Force Base, Ala.: Air University Press, 1986. Compiles papers from a military symposium that concluded that publicizing the nuclear-winter scenario would lead to increased international pressure for nuclear disarmament, thus affecting the capability of the military to respond to threats. Reflects an unwillingness on the part of military planners to accept the predictions of the nuclear-winter model.

Robock, A., et al. "Climatic Consequences of Regional Nuclear Conflict." *Atmospheric Chemistry and Physics* 7 (2007): 2003-2012. Discusses the potential effects of a regional nuclear war in the subtropics on the global climate. Based on an arsenal of 50 Hiroshima-sized bombs for each country. Recognizes that the results of this scenario are less dramatic than previous studies, but are no less relevant. Clearly explained and useful to readers across multiple disciplines.

Sagan, Carl. "Nuclear Winter and Climatic Catastrophe: Some Policy Implications." *Foreign Affairs* 62 (Winter, 1983): 257-292. Represents a major contribution toward introducing the nuclear-winter issue into foreign-policy debates among the informed public, by the most prominent science writer on the TTAPS team.

Sagan, Carl, and Richard Turco. *A Path Where No Man Thought: Nuclear Winter and the End of the Arms Race.* New York: Random House, 1990. Tells the story of the TTAPS research program and the campaign to bring the results to the attention of political and military authorities. Offers interesting speculations on how the findings may have influenced geopolitical and military thinking of major powers other than the United States.

Sutton, Gerard K., and Joseph A. Cassalli, eds. *Catastrophe in Japan: The Earthquake and Tsunami of 2011.* Hauppauge, N.Y.: Nova Science Publishers, 2011. Compiles a number of reports on the effects of the earthquake and resulting tsunami, including the impact on agriculture and economics. Focuses on the nuclear crisis following the earthquake and tsunami. Describes the events within the nuclear power plant resulting from the natural disaster.

Takada, Jun. *Nuclear Hazards in the World.* New York: Springer, 2010. Discusses the risks of nuclear energy including nuclear hazards, pollution, fallout, the Chernobyl accident, and nuclear testing. Written for the layperson and contains information from field work conducted between 1995 and 2002.

Toon, Brian, Alan Robock, and Rich Turco. "Environmental Consequences of Nuclear War." *Physics Today* 61 (Dec. 2008): 37-42. Summarizes research conducted on the effects of nuclear warfare from the 1980's through 2008. Also discusses the future of nuclear warfare.

Turco, Richard P., et al. "The Climatic Effects of Nuclear War." *Scientific American* 251, no. 2 (August, 1984): 33-43. An extraordinarily important article for its introduction of the TTAPS research to the scientifically educated public.

_____. "Nuclear Winter: Global Consequences of Multiple Nuclear Explosions." *Science* 222 (1983): 1283-1292. The key article announcing the TTAPS team findings to the scientific community.

**See also:** Acid Rain and Acid Deposition; Air Pollution; Atmosphere's Evolution; Atmosphere's Global Circulation; Atmosphere's Structure and Thermodynamics; Atmospheric Properties; Auroras; Climate; Climate Change Theories; Cosmic Rays and Background Radiation; Earth-Sun Relations; Geochemical Cycles; Global Warming; Greenhouse Effect; Hydrologic Cycle; Ocean-Atmosphere Interactions; Ozone Depletion and Ozone Holes; Radon Gas; Rainforests and the Atmosphere